Engineering Thermodynamics

Engineering Thermodynamics
Fundamental and Advanced Topics

Kavati Venkateswarlu

CRC Press
Taylor & Francis Group
Boca Raton London New York

CRC Press is an imprint of the
Taylor & Francis Group, an informa business

First edition published 2021
by CRC Press
6000 Broken Sound Parkway NW, Suite 300, Boca Raton, FL 33487-2742

and by CRC Press
2 Park Square, Milton Park, Abingdon, Oxon, OX14 4RN

First edition published by CRC Press 2021

CRC Press is an imprint of Taylor & Francis Group, LLC

ISBN: 978-0-367-64628-8 (hbk)
ISBN: 978-0-367-65210-4 (pbk)
ISBN: 978-1-003-12836-6 (ebk)

Typeset in Times
by codeMantra

Access the Support Material: https://www.routledge.com/9780367646288

Dedicated to my beloved father Late Sri. Kavati Sailu

Contents

Foreword

This textbook fosters an understanding of fundamental and advanced concepts of thermodynamics based on real-world problems. Design-oriented problems provide an opportunity for the readers to understand the practical problems and develop new ideas and concepts based on the fundamental theory presented herein.

Distinctive features of the book include a detailed discussion of energy efficiency, energy sustainability, energy security, and exergy efficiency, vital in everyday applications such as energy-efficient buildings, compressors, and motors. Advances in power and refrigeration cycles such as organic Rankine cycle (ORC), combined cycle power plants, combined cycle power plant integrated with ORC and absorption refrigeration system, the effect of operating parameters on combined cycle performance, integrated coal gasification combined cycle power plants, innovative refrigeration techniques, and next-generation low-global warming potential refrigerants with the current refrigerant options are presented elaborately. Power cycles for nuclear power generation and advanced combustion technologies in gas turbines are also included. In addition, the chemical potential of ideal Fermi and Bose gases, irreversible thermodynamics with an emphasis on Onsager reciprocal relations, and the application of irreversible thermodynamics to the thermo-electric phenomenon are presented elaborately.

Dr. Venkateswarlu has the experience of teaching both basic and advanced concepts of thermodynamics for undergraduate and post-graduate students at various engineering colleges over a long period. He has put all his efforts using that experience in making this book fruitful. I take this opportunity to congratulate the author on his admirable success and recommend this book with great enthusiasm to the engineering students all over the world.

Dr. A.V.S.S.K.S. Gupta
Professor and Head, Department of Mechanical Engineering
JNTUH College of Engineering Hyderabad

Foreword

Preface

This textbook explains the most up-to-date advanced concepts in steam power plants, refrigeration, air-conditioning, and waste heat recovery systems. The detailed discussion of energy efficiency, energy sustainability, and energy security will be useful for the readers in the design of energy-efficient buildings. It reinforces concepts presented with exercises, review questions, multiple-choice questions, and a discussion of experimental design and recent findings. It presents irreversible thermodynamics of equilibrium and non-equilibrium systems and introduces Onsager's reciprocal relations that connect thermodynamics, transport theory, and statistical mechanics. Infused with design and experimental problems, this text enables undergraduate students to perform case studies, design equipment and devices, and apply new concepts related to the industry and real-world.

Chapter 1 presents the introduction and basic concepts of thermodynamics such as system and different systems, surroundings, and the universe. It distinguishes between macroscopic and microscopic viewpoints, homogeneous and heterogeneous systems, and intensive and extensive properties. It covers the thermodynamic process, state, thermodynamic cycle, the state postulate-I, and the state postulate-II, and pressure measurement. It introduces the thermodynamic equilibrium concept and presents the mechanical, chemical, and thermal equilibrium.

Chapter 2 presents the property temperature and zeroth law of thermodynamics and its significance for measurement of temperature. It includes various temperature-measuring devices and reference points used in the measurement of temperature and thermometric property used in each of the temperature-measurement devices. It also covers temperature scales such as the ideal gas temperature scale and international temperature scale.

Chapter 3 presents the energy, different forms of stored energy, and the first law of thermodynamics also known as the principle of conservation of energy. It distinguishes between heat and work and different forms of work. It also presents various mechanical forms of work. It covers the energy analysis of closed systems, first law for a closed system undergoing a cycle, and first law for a closed system undergoing a change of state. Also presented are energy conversion efficiency and energy-efficient buildings, motors, and compressors.

Chapter 4 presents pure substances and their phases and phase change processes of pure substances such as water and others on p-v, T-s, p-t, and h-s diagram or Mollier diagram. It also presents the quality or dryness fraction—property tables for measuring the properties of steam such as specific volume, enthalpy, and entropy.

Chapter 5 presents the core concepts related to control volume analysis and distinguishes between steady-state and unsteady (transient) analysis. It covers the development of a steady-flow energy equation using mass conservation and energy conservation principles. It includes first law analysis of steady-flow processes, steady-flow energy equation needs, and steady-flow devices. It also presents the first law analysis of unsteady-flow processes.

Chapter 6 presents the limitations of the first law of thermodynamics and introduces the second law statements, namely Kelvin–Planck and Clausius statements and their equivalence including second law application to power cycles, refrigeration, and heat pump cycles. It presents the concepts of reversible and irreversible processes. It covers the Carnot theorem and Carnot cycle with the help of p-v and T-s diagrams and thermodynamic temperature scale.

Chapter 7 presents the inequality of Clausius, which is the basis for determining whether a process is reversible or irreversible or impossible, including entropy—a property of a system and principle of entropy. It also presents the TDS equations, entropy change of pure substances, ideal gases, solids, and liquids, entropy balance equation, entropy transfer by heat and mass transfer, and entropy generation. The isentropic process, isentropic efficiency, and isentropic efficiency of a turbine, compressor, pump, and nozzle are also covered.

Chapter 8 presents the properties of gases and gas mixtures, including the ideal gas equation of state and other equations of state such as Van der Waals. It also presents the compressibility factor—the deviation of real gases from the ideal gas behavior and gas compression–reducing the work of compression.

Chapter 9 presents the concept of availability (exergy), reversible work, irreversibility, and useful work. It covers exergy of a flow stream (open system) and non-flowing fluids (closed systems) and exergy transfer by heat, work, and mass. Also covered are second-law efficiency of work-producing and work-consuming devices, exergy destruction, and exergy balance.

Chapter 10 presents the vapor and advanced power cycles. It presents the Carnot cycle and its limitations as a practical case, the basic Rankine cycle, and its improvement of efficiency. Co-generation and combined cycle power plants are included. In addition, the advanced power cycles such as organic Rankine cycle (ORC), combined cycle power plant integrated with ORC and absorption refrigeration system, the effect of operating parameters on combined cycle performance, integrated coal gasification combined cycle power plants, and power cycles for nuclear power generation are presented elaborately.

Chapter 11 presents the gas power cycles Carnot cycle (idealized cycle) and Otto cycle, Diesel cycle, dual cycle for internal combustion engines, and Stirling and Ericsson cycles for external combustion engines including the advents in both spark Ignition and compression ignition engines. It also covers the Brayton cycle for gas turbine power plants and jet propulsion. In addition, advanced combustion technologies in gas turbines and compressors used in jet propulsion systems are also included.

Chapter 12 presents the reversed Carnot cycle, vapor compression refrigeration cycle, air refrigeration cycle, and vapor absorption refrigeration cycle including various refrigerants including next-generation low-global warming potential refrigerants with the current refrigerant options. Innovative refrigeration techniques are also presented elaborately. In addition, energy conservation in domestic refrigerators is also presented.

Chapter 13 presents the important mathematical relations necessary for the development of Maxwell relations and then other relations developed based on Maxwell relations such as the Clausius–Clapeyron equation and the Joule–Thomson coefficient.

Chapter 14 presents the properties of dry air and water vapor such as specific humidity, relative humidity, dew-point temperature, wet-bulb temperature, and dry-bulb temperature including the adiabatic saturation process. It also covers the psychrometric chart and its application to air-conditioning processes.

Chapter 15 presents the concepts of chemical potential and fugacity, Gibbs free energy, and chemical potential application to thermal radiation. It also covers the chemical potential of ideal Fermi-Dirac and Bose-Einstein gases, low-temperature behavior of physical systems, and Fermi and Bose low-temperature expansions.

Chapter 16 presents the irreversible thermodynamics with an overview of equilibrium and non-equilibrium thermodynamics, coupled phenomena, maximum entropy production principle, and minimum entropy production principle. It presents an emphasis on Onsager reciprocal relations and the application of irreversible thermodynamics to a thermo-electric phenomenon including the Seebeck effect, Peltier effect, Joule effect, and Kelvin effect.

Property tables, charts, and multiple-choice questions comprise appendices of the book and are available at https://www.routledge.com/9780367646288.

Acknowledgments

I take this opportunity to thank Dr. A.V.S.S.K.S. Gupta, Professor, Department of Mechanical engineering, JNTU Hyderabad for writing the Foreword for this book who happens to be my teacher for the Advanced Thermodynamics course during my master's at the University. I wish to thank Dr. K. Ramakrishna, a professor in the Department of Mechanical Engineering, K.L University, Guntur, who is an inspiration for me to write this book. My special thanks are due to Gokaraju Rangaraju Institute of Engineering and Technology, Hyderabad college administration.

I also wish to thank my colleagues Dr. R. Karthikeyan, Mr. D. Suresh Kumar, Mr. L Gopinath and few students of UG and PG in the department for their assistance in completing the book.

I extend my gratitude to the editorial team of CRC Press, Taylor & Francis Group, senior commissioning editor (engineering) Gauravjeet Singh Reen and editorial assistant Lakshya Gaba, and others for their relentless efforts in the improvisation of the book. I am indebted to all authors and publishers whose work is consulted freely and referred to in shaping this book.

Above all, I wholeheartedly thank my parents and family members for their patience and cooperation throughout the completion of the book.

Author

K. Venkateswarlu is currently working as a Visiting Postdoctoral Research Fellow, RAK Research and Innovation Center, American University of Ras Al Khaimah, Ras Al Khaimah, United Arab Emirates and a former Professor, department of Mechanical Engineering at Gokaraju Rangaraju institute of Engineering and Technology, Hyderabad, India. He received both his B.Tech in Mechanical Engineering and M.Tech in Thermal Engineering from JNTU College of Engineering, Hyderabad and Ph.D in Mechanical Engineering (IC Engines) from JNTU Kakinada. His research interests include Fuel efficiency and emission improvement of diesel engines; Phase change materials (PCMs) for thermal energy storage; and Nanofluids in heat transfer and solar energy applications. He has twenty-four years of professional experience with twenty years in teaching and research in the faculty of Mechanical engineering including two years of foreign experience.

Dr. Venkateswarlu has published more than 25 research papers in national and international journals and conferences of repute. He has guided 12 M.Tech theses. He has authored one text book "Alternative Fuels and Energy Technologies".

He is a Fellow, Institution of Engineers India (FIE) and Life Member, Combustion Institute Indian Section.

Dr. Venkateswarlu is the Associate Editor, SAE International Journal of Engines and the technical reviewer for journals SAE International Journal of Engines, Journal of the Brazilian society of Mechanical sciences and engineering, Institution of Engineers and African journal of Agricultural research.

Author

1 Introduction and Basic Concepts

LEARNING OUTCOMES

After learning this chapter, students should be able to

- Form a sound base for the development of the principles of thermodynamics with a thorough understanding of the definition of thermodynamics.
- Explain the basic concepts of thermodynamics such as system, process, state, cycle, and equilibrium.
- Demonstrate the knowledge of control volume and control mass for distinguishing the thermodynamic systems.
- Understand the principles of pressure-measuring devices and evaluate their merits and demerits.
- Demonstrate the knowledge of path function and point function.
- Understand the importance of thermodynamics in various fields of mechanical engineering such as heat transfer, combustion, refrigeration, air conditioning, and cryogenics.

1.1 INTRODUCTION TO THERMODYNAMICS

The formal study of thermodynamics started in the early 19th century in the application to convert heat into work, though the aspects of thermodynamics are quite old. Nowadays, its scope is much broader to provide solutions to a great diversity of problems in many fields, which involve the transfer of energy. The concepts of thermodynamics play a vital role in the present-day issues, such as effective usage of fossil fuels, development of renewable and new energy sources, and improvement of thermal system performance.

Thermodynamics is a branch of science and engineering. A scientist deals with the physical and chemical behavior of fixed quantities of matter and applies thermodynamic principles to relate the properties of matter, whereas an engineer deals with the design and analysis of systems and their interactions with the surroundings. Thermodynamics is thus the study of systems through which matter flows. That is the science of energy interactions and their effect on the surroundings. In daily life, more often we have a feeling of what energy is and it is defined as the ability to cause changes. The name thermodynamics is derived from the Greek words therme (heat) and dynamis (power), which were used to describe the conversion of heat into power. The energy transformations occur in so many engineering applications such as power generation, refrigeration and air conditioning, and relationships among the properties of matter. In all the activities, there is certainly some interaction between energy and matter.

Engineering thermodynamics plays a vital role in the design of thermal systems. For example, for a steam power plant to generate the electric power, the steam turbine must produce a net work output to drive the generator. This is possible when the steam generated by the boiler expands in the turbine by rotating the rotor. For a specific net work output of the turbine, the parameters, such as the rate of steam flow, steam pressure, and speed of the rotor, require a thorough understanding of principles of thermodynamics in the design of a power plant.

Thermodynamics is also encountered in many aspects of life, and one does not need to go very far to see some of its application areas. A person standing in a breezy room (exposed to the wind) loses heat in the form of thermal radiation; it is a common phenomenon related to thermodynamics that occurs in everyday life.

1.2 THERMODYNAMIC SYSTEMS

A *thermodynamic system* is defined as any quantity of matter or any region of space within a prescribed boundary on which we focus our attention for the purpose of analysis. It may be a simple device such as a small rotating fan or a combination of devices such as a large power plant. We can consider some quantity of matter contained in a closed cylinder or the flow of steam through a pipeline. The system may contain fixed matter or changing composition through chemical reactions. The volume of the system being analyzed may even change as the gas is compressed in a piston-cylinder arrangement. Everything external to the system is called *surroundings*. The system and its surroundings are distinguished by a specified boundary, which may be at rest or in motion. The system can interact with its surroundings in many ways and these interactions take place across its boundary. The real or imaginary surface that separates the system and its surroundings is called *boundary.*

Types of Systems

There are three kinds of thermodynamic systems: closed, isolated, and open. A *closed system* refers to a fixed quantity of matter and it is also called *control mass*, as mass cannot cross its boundary, only energy can cross. Examples of closed systems are air trapped in a piston-cylinder device and gas inside a closed balloon. An *isolated system* is the one in which neither energy nor mass can cross the boundaries of the system. It is a special case of closed system. Although it seems that a system that doesn't interact with the surroundings has no significance, if the combination of two or more systems, interacting with each other, is surrounded by a boundary, then they can be regarded as an isolated system for the analysis. A thermos with a lid used to keep the things either cold or hot is an example of an isolated system since it doesn't allow either mass or energy transfer across it. An *open system*, also called as control volume, is a region of space through which mass and energy can cross the boundaries of the system. Examples of open systems are air and fuel entering and exhaust gases leaving an internal combustion engine, and the engineering devices that involve mass flow such as turbines, compressors, and nozzles can be considered as control volume. Figures 1.1–1.3 show how the closed system, isolated system, and open system interact with their surroundings, respectively.

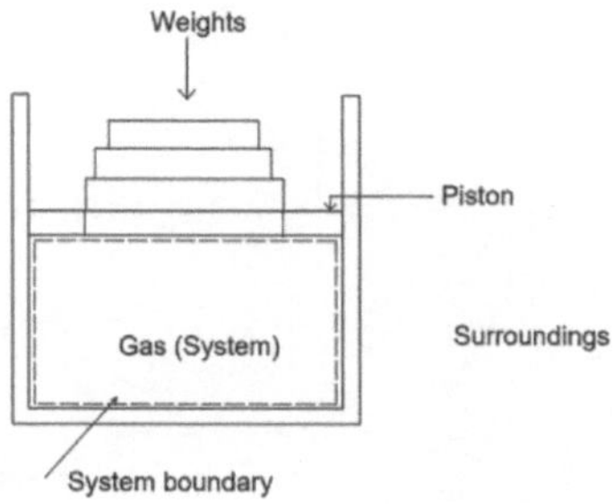

FIGURE 1.1 A closed system.

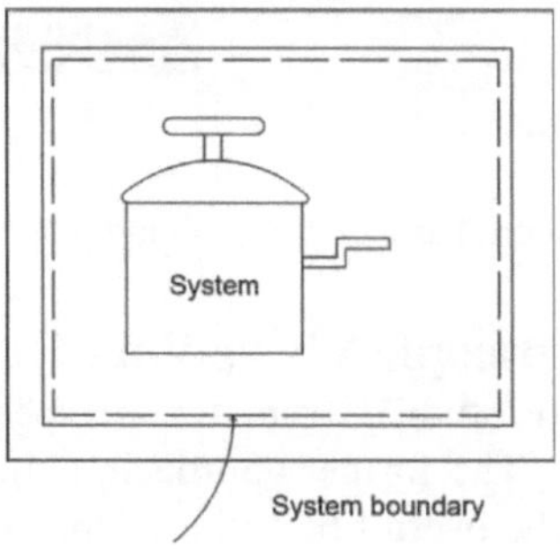

FIGURE 1.2 An isolated system.

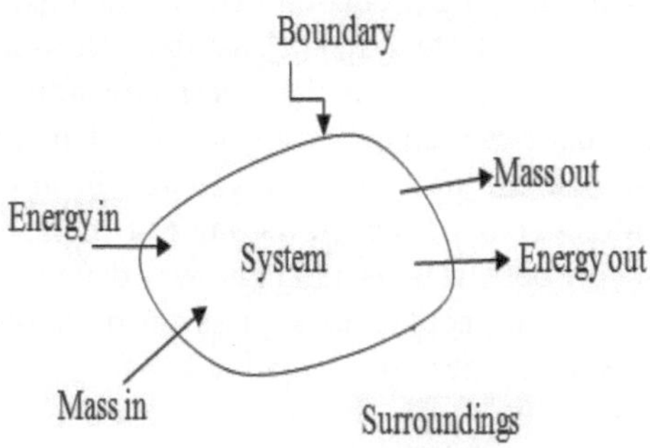

FIGURE 1.3 An open system.

The *control volume* approach is used in the thermodynamic analyses of the systems that involve the mass flow such as turbines, compressors, pumps, and nozzles. A control volume is defined as a certain volume in space surrounding the system through which mass can cross. The surface which bounds the region is called the control surface. Figure 1.4 shows the control volume. The system boundary is frequently referred to as a control surface when the terms control mass and control volume are used.

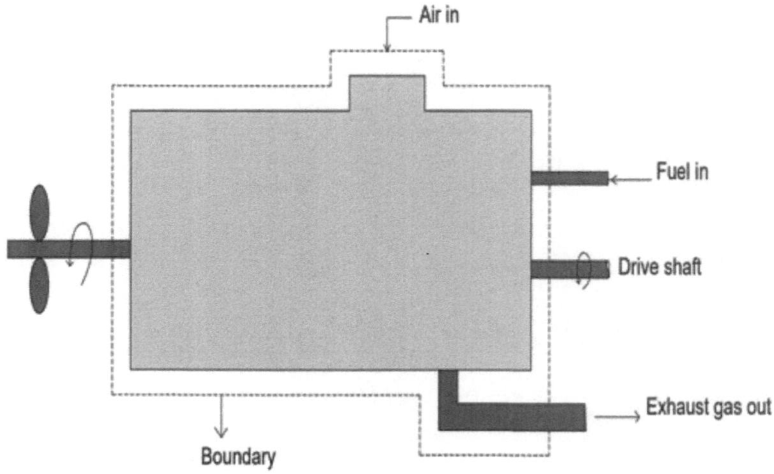

FIGURE 1.4 An engine considered as a control volume.

Macroscopic and Microscopic Viewpoints

Thermodynamic systems are studied from the perspective of both macroscopic and microscopic viewpoints. The former approach, called classical thermodynamics, deals with the study of the overall behavior of the systems without considering the interactions at the molecular level. The classical thermodynamics describes the behavior of a system based on the observations of the overall system. The latter approach, known as statistical thermodynamics, however, deals with the structure of matter. This approach, applying probability considerations, analyzes the behavior of particles that make up a system and relates this information to the observed macroscopic behavior of the system. Statistical thermodynamics plays a vital role in analyzing the physical processes such as cryogenics (low-temperature technologies), compressible fluid flows (high gas flows) encountered in aircrafts, and chemical kinetics in combustion reactions. However, the laws of thermodynamics are based exclusively on macroscopic observations and in no way depend on molecular theory. In a wide range of engineering applications, classical thermodynamics provides a direct approach for analysis and design.

1.3 THERMODYNAMIC PROPERTIES

A property of a system is defined as a characteristic of the system which describes its behavior. It depends upon the state of the system but not on how the state is reached. There are two types of properties: intensive and extensive. *Intensive properties* are independent of the mass of the system, e.g., temperature, pressure, and density. For example, the density of water is 1000 kg/m^3 irrespective of its mass and hence the property density is an intensive property. Extensive properties are related to the size or extent of the system, e.g., total volume, total mass, and total entropy. *Extensive properties* are often divided by mass related to them to obtain the specific extensive properties. For example, if the entropy of a system of mass m is S, then the specific entropy s of matter within the system is

$$s = \frac{S}{m}\left(\text{kJ/kg K}\right) \tag{1.1}$$

Specific entropy is a specific extensive property.

1.4 STATE, PROCESSES, AND CYCLES

State

A state of a system is a unique condition at an instant of time, which is described by certain noticeable properties such as temperature and pressure. Each property of a substance will have a definite value at a particular state and therefore the properties are point functions. A variable is a property, if it has a single value at each equilibrium state. A variable is a property, if and only if the change in its value between any two prescribed equilibrium states is single valued. Therefore, any variable whose change is fixed by the end states is a property.

The State Postulate-I

As stated in the preceding section, a state of a system is described by its properties; however, from experience, it is observed that it is not necessary to specify all the properties in order to fix the state. Whenever the required number of properties are specified, the rest can assume certain values automatically. To fix the state of a system, only a certain number of properties is sufficient and it is called *state postulate.*

A simple compressible substance doesn't involve the effects of electricity, magnetism, and surface tension. A closed system, composed of a simple compressible substance without involving the effects of motion or gravity, is called a *simple compressible closed system.* The state of a simple compressible system is completely specified by two independent intensive properties. Two properties are said to be independent if one property can be varied by keeping the other one constant. Temperature and specific volume are two independent properties that can fix the state of a simple compressible system. Temperature and pressure, however, are two independent properties for a single-phase system and dependent properties for multiphase systems. For example, if we consider the two-phase process of boiling of water, boiling temperatures are different at different pressures, i.e., 100°C at 1 atm pressure and 120°C at 2 atm pressure. Therefore, to fix the state of a multiphase system, temperature and pressure are not sufficient, but one more property is required to be specified.

The State Postulate-II

Postulate I ensures that the state of a thermodynamic system is completely specified by independent variables. Now it is desirable to see how these variables change in a change of state, i.e., changing conditions resulting in a new equilibrium state. This can be specified by considering a simple system consisting of two subsystems called a composite system. The walls of the composite system and the wall dividing the system into two subsystems play a role in determining the consequences that the changing conditions will have on the system. Finally, this arrangement provides an understanding of the basic problem of thermodynamics related to the calculation of the properties of the new equilibrium state in a change of state. To develop the criterion of thermodynamic equilibrium, an extremum principle is formulated in postulate II of thermodynamics; according to it, there exists a function, known as entropy, of any composite system defined for all equilibrium states with the property that the values assumed by extensive parameters in the absence of an internal constraint are those which maximize the entropy over the manifold of constrained equilibrium states.

Process

A process takes place when a system undergoes a change of state or an energy transfer takes place at a steady state. Figure 1.5 shows the process in which a system undergoes a change of state from 1 to 2. A process may be a flow process or a non-flow process. A flow process is the one in which mass enters and leaves through the boundary of a control volume (open system). Example: A certain mass of working fluid (water) enters a steam power plant and undergoes a change of state in the flow process. In an open system, it is necessary to take account of the work delivered from the surroundings to the system at the entry to cause the mass to enter, and also of the work delivered from the system at surroundings to cause the mass to leave, as well as any heat or work crossing the boundary of the system. For example, a substance that is being heated in a closed cylinder undergoes a non-flow process. In a steady flow process, a certain quantity of mass enters the boundary from surroundings at entry and an equal mass leaves the boundary at the exit so that the total mass of the system remains constant.

Quasi-Static or Quasi-Equilibrium Process

If a system passes through a series of equilibrium states during a process, in such a way that at any instant the system is in equilibrium or infinitesimally close to being on equilibrium, then that process is called a quasi-static or quasi-equilibrium process. A closed system will be in quasi-equilibrium if there are no frictional effects and at any instant, the properties are uniform throughout; for an open system, quasi-equilibrium

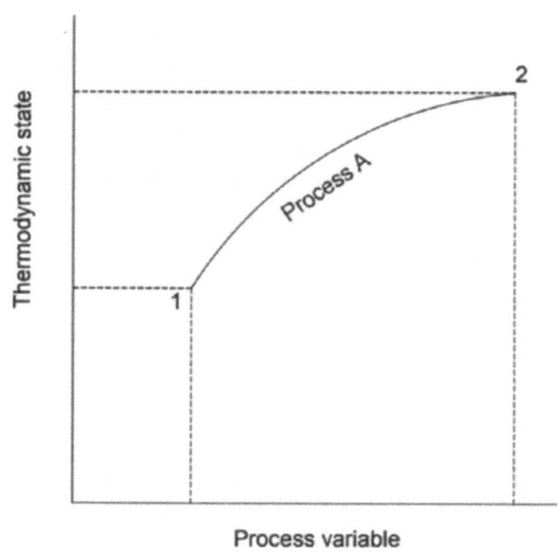

FIGURE 1.5 A thermodynamic process.

requires that there should not be friction but the properties may vary from point-to-point throughout. A quasi-static process is considered to take place infinitely slowly, so that the system adjusts to itself internally and the properties in one part do not change any faster than those at other parts of the system.

Let us consider a very slow frictionless isothermal process in which a gas is contained in a well-conducting cylinder having a gas-tight frictionless piston. If the gas is compressed very slowly, its temperature rises so that there is a temperature difference between the gas and surroundings allowing heat to transfer from the gas to surroundings, so that the gas is maintained at a constant temperature as work is done on it. Similarly, when the gas is expanded very slowly, its temperature decreases slightly and heat in this case transfers into the gas to maintain it at a constant temperature as the work is done by it. Only a minute temperature difference (almost it approaches zero) is allowed to keep the compression or expansion to occur a long time. The temperature inside the cylinder is uniform and it will rise uniformly at all loactions. Since equilibrium is maintained at all times, the process is a quasi-static process. If the compression or expansion is carried out suddenly, there is a large temperature difference, and the system can no longer be in equilibrium, which makes the entire process non-quasi-static.

Cycle

When a system, from its initial state, undergoes a series of state changes or a series of processes and finally returns to its initial state, then it is said to have undergone a *cycle*. Figure 1.6 shows the cycle in which a system undergoes a change of state from 1 to 2 and restores to the initial state from state 2 to 1; the final state is identical with the initial state; thus, it is called a cycle. All the properties of the system will have the same values at the conclusion of the cycle.

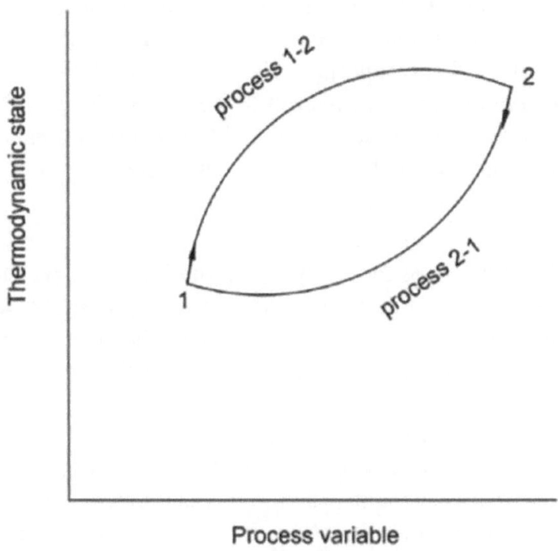

FIGURE 1.6 A thermodynamic cycle.

In steam power plants, the working fluid water undergoes a cycle. It is important here to distinguish between the thermodynamic cycle and the mechanical cycle. Water in steam power plants undergoes a thermodynamic cycle since it is reused after each cycle, while working fluid that is a fuel–air mixture in internal combustion engines undergoes a mechanical cycle since it is thrown out so that a fresh charge is inducted in each cycle.

1.5 HOMOGENEOUS AND HETEROGENEOUS SYSTEMS

A phase is defined as a quantity of matter which is homogeneous throughout its chemical composition and physical structure. There are three phases, namely, solid, liquid, and gas, and a substance can exist in any one of these three phases. A *homogeneous system* is the one that has a single-phase, a blend of gasoline and alcohol, for instance, is a homogeneous system since both are in the same phase, while a *heterogeneous system* has more than one phase, an ice cube in water, for instance, is a heterogeneous system since there are two phases(liquid and solid).

1.6 THERMODYNAMIC EQUILIBRIUM

A system is said to be in thermodynamic equilibrium if it is in thermal, mechanical, and chemical equilibrium with itself and with the surroundings. Temperature and pressure at all points are the same and there should be no velocity gradient. Systems under temperature and pressure equilibrium but not under chemical equilibrium are sometimes said to be in metastable equilibrium conditions. It is only under thermodynamic equilibrium conditions that the properties of a system can be fixed. Thus for attaining a state of thermodynamic equilibrium, the following three types of equilibrium states must be achieved: (i) *Thermal equilibrium:* the temperature of the system does not change with time and has the same value at all points of the system. (ii) *Mechanical equilibrium:* there are no unbalanced forces within the system or between the system and surroundings. The pressure in the system is the same at all points and does not change with respect to time. (iii) *Chemical equilibrium:* no chemical reaction takes place in the system and the chemical composition, which is the same throughout the system, does not vary with time.

Continuum

In the macroscopic approach to the study of thermodynamics, the substance is considered to consist of myriads of molecules that are widely spaced apart, and volumes are very large compared to molecular dimensions. The behavior of the matter is analyzed conveniently by disregarding the molecular-level interactions. So the matter is assumed to be continuous and homogeneous with no holes and hence treated as a continuum. The continuum theory is only an idealization that loses its validity when the mean free path of the molecules approaches the order of extent of the dimensions of the vessel, the one that is encountered in high-vacuum technology. In engineering practice, the concept of the continuum is valid, closely connected with the macroscopic view.

1.7 SPECIFIC VOLUME AND DENSITY

Volume V is the space occupied by a substance and is expressed in m^3. The *specific volume* is the volume per unit mass and is indicated by the symbol ν. The reciprocal of the specific volume is *density*, which is defined as the mass per unit volume. Density is designated by the symbol ρ.

$$\nu = \frac{V}{m}\left(m^3/kg\right) \tag{1.2}$$

$$\rho = \frac{m}{V}\left(kg/m^3\right) \tag{1.3}$$

or

$$\rho = \frac{1}{\nu}$$

Density, in case of a differential volume element of mass δm and volume δV, can be written as

$$\rho = \frac{\delta m}{\delta V}$$

Volume is also specified by another commonly used unit liter (l).

$$1l = 10^{-3}\,m^3$$

Specific volume or density may be expressed on the basis of either mass or mole. A mole of a substance has a mass numerically equal to its molecular weight. One g mol of carbon has a mass of 12 g and 1 kg mol of hydrogen has a mass of 2 kg. Molar specific volume is indicated by the symbol, $\bar{\nu}$, and is expressed in m^3/k mol.

Specific Gravity

If the density of a substance is denoted with respect to the density of a standard substance, then it is called *specific gravity (S.G)* or relative density. It is defined as the ratio of the density of a substance to the density of some standard substance at a specified temperature (usually, water at 4°C, with a density of 1000 kg/m^3 is considered a standard substance). It is given by

$$S.G = \frac{\rho}{\rho_{H_2O}} \tag{1.4}$$

Specific gravity of a substance is a dimensionless quantity. The numerical value of specific gravity of a substance in SI units is numerically equal to its density in gm/cm^3. Water, for example, has a specific gravity of 1 meaning that its density at 4°C is 1 gm/cm^3 and mercury, however, has a specific gravity of 13.6 and its density at 0°C is 13.6 gm/cm^3.

1.8 PRESSURE

Generally, we speak of pressure when we are dealing with liquids and gases while stresses in dealing with solids. According to Pascal's law, the pressure in a fluid at rest at a given point is the same in all directions. Pressure is defined as the normal component of force exerted per unit area. When a uniform pressure acts on a flat plate of area A and a force F pushes the plate, then

$$\text{Pressure} = \frac{\text{Normal force}}{\text{Area}}$$

$$p = \frac{F}{A}\left(N/m^2\right) \tag{1.5}$$

where p is the pressure and F is the pressure force. If the pressure is not uniform, the variable pressure ΔP acting on the minute area ΔA is expressed by the following equation:

$$p = \lim_{\Delta A \to 0} \frac{\Delta P}{\Delta A} = \frac{dP}{dA} \tag{1.6}$$

Units of Pressure

The unit of pressure is pascal (Pa), but it can also be expressed in bars or mm of mercury column (mm Hg) or meters of water column (m H_2O). The conversion table of pressure units is given in Table 1.1. The pressure of 1 atm = 760 mm Hg (at 273.15 K, $g = 9.807\ m/s^2$) = 101.325 kPa. One atmospheric pressure is standard atmospheric pressure in metrology and is called the standard atmospheric pressure. Table 1.1 shows the conversion units of pressure.

Types of Pressure

i. Absolute pressure
ii. Gauge pressure
iii. Vacuum pressure

TABLE 1.1
Pressure Conversion Units

Name of the Unit	Unit	Conversion
Pascal	Pa	$1\ Pa = 1\ N/m^2$
Bar	bar	$1\ bar = 10^5\ N/m^2$
Atmospheric pressure	atm	1 atm = 101.325 kPa
Water column meter	m H_2O	1 m H_2O = 9806.65 Pa
Mercury column mm	mm Hg	1 mm Hg = 1/760 atm
Torr	torr	1 torr = 1 mm Hg
Pound-force per square inch	psi	$1\ psi = 1/14.223\ kgf/cm^2$

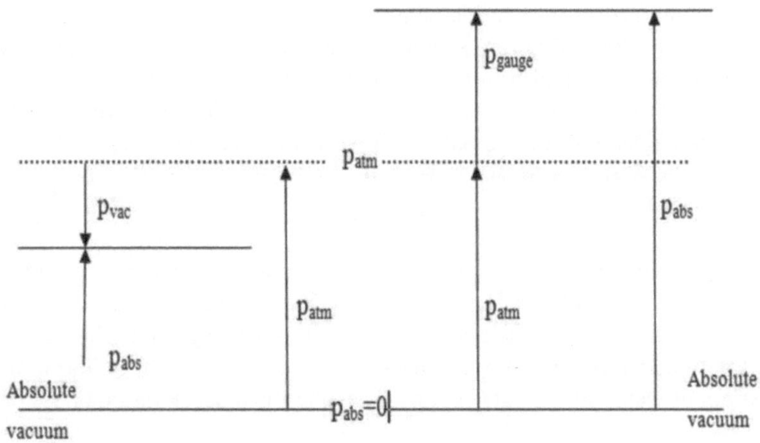

FIGURE 1.7 Absolute, gauge, and vacuum pressures.

The actual pressure at a given position is called *absolute pressure*, p_{abs}, and is measured with respect to absolute zero pressure (absolute vacuum). The difference between absolute pressure and local atmospheric pressure is called *gauge pressure*, p_{gauge}. It is the difference between the pressure of a fluid and the pressure of the atmosphere. Most pressure-measuring devices are calibrated to read zero in the atmosphere; therefore, they indicate the difference between absolute and local atmospheric pressures. The pressure below atmospheric pressure is called *vacuum pressure*, p_{vac}, and it is measured by vacuum gauges. For example, a gauge pressure of 50 kPa is spoken of as a vacuum of 50 kPa. The vacuum is created in steam condensers; to measure this vacuum pressure, vacuum gauges are used. There are two methods based on which the pressure is measured: one is based on the perfect vacuum and the other on the atmospheric pressure. The former is called the absolute pressure and the latter is called the gauge pressure.

If the pressure measured is above the atmospheric pressure, then

$$\text{Gauge pressure} = \text{absolute pressure} - \text{atmospheric pressure}$$

If the pressure measured is less than the atmospheric pressure, then

$$\text{Vacuum pressure} = \text{atmospheric pressure} - \text{absolute pressure}$$

The above relation is shown in Figure 1.7. Most gauges are constructed to indicate the gauge pressure only. For example, the gauge connected to an automobile tire indicates the gauge pressure of 32 psi, that is, 32 psi (1 kgf/cm^2 = 14.223 psi) above the atmospheric pressure.

1.9 PRESSURE-MEASURING DEVICES

Atmospheric pressure, p_{atm}, is measured using a barometer and hence atmospheric pressure is also called barometric pressure. Italian scientist E. Torricelli measured atmospheric pressure by inverting a tube filled with mercury into a container that is open to the atmosphere. Figure 1.8 shows the barometer used for this purpose. The pressure at point L is equal to atmospheric pressure, and pressure at M is zero as there is only mercury vapor above M; as it is very low relative to atmospheric pressure, it can be neglected. A force balance in vertical direction gives

$$p_{atm} = \rho g z \tag{1.7}$$

where ρ is the density of mercury, g is the gravitational acceleration, and z is the height of the mercury column above the free surface; the effects of tube length and cross-sectional areas on the height of fluid column are negligible. Standard atmosphere is defined as the pressure produced by a mercury column of 760 mm height (or a column of water 10.3 m) at 0°C under gravitational acceleration of 9.8 m/s^2.

Pressure Measurement

There are two types of pressure-measuring devices: manometers and mechanical gauges. Manometers measure the pressure at a single point or multiple points in a single pipeline or multiple pipelines by balancing the fluid column by the same or another column of fluid. Mechanical gauges operate on the principle of elastic deformation of measuring element under the influence of pressure and this motion is coupled to a pointer mechanism. The most commonly used mechanical pressure-measuring device is the Bourdon tube. It consists of a hollow metal tube bent into a hook shape, whose end is closed and connected to a dial indicator needle. The tube is undeflected when it is open to the atmosphere, and at this state, the needle on the dial is calibrated to zero (gauge pressure). When the pressure is applied to the fluid inside the tube, the tube stretches and moves the needle in proportion to the applied pressure. Figure 1.9 shows the Bourdon tube pressure gauge.

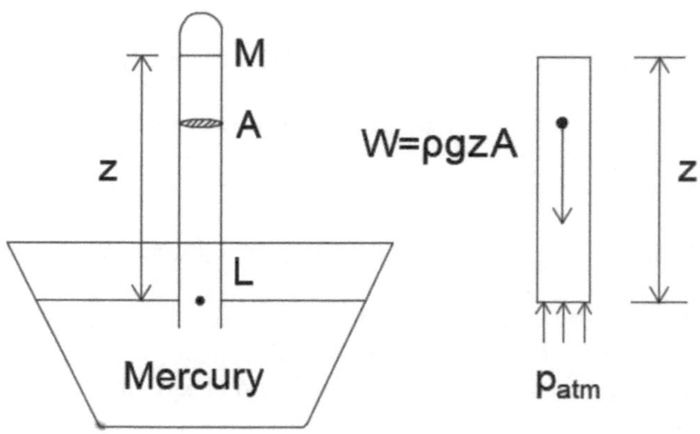

FIGURE 1.8 The barometer.

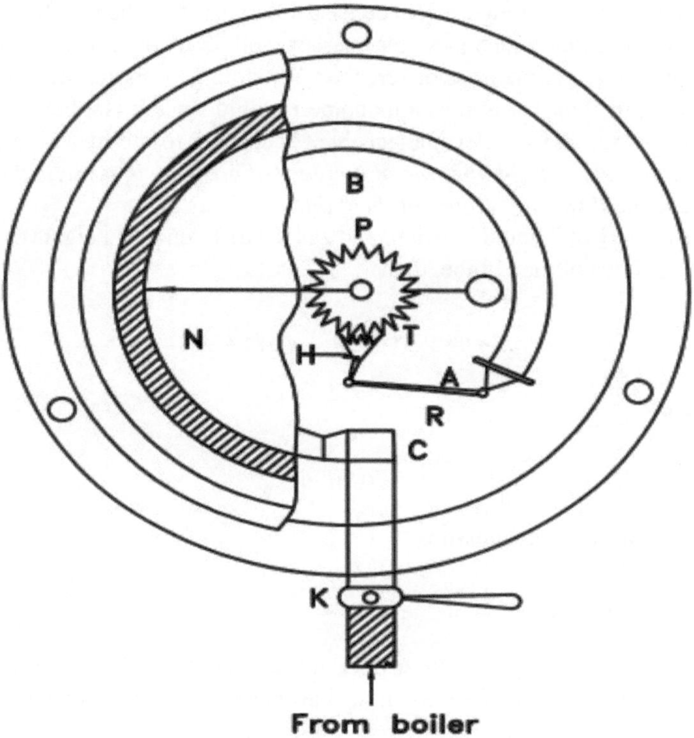

FIGURE 1.9 Bourdon tube pressure gauge.

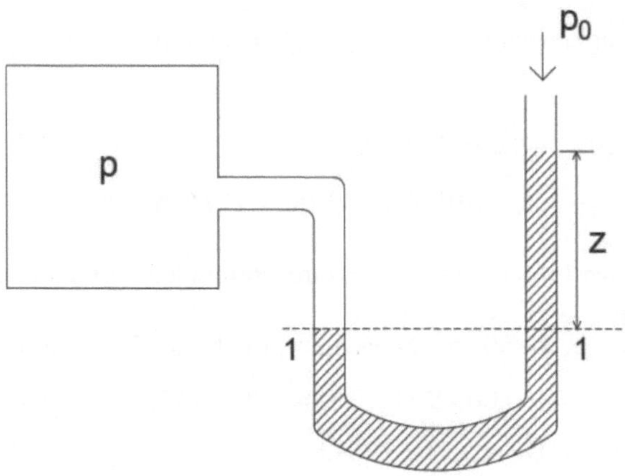

FIGURE 1.10 U-tube manometer.

Manometer as shown in Figure 1.10 consists of a glass tube bent in U shape containing one or more fluids such as water, mercury, oil, or alcohol. It is used to measure small and moderate pressure differences. When large pressure differences are to be measured, mercury is used as a manometric fluid since it is a heavy fluid and keeps the size of the manometer manageable. According to the hydrostatic law, *at any point inside a static fluid, the rate of increase of pressure in a vertical direction must equal the local specific weight of the fluid.*

If ρ is the density of a liquid in a tank and z is the difference in heights of the fluid column in two limbs of the U-tube, then

$$\text{gauge pressure } p_{gauge} = \rho g z$$

Then pressure,

$$p = p_0 + \rho g z \tag{1.8}$$

where p_0 is the atmospheric pressure.

EXAMPLE PROBLEMS

Example 1.1 Determine the absolute pressure at a depth of 5 m below the free surface of a tank of oil (S.G = 0.8), when the barometer reads 760 mm Hg.

Solution The pressure at a location is estimated with $p = \rho g z$

$p_{abs} = p_{atm} + p_{gauge}$

$p_{atm} = \rho g z$ (where z = 760 mm or 0.76 m)

$= 13{,}600 \times 9.81 \times 0.76 = 101.396$ kPa

$p_{gauge} = \rho g z$, where $\rho_{oil} = S.G_{oil} \times \rho_{water} = 0.8 \times 1000 = 800\ \text{kg/m}^3$

$z = 5$ m

$= 800 \times 9.81 \times 5 = 39.240$ kPa

$p_{abs} = p_{atm} + p_{gauge} = 101.396 = 39.240 = 140.636$ kPa Ans.

Example 1.2 Convert the following pressures into kPa assuming that the barometer reads 760 mm Hg.
(i) 850 mm Hg gauge, (ii) 50 cm Hg vacuum, (iii) 1.3 m H_2O, and (iv) 2.5 bar.

Solution i. $p = \rho g z = 13{,}600 \times 9.81 \times 0.85 = 113.403$ kPa Ans.

ii. $13{,}600 \times 9.81 \times 0.5 = 66.708$ kPa Ans.

iii. $1000 \times 9.81 \times 1.3 = 12.753$ kPa Ans.

iv. when 1.01325 bar = 101.325 kPa

then 2.5 bar = 250 kPa Ans.

Example 1.3 What will be the gauge and absolute pressures of crude oil at a depth of 32 m below the earth's crust. What depth, below the surface of the oil, will produce a pressure of 132 kPa (density of crude oil: 782 kg/m³, atmospheric pressure: 101.32 kPa).

Solution Given: $\rho_{oil} = 782$ kg/m³, $p_{atm} = 101.32$ kPa

i. Gauge pressure, $p_{gauge} = \rho gz = 782 \times 9.81 \times 32 = 245.485$ kPa

   Absolute pressure $p_{abs} = p_{atm} + p_{gauge} = 101.32 + 245.485 = 346.805$ kPa

ii. $p_{gauge} = \rho gz = 132 = 782 \times 9.81 \times z \Rightarrow z = 17.20$ m

   The depth below the surface of the oil is 17.20 m which produces a pressure of 132 kPa.

Example 1.4 Determine the absolute pressure at a depth of 4.5 m below the free surface of a tank of water where the gauge pressure is 70 kPa (take barometric pressure as 760 mm Hg).

Solution $p_{atm} = \rho gz$, where h is the barometric height, i.e., 760 mm Hg

$13{,}600 \times 9.81 \times 0.76 = 101.396$ kPa

$p_{abs} = p_{atm} + p_{gauge} = 101.396 + 70 = 171.396$ kPa Ans.

Example 1.5 A vacuum gauge connected to a condenser reads 650 mm Hg. What is the absolute pressure in the condenser when the barometer reading is 760 mm Hg?

Solution $p_{abs} = p_{atm} - p_{gauge}$ (where p_{gauge} is the vacuum pressure)

$p_{atm} = \rho gz = 13{,}600 \times 9.81 \times 0.76 = 101.396$ kPa (due to barometer)

$p_{gauge} = \rho gz = 13{,}600 \times 9.81 \times 0.65 = 86.720$ kPa

$\therefore p_{abs} = 101.396 - 86.720 = 14.676$ kPa Ans.

Example 1.6 A mercury manometer is used to measure the pressure inside a container as shown in Figure Ex. 1.6. The density of mercury is 13,600 kg/m³ and the difference between the heights of the two columns of the manometer is measured to be 250 mm. Determine the pressure inside the container when the barometer reads 760 mm Hg.

Solution

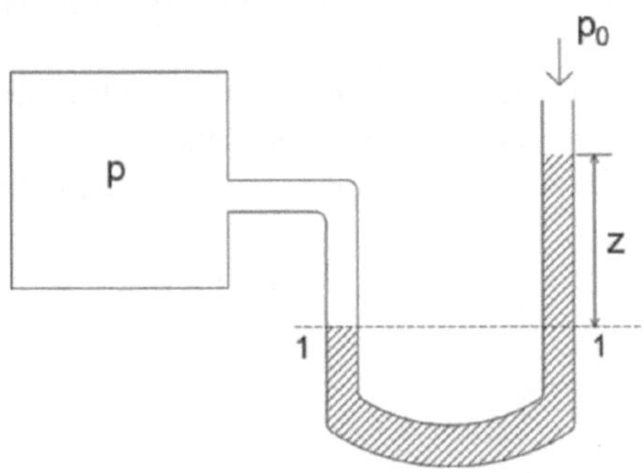

FIGURE EX. 1.6

At plane 1-1, we have from Eq. 1.8,

$$p = p_0 + \rho g z \qquad \text{(pressure is same in the horizontal direction)}$$

where p_0 is atmospheric pressure $= \rho g z_0$.

z_0 is the barometric height $= 760$ mm and $z = 250$ mm.

$\rho_{Hg} = 13{,}600$ kg/m^3

Therefore $$p = \rho g z_0 + \rho g z = \rho g(z + z_0)$$

$$13{,}600 \times 9.8 \times (0.250 + 0.760) = 134.612 \text{ kPa or } 1.346 \text{ bar} \qquad \text{Ans.}$$

REVIEW QUESTIONS

1.1 What is a quasi-static process? What is its importance in engineering?
1.2 Distinguish between intensive and extensive properties. Give examples.
1.3 What is thermodynamic equilibrium? Explain how it is achieved.
1.4 Distinguish between macroscopic and microscopic viewpoints.
1.5 Define control volume and control mass. Give examples.
1.6 Define system and what are the different means in which a system can interact with its surroundings?
1.7 How can an isolated system interact with its surroundings? Give some examples for isolated systems.
1.8 Define state, process, and cycle.
1.9 Distinguish between homogeneous and heterogeneous systems.
1.10 Distinguish between absolute, gauge, and vacuum pressures.
1.11 A pressure gauge attached to an automobile tire shows a reading of 33 psi, what does it mean and what is its value in kgf/cm^2?
1.12 How does the pressure change with the increase in elevation and increase in depth into the sea?
1.13 Convert the pressure of 750 mm Hg into mm of water column.
1.14 Convert the pressure of 200 kPa into pressure in bar and kgf/cm^2.
1.15 What is the difference between classical and statistical thermodynamics?
1.16 Is the density of a fluid an intensive or extensive property?
1.17 Is the weight of a body an intensive or extensive property?
1.18 What is state postulate?
1.19 Consider the case of steam expansion in a steam turbine, what would you choose as the system and what type of system it is?
1.20 What type of system is the expansion of hot gases in a piston-cylinder device?

EXERCISE PROBLEMS

1.1 A 4.5 kg plastic tank having a volume of 0.32 m^3 is filled with an oil of density 800 kg/m^3. Determine the weight of the combined system.

1.2 If Superman has a mass of 120 kg on his birth planet Krypton, where the acceleration of gravity is 22 m/s^2, determine (i) his weight on Krypton, in N, and (ii) his mass, in kg, and weight, in N, on Earth where g is 9.81 m/s^2.

1.3 A spring compresses in length by 0.24 cm for every 1 kg of applied force. Determine the mass of an object, in kg mass, which causes a spring deflection of 2 cm. The local acceleration of gravity is 9.81 m/s^2.

1.4 A closed system consists of 0.45 kmol of ammonia occupying a volume of 6.4 m^3. Determine (i) the weight of the system, in N, and (ii) the specific volume, in m^3/kmol and m^3/kg. Let $g = 9.81$ m/s^2.

1.5 Determine the atmospheric pressure at a location where the barometer reads 755 mm Hg. Take the density of mercury to be 13,600 kg/m^3.

1.6 The value of the gravitational acceleration decreases with elevation from 9.8 m/s^2 at sea level to 9.7 m/s^2 at an altitude of 12,000 m, where large passenger planes cruise. Determine the percent reduction in the weight of an airplane cruising at 12,000 m relative to its weight at sea level.

1.7 In water, the absolute pressure at a depth of 300 mm is 1.2 bar. Determine (i) the local atmospheric pressure and (ii) the absolute pressure at a depth of 300 mm in a liquid whose specific gravity is 0.88 at the same location.

1.8 A vacuum gauge is connected to a tank at a location where the barometer reads 760 mm Hg. Determine the absolute pressure in the tank when the vacuum gauge reads 35 kPa. Take the density of mercury as 13,600 kg/m^3.

1.9 The difference between the heights of two columns of a manometer is 190 cm, when a fluid of density 878 kg/m^3 is used. Determine the pressure difference. What is the height difference if the same pressure difference is measured using mercury ($\rho = 13{,}600$ kg/m^3) as a fluid in the manometer.

1.10 The pressure gauge attached to an air tank shows 78 kPa when the diver is 12.1 m down in the river. At what depth will the gauge pressure be zero?

1.11 A vacuum gauge connected to the condenser of a steam power plant records 725 mm of Hg. Determine the absolute pressure in the condenser when the barometric reading is 755 mm of Hg.

1.12 Convert the following pressures into kPa assuming that the barometer reads 760 mm Hg; (i) 70 mm Hg gauge, (ii) 35 psi, (iii) 1.3 atm, and (iv) 5 kgf/cm^2.

1.13 A vacuum gauge attached to a steam condenser records 650 mm Hg, what does it mean and what is its value in kgf/cm^2?

2 Temperature: Zeroth Law of Thermodynamics

LEARNING OUTCOMES

After learning this chapter, students should be able to

- Understand the importance of zeroth law of thermodynamics in temperature and its measurement.
- Interpret the basic concepts of temperature-measuring scales.
- Evaluate the relative performance of temperature-measuring instruments.
- Define international temperature scale.

2.1 TEMPERATURE

Temperature is a familiar property that we come across in daily life; however, it is difficult to define it exactly. Temperature is a sense of hotness or coldness when we touch an object. When a hot body is brought in contact with a cold body, the hot body becomes cooler and the cold body becomes warmer owing to heat exchange resulting from the temperature difference. If we keep these two bodies in contact for some time, they attain a common temperature and are said to be in thermal equilibrium with each other. However, we also realize that our sense of hotness or coldness is very unreliable. Sometimes very cold bodies may seem hot, and bodies of different materials that are at the same temperature appear to be at different temperatures.

The temperature is a property that plays an important role in thermodynamics. Based on experimental evidence, it is found that certain measurable properties of materials change with a change in temperature. It forms the basis for measurement of temperature by measuring devices known as thermometers. For example, the volume of the hot body decreases slightly with time while that of the cold body increases slightly. This will be continued till all changes in such observable properties stop, then the thermal interaction ceases. The two bodies are said to have reached a state of thermal equilibrium. The only required thing for attainment of thermal equilibrium is the equality of a property known as temperature. Hence we assume that when the two blocks are in thermal equilibrium, their temperatures are equal.

2.2 ZEROTH LAW OF THERMODYNAMICS

The zeroth law of thermodynamics can be stated as follows: *when two objects are in thermal equilibrium with a third object, those two are in thermal equilibrium with one another.* This law forms the basis of temperature measurement. So many properties of the materials vary with variation in temperature and make it convenient to accurately measure the temperature. To find out whether the two objects are in thermal equilibrium, i.e., at the same temperature, it is not required that they be brought in contact with each other and to find whether their properties vary with time. But it is required that whether they are individually in thermal equilibrium with a third object, which is obviously a thermometer.

The zeroth law is not derived based on other laws and it precedes the first and second laws of thermodynamics; therefore, it is called the zeroth law of thermodynamics. Every time a body has equality of temperature with the thermometer, we can say that the body has the temperature we read on the thermometer. The problem remains of how to relate temperatures that we might read on different mercury thermometers or obtain from different temperature-measuring devices, such as thermocouples and resistance thermometers. This observation suggests the need for a standard scale for temperature measurements.

2.3 THERMOMETERS—TEMPERATURE MEASUREMENT

In temperature measurements, a measurable property that changes with a change in temperature is chosen and this property is called a thermometric property. The most commonly used temperature-measuring device is a liquid-in-glass tube thermometer in which the change in length of liquid, usually mercury, which changes with change in temperature, is considered as the thermometric property. There are other such devices as well for the measurement of temperature: constant volume gas thermometers, constant pressure gas thermometers, electrical resistance thermometers, and thermocouples with their respective thermometric properties as pressure, volume, resistance, and thermal electromotive force (e.m.f.).

2.3.1 Reference Points

For measurement of temperature, one common basis is required; the temperature scales provide us such a common basis. So many temperature scales have been introduced so far. These scales earlier relied on some easily reproducible states: ice point (the freezing point of water) and steam point (boiling point of water). Ice point can be defined as an equilibrium state of a mixture of ice and water with air saturated with vapor at 1 atm pressure. Steam point can be defined as an equilibrium state of a mixture of liquid water and water vapor (with no air) at 1 atm pressure. To measure the temperatures between the ice point and steam point, the distance between two marks is divided into 100 equal parts and each graduation is marked as 1°C on the Celsius scale. However, there are certain difficulties in using two fixed points, one is achieving equilibrium between pure ice and air-saturated vapor and the other is the sensitiveness of steam point to the changes in pressure. Owing to these difficulties, the use of two fixed points was discarded. In 1954, at the tenth General Conference on Weights and Measurements (CGPM) meeting, it was decided that the triple point of water, i.e., 273.16 K or 0°C, should be considered as the single fixed reference point. When a body whose temperature θ is to be measured is placed in contact with water at its triple point, the temperature can be obtained from the equation shown below,

$$\theta = 273.16\frac{X}{X_t} \tag{2.1}$$

where θ is the temperature to be measured, X is the thermometric property, and X_t is the thermometric property at the triple point. If the thermometric property is pressure (in the case of constant volume gas thermometer), then the above expression can be

$$\theta(P) = 273.16\frac{P}{P_t}$$

Table 2.1 shows the temperatures of fixed points starting from oxygen point to gold point, which have been divided into three groups. From 0°C to 660°C, from −190°C to 0°C, and from 660°C to 1063°C.

2.3.2 Liquid-in-Glass Tube Thermometer

It consists of a capillary glass tube connected to a bulb filled with a liquid such as mercury or alcohol, the space above which is filled with the vapor of the liquid used. The other end of the glass is sealed. As the temperature rises, the liquid in the tube expands, the rise in length is proportional to the rise in temperature. The length of the liquid is the thermometric property in this device. This can be preferred for measuring ordinary temperature measurements. The main drawback of this device

TABLE 2.1
Temperatures of Fixed Points

Fixed Point	Temperature, °C
Normal boiling point of oxygen	−182.97
Triple point of water	0.01
Normal boiling point of water	100.00
Normal boiling point of sulfur	444.60
Normal melting point of zinc	419.50
Normal melting point of antimony	630.50
Normal melting point of silver	960.80
Normal melting point of gold	1063.00

is that it is not suitable for temperature measurements where greater accuracy is needed.

2.3.3 Gas Thermometers

This type of thermometers utilize gas as the thermometric substance. These are better devices than liquid-in-glass tube thermometers in the sense that they use gases that have a higher coefficient of expansion than that of liquids and are more sensitive to the changes in temperatures. There are two types of gas thermometers: constant volume gas thermometer and constant pressure gas thermometer. Constant volume gas thermometers are widely used because of their simpler construction and ease of use.

Constant Volume Gas Thermometer

It consists of a bulb in which a small amount of gas is filled. The tube is connected to one end of the manometer through a capillary tube. The other end of the manometer is open to the atmosphere, the mercury level is adjustable so that it touches the lip of the capillary tube. Figure 2.1 shows the constant volume gas thermometer.

In this device, the pressure in the bulb is used as the thermometric property that is given by

$$p = p_0 + \rho g z \tag{2.2}$$

where p_0 is the atmospheric pressure, ρ is the density of mercury, and z is the difference in heights of the mercury column of two limbs. The bulb, whose temperature is to be measured, is brought in contact with the system so that it comes in thermal

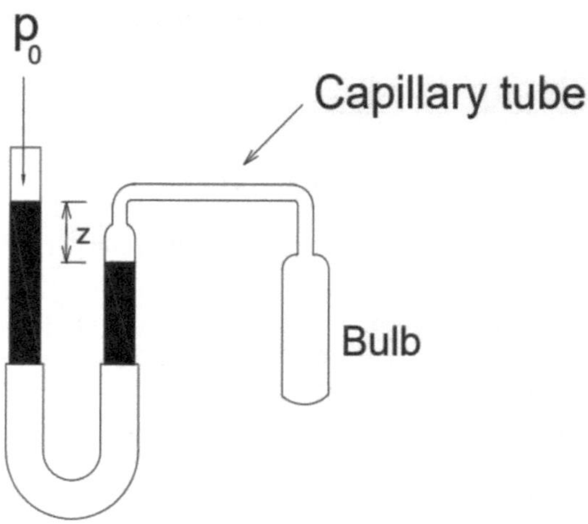

FIGURE 2.1 Constant volume gas thermometer.

equilibrium with the system in the meantime. The expanded gas in the bulb, being heated, pushes down the mercury, making the manometer limbs to be adjusted so that the mercury touches the capillary lip. The difference in mercury levels, z, is recorded and pressure in the bulb is estimated. From the ideal gas equation,

$$\Delta T = \frac{V}{R}\Delta p \tag{2.3}$$

The rise in temperature is proportional to the rise in pressure, since R is a constant, and also the volume of the gas trapped is constant.

2.3.4 Electrical Resistance Thermometer

In this, the electrical resistance of a metal wire, which changes with change in temperature, is chosen as the thermometric property. The platinum wire is commonly used in the Wheatstone bridge circuit of resistance thermometers. Figure 2.2 shows an electrical resistance thermometer.

For the measurement of temperature, the equation given below is used

$$R = R_0\left(1 + At + Bt^2\right) \tag{2.4}$$

where R_0 is the resistance of the platinum wire while A and B are constants.

The platinum electrical resistance thermometers are highly accurate and sensitive for the measurement of temperature and hence are used in the calibration of other thermometers.

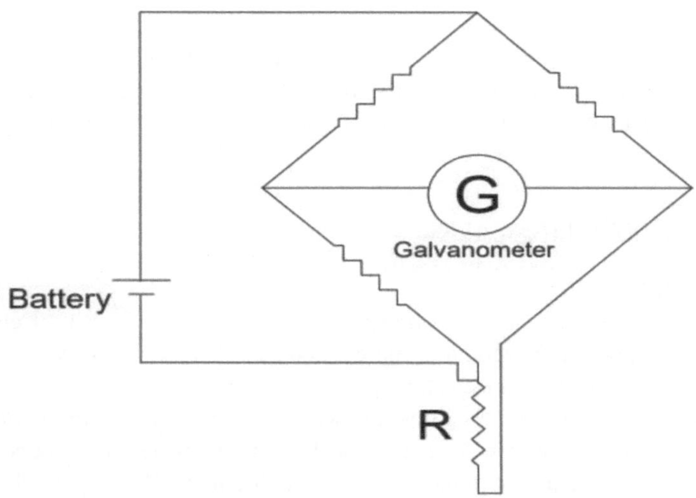

FIGURE 2.2 Electrical resistance thermometer.

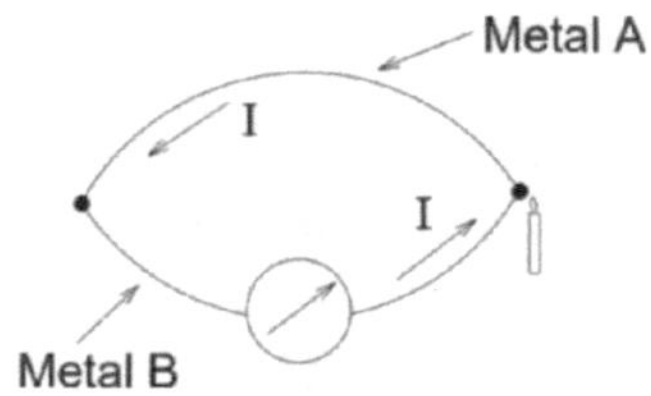

FIGURE 2.3 Thermocouple.

2.3.5 Thermocouple

Gas thermometers typically are large and cumbersome and not suitable for routine work as they are fixed in certain places. They are, however, used for the purpose of calibration and standardization of other temperature-measuring devices. A thermocouple is a simple device comparatively and it comes to thermal equilibrium with the system, whose temperature is to be measured very rapidly. It is the most widely used temperature-measuring device owing to the above advantages.

A thermocouple works on the principle of *Seebeck effect*, named after the German physicist Thomas Johann Seebeck (1770–1831) who discovered this principle in 1821, which is stated as follows: *when two wires made up of different metals are joined together to form a closed circuit and one of the junctions is heated, current flows continuously in the circuit.* Seebeck effect has two prominent applications, one is temperature measurement and the other is power generation.

Figure 2.3 shows a thermocouple made up of two dissimilar metal wires. A net e.m.f. is generated in the circuit due to the Seebeck effect and it is proportional to the temperature difference between the hot and cold junctions. This circuit is called the thermoelectric circuit since it integrates the thermal and electrical effects. When this circuit is broken, the current stops to flow and the e.m.f. or voltage generated in the circuit can be measured using a voltmeter.

The commonly used metal combinations for thermocouples are copper–constantan and chromel–alumel. The thermocouple made of copper–constantan wires is known as a T-type thermocouple, which produces nearly 40 μV of voltage per degree centigrade difference in temperature.

2.4 TEMPERATURE SCALES

Unit of Temperature

In SI units, temperature has units *degree Celsius* abbreviated as °C. Kelvin is the other unit of temperature in SI units abbreviated as K. The use of Kelvin is explained after the second law of thermodynamics.

The predominantly used temperature scales are the *Fahrenheit scale* and the *Celsius scale*. The Fahrenheit scale, used with the English system of units, is named after the German instrument maker Gabriel D. Fahrenheit (1686–1736). The Celsius scale, used with the SI system of units, was formerly known as the centigrade scale and renamed after the Swedish astronomer, Anders Celsius (1702–1744). Until 1954, both these scales were based on two fixed reference points: ice point and steam point.

The symbols °F and °C are used to indicate the units of temperature for both the Fahrenheit scale and Celsius scale, respectively. On the Celsius scale, the ice point and steam point were assigned the values of 0°C and 100°C, respectively, and on the Fahrenheit scale, the values are 32°F and 212°F respectively for both ice point and steam point. Since the temperature values were assigned at two different points, these are termed two-point scales.

A thermodynamic temperature scale is a temperature scale that is independent of the properties of any substance or substances, which is developed in concurrence with the second law of thermodynamics. *Kelvin scale* is the thermodynamic temperature scale in the SI system of units, which is named after Lord Kelvin (1824–1907). On the Kelvin scale, the unit of temperature is indicated by K.

Rankine scale, named after William Rankine (1820–1872), is the thermodynamic temperature scale in the English system of units. On this scale, Rankine is the temperature unit, designated by R.

2.4.1 Ideal Gas Temperature Scale

A constant volume gas thermometer as shown in Figure 2.4 is used for measuring the temperatures on this scale. It is based on the principle that as the pressure of a gas approaches zero, it obeys the ideal gas equation of state given as

$$Pv = RT$$

To maintain the constant volume of the gas, the mercury level in the tube is adjusted in such a way that it stands at a reference mark. The gas bulb is then placed in a location where the temperature is to be measured. It is assumed that the gas in the tube is

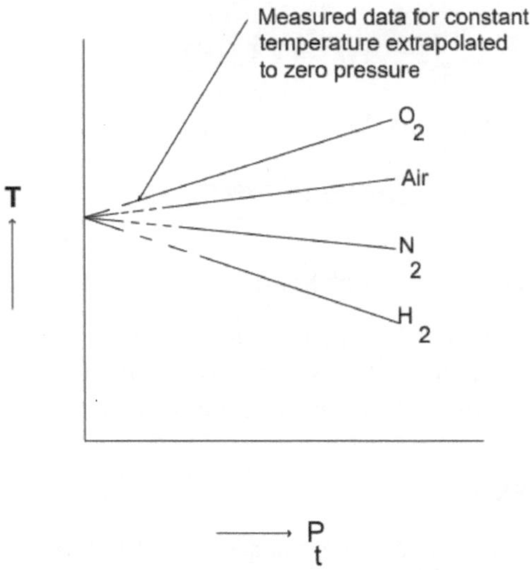

FIGURE 2.4 Determination of ideal gas temperature.

the same as the gas in the bulb, then the measurement of pressure of the gas can be used to determine the temperature.

Let pressure associated with the triple point of water be P_t, and by measuring this pressure, the unknown temperature T is measured by measuring the pressure P given by the relation

$$T = 273.16\left(\frac{P}{P_t}\right) \tag{2.5}$$

Equation 2.5 is obtained from Eq. 2.1 by replacing θ with T and thermometric property X with pressure P.

If the above experiment is repeated with varying amounts of gases in the bulb, the pressure measured at triple point and any other will vary. If we assume the gas as an ideal gas and the corresponding temperature T is plotted against the pressure corresponding to a triple point and the curve is extrapolated to absolute zero pressure, the ideal gas temperature can be obtained. If different gases are used, different curves will result as shown in Figure 2.4 and they meet at zero pressure, and the temperature of all the gases is the same at this zero pressure.

The temperature scales in all the above systems are related.

The Kelvin scale is related to the Celsius scale by

$$T(°C) = T(K) - 273.14 \tag{2.6}$$

The Rankine scale is related to the Fahrenheit scale by

$$T(°F) = T(R) - 459.67 \tag{2.7}$$

The Rankine scale is proportional to the Kelvin scale according to

$$T(R) = 1.8\ T(K) \tag{2.8}$$

Substituting Eqs. 2.5 and 2.6 into Eq. 2.7, we get

$$T(°F) = 1.8\ T(°C) + 32 \tag{2.9}$$

2.4.2 International Temperature Scale

In the seventh CGPM held in 1927, the international temperature scale is established and adopted, which forms the basis for easy and rapid calibration of scientific and industrial instruments. This is developed in compliance with the Celsius scale.

EXAMPLE PROBLEMS

Example 2.1 The temperature of water changes by 20°F during the heating process. Express this change in Celsius (°C), Kelvin (K), and Rankine (R) units.

Solution The temperature changes are identical in Celsius and Kelvin scales as

$$\Delta T(K) = \Delta T(°C)$$

The temperature changes are also identical in Fahrenheit and Rankine scales as

$$\Delta T(R) = \Delta T(°F)$$

Therefore, the temperature change in the Rankine scale is = 20 R
The temperature scales in two-unit systems are related as

$$\Delta T(R) = 1.8\ \Delta T(K)$$

Therefore, $\Delta T(K) = \frac{20}{1.8} = 11.11\,K$; this is also equal to the temperature change in the Celsius scale.

Example 2.2 The fire point of certain fuel is 350°F. What is the absolute fire point in Celsius (°C), Kelvin (K), and Rankine (R) units?

Solution The Rankine scale is related to Fahrenheit and as $\Delta T(R) = \Delta T(°F) + 459.67$

$$T(R) = 350 + 459.67 = 809.67\,R$$

$$T(R) = 1.8T(K) \Rightarrow T(K) = \frac{T(R)}{1.8}$$

$$\therefore T(K) = \frac{809.67}{1.8} = 449.82\,K$$

$$T(°C) = T(K) - 273.15 = 449.82 - 273.15 = 176.67°C \qquad \text{Ans.}$$

Example 2.3 A thermometer has a temperature scale of the relation $t = x\log_e k + y$, where x and y are constants and k is the thermometric property of the fluid used in the thermometer. The thermometric properties of the fluid respectively at ice and steam points are 1.75 and 8.5. Determine the temperature corresponding to the thermometric property value of 4.25 on the scale.

Solution Ice point temperature = 0°C, and thermometric property at the ice point, k = 1.75.
Steam point temperature = 100°C, and thermometric property at the steam point, k = 8.5.

Now $t = x\log_e k + y$

$$\text{At ice point} \Rightarrow \quad 0 = x\log_e 1.75 + y \Rightarrow y = -0.559x \tag{2.10}$$

$$\text{At steam point,} \quad 100 = x\log_e 8.5 + y \Rightarrow y = 100 - 2.14x \tag{2.11}$$

On solving Eqs. 2.10 and 2.11 $\Rightarrow x = 63.25$ and $y = -35.35$
By substituting x and y values in Eq. 2.5 $\Rightarrow t = 63.25\log_e 4.25 - 35.35$

$$t = 56.16°C \qquad \text{Ans.}$$

Example 2.4 The resistance of a platinum wire used in an electrical resistance thermometer is estimated to be 10.805, 14.583, and 29.332 Ω at ice point, steam point, and sulfur point respectively. Determine the constants A and B in $R = R_0(1 + At + Bt^2)$.

Solution At ice point, $t_1 = 0°C, \quad R_1 = 10.805\Omega$

At steam point $t_2 = 100°C, \quad R_2 = 14.583\Omega$

At sulfur point, $t_3 = 444.6°C, \quad R_3 = 29.332\Omega$

$$R = R_0\left(1 + At + Bt^2\right)$$

$$\text{Now at } t_1 = 0°C \Rightarrow 10.805 = R_0(1 + 0 + 0) \Rightarrow R_0 = 10.805 \quad (2.12)$$

$$\text{At } t_2 = 100°C \Rightarrow 14.583 = 10.805\left(1 + A \times 100 + B \times 100^2\right) \quad (2.13)$$

$$\text{At } t_3 = 444.6°C, \Rightarrow 29.332 = 10.805\left(1 + A \times 444.6 + B \times 444.6^2\right) \quad (2.14)$$

$$14.583 = 10.805 + A \times 1080.5 + B \times 10.805 \times 10^4 \quad (2.15)$$

$$29.332 = 10.805 + A \times 10.805 \times 444.6 + B \times 10.805 \times 444.6^2 \quad (2.16)$$

$$\text{Multiplying Eq. 2.15 with 4.46} \Rightarrow 65.040 = 48.19 + 48.19 \times 10^4 B \quad (2.17)$$

$$\text{Rearranging Eq. 2.16} \quad 29.332 = 10.805 + 10.805 \times 444.6^2 B \quad (2.18)$$

$$\text{Subtracting 2.18 from 2.17} \quad 72.708 = \left(4819 \times 10^2 - 2{,}135{,}815\right)B$$

$$\Rightarrow B = -4.35 \times 10^{-5} \quad \text{Ans.}$$

$$A = 7.84 \times 10^{-3} \quad \text{Ans.}$$

REVIEW QUESTIONS

2.1 How do you define temperature?
2.2 What is the zeroth law of thermodynamics?
2.3 What is a thermometric property?
2.4 What is a thermocouple and what is its purpose?
2.5 What are the reference points for temperature measurement in the past and at present?
2.6 What are the ordinary and absolute temperature scales in the SI and the English system?
2.7 The body temperature of a healthy person is 37°C, what is it in Fahrenheit?
2.8 Ambient temperature is 30°C. Express this temperature in R, K, and °F.
2.9 The temperature of a hot metallic bar drops by 20°C when it is cooled. Express this drop in temperature in Kelvin.
2.10 Explain how a thermometer measures the temperature of a human body.
2.11 What are the various temperature-measuring devices?
2.12 What are the thermometric properties in various temperature-measuring devices?
2.13 How is the thermodynamic temperature scale different from other temperature scales?

2.14 What is the principle used in the ideal gas temperature scale?
2.15 What is ice point?
2.16 What is steam point?

EXERCISE PROBLEMS

2.1 The 30-year average temperature in Toronto, Canada, during summer is 19.5°C and during winter is −4.9°C. What are the equivalent average summer and winter temperatures in °F and °R?
2.2 Convert the following temperatures from °F to °C: (i) 86°F, (ii) −22°F, (iii) 50°F, (iv) −40°F, (v) 32°F, and (vi) −459.67°F. Convert each temperature to K.
2.3 Natural gas is burned with air to produce gaseous products at 1985°C. Express this temperature in K, °R, and °F.
2.4 The temperature of a child ill with a fever is measured as 40°C. The child's normal temperature is 37°C. Express both temperatures in °F.
2.5 Does the Rankine degree represent a larger or smaller temperature unit than the Kelvin degree? Explain.
2.6 A thermocouple has its test junction in gas and reference point at the ice point. The temperature of gas is found to be 45°C with the gas thermometer. The thermocouple is calibrated with its e.m.f. varying linearly between ice and steam points. The e.m.f. produced when the thermocouple has its test junction in gas at t°C and reference point at ice point is given by $e = 0.16t - 4.8 \times 10^{-4} \times t^2$ millivolts. Estimate percent variation in temperature reading.
2.7 The temperature of 0°C on a temperature scale is equivalent to 200°X and 100°C is equivalent to 400°X. Compute the temperature corresponding to 300°X.

DESIGN AND EXPERIMENT PROBLEMS

2.8 Using different temperature-measuring instruments such as thermometer, thermocouple, and thermistors, measure the temperature of the flames near 1000 K. Suggest the most suitable measuring instrument for this measurement.
2.9 The refrigerant fluid in a household refrigerator changes its phase from liquid to vapor at the low temperature in the refrigerator. It changes phase from vapor to liquid at a higher temperature in the heat exchanger that gives the energy to the room air. Measure or otherwise estimate these temperatures. Based on these temperatures, make a table with the refrigerant pressures for the refrigerants. Discuss the results and the requirements for a substance to be a potential refrigerant.

3 Energy and the First Law of Thermodynamics

LEARNING OUTCOMES

After learning this chapter, students should be able to

- Define the three mechanisms of heat transfer: conduction, convection, and radiation.
- Analyze several forms of mechanical work and electrical work.
- Apply the first law of thermodynamics for closed systems and construct conservation of mass and energy equations.
- Define the first law of thermodynamics for a closed system undergoing a cyclic process and a change of state.
- Deduce the relationship among internal energy, enthalpy, and specific heats of solids, liquids, and gases.

3.1 ENERGY ANALYSIS

The two principles of classical thermodynamics that need much emphasis in the engineering perspective are the first and second laws of thermodynamics. The first law deals with the equivalence of work transfer and heat transfer, which are the two means of energy transfer. The first law of thermodynamics, also known as the principle of conservation of energy, and the energy equation are alternative expressions for the same fundamental principle. However, the difference lies in the expression of both the first law and the energy equation. The energy equation is stated for a control mass undergoing a process with a change of state of the system with time. It is then expressed for a complete cycle and recognized as the first law of thermodynamics. The energy equation can be used to establish a relationship between the change of state in a process inside a control volume and the amount of energy transferred as work or heat. For example, if a steam turbine provides a certain amount of work to a rotor, it will rotate; in this, the increase in kinetic energy can be related to the work transfer; similarly, if heat is added to the water in the boiler, the water evaporates, and the rise in water temperature can be related to the heat transfer. The second law is presented in Chapter 6.

Energy exists in several forms such as mechanical, thermal, electrical, chemical, and nuclear energy. Mass is also regarded as a form of energy. Energy transfer either into or out of a closed system takes place by two mechanisms: heat transfer and work transfer, and it involves no mass crossing its boundary. For open systems, it is by mass flow also. An energy transfer either into or out of a closed system is heat, if it

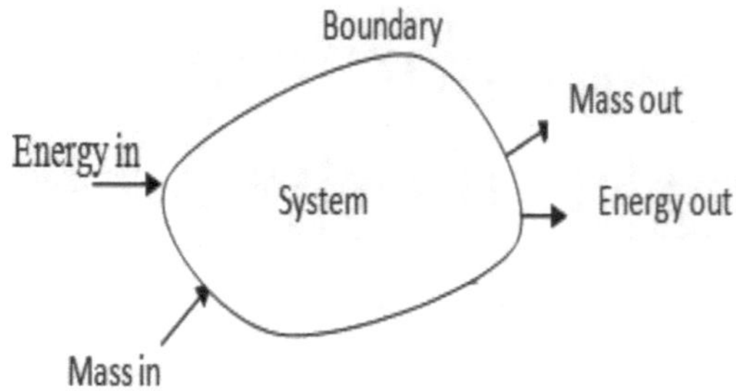

FIGURE 3.1 An open system interacting with its surroundings.

is as a result of a temperature difference and it is work, if it is as a result of a force acting through a distance. Figure 3.1 shows how an open system interacts with its surroundings.

3.2 DIFFERENT FORMS OF STORED ENERGY

The word energy was first used by Thomas Young in 1807, whereas its use in thermodynamics was suggested by Lord Kelvin in 1852. The total of all the energies—mechanical, thermal, electrical, chemical, and nuclear energies—is called the total energy of a system, indicated by E. The total energy on a unit mass basis is indicated as

$$e = \frac{E}{m} \tag{3.1}$$

The total energy of a system can be divided into macroscopic and microscopic forms. The macroscopic form of energy is the energy a system possesses as a whole with respect to some reference point. Kinetic and potential energies are of the macroscopic form. The microscopic form, on the other hand, is concerned with the molecular structure of a system and molecular activity. The internal energy is the sum of all the microscopic forms of energy, which is indicated by U.

The macroscopic form of energy caused by motion is called kinetic energy and it is influenced by gravity, electricity, magnetism, and so on. This type of energy is the energy possessed by a moving device. The kinetic energy is expressed as

$$KE = \frac{1}{2} mV^2 \text{ in kJ} \tag{3.2}$$

where m is mass in kg and V is the velocity of the system in m/s. In the case of a solid body rotating with an angular velocity ω, it is indicated by

$$\frac{1}{2} I \cdot \omega^2 \tag{3.3}$$

where I is the moment of inertia of the body.

The energy possessed by a system due to its elevation in the gravitational field is called the potential energy, which is expressed as PE = mgh, where g is the gravitational acceleration and h is the height or elevation of the center of gravity of a system with respect to some standard reference datum.

The total energy consists of kinetic, potential, and internal energies only in the absence of magnetic, electric, and surface tension effects (which are considered in some special cases only). In such cases, the total energy is given by

$$E = \underbrace{U}_{\text{micro}} + \underbrace{KE + PE}_{\text{macro}} \tag{3.4}$$

where U is the total internal energy, $KE = \frac{1}{2} mV^2$, and PE = mgh.

On a unit mass basis, it can be expressed as

$$e = u + ke + pe = u + \frac{V^2}{2} + gh \tag{3.5}$$

For closed stationary systems in which there are no changes in kinetic and potential energies, the change in total energy is equal to the change in internal energy. Usually, a closed system is treated as stationary when its velocity and elevation with respect to a standard datum remain unchanged. Thus, for a closed stationary system

$$\Delta E = \Delta U \tag{3.6}$$

Open systems, unlike closed systems, essentially involve fluid and hence, energy flow related to the fluid flow can be expressed in the rate form with the use of mass flow rate $\dot{m}$ and volume flow rate $\dot{V}$. Mass flow rate is defined as the amount of mass flowing across a cross-section per unit time (kg/s), whereas volume flow rate is the volume of fluid flowing across a cross-section per unit time (m^3/s). Then the mass flow rate is given by

$$\dot{m} = \rho\dot{V} = \rho AV \tag{3.7}$$

where ρ is the fluid density (kg/m^3), A is the cross-sectional area of flow (m^2), and V is the average velocity of flow normal to flow area. Then the energy flow rate (kJ/s or kW) of fluid is given by

$$\dot{E} = \dot{m}e \tag{3.8}$$

3.3 POINT FUNCTION AND PATH FUNCTION

In order to understand clearly the differences between the properties, such as temperature, pressure, and specific volume, and quantities, such as heat and work, it is essentially important to understand both point and path functions. If a thermodynamic system in a change of state depends on initial and final states and is independent of

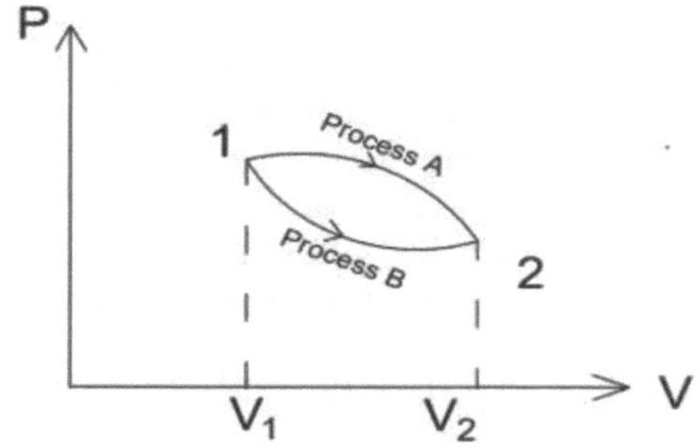

FIGURE 3.2 Diagram representing that properties are point functions.

the path followed, then it is termed *point function* and if it depends on the path followed, then it is termed *path function.* Properties are point functions and quantities such as heat and work are path functions.

Point functions have exact differentials designated by the symbol 'd', for example, a change in volume is indicated by 'dV'. Path functions, on the other hand, have inexact differentials designated by the symbol 'δ'. A differential change in heat or work is indicated by δQ or δW and not by dQ or dW. The change in volume between states 1 and 2 is the volume between state 2 minus state 1 irrespective of the path the process followed whether A or B as shown in Figure 3.2.

$$\int_1^2 dV = V_2 - V_1 = \Delta V \tag{3.9}$$

Total work done when a process undergoes a change of state from 1 to 2 is given as

$$\int_1^2 \delta W = W_{12} \neq \Delta W \tag{3.10}$$

Total work is the sum of the differential amounts of work δW along the process path. Work is not a property, it does not have a value at each state, and hence, it is a path function not a point function.

3.4 HEAT TRANSFER

Heat and work are the two means of energy transfer that occur at the boundaries of a system. They are the two forms of energy, yet they are not the same, and thus it is essential to differentiate between these two. If a hot iron bar is placed in a mug of cold water, the iron bar cools down eventually and the water warms up until the iron and water attain the common temperature. Both the decrease in the temperature of the iron bar and the increase in the temperature of the water are caused by the transfer of energy, in the form of heat from the iron bar to the water.

Heat is defined as a form of energy that is transferred across the boundary of a system by virtue of temperature difference. Heat transfer takes place from the system at a higher temperature to the system at a lower temperature, and it occurs exclusively due to the temperature difference between the two systems. Heat is energy in transition and is recognized only as it crosses the boundary of a system. In thermodynamics, the term heat simply means heat transfer. For example, a hot iron bar contains energy, but this energy can be heat transfer as and when this heat crosses the system boundary during the cooling process.

An adiabatic process is defined as a process during which there is no heat transfer. The word 'adiabatic' comes from the Greek word adiabatos, which means not to be passed. An adiabatic process can be made by insulating the system so that only an insignificant amount of heat passes through the boundary. It is important to note that though an adiabatic process does not involve any heat transfer, it involves work interaction between the system and surroundings, which can change the energy content and thus the temperature of the system.

Heat is denoted by the symbol Q and since it is a form of energy, it has energy units kJ. Heat transfer rate is the amount of heat transferred per unit time and has the unit kJ/s (kW). Heat transferred into a system is taken to be positive and that transferred out is negative. Heat is a path function and is considered an inexact differential. That is, the amount of heat transferred, when a system changes its state from 1 to 2, depends on the path the system follows during that change of state. The amount of heat transfer during a process is determined by integrating the inexact differential, δQ

$$\int_1^2 \delta Q = Q_{1\text{-}2} \tag{3.11}$$

where $Q_{1\text{-}2}$ is the heat transferred during the process between states 1 and 2.

Heat transfer per unit mass of a system is denoted by q and is determined from

$$q = \frac{Q}{m}(\text{kJ/kg}) \tag{3.12}$$

The rate of heat transfer per unit area normal to the direction of heat transfer is called heat flux, and the average heat flux is expressed by

$$\dot{q} = \frac{\dot{Q}}{A}\left(\text{W/m}^2\right) \tag{3.13}$$

where $\dot{Q}$ is the rate of heat transfer in kJ/s or kW, and A is the heat transfer area in m^2.

Modes of Heat Transfer

Basically, heat is transferred by three modes: conduction, convection, and radiation. *Conduction* is the type of heat transfer from the particles of higher energy levels of a substance to the adjacent particles of lower energy levels resulting from the interaction between particles. *Convection* is the type of heat transfer between a solid surface and an adjacent fluid that is in motion, and it involves the combined effects of conduction and fluid motion. Examples include air circulation with a fan or blower in a room or flow through heat exchangers such as a condenser in which water/air is circulated through piping to cool the working fluid. *Radiation*, on the other hand, is the type of heat transfer that takes place due to the emission of electromagnetic waves (or photons). Thermal radiation is the energy emitted by matter that is at a non-zero temperature. Radiation, unlike the other two modes, can even happen in vacuum (most efficiently) and does not require any medium; however, the emission of the radiation and the absorption require a substance to be present. Although the radiation from solid surfaces is of importance all the time, liquids and gases also emit radiation.

Heat conduction is governed by Fourier's law of heat conduction, which states that the heat conducted is proportional to the area normal to the direction of heat transfer and temperature gradient in that direction. Figure 3.3 shows the one-dimensional plane wall which has a temperature distribution T(x); the rate equation of heat conduction for this plane wall is given by

$$\dot{Q}_{cond} = -kA\frac{dT}{dx} \tag{3.14}$$

where $\dot{Q}_{cond}$ is the rate of heat conduction in the x-direction normal to the direction of heat transfer and dT/dx is the temperature gradient in that direction. The parameter k is a transport property known as thermal conductivity (W/m K). The negative sign is a consequence of the fact that heat always flows in the direction of decreasing temperature (temperature gradient is negative in the direction of heat transfer).

Convection is termed *forced convection*, if the fluid motion is caused by a fan, pump, or blower, whereas it is called *free or natural convection*, if the fluid motion is caused by buoyancy forces induced by density differences due to the temperature difference in the fluid. Figure 3.4 shows the heat transfer by natural convection from a hot fluid as the fluid motion (in the absence of a pump) is caused by buoyancy forces, i.e., the rise of lighter fluid near the surface and the fall of the heavier fluid to fill its place due to the temperature difference in the fluid.

The rate of heat transfer by convection is governed by Newton's law of cooling, given as

$$\dot{Q}_{conv} = hA(T_S - T_\alpha) \tag{3.15a}$$

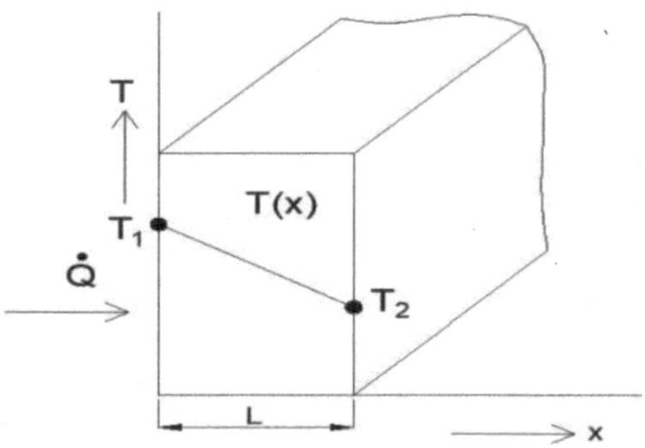

FIGURE 3.3 Heat transfer by conduction.

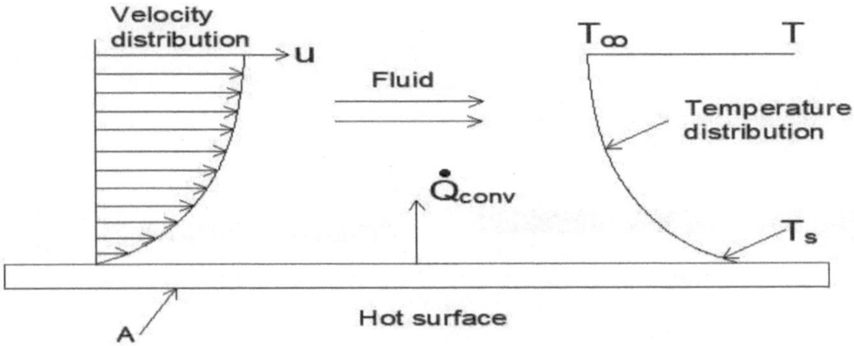

FIGURE 3.4 Heat transfer by convection from a hot fluid.

where h is a parameter (not the property of the fluid) called convective heat transfer coefficient, A is the surface area through which heat transfer takes place, T_S is the surface temperature, and T_α is the bulk fluid temperature away from the surface. The heat transfer coefficient depends on the conditions that exist in the boundary layer that are influenced by the geometry of the surface, nature of fluid motion, and transport properties. In Eq. 3.15a, convection heat transfer is taken to be positive if heat is transferred from the surface, whereas it is negative when the heat is transferred to the surface. In such a case, we can write Eq. 3.15a as

$$\dot{Q}_{conv} = hA(T_\alpha - T_S) \tag{3.15b}$$

The radiation that is emitted by the surface, as shown in Figure 3.5a, originates from the thermal energy of matter bounded by surface, and the rate at which energy is released per unit area (W/m^2) is called the surface *emissive power*, E. Black body is an ideal radiator which emits maximum energy, which is given by the Stefan–Boltzmann law

$$E_b = \sigma T_s^4 \tag{3.16}$$

where T_S is the absolute temperature (K) of the surface and σ is the Stefan–Boltzmann constant ($\sigma = 5.67 \times 10^{-8}$ W/m^2 K^4). A black body is an idealized body that emits maximum radiation. The radiation emitted by all real surfaces is less than that emitted by a black body at the same temperature, given by

$$E = \varepsilon \sigma T_s^4 \tag{3.17}$$

where ε is the *emissivity* of a surface, defined as the ratio of emissive power of a surface to the emissive power of a black surface. Absorptivity (α) is another property defined as the fraction of radiant energy incident on a surface that is absorbed by the surface. The rate at which a surface absorbs radiation, G_{abs}, is given by

$$G_{abs} = \alpha G \tag{3.18}$$

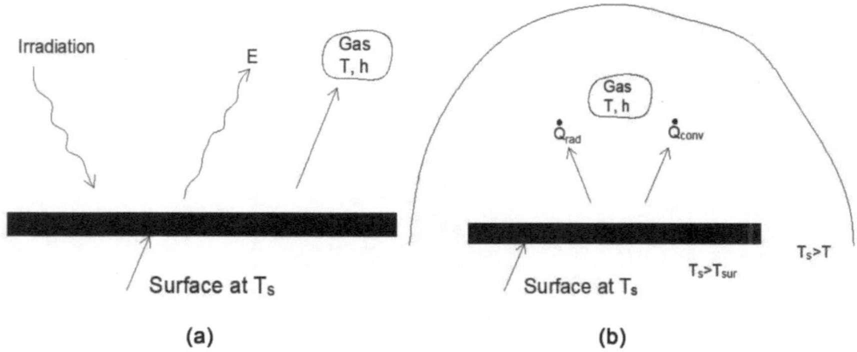

FIGURE 3.5 Radiation exchange (a) at a surface and (b) between a surface and a large enclosure.

where G is the rate at which radiation is incident on a surface per unit area termed as irradiation as shown in Figure 3.5b. A more frequently occurring radiation exchange is the one that takes place between a small surface of emissivity ε at T_s and a much larger, isothermal surface that completely surrounds the smaller one, like a furnace whose temperature is T_{surr}. For a gray surface ($\alpha = \varepsilon$), the net rate of radiation heat transfer is the difference between these two surfaces given by

$$\dot{Q}_{rad} = \varepsilon \sigma A \left(T_s^4 - T_{surr}^4 \right) \tag{3.19}$$

and per unit area of the surface, the net radiation heat transfer is given by

$$\dot{Q}_{rad} = \varepsilon \sigma \left(T_s^4 - T_{surr}^4 \right)$$

Eq. 3.19 gives the difference between thermal energy released due to radiation emission and that gained due to radiation absorption.

3.5 WORK TRANSFER

The concept of work in thermodynamics is a normal extension of the concept of work in mechanics. In mechanics, work can be defined as a force F acting through a displacement s, so

$$\delta W = F ds$$

$$W_{1\text{-}2} = \int_1^2 F\, ds \tag{3.20}$$

We can evaluate the work if the force F as a function of displacement s is known.

In thermodynamics, work can be defined as follows: "work is said to be done by a system on its surroundings, if the sole effect on the things external to the system is the raising of a weight". Thus the raising of a weight, in effect, is a force acting through a distance.

Work is also an energy interaction between a system and its surroundings. Energy can cross the boundary of a closed system in the form of heat or work. If the energy interaction between a system and its surroundings is not by heat, it must be by work. The driving force for heat transfer is the temperature difference between the system and its surroundings. Then an energy interaction that is not caused by a temperature difference between a system and its surroundings is work. More clearly, work is the energy transfer associated with a force acting through a distance. A rotating rotor as in a turbo-generator system, electrical work from a battery, and chemical work are some of the examples of work transfer. Work, like heat, has energy units kJ. The work done during a process between states 1 and 2 is denoted by $W_{1\text{-}2}$, or simply W.

$$\int_1^2 \delta W = W_{1\text{-}2} \tag{3.21}$$

The value of this integral should not be indicated as $W_2 - W_1$. The differential of work, δW, is said to be inexact, because, in general, the following integral cannot be evaluated without specifying the details of the process.

On a unit mass basis, the work done by a system is denoted by w and is expressed as

$$w = \frac{W}{m} (\text{kJ/kg}) \tag{3.22}$$

Power is the rate of work done per time and is designated by the symbol $\dot{W}$

$$\dot{W} = \frac{\delta W}{dt} \tag{3.23}$$

The unit of power is watt (W), that is the rate of work of one joule per second, 1 W = 1 J/s.

The differential of a property is said to be exact, because the change in a property in a change of state depends on the initial and final states and is not on the path of the process linking the two states. For instance, the change in volume from state 1 to state 2 can be found by integrating the differential dV, without considering the details of the process, that is

$$\int_{V_1}^{V_2} dV = V_2 - V_1 \tag{3.24}$$

where V_1 and V_2 are the volumes at state 1 and state 2, respectively. The differential of every property is exact and exact differentials are written using the symbol d. To emphasize the difference between exact and inexact differentials, the differentials of work and heat are indicated as δW and δQ, respectively.

3.6 DIFFERENT FORMS OF WORK

Some common forms of work including both mechanical and non-mechanical forms are discussed in the following sections.

Electrical work: Let us consider a system in which there is a flow of current (I) through the resistor as shown in Figure 3.6. There is a work transfer into the system, since the current can drive a motor, the motor in turn can drive the pulley which can raise a weight. The current flow is represented by

$$I = \frac{dC}{dt} \tag{3.25}$$

where dC is the charge (in coulombs) crossing the boundary in time interval dt. Then the work is

$$\delta W = E \cdot dC \, (\text{where E is the voltage}) \tag{3.26}$$

$$= E \cdot Idt$$

On integration

$$\Rightarrow W = \int_1^2 EI\,dt \tag{3.27}$$

The electrical power is given by

$$\dot{W} = \lim_{dt \to 0} \frac{dW}{dt} = EI \tag{3.28}$$

and it is the rate of work transfer.

Stirring work: Let us consider the case of stirring of a fluid system when there is a work transfer into it due to the turning of the paddle wheel as the weight is lowered as shown in Figure 3.7. If m is the mass of the weight lowered through a distance dl and T is the torque transmitted by the shaft due to its rotation through an angle dΘ, then the differential amount of work transfer to the fluid system is

$$\delta W = mgdl = T \cdot d\Theta$$

The total work transfer is

$$W = \int_1^2 mg\,dl = \int_1^2 W^1 dl = \int_1^2 T\,d\theta \left(W^1 \text{ is the weight lowered}\right) \tag{3.29}$$

Shaft work: Let us consider an energy transmission with a rotating shaft as shown in Figure 3.8. The work done during 'n' revolutions for a constant torque T and force F is determined as

$$T = F \cdot R \text{ then } F = \frac{T}{r} \tag{3.30}$$

When s is the distance through which a force acts which is related to radius r as $s=(2\pi r)n$, then the shaft work

$$W = F \cdot s = \left(\frac{T}{r}\right)(2\pi r)n \tag{3.31}$$

The power transmitted through the shaft is the work done per unit time.

$$W = 2\pi nT \tag{3.32}$$

where n is the number of revolutions per unit time.

Spring work: Whenever a force is applied on a spring, its length changes. The work done when a force F is applied, which causes a differential change in length dx, as shown in Figure 3.9, is given by

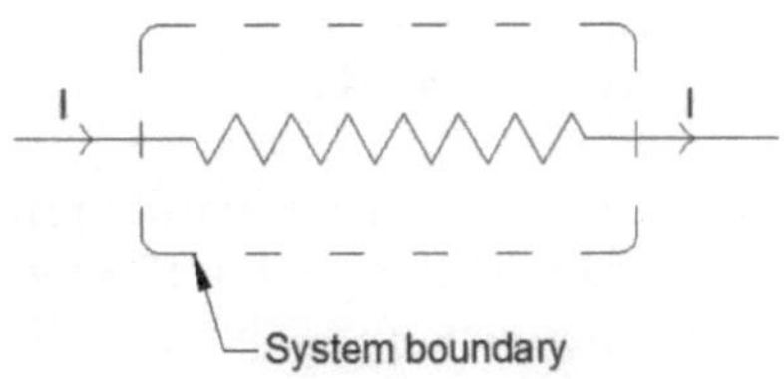

FIGURE 3.6 Electrical work.

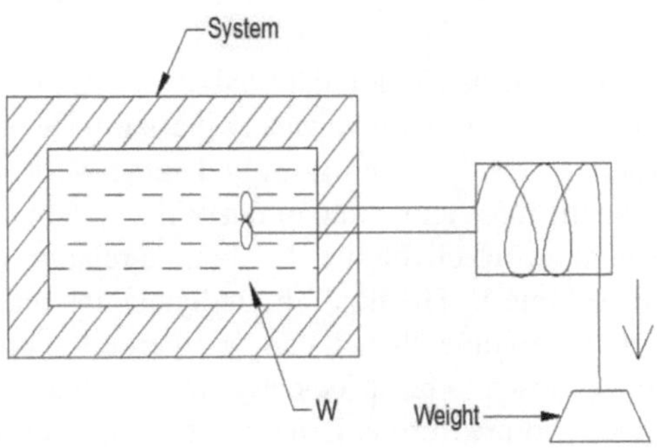

FIGURE 3.7 Paddle-wheel work.

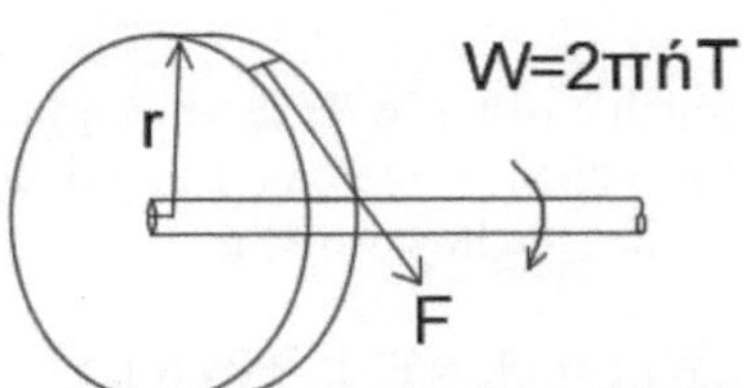

FIGURE 3.8 Shaft work.

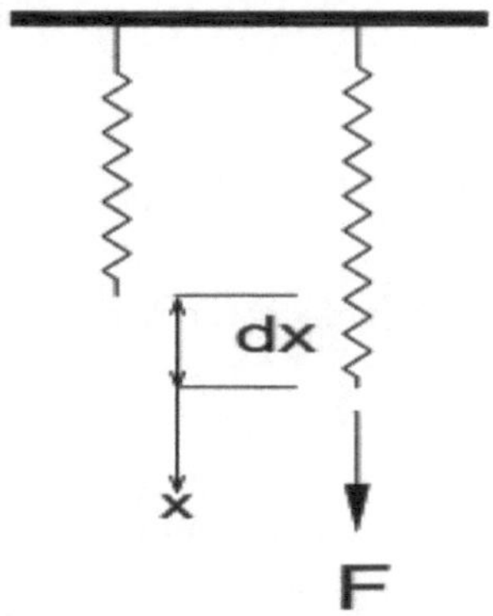

FIGURE 3.9 Spring work.

$$\delta W = Fdx \tag{3.33}$$

To calculate the total spring work, for the linear elastic springs, the displacement x is proportional to the applied force F. That is F = Kx, K is the spring constant in N/m and x is measured from the rest position of spring, i.e., at F = 0. Substituting this in Eq. 3.33 and integrating it

$$W = \frac{1}{2}k\left(x_2^2 - x_1^2\right) \tag{3.34}$$

where x_1, x_2 are the displacements (initial and final).

Flow work: Closed systems do not involve any mass transfer across their boundaries while control volumes involve mass flow, and some work, known as the flow work, is required to push the mass either into or out of the system and for maintaining a continuous flow through a control volume. To obtain a relation for flow work, consider a fluid element of volume V. The fluid, immediately upstream, forces this fluid element to enter the control volume; thus, it can be regarded as an imaginary piston. The fluid element can be chosen to be sufficiently small so that it has uniform properties throughout. If the fluid pressure is P and the cross-sectional area of the fluid element is 'A', then the force applied on the fluid element by the imaginary piston is

$$F = PA \tag{3.35}$$

This force 'F' must act through a distance 'L' as shown in Figure 3.10 to push the fluid element into the control volume. Therefore, the work done in pushing the fluid element across the boundary (i.e., the flow work) is

$$W_{flow} = FL = PAL = PV\,(kJ) \tag{3.36}$$

The flow work per unit mass is obtained by dividing both sides of this equation by the mass of the fluid element:

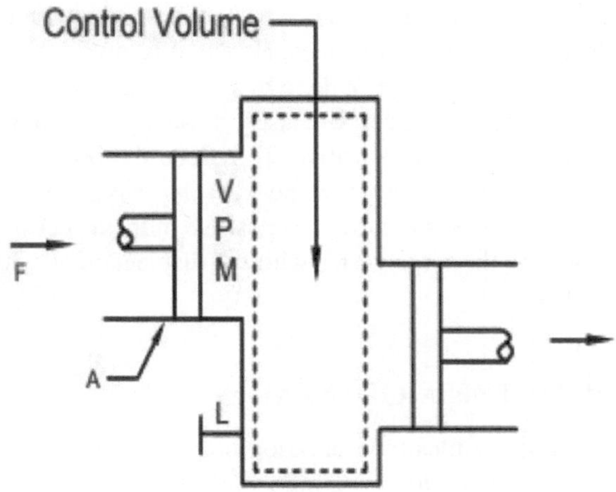

FIGURE 3.10 Flow work.

$$w_{flow} = Pv \text{ (kJ/kg)} \tag{3.37}$$

Free Expansion: Work transfer is basically a boundary phenomenon that is recognized at the boundary of the system. Free expansion is the expansion of gas against the vacuum. If we consider a gas separated from the vacuum by a partition, after the partition is removed, gas escapes to fill the entire volume. The work is said to be done when a force acts through a distance. In this case, though there is a change in the volume, work done will be zero, disregarding the work done for removal of the partition, since the force applied is zero though there is a displacement.

$$\int_1^2 \delta W = 0 \tag{3.38}$$

3.7 RELATIONSHIP BETWEEN HEAT AND WORK

Heat and work are the two means of energy transfer between a system and its surroundings and they are the two forms of energy. Certain similarities between the two are given below.

1. Both heat and work are boundary phenomena and occur at the boundaries of a system as they cross the boundaries.
2. We can neither store heat nor work in a system, though systems possess energy.
3. Both are path functions, that is, the change of work or heat in a change of state depends on the path followed during a process.
4. Both heat and work, unlike properties such as energy and entropy, are associated with a process, not a state.

In engineering thermodynamics, the focus will be on work-producing devices such as internal combustion engines and turbines. Hence, it is always suitable to consider such work as positive. That is, the work done by the system is positive, and the work done on the system is negative. In certain cases, however, it is convenient to regard the work done on the system to be positive. To reduce the possibility of misunderstanding in any such case, the direction of energy transfer is shown by an arrow on a sketch of the system, and work is regarded as positive in the direction of the arrow. It is necessary to know how the force varies with the displacement. This brings out an important idea about work.

3.8 FIRST LAW OF THERMODYNAMICS

The first law of thermodynamics (conservation of energy principle) establishes the relationships among the various forms of energy and energy interactions such as heat Q and work W during a process. The first law of thermodynamics states that energy is neither created nor destroyed during a process; it gets transformed from one form to the other. When an electrical energy is supplied to a motor, it will rotate a pump circulating the fluid. In this case, electrical energy is converted to kinetic energy (rotation of the shaft). A speeding vehicle possesses kinetic energy, if the brake is applied it comes to rest slowly and at rest, it possesses potential energy. The loss in kinetic energy equals the increase in potential energy when the air resistance is negligible, thus confirming the conservation of energy principle for mechanical energy.

The conservation of energy principle can be demonstrated with the help of the processes that involve either heat or work interactions or simultaneous heat and work interactions as well. In one case, we consider a process that involves heat transfer but no work interactions such as heating of a metal bar in a furnace. Owing to heat transfer to the metal bar, its energy increases. The increase in the total energy of the metal bar will be equal to the amount of heat transferred to it. Similarly, in the case of heating of water in a steam boiler, if 100 kJ of heat is transferred to the water from the furnace and 10 kJ of it is lost from the water to the ambient surroundings, the increase in energy of the water becomes equal to the net heat transfer to water, i.e., 90 kJ.

In the other case, we consider the process that involves work interactions without heat transfer. Compression of air in an insulated air compressor is an example in this case. The work transfer equals the increase in the energy of the system (air); the compressor is insulated, and therefore, it involves no heat transfer.

Energy Balance

The energy conservation principle can be stated as follows: the net change (increase or decrease) in the total energy of a system during a process is equal to the difference between the total energy entering and the total energy leaving the system during that process. It can be expressed as

$$\underbrace{E_{in} - E_{out}}_{\substack{\text{Total energy entering system} \\ -\text{ total energy leaving system}}} = \underbrace{\Delta E_{system}}_{\text{Total energy change of system}} \qquad (3.39)$$

Equation 3.39 is termed the energy balance equation which can be applied to any kind of system undergoing any kind of process. This equation can be effectively applied to solve engineering problems by understanding the different forms of energy and recognizing the forms of energy transfer.

Energy Change of a System

The energy change of a system during a process can be determined by evaluating the difference in energy of the system at the beginning and at the end of the process, expressed as

$$\Delta E_{system} = E_{final} - E_{initial} \tag{3.40}$$

Since energy is a property, the value of a property does not change when there is no change in the state of the system. Thus, the change in energy of a system can be zero if the state of the system does not change during the process. Moreover, energy exists in several forms such as internal, kinetic, potential, electrical, and magnetic energy. The sum of all these energies constitutes the total energy, E, of a system. The change in the total energy ΔE of a system during a process is the sum of the changes in its internal, kinetic, and potential energies when electrical magnetic, and surface tension effects are negligible, which is expressed as

$$\Delta E = \Delta KE + \Delta PE + \Delta U \tag{3.41}$$

Where $$\Delta KE = \frac{1}{2}m\left(v_2^2 - v_1^2\right)$$

$$\Delta PE = mg(h_2 - h_1)$$

$$\Delta U = m(u_2 - u_1)$$

Subscripts 1 and 2 are specified initial and final states. The values of the specific internal energies u_1 and u_2 can be taken directly from the property tables. Generally, most of the systems we come across in practice are stationary and the changes in their velocity or elevation during a process are zero. For a stationary system, when ΔKE and ΔPE are zero, Eq. 3.41 becomes

$$\Delta E = \Delta U$$

Also, the energy of a system during a process will change even if one form of its energy changes without the other forms of energy changing.

Energy Transfer

Energy change of a system occurs when there is an energy transfer to or from a system. For a control volume, energy transfer takes place through three possible means: heat, work, and mass flow. For a closed system, heat transfer and work transfer are the only two forms of energy interactions. For a process, energy interactions are recognized at the system boundary as they cross it. The energy gain or loss for a system during a process can be determined by the type of interaction that is taking place. The three different forms of energy transfer and their effect on both the system and surroundings are discussed below.

An energy interaction in which heat is transferred to a system increases the internal energy of the system as the heat added results in an increase of molecular energy. Addition of heat to water to convert it to steam is an example of this case. While heat is transferred from a system, it decreases internal energy, since the energy transferred out as heat comes from the energy of the molecules of the system.

Work transfer is another energy interaction that is not as a result of temperature difference between a system and its surroundings. A rising piston and a rotating shaft are associated with work interactions. There are work-producing devices such as car engines and turbines, which produce work, and work-consuming devices such as compressors, pumps, and mixers, which consume work. Work transfer to a system increases the energy of the system, and work transfer from a system decreases it, since the energy transferred out as work comes from the energy contained in the system.

Mass flow in and out of the open system is another mechanism of energy transfer. When mass enters a system, it carries energy with it causing the energy of the system to increase. Similarly, when mass leaves the system, it reduces the energy contained within the system because the leaving mass carries some energy with it. For example, if some compressed air is taken out of an air compressor and is replaced by the same amount of atmospheric air, the energy content of the air compressor decreases as a result of this mass interaction.

3.9 MOVING BOUNDARY WORK (pdV work)

A simple compressible substance does not involve the effects of electricity, magnetism, and surface tension. A closed system, composed of a simple compressible substance without involving the effects of motion or gravity, is called a *simple compressible closed system.* The compression or expansion of a gas or air in a piston-cylinder device is most commonly encountered in applications such as automobiles and reciprocating air compressors. This type of compression or expansion is called displacement work or boundary work. In automobiles, the hot gases are compressed by piston motion, and the expansion of gases forces the piston to reciprocate in the cylinder and forces the crankshaft to rotate.

Let us consider the gas in the cylinder as a system having the initial pressure p_1 and volume V_1 as shown in Figure 3.11. The system is in quasi-equilibrium, a process during which the system is in equilibrium at all states, quasi-static process as shown in Figure 3.12.

The piston is moved due to the gas pressure to a new equilibrium state specified by p_2 and V_2. Let the pressure and volume be p and V at any intermediate point. If the piston moves through a distance ds in equilibrium manner, then the work done during this process is $\delta W = F\,ds = P\,A\,ds = pdV$, where A is the area of the piston, F is the force acting on the piston ($F = p\,A$), and $dV = A\,ds$, where dv is the displacement volume.

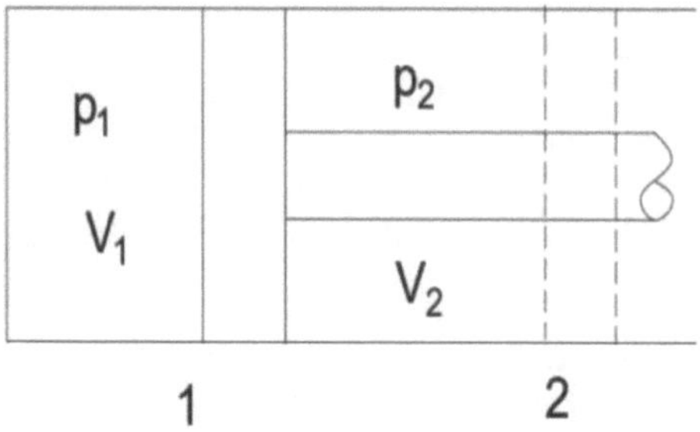

FIGURE 3.11 pdV work.

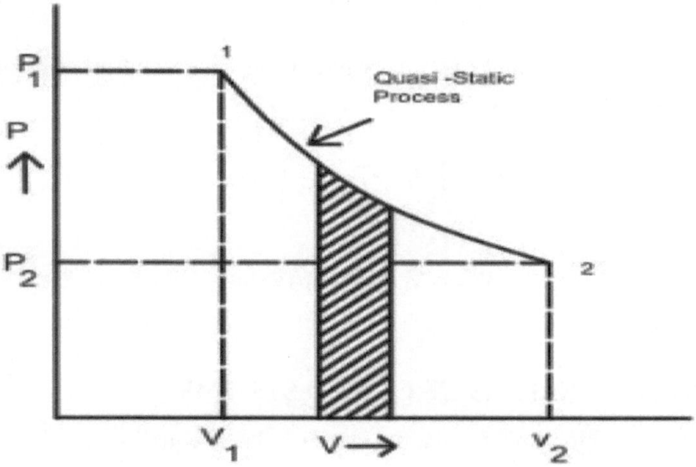

FIGURE 3.12 Quasi-static pdV work.

If the piston moves from an initial position (p_1V_1) to another position (p_2V_2), the amount of work done by gas

$$W_{1\text{-}2} = \int_{V_1}^{V_2} p\,dV \tag{3.42}$$

The work done is equal to the area under the pV diagram (path 1-2). However, p is the absolute pressure which is always positive, while volume change dV is positive for expansion and negative for compression.

Polytropic Process

The pressure and volume are related to each other through the equation $pV^n = C$ in the compression or expansion process, where n and C are constants. This is called a polytropic process. The expression for work done in a polytropic process is derived from the above expression (Eq. 3.42).

$$p = CV^{-n}$$

$$\text{Work done } W = \int_1^2 p dV = \int_1^2 CV^{-n} dv$$

$$\Rightarrow \left[\frac{CV^{-n+1}}{-n+1}\right]_{V_2}^{V_2} = c\frac{\left[V_2^{-n+1} - V_1^{-n+1}\right]}{-n+1} = c\frac{\left[V_2 V_2^{-n} - V_1 V_1^{-n}\right]}{1-n}$$

$$\text{Now} \qquad C = p_1 V_1^n = p_2 V_2^n$$

Substituting this in the above equation, work done

$$W = \frac{p_2 V_2 - p_1 V_1}{1-n} \text{ or } \frac{p_1 V_1 - p_2 V_2}{n-1} \tag{3.43}$$

If $n = 1$, then $pV^n = C$ becomes $pV = C$, which is an isothermal process for an ideal gas; in this case, work done is

$$W = \int_1^2 C \cdot V^{-1} dv = C \cdot \ln\frac{V_2}{V_1}$$

$$W = p_1 V_1 \ln\frac{V_2}{V_1} \text{ or } p_2 V_2 \ln\frac{V_2}{V_1} \tag{3.44}$$

3.10 ENERGY ANALYSIS OF CLOSED SYSTEMS

For any system and for any process, the energy balance equation is

$$E_{in} - E_{out} = \Delta E_{system} \tag{3.45}$$

On a unit mass basis, it is expressed as $e_{in} - e_{out} = \Delta e_{system}$.

In the differential form, it can be expressed as $\delta E_{in} - \delta E_{out} = dE_{system}$

$$\delta e_{in} - \delta e_{out} = de_{system}$$

The above energy balance equation can be applied to two cases, a closed system undergoing a cycle and a closed system undergoing a change of state.

3.10.1 First Law for a Closed System Undergoing a Cycle

For a closed system undergoing a cycle, the initial and final states are identical and change in energy $\Delta E = 0$. Then the energy balance equation $E_{in} - E_{out} = 0$ and hence $E_{in} = E_{out}$.

Since a closed system involves no mass transfer across its boundaries, the energy balance equation is expressed in terms of heat and work interactions. The net work output is equal to the net heat input during a cycle, that is

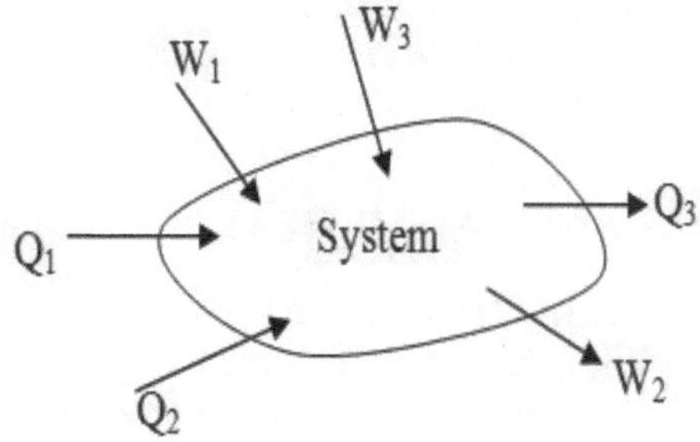

FIGURE 3.13 System–surroundings interactions with many energy fluxes.

$$W_{net,out} = Q_{net,in} \tag{3.46}$$

The above equation is the first law for a closed system undergoing a cycle.

In the expression $E_{in} - E_{out} = \Delta E_{system}$, taking the heat transfer to the system and work transfer from the system to be positive,

$$Q_{in} - W_{out} = 0 \Rightarrow Q = W \ (\Delta E = 0 \text{ for cycle})$$

3.10.2 First Law for a Closed System Undergoing a Change of State

The equation $W_{net,out} = Q_{net,in}$ is applied to systems undergoing a cycle. For a cycle, since $\Delta E = 0$, $Q = W$.

In the equation $E_{in} - E_{out} = \Delta E$, taking the heat transfer to the system and work transfer from the system as positive.

If a system undergoes a change of state during which energy transfer takes place due to heat transfer and work transfer, this energy transfer will be stored within the system. If Q is the heat transfer to the system and W is the work transfer from the system, then $Q - W$ is the net energy transfer stored within the system.

$$Q - W = \Delta E \tag{3.47}$$

For a system involving more number of heat and work transfer quantities as shown in Figure 3.13,

$$(Q_1 + Q_2 - Q_3) - (-W_1 + W_2 - W_3) = \Delta E$$

$$\text{or} \quad (Q_1 + Q_2 - Q_3) = \Delta E + (W_2 - W_1 - W_3)$$

$$Q_{net} - W_{net} = \Delta E_{system} \tag{3.48}$$

where $Q_{net,in} = Q_{in} - Q_{out}$, and $W_{net,out} = W_{out} - W_{in}$.

$Q_{net,in}$ is the net heat input, and $W_{net,out}$ is the net work output.

Though the first law is not proved mathematically, since no violation of it is proved so far and no process in nature is known to have happened in violation of the first law, it can be considered as a proof.

3.11 SPECIFIC HEAT AND LATENT HEAT

Specific Heat at Constant Pressure, c_p

Specific heat can be defined as the energy required to raise the temperature of a unit mass of a substance by 1°. However, the energy required to raise 1° temperature of identical masses of different substances differs. There are two types of specific heats: (i) specific heat at constant pressure and (ii) specific heat at constant volume. Specific heat at constant pressure (c_p) is the energy required to raise the temperature of a unit mass of a substance by 1° when the pressure is held constant.

$$c_p = \left(\frac{\partial h}{\partial T}\right)_P \tag{3.49}$$

c_p can also be defined as the rate of change of enthalpy with respect to temperature when pressure is held constant.

Specific Heat Constant Volume, c_v

$$c_v = \left(\frac{\partial u}{\partial T}\right)_v \tag{3.50}$$

c_v is the rate of change of internal energy with respect to temperature when volume is held constant. Specific heat is different for dissimilar substances of the same mass; for instance, the specific heat needed to raise the temperature of 1 kg of water from 27°C to 37°C is 41.8 kJ/kg K, while for steam it is 18.723 kJ/kg K. c_p is always greater than c_v, because in the constant pressure process, the system is allowed to expand and the energy required for this expansion must also be supplied to the system.

Latent Heat

The amount of heat required to cause a phase change is called *latent heat*. Latent heat is also defined as the amount of energy absorbed or released during a phase change process. The amount of heat absorbed during melting or the amount of heat released during freezing is called the *latent heat of fusion*. The amount of heat absorbed during vaporization or the amount of heat released during condensation is called the *latent heat of vaporization*. The amount of heat required to convert a solid to vapor or vapor to solid is called the *latent heat of sublimation*.

3.12 INTERNAL ENERGY, ENTHALPY, AND SPECIFIC HEATS OF IDEAL GASES

Enthalpy can be defined as the energy of a flowing fluid and is a quantity that occurs regularly in thermodynamics. It is the sum of internal energy u and the product of pressure and volume $p\nu$. It is given as

$$h = u + pv \tag{3.51}$$

For a fluid of mass m, enthalpy is

$$H = U + PV$$

For an ideal gas, pressure and specific volume are related by the equation $pv = RT$, and also internal energy is a function of absolute temperature only, that is

$$h = u + RT$$

$$u = u(T) \tag{3.52}$$

Since R is a constant, enthalpy is also a function of absolute temperature given as

$$h = h(T) \tag{3.53}$$

Enthalpy is a property of a fluid, since internal energy, pressure, and specific volume are properties. Internal energy is the energy of a non-flowing fluid.

For an ideal gas, internal energy and enthalpy are functions of absolute temperature only; therefore, the specific heats c_p and c_v are also functions of absolute temperature only. For ideal gases, the differential changes in internal energy and enthalpy based on the Eqs. 3.52 and 3.53, replacing the partial derivatives by ordinary derivatives, are

$$du = c_v(T)dT \tag{3.54}$$

$$dh = c_p(T)dT \tag{3.55}$$

For a process, for an ideal gas, the changes in internal energy and enthalpy are obtained by integrating the Eqs. 3.54 and 3.55 respectively given as

$$\Delta u = \int_1^2 c_v(T)dT \tag{3.56}$$

$$\Delta h = \int_1^2 c_p(T)dT \tag{3.57}$$

3.13 PERPETUAL MOTION MACHINE OF THE FIRST KIND—PMM1

The first law of thermodynamics states the general principle of conservation of energy. Energy is neither created nor destroyed, but only gets transformed from one form to another. There can be no machine, which would continuously supply mechanical work without some form of energy disappearing simultaneously. Any such imaginary machine that creates energy in violation of the first law of thermodynamics is called a perpetual motion machine of the first kind (PMM1). A PMM1 is thus impossible. Figure 3.14 shows the PMM1.

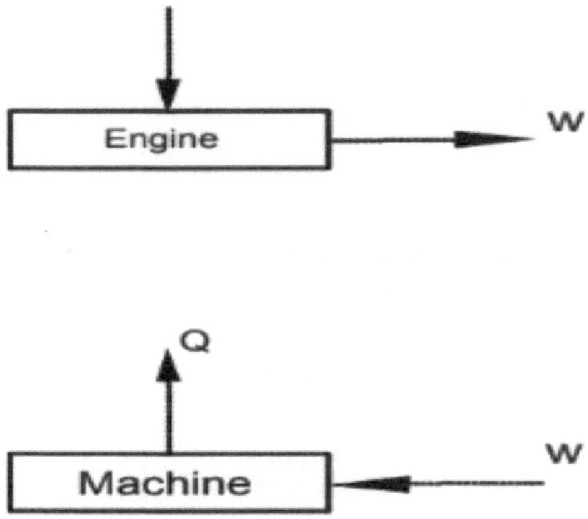

FIGURE 3.14 Perpetual motion machine of the first kind (PMM1).

3.14 ENERGY EFFICIENCY

Energy plays a vital role in wealth generation, economic improvement, and social development in all countries. Energy efficiency is basically using less energy to perform the same task by eliminating the wastage of energy. According to the International Energy Agency, "*energy efficiency* is a way of managing and restraining the growth in energy consumption. Something is more energy efficient if it delivers more services for the same energy input, or the same services for less energy input." It is the easiest and cost-effective means of reducing the greenhouse gas emissions, improving the competitiveness of the businesses and cost of the energy utilities. Energy efficiency offers a lot of benefits such as reducing dependency on energy imports and lowering our costs on a household and economy-wide level. Therefore, the promotion of energy efficiency will contribute to energy conservation and is treated as an integral part of energy conservation promotional policies. Energy efficiency is often regarded as a resource option like coal, oil, or natural gas. It improves the economy by preserving the resource base and reducing pollution. For example, replacing conventional light bulbs with compact fluorescent lamps for lighting a room will reduce the energy consumption by about 70%.

3.14.1 Energy Conversion Efficiency

An energy conversion device converts one form of energy into another. An automobile engine converts chemical energy of fuel into mechanical energy (rotation of shaft). The most important energy conversion device is an electric power plant in which chemical energy of fuel is converted into thermal energy in the boiler, thermal energy, in turn, is converted into mechanical energy in the turbine, and finally mechanical energy is converted into electrical energy in the generator. The efficiency of an energy conversion device can be expressed in terms of the energy input and energy output given as

$$\eta = \frac{\text{Energy output}}{\text{Energy input}} \tag{3.58}$$

An electric motor converts electrical energy into mechanical energy. *Motor efficiency* is the ratio of mechanical power delivered by the motor (output) to the electrical power supplied to the motor (input).

$$\eta_{\text{motor}} = \frac{\text{Mechanical power output}}{\text{Electrical power input}} \tag{3.59}$$

A generator is a device that converts mechanical energy to electrical energy, and the effectiveness of a generator is characterized by the *generator efficiency*, which is the ratio of the electrical power output to the mechanical power input.

$$\eta_{\text{generator}} = \frac{\text{Electrical power output}}{\text{Mechanical power input}} \tag{3.60}$$

The thermal efficiency of a power plant is defined as the ratio of the net work output of the turbine to the heat input to the working fluid.

$$\eta_{\text{thermal}} = \frac{\text{Net work output}}{\text{Total heat input}} \tag{3.61}$$

Combustion efficiency is the ratio of the amount of heat released during combustion to the heating value of the fuel given as

$$\eta_{\text{combustion}} = \frac{Q}{CV} \tag{3.62}$$

Power plant efficiency is expressed by defining an *overall efficiency*, which takes into account the effect of other factors. It is defined as the ratio of the net electrical power output to the rate of fuel energy input given as

$$\eta_{\text{overall}} = \frac{\dot{W}_{\text{net}}}{\dot{m}_f \times CV} \tag{3.63a}$$

where $\dot{W}_{\text{net}}$ is the net electrical power output (kJ/h), $\dot{m}_f$ is the fuel burning rate (kg/h) in the boiler, and CV is the calorific value of fuel (kJ/kg).

Overall efficiency is also expressed as

$$\eta_{\text{overall}} = \eta_{\text{combustion}} \times \eta_{\text{thermal}} \times \eta_{\text{generator}} \tag{3.63b}$$

Table 3.1 shows the efficiencies of thermal power plants operating on various fuels and combustion technologies.

Energy Conversion Efficiency of Biological Systems

For most of the plants and animals, either incoming solar radiation is directly converted into energy or direct body warming by incoming solar radiation results in

TABLE 3.1
The Efficiencies of Various Thermal Power Plants

Thermal Power Plant Type	Efficiency Range (%)
Oil-fired	38–44
Coal-fired	39–47
Natural-gas fired	Up to 39
Pressurized fluidized bed combustion	>40
Combined gas–vapor cycle	>40
Binary vapor cycle	>50
Coal-fired integrated gasification combined cycle	>43

metabolic reduction. In this case, the incoming solar radiation (which is a mode of heat transfer) $\dot{Q}$ is considered as part of the total energy input. The energy conversion efficiency of plant photosynthesis is defined as

$$\eta_{\text{photosynthesis}} = \frac{\text{Energy converted to organic molecules by photosynthesis (per unit area)}}{\text{Solar energy input to earth (per unit area)}} \quad (3.64)$$

The photosynthesis efficiency is in the range of 0.01%–1%, which is too low. This may be due to the reason that only a part of the incident radiation on the earth's surface (per unit area) is intercepted by the plant. The energy conversion efficiency of animals is defined as

$$\eta_{\text{animal}} = \frac{\text{Rate of food energy stored in the body as complex organic molecules}}{\text{Rate of energy taken into the body as food}} \quad (3.65)$$

3.14.2 Energy-Efficient Buildings

The energy consumed by buildings constitutes a substantial share in total energy use globally, and hence, it has an insightful impact on the environment. Energy is used in several stages of the building life cycle, which includes the choice of locality, architectural design, structural systems and material selection, building construction, usage and maintenance, demolition, reuse-regain-recycle, and waste disposal. As per the estimates of the World Watch Institute, the annual energy consumption of buildings is around 40% of the total world's energy consumption. This can be reduced considerably in every stage of a building life cycle by different strategies.

The building life cycle is typically divided into three main phases: (i) the prebuilding phase, (ii) building phase, and (iii) postbuilding phase. The first phase includes the suitable site selection, site planning, building form, building plan, and suitable space organization, building envelope design selecting energy-efficient building materials, energy-efficient landscape design, obtaining raw materials for building material, manufacturing, and transporting them. The second phase includes the construction and usage processes of the building. The third phase is the phase following the completion of building usage. This phase includes the demolition, recycling, and wipe-out of the building.

3.14.3 Cost-Effectiveness of Reflective White Materials

There are two properties that will influence the surface temperature of opaque materials under the sun, one is solar reflectance and the other is thermal emittance. Solar reflectance is the portion of solar radiation that is reflected from a surface. The thermal emittance (ε), on the other hand, is the rate at which a surface radiates energy as compared to that of a black body operating at the same temperature.

The use of reflective white materials applied to building roofs can reduce up to 80% of cooling energy consumption depending on climate, thermal insulation, internal gains, and previous condition of the roof and can be an energy-efficient retrofitting measure for building roofs. The savings in energy expenditure ensued depend on several factors such as local climate, the amount of insulation in the roof, purpose to which the building is used, cost of energy, and the type and efficiency of the heating and cooling systems. It is therefore important to assess the costs and benefits before one can decide whether to use reflective white materials for buildings.

3.14.4 Energy-Efficient Motors

Electricity consumption of motor systems in the industrial and in the service sectors accounts for nearly 69% and 38% of the total electricity consumption respectively in the European Union, which represents about 575 and 186 TWh/year (terawatt-hour per year). Moreover, of the total motor electricity consumption, pumps, fans, and compressors account for 62% and 82% in the industry and in the service sectors, respectively. Energy efficiency should be of primary importance in the purchase or rewind of a motor since the annual energy cost for running a motor is several times higher than the initial cost of motor. As per the estimates of the U.S. Department of energy, at an energy rate of $0.04/kWh, the annual electricity cost of a typical 20-horsepower motor is $6000, which is nearly six times its initial price. There is a considerable variation between the performances of a standard and an energy-efficient motor. Hence, improving the efficiency of electric motors and the equipment they drive not only saves the energy but also reduces the running costs, thereby improving the Nation's productivity. This can be possible with the improved design, superior materials, and better manufacturing techniques which help energy-efficient motors to achieve more work per unit of electrical energy consumed.

A power electronic drive typically consumes 25% less energy than a classic motor system on an average. The advents in solid-state technologies, on the other hand, are enabling the manufacturers to build power electronic converters for electric drive systems allowing the motors to be used in more precise applications with the effective speed and position control. The usage of more number of advanced electric motor drives, with better performance and precision, is increasing, which replace the older ones. These advanced drives also have the flexibility of utilizing superior switching schemes to boost the performance. Induction motors used for irrigation in the rural sector and industrial purpose in the urban sector consume a large share of electrical energy. It is found that even a 5% improvement in the overall efficiency of induction motor can save enough energy that is equivalent to the energy produced by a new power plant of few hundred megawatts.

A motor can be energy efficient, if the design improvements are incorporated explicitly to increase operating efficiency over standard motors. These design improvements particularly focus on reducing intrinsic motor losses. Improvements include the use of superior materials such as lower-loss silicon steel, a longer core, and thicker wires (to reduce resistance). Others include thinner laminations, smaller air gap between stator and rotor, copper instead of aluminium bars in the rotor, superior bearings, and a smaller fan.

Energy-efficient motors, in comparison with the NEMA (National Electrical Manufacturers Association) B motors of the same size, possess several advantages such as less electric energy consumption, cool running, and are long lasting. An 85% efficient motor converts 85% of the electrical energy input into mechanical energy. The remaining 15% of the electrical energy is dissipated as heat, which is evident from a rise in temperature of the motor. In addition, energy-efficient electric motors utilize improved motor design and high-quality materials to reduce motor losses, thereby improving motor efficiency with the consequent decrease in heat dissipation and reduction in noise output. There was a compromise in performance levels of majority of electric motors manufactured prior to 1975 in order to reduce the purchase price. Efficiency was maintained only at levels high enough to meet the temperature-rise restrictions of the particular motor. As per the recommendations of NEMA, a procedure was followed for labelling standard three-phase motors with an average nominal efficiency. These efficiencies are a benchmark of industry average for a large number of motors of the same design. Energy-efficient motors are only marketed with NEMA B speed-torque characteristics.

3.14.5 Energy-Efficient Compressors

Air compressors are widely used in industrial applications such as supplying process requirements, operating pneumatic tools and equipment. Energy-efficient use of compressors is essentially required since only 10%–30% of total energy is useful and the rest, i.e., 70%–90% of energy of the prime mover, is converted to heat energy which is unusable and to a small extent lost as friction. The following section presents the energy efficiency improvement methods of the compressors.

Power Drive Improvement

The energy efficiency of compressors can be improved by integrating variable speed drives (VSD) into compressors, because the VSD compressors allow the unit to operate at maximum efficiency for varying weather conditions and heating or cooling requirements as determined by the microprocessor. Usage of high-efficiency drives renders in large savings to new systems.

Optimal Selection of Compressor Type

The selection of compressor type used for a particular application plays a dominant role in power savings. For example, rotary screw compressors with oil injection are predominantly used in the power range from 10 to 300 kW in the European Union. There are other types of compressors as well with their specific advantages. Use of multiple compressors is another option, because the efficiency of a compressor changes as it is unloaded leading to reduced efficiency. The entire system efficiency is affected by the part-load characteristics of individual compressors. It is therefore desirable to find the operating strategies, suitably accounting for compressor unloading characteristics so that the efficiency of the entire system is maximized. An optimum point can be identified at which it is best to switch from each compressor equally sharing the load to one compressor operating at full load and the other unloaded to match the remaining system load.

Application of Sophisticated Control Systems

Advanced control systems with the aid of microprocessors can effectively adjust the compressor outlet flow to the requirements of the consumers. Control systems can optimize the transition among non-loaded working state, loaded working state, and non-operating state of compressor, thereby reducing the energy requirements. The advantage with control systems is that they can be integrated to the existing machines instead of purchasing new ones. Moreover, the price of control technologies is decreasing and awareness of its usage in applications is simultaneously increasing, leading to their rapid expansion of usage.

Reduction of Air Pressure Setting

Compressing the air to a higher pressure than that required by the air-driven equipment is an important source of energy waste in compressed air systems (CAS) as the energy requirement increases to raise the air pressure to higher values. Energy savings in such cases can be achieved by identifying the minimum required pressure and then reducing the air pressure control setting on the compressor.

Reduction of Pressure Losses

There are several reasons for pressure losses in CAS such as geometrical errors (pipeline and radius of curvature) and materials used, which can lead to the frictional losses. To address this problem, the possible solution is proper design and realization of distribution network.

Air Leakage Elimination

Air leakage also contributes largely to the energy loss in industrial facilities using CAS, since nearly 40% of the compressed air is wasted in the form of leakage. Energy is needed to compress the air whether the compressed air is useful or wasted and hence loss of compressed air is a loss of energy. Air leakage accounts for nearly 40% of the outlet air.

Regeneration of the Dissipated Heat

In most of the cases, the heat dissipated by the compressor is extremely low in temperature, or too low in its quality to sufficiently respond to the industrial needs such as their main processes or heating. Moreover, the influence of climate and seasonal changes is also considerable on the ratio between investments and yields. There are certain possibilities for using the compressor-recycled energy such as water heating, building heating, and heating and drying processes.

3.15 ENERGY SUSTAINABILITY

Sustainability is most widely known in relation to sustainable development. It is defined as a behavior that *meets the needs of the present without compromising the ability of future generations to meet their own needs*. Sustainable housing development provides a background for the assimilation of environmental policies and developmental strategies.

In order to achieve sustainable development, there appears to be various energy efficiency parameters in the design of housing projects that are to be considered. These parameters can contribute to a reduction in energy consumption. The most effective energy efficiency parameters include insulation, application of lighting choices to save energy, application of passive solar energy, application of natural ventilation, and making clean electricity. To improve sustainable housing development, it is the responsibility of designers to consider the optimization of energy parameters. The designers in Southern California could reduce energy consumption by nearly 60% by using different methods such as improving the HVAC system, efficient water heating, introducing daylight, building orientation and insulation.

3.16 ENERGY SECURITY

Energy security has been defined by the United Nations as *the continuous availability of energy in varied forms, in sufficient quantities, and at affordable prices*. There are two cases related to energy security: long-term security and short-term security. The primary factor as far as energy security is concerned is resource availability, which is the actual physical amount of the resource available around the world, and it can be considered as long-term security. The need for system reliability, such as the continuous supply of energy, particularly electricity, to meet consumer demand at any given time is termed short-term security. The economic growth of a nation largely depends on its resources and their utilization. Therefore, the primary objective of energy security for a nation is to reduce its dependency on the imported energy sources.

As per the estimates of the U.S. Energy Information Administration (EIA), the global energy consumption would grow by about 50% between 2019 and 2050. Most of this growth is expected to come from the regions outside of the Organization for Economic Cooperation and Development (OECD) that is from the Asian countries where strong economic growth is the driving force for this demand. Industrial sector, consisting of manufacturing, mining, agriculture, processing, and construction industries, constitutes the largest share of the energy consumption globally. This is expected to grow by nearly 30% between 2018 and 2050. While the energy consumption of the transportation sector is expected to grow by nearly 40%, the energy consumption of the building sector (both residential and commercial) is expected to grow by 65% between the same periods. Figures 3.15 and 3.16 show the global primary energy consumption by region and by sector respectively during 2010 and 2050.

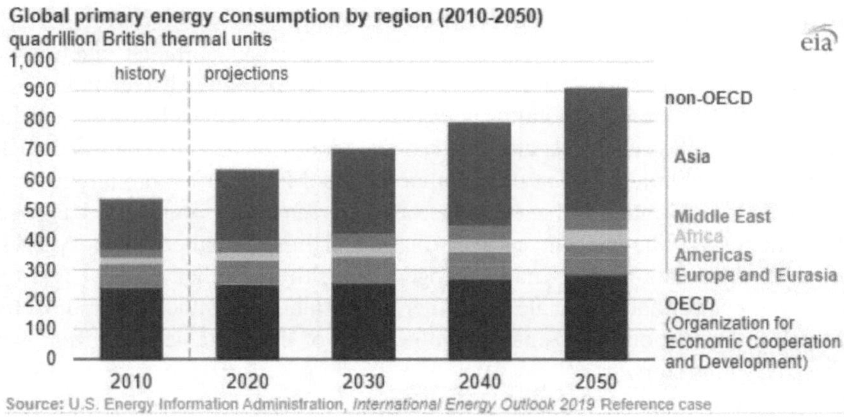

FIGURE 3.15 Global primary energy consumption by region (2010–2050). (Obtained with permission from the U.S. Energy Information Administration (EIA).)

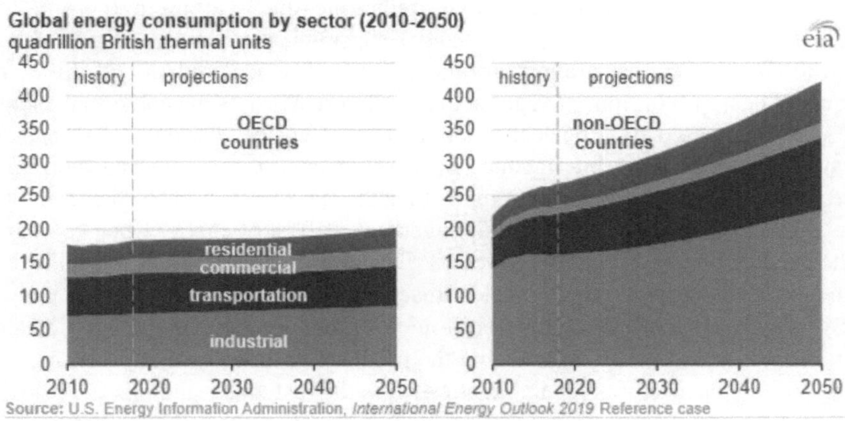

FIGURE 3.16 Global primary energy consumption by sector (2010–2050). (Obtained with permission from the U.S. Energy Information Administration (EIA).)

3.17 ENERGY CONSERVATION

Though the words energy conservation and energy efficiency seem to be separate, they are related concepts. Energy conservation is basically reducing the consumption of energy by using less of an energy service. Energy can be conserved either by using it more efficiently, for example, driving the car sparingly or by reducing the amount of service used. Energy conservation depends primarily on many processes or developments such as productivity increase or technological progress. It could be possible when the growth of energy consumption can be reduced. Energy efficiency, on the other hand, is possible if energy intensity in a specific product, process, or area of production or consumption can be reduced without upsetting the output, consumption, or comfort levels.

Energy Conservation in Nuclear Reactions

Except for radioactivity, most of the transformations of states of matter that occur at terrestrial temperatures are chemical. At very high temperatures (exceeding 10^6 K) that are usually attained in the stars, nuclei collide and undergo nuclear reactions just like molecules collide and react at terrestrial temperatures. The electrons and nuclei of atoms are totally uncertain at these temperatures. Matter turns into an unknown state and the transformations that take place are between nuclei, and hence it is called nuclear chemistry. The nuclear reactions that occur in stars called nucleosynthesis result in the elements heavier than hydrogen on earth and other planets. Just like unstable molecules that dissociate into other more stable molecules, the radioactive elements are formed due to disintegration of some of the unstable nuclei that were synthesized in the stars.

The source of heat for the earth's interior is the energy released by radioactive elements turning into heat due to the radioactive decay that is mass is converted to energy. This energy is set free as kinetic energy, emitting α- and β-particles and γ-radiation. The most of the decay is converted to heat. Thus, the heat produced by naturally radioactive elements in the earth's crust is a key source for geothermal energy. For example, natural radioactivity in granite, due to uranium, thorium, and potassium isotopes (U^{238}, U^{235}, Th^{232}, and K^{40}), produces a small amount of heat equal to about 5 μcal per gram of granite per year. The accumulation of this small amount of heat over billions of years in the earth's crust results in geothermal energy. The energy released by a nuclear reaction such as nuclear fission (heavy nucleus decays into two or more intermediate mass fragments) of uranium and thorium is enormous, which is a million times higher than the energy released by chemical reactions. Nuclear fission of 1 kg of uranium generates 10^{14} J of energy when compared to the burning of 1 kg of coal that gives 10^7 J. Nuclear fusion (two light nuclei fuse to form a larger nucleus) is another nuclear reaction.

By using Einstein's relation $E^2 = p^2c^2 + m^2c^4$, the energy released can be calculated from the rest mass of the reactants and the products. In this equation, E is the total energy of the particle, p is the momentum, m is the rest mass, and c is the velocity of light. If the total rest mass of the products is less than the total rest mass of the reactants, then this excess rest mass turns into kinetic energy of the products, which in turn turns into heat due to collisions. If there is a negligible difference between the kinetic energy of the reactants and that of the products, then the heat liberated is estimated using the relation $\Delta E = \Delta mc^2$, in which Δm is the difference between the rest mass of the reactants and the products. The following nuclear fusion reaction, in which deuterium and tritium (D–T) react to form helium isotope and neutron (n), is used to estimate the heat released.

$$ {}^2_1H + {}^3_1H \rightarrow {}^4_2He + n $$

The atomic mass units ($1\ \text{amu} = 1.6 \times 10^{-27}$ kg) of deuterium, tritium, neutron, and helium are given as

$${}^2_1H = 2.014102\,\text{amu};\ {}^3_1H = 3.016049\,\text{amu};\ \text{neutron} = 1.0087\ \text{amu and}$$

$${}^4_2He = 4.002602\,\text{amu}$$

$$\Delta m = 1\,\text{mol of } {}^2\text{H} + 1\,\text{mol of } {}^3\text{H} - \left(1\,\text{mol of } {}^4\text{He} + 1\,\text{mol of n}\right)$$

$$= 2.014102 + 3.016049 - (4.002602 + 1.0087) = 0.018849\,\text{amu} = 3.016 \times 10^{-29}\,\text{kg}$$

$$\Delta E = \Delta mc^2 = 3.016 \times 10^{-29} \times \left(3 \times 10^8\right)^2 = 2.714 \times 10^{-12}\,\text{J}$$

In a g mol of a substance, there are 6.023×10^{23} molecules (Avogadro's law).

$$\therefore \text{Heat released per g mol of a substance} = 2.714 \times 10^{-12} \times 6.023 \times 10^{23}$$

$$= 1.635 \times 10^{12}\,\text{J/mol}$$

In chemical reactions, the *additivity* of enthalpy (H) and energy (U) is a state function, because U and H are state functions. Additivity is expressed in Hess's law, which states that the total enthalpy change for the reaction is the sum of all changes irrespective of multiple steps of the reaction. Hess's law is useful in predicting the heat of reaction. In a nuclear reaction, the heat released is equal to the enthalpy, if the nuclear process occurs at constant pressure; in such a case, all the thermodynamic formalism that applies to the chemical reactions equally applies to nuclear reactions as well. Hess's law is also valid for nuclear reactions in accordance with the first law of thermodynamics.

EXAMPLE PROBLEMS

Example 3.1 The cylinder of an engine contains a gas at a pressure of 1600 kPa, which expands in the cylinder according to a process represented by a straight-line on p–V co-ordinates. Estimate the work done by the gas on the piston when the final pressure is 175 kPa. The stroke of the piston is 250 mm and the area of the cylinder is 0.118 m².

Solution

Figure Ex. 3.1

$$\text{Work done } w = \int p\,dV \quad (\text{The area under p} - \text{V diagram})$$

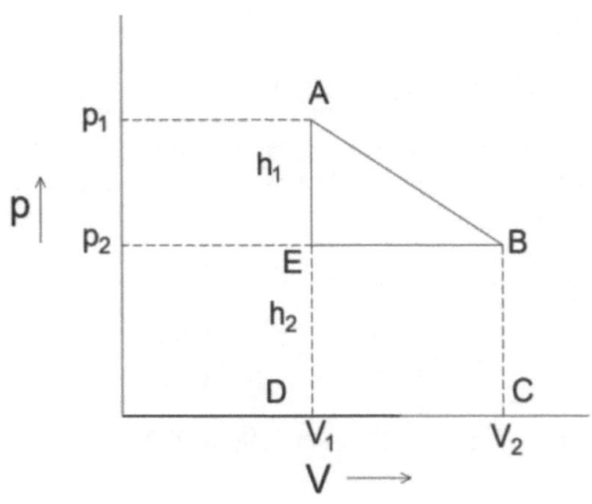

$= \text{Area of rectangle BCDE} + \text{area of triangle ABE}$

$\text{Area of rectangle} = lh_2$

$\text{where } l = \Delta V = \text{area} \times \text{length}$

$= 0.118 \times 0.25 = 0.0295\,m^3$

$\text{and } h_2 = p_2 = 175\,kPa$

Area of rectangle $= 0.0295 \times 175 = 5.162$ kJ

$$\text{Area of triangle} = \frac{1}{2} lh_1$$

$\text{where } h_1 = p_1 - p_2$

$= 1600 - 175 = 1425\,kPa$

$$\text{Area of triangle} = \frac{1}{2} \times 0.0295 \times 1425 = 21.018 \text{ kJ}$$

$\therefore$ Work done $= 5.162 + 21.018 = 26.18\,kJ$ Ans.

Example 3.2 In a cyclic process, the working fluid in a steam engine involves in two heat interactions: 50 kJ to the fluid and 20 kJ from the fluid, and three work interactions: 20 kJ to the fluid, 40 kJ from the fluid, and third is unknown. Determine the magnitude and direction of the third work transfer.

Solution For a cyclic process, according to the first law of thermodynamics, $\sum Q_{cycle} = \sum W_{cycle}$

$$Q_1 + Q_2 = w_1 + w_2 + w_3$$

$$Q_1 - Q_2 = w_2 - w_1 + w_3$$

$$\therefore 50 - 20 = 40 - 20 + W_3$$

$$w_3 = 30 - 20 = 10 \text{ kJ}$$

$\therefore$ Third work transfer is 10 kJ from the fluid. Ans.

Example 3.3 Determine specific heats c_p and c_v for a fluid whose properties are related as $u = 205 + 0.827t$ and $pv = 0.279(t + 273)$, where t is in °C, v is in m^3/kg, and 'u' is the specific internal energy.

Solution Specific heats at constant pressure and constant volume are

$$c_p = \left(\frac{\partial h}{\partial T}\right)_P \text{ and } c_v = \left(\frac{\partial u}{\partial T}\right)_v$$

where $h = u + pv \Rightarrow h = (205 + 0.827t) + 0.279(t + 273)$

$$c_p = \left(\frac{\partial h}{\partial T}\right)_P = \frac{\partial}{\partial t}(205 + 0.827t) + 0.279(t + 273)$$

$0.827 + 0.279 = 1.106$ kJ/kg K

$$c_v = \left(\frac{\partial u}{\partial T}\right)_v = \frac{\partial}{\partial t}[205 + 0.827t] = 0.827\,\text{kJ/kg K}$$

$\therefore c_p = 1.106$ kJ/kg K and $c_v = 0.827$ kJ/kg K Ans.

Example 3.4 A gas of mass 2 kg undergoes a quasi-static expansion, which follows $p = a + 2bV$. The initial pressure and volume are 1200 kPa and 0.25 m^3, and the final pressure and volume are 250 kPa and 1.35 m^3, respectively. The specific internal energy of the gas is given by $u = 2pv - 70$ kJ/kg. Estimate the net heat transfer.

Solution $$p = a + 2bV \Rightarrow 1200 = a + 2b \times 0.25 \quad (3.66)$$

$$250 = a + 2b \times 1.35 \quad (3.67)$$

Solving Eqs. 3.46 and 3.47 $\Rightarrow a = 984.09,\ b = -431.82$

$$\text{Work transfer } W_{1\text{-}2} = \int_{V_1}^{V_2} p\,dV = \int_{V_1}^{V_2} (a + 2bV)\,dV$$

$$= a(V_2 - V_1) + 2b\left(\frac{V_2^2 - V_1^2}{2}\right) = (V_2 - V_1)\left[a + \frac{2b}{2}(V_1 + V_2)\right]$$

$$= 1.1[984.09 + (-431.82)(1.6)] = 322.495\ \text{kJ}$$

Δu = Change in internal energy $= u_2 - u_1 = 2(p_2v_2 - p_1v_1)$

$\Delta U = U_2 - U_1 = 2(p_2V_2 - p_1V_1)$ (Since $V = vm$ and $U = mu$)

$\Delta U = 2[250 \times 1.35 - 1200 \times 0.25]$

$= 2(337.5 - 300) = 75$ kJ

$Q_{1\text{-}2} = W_{1\text{-}2} + \Delta U = 322.495 + 75$

The net heat transfer is 397.495 kJ Ans.

Example 3.5 Gas (5 kg) expands in an engine in which the p–V relationship is $pV^{1.3}$ = const. The initial and final pressures are 980 and 2.5 kPa, respectively. The initial volume is 0.93 m^3. Determine the heat transfer, if the specific internal energy of gas reduces by 50 kJ/kg.

Solution $$W_{1\text{-}2} = \frac{p_1V_1 - p_2V_2}{n - 1} \quad \text{when } pV^n = \text{constant}$$

$$\text{Now } p_1V_1^n = p_2V_2^n \Rightarrow V_2^n = \left(\frac{p_1}{p_2}\right)V_1^n$$

$$\therefore V_2^n = \left(\frac{980}{2.5}\right) \times (0.93)^{1.3} = 392 \times 0.909 = 356.70$$

$$V_2 = (356.70)^{1/1.3} = 91.77\,\text{m}^3$$

$$\therefore W_{1\text{-}2} = \frac{980 \times 0.93 - 2.5 \times 91.77}{0.3} = 2273.25\,\text{kJ}$$

$$\therefore Q_{1\text{-}2} = W_{1\text{-}2} + \Delta U$$

$$\Delta U = U_2 - U_1 = m(u_2 - u_1) = 5(50) = 250\ \text{kJ/kg}$$

$$\therefore Q_{1\text{-}2} = 2273.25 - 250 = 2023.25\,\text{kJ}$$ Ans.

Example 3.6 A 700 kg of fish, initially at 7°C, is to be frozen at −15°C. How much is the heat to be removed to cool it? Take specific heat c_p of fish above and below the freezing point as 3.2 and 1.699 kJ/kg K, respectively. The freezing point of fish is −2°C and the latent heat of fusion is 232 kJ/kg. Also find the total latent heat to be removed.

Solution Total heat to be removed from fish = heat removal above the freezing point + latent heat + heat removal below the freezing point

Mass of fish $m = 700$ kg

Initial temperature $t_1 = 7°\text{C}$ freezing point $t_f = -2°\text{C}$

Temperature required, $t_2 = -15°\text{C}$

Specific heat of fish above the freezing point $c_{p1} = 3.2$ kJ/kg K

Specific heat of fish below the freezing point $c_{p2} = 1.699$ kJ/kg K

Heat removal $Q = mc_p\,\Delta t$

Latent heat $h_{fg} = 232$ kJ/kg

$$Q = m\left[c_{p1}(t_1 - t_f) + h_{fg} + c_{p2}(t_f - t_2)\right]$$

$$= 700[3.2 \times [7 - (-2)] + 232 + 1.699 \times [-2 - (-15)]$$

Total heat to be removed from fish = 198,020.9 kJ or 198.0209 MJ Ans.

Total latent heat to be removed = 700 × 232 = 162,400 kJ or 162.4 MJ Ans.

Example 3.7 The capacity of the milk chilling unit to remove heat from milk is 15 kJ/s. The rate of heat leakage into the milk is 2 kJ/s. Estimate the time required for cooling 1 ton of milk from 30°C to 7°C. The specific heat of milk is 4.18 kJ/kg K.

Solution Mass of milk = 1000 kg c_p of milk = 4.18 kJ/kg K

$$\Delta t = 30 - 7 = 23°\text{C}$$

Total heat to be removed from milk $Q = mc_p\,\Delta t$

$$= 1000 \times 4.18 \times 23 = 96{,}140\ \text{kJ}$$

Net capacity of the plant = 15 − 2 = 13 kJ/s

$$\text{Time required for cooling the milk} = \frac{\text{Total heat to be removed}}{\text{Net capacity of the unit}}$$

$$= \frac{96{,}140}{13} = 7395\,\text{s} = 2.05\,\text{hours}$$ Ans.

Example 3.8 A gas of 1 kg mass with an initial pressure of 250 kPa and a volume of 0.035 m^3 is contained in a cylinder fitted with a piston on which some weights are placed. Estimate the work done when (i) the gas expands to a volume of 0.09 m^3 when the pressure remains constant, (ii) gas expands isothermally when the weights are removed, and (iii) the expansion follows $pV^{1.4} = \text{constant}$.

Solution Work done by the gas is to be estimated in three different cases.

$$p_1 = 250\text{ kPa} \qquad m = 1\text{ kg}$$

$$V_1 = 0.035\text{ m}^3 \qquad V_2 = 0.09\text{ m}^3$$

Case 1: Work done in isobaric process, $W_{1\text{-}2} = \int_1^2 p\,dV = p(V_2 - V_1)$

$$= 250(0.09 - 0.035) = 13.75\text{ kJ}$$

Case 2: Work done in isothermal process, $W_{1\text{-}2} = p_1V_1 \ln\dfrac{V_2}{V_1} = 250 \times 0.035$

$$\times \ln\frac{0.09}{0.035} = 8.264\text{ kJ}$$

Case 3: Work done in polytropic process, $W_{1\text{-}2} = \dfrac{p_1V_1 - p_2V_2}{n-1}$

Again $\quad p_1V_1^n = p_2V_2^n \Rightarrow p_2 = p_1\left(\dfrac{V_1}{V_2}\right)^n$

$$p_2 = 250\left(\frac{0.035}{0.09}\right)^{1.4} = 67.85\text{ kPa}$$

$$W_{1\text{-}2} = \frac{250 \times 0.035 - 67.85 \times 0.09}{1.4 - 1} = 6.608\text{ kJ}$$

Example 3.9 A gas undergoes a thermodynamic cycle consisting of three processes: (i) process 1-2: constant pressure process, $p_1 = 1.4$ bar, $v_1 = 0.028\text{ m}^3$, $W_{1\text{-}2} = 10.5$ kJ, (ii) process 2-3: compression with $pV = C$, $U_3 = U_2$, and (iii) process 3-1: constant volume $U_1 - U_3 = -26.4$ kJ. (i) Sketch a cycle on the p–V diagram, (ii) calculate W_{net} for the cycle, (iii) calculate $Q_{1\text{-}2}$, and (iv) show that $\sum Q_{cycle} = \sum W_{cycle}$.

Solution

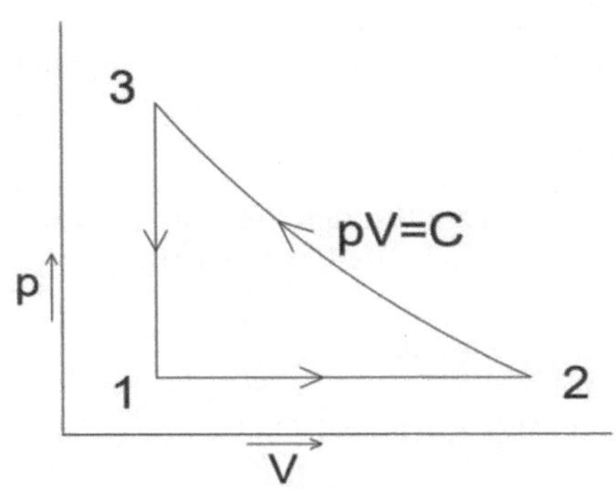

Figure Ex. 3.9

Process 1-2 is a constant pressure process, $W_{1\text{-}2} = p_1(V_2 - V_1)$

$$10.5 = 1.4 \times 10^2(V_2 - 0.028)$$

$$V_2 = 0.103\ m^3 \text{ and } V_1 = V_3 = 0.028\ m^3$$

$$\text{Process 2-3} \Rightarrow W_{2\text{-}3} = p_2V_2 \ln\frac{V_3}{V_2} = 1.4 \times 10^2 \times 0.103 \times \ln\left(\frac{0.028}{0.103}\right) = -18.8\,\text{kJ}$$

$$Q_{2\text{-}3} = W_{2\text{-}3} + U_3 - U_2 = -18.8 + 0 = -18.8\ \text{kJ}$$

$$\text{Process 3-1} \Rightarrow W_{3\text{-}1} = 0,\ Q_{3\text{-}1} = -26.4\,\text{kJ}$$

$$W_{net} \text{ for cycle, } \sum W_{cycle} = W_{1\text{-}2} + W_{2\text{-}3} + W_{3\text{-}1}$$

$$= 10.5 - 18.8 + 0 = -8.3\ \text{kJ}$$

$$Q_{1\text{-}2} = W_{1\text{-}2} + U_2 - U_1 = 10.5 + 26.4$$

$$Q_{1\text{-}2} = 36.9\ \text{kJ}$$

$$\sum Q_{cycle} = Q_{1\text{-}2} + Q_{2\text{-}3} + Q_{3\text{-}1}$$

$$36.9 + (-18.8) + (-26.4) = -8.3\ \text{kJ}$$

$$\text{Therefore} \sum Q_{cycle} = \sum W_{cycle}$$

Example 3.10 The wall of an industrial furnace is made of 20-cm thick fireclay bricks with a thermal conductivity of 1.7 W/m K. The inner surface of the wall is maintained at 1000°C. Determine the outer surface temperature when there is a heat loss of 1500 W through a wall of 0.5 m × 1.5 m.

Solution Since the heat transferred through the wall is by conduction, the heat loss is determined by Fourier's equation given as

$$\dot{Q}_{cond} = kA\frac{\Delta T}{L}$$

where $k = 1.7$ W/m K, $L = 0.20$ m, $T_1 = 1273$ K and $\dot{Q} = 1500$ W.

$$\therefore 1500 = 1.7 \times 0.5 \times 1.5\left(\frac{1273 - T_2}{0.20}\right)$$

Outer surface temperature, $t_2 = 764.70°C$ Ans.

Example 3.11 Two surfaces of a 25-mm thick plate are maintained at 0°C and 80°C, respectively. If the heat is transferred through the plate at a rate of 450 W/m², find the thermal conductivity of the plate.

Solution Since the heat transferred through the plate is by conduction, the heat flux in W/m² through the plate is given by Fourier's equation,

$$\ddot{q}_{cond} = k\frac{\Delta T}{L}$$

where $L = 0.025$ m, $\dot{q} = 450$ W/m², $\Delta T = 80°C–0°C = 80°C$ and $A = 1.5 \times 3$ m².

$$450 = \frac{k \cdot (80 - 0)}{0.025}$$

∴ Thermal conductivity of the plate, k = 0.1406 W/m K. Ans.

Example 3.12 Hot air at a temperature of 100°C is blown over a plate of 1.5 m × 3 m flat surface at 25°C. Find the rate of heat transfer from the air to the plate, with a heat transfer coefficient of 50 W/m² K.

Solution Since the air is blown over a plate, heat transfer by forced convection takes place from hot air to the plate. The rate of heat transfer by convection is governed by Newton's law of cooling.

Convective heat transfer coefficient, $h = 50$ W/m² K and $A = 1.5 \times 3$ m²

Heat transfer by convection can be determined by using Eq.3.15

$$\dot{Q}_{conv} = hA(T_\alpha - T_S)$$

$$= 50 \times 1.5 \times 3(100 - 25) = 16.875 \text{ kW}$$ Ans.

REVIEW QUESTIONS

3.1 What are the macroscopic and microscopic forms of energy?
3.2 What is total energy? Identify the different forms of energy that constitute the total energy.
3.3 List the forms of energy that contribute to the internal energy of a system.
3.4 How are heat, internal energy, and thermal energy related to each other?
3.5 What is mechanical energy? How does it differ from thermal energy? What are the forms of mechanical energy of a fluid stream?
3.6 Define specific heats at constant pressure and constant volume.
3.7 Why is the specific heat at constant pressure greater than the specific heat at constant volume?
3.8 What is a perpetual motion machine of the first kind (PMM1)? Does it exist?
3.9 Is energy a point function? Justify your answer.
3.10 What is a path function?
3.11 What are the similarities between heat and work?
3.12 Explain why heat and work are path functions.
3.13 Does a system contain heat? Explain.
3.14 Does a system contain work? Explain.
3.15 For a closed system, energy balance in the differential form is $dE = \delta Q + \delta W$. Explain why is d and not δ used for the differential on the left.
3.16 What is energy efficiency and what is energy conservation?
3.17 What is energy sustainability?
3.18 What is energy security?
3.19 Define thermal conductivity.
3.20 Define the convection heat-transfer coefficient. What are the factors that influence the coefficient?

3.21 What is the governing law of heat conduction?
3.22 What is the governing law of heat convection?
3.23 Define emissivity.
3.24 Define emissive power.
3.25 What is a black body?

EXERCISE PROBLEMS

3.1 A piston-cylinder device contains air at 5.5 bar, 27°C, and a volume of 0.008 m^3. A constant-pressure process gives 48 kJ of work output. Find the final volume, the temperature of the air, and the heat transfer.

3.2 A piston-cylinder device contains 2.45 kg of water at 3 bar, 180°C. It is now heated by a process in which pressure is linearly related to volume to a state of 6.4 bar, 300°C. Find the final volume, heat transfer, and work in the process.

3.3 A piston-cylinder device contains 1.5 kg of water at 5.6 bar, 300°C. It is now cooled in a process wherein pressure is linearly related to volume to a state of 2.3 bar, 135°C. Determine the work and the heat transfer in the process.

3.4 Helium gas expands from 1.4 bar, 90°C and 0.198 m^3 to 1.02 bar in a polytropic process with $n=1.45$. How much heat transfer is involved?

3.5 A piston-cylinder device contains 0.15 kg of air at 30°C and 1.03 bar. The air is now slowly compressed in an isothermal process to a final pressure of 2.56 bar. Show the process in a P–V diagram, and find both the work and heat transfer in the process.

3.6 A piston-cylinder device has an initial volume of 0.08 m^3 and contains nitrogen at 1.45 bar, 25°C. The piston is moved, compressing the nitrogen until the pressure is 8 bar and the temperature is 135°C. During this compression process, heat is transferred from the nitrogen, and the work done on the nitrogen is 18 kJ. Determine the amount of this heat transfer.

3.7 A 1000 kg car is accelerated from 0 to 120 km/h over a distance of 500 m. The road at the end of the 500 m is at 12 m higher elevation. Determine the total increase in the car's kinetic and potential energy.

3.8 A car of a mass of 700 kg, with an initial velocity of 80 m/s, decelerates to a final velocity of 15 m/s. Estimate the kinetic energy of the car in kJ.

3.9 An object of 500 kg mass is at an elevation of 30 m above the earth. Estimate the potential energy of the object relative to the surface of the earth. Take $g=9.8$ m/s^2.

3.10 An object of 70 kg mass is projected upward from the earth with an initial velocity of 50 m/s. Estimate the elevation of the object at which its velocity becomes zero. Take $g=9.8$ m/s^2. Assume that the force acting on the object is the force of gravity only.

3.11 A 12-V battery supplies a constant current of 0.70 amp to a resistance for 1 hour. Determine (i) the resistance, in ohms, and (ii) the amount of energy transfer to the battery, by work, in kJ.

3.12 An automobile engine's alternator supplies charge constantly to a 12-V battery. The voltage across the terminals is 12.5 V dc, and the current is 8 A. Determine the electrical work energy transport rate from the automobile engine to the battery in both watts and horsepower.

3.13 A mass of 12 kg undergoes a process during which there is heat transfer from the mass at a rate of 5 kJ/kg, an elevation decrease of 35 m, and an increase in velocity from 20 to 40 m/s. The specific internal energy decreases by 7.5 kJ/kg and the acceleration of gravity is constant at 9.81 m/s^2. Determine the work for the process.

3.14 Water, in a closed pan on top of a range, is heated by the stirring process by a paddle wheel. Heat (25 kJ) is transferred to the water and 7 kJ of heat is lost to the surrounding air during the process. The paddle-wheel work is 450 N m. Determine the final energy of the system if its initial energy is 12.5 kJ.

3.15 A rigid, closed, stationary tank contains a unit mass of saturated liquid water at a pressure of 3 kgf/cm^2. A paddle wheel does 350 N m of the work on the system, while heat is transferred to or from the system. The final pressure of the system is 3 kgf/cm^2. Determine the amount of heat transferred and indicate its direction.

3.16 A pulley of 20 cm diameter turns a belt rotating the driveshaft of a power plant pump. The torque applied by the belt on the pulley is 180 N m, and the power transmitted is 8.5 kW. Estimate the net force applied by the belt on the pulley and the rotational speed of the driveshaft.

3.17 The inner and outer surfaces of a 45-mm thick 2-m × 2-m window glass in winter are 15°C and 5°C, respectively. If the thermal conductivity of the glass is 0.85 W/m °C, evaluate the amount of heat loss through the glass over a period of 3 hours.

3.18 A 45-mm external diameter, 10-m long hot water pipe at 100°C is losing heat to the surrounding air at 10°C by natural convection with a heat transfer coefficient of 25 $W/m^2 \cdot °C$. Determine the rate of heat loss from the pipe by natural convection.

3.19 A 5-cm-diameter spherical ball whose surface is maintained at a temperature of 60°C is suspended in the middle of a room at 25°C. If the convection heat transfer coefficient is 12 $W/m^2 \cdot °C$ and the emissivity of the surface is 0.8, determine the total rate of heat transfer from the ball.

3.20 Two very large parallel planes having surface conditions that are closely approximated as a black body are maintained at 1000°C and 500°C, respectively. Calculate the heat transfer by radiation between the planes per unit time and per unit surface area.

DESIGN AND EXPERIMENT PROBLEMS

3.21 Design an experiment complete with instrumentation to determine the specific heats of a gas using a resistance heater. Discuss how the experiment will be conducted, what measurements need to be taken, and how the specific heats will be determined. What are the sources of error in your system? How can you minimize the experimental error?

3.22 Design an experiment complete with instrumentation to determine the mass flow rates and temperatures of exhaust gases coming out of a diesel engine in your IC engines laboratory. Write the procedure for the conduct of the experiment and measurements need to be taken.

3.23 In an air compressor experiment of thermal engineering laboratory, compressed air from a storage tank is discharged to the atmosphere while temperature and pressure measurements of the air inside the vessel are recorded. The expansion on compression of a gas can be described by the polytropic relation $pV^n = constant$, where p is pressure, v is specific volume, c is a constant, and the exponent n depends on the thermodynamic process. Obtain the polytropic exponent n for the process using these measurements, along with the first law of thermodynamics.

4 Properties of Pure Substances

LEARNING OUTCOMES

After learning this chapter, students should be able to

- Understand the concept of a pure substance.
- Analyze the phase change phenomena useful in so many applications such as steam power plants.
- Illustrate the p-v, T-v, and p-T property diagrams and p-v-T surfaces of pure substances.
- Demonstrate the procedures for determining the thermodynamic properties of pure substances from tables of property data.
- Describe the hypothetical substance "ideal gas" and the ideal-gas equation of state.
- Apply the ideal-gas equation of state for typical problems.
- Introduce the compressibility factor, which accounts for the deviation of real gases from ideal-gas behavior.

4.1 PURE SUBSTANCES AND THEIR PHASES

A substance that has a fixed chemical composition throughout its volume is called a pure substance. A substance may be a single element or compound or a mixture of various chemical elements or compounds. A pure substance can exist in more than one phase, but its chemical composition must be the same in each phase. For example, water is a pure substance when it is in solid (ice), liquid (water), and vapor (gaseous) states separately and combinedly as water possesses the fixed ratio of hydrogen and oxygen. Similarly, air is also a pure substance as it possesses a uniform mixture of its constituents throughout its volume. Liquid air and air (gaseous) are not considered as pure substances because their compositions vary in the different phases.

A substance exists in three different phases: solid, liquid, and gas. In a solid phase, the molecules are closely arranged and hence the attractive forces of molecules on each other are so high that they keep the molecules at fixed positions. There is no relative motion of molecules in solids; however, there is a continuous oscillation of molecules about their equilibrium positions. Temperature plays an important role during these oscillations. Liquids, like solids, have almost similar molecular structures as molecules are closely spaced; however, they differ in that molecules can rotate and translate freely apart from not having fixed positions relative to each other. Gases, on the other hand, have molecules spaced far apart when compared to both solids and liquids and molecules move randomly and collide continually with each other and with the walls in which they are contained. The molecules in the gas phase have higher energy levels when compared to the same molecules in solid or liquid phases.

4.2 PHASE CHANGE PROCESSES OF PURE SUBSTANCES

For a pure substance, two of its phases, liquid and vapor that are commonly encountered in many of the practical situations such as boiling and condensation applications, coexist in equilibrium. For example, water, in steam power plants, exists as a mixture of liquid and vapor. As another example, a refrigerant changes its phase from liquid to vapor in the evaporator section of the refrigerators.

Saturation Temperature and Saturation Pressure

A compressed liquid or subcooled liquid is a substance that is not about to vaporize. *A saturated liquid* is the one which is about to vaporize and doesn't contain any vapor with it. A vapor is said to be *saturated vapor* when it is about to condense and doesn't contain any liquid with it, whereas a vapor is said to be *superheated vapor* when its temperature is higher than the saturation temperature corresponding to the given pressure. A superheated vapor is also defined as a vapor that is not about to condense.

Saturation temperature, t_{sat}, is defined as the temperature at which a pure substance changes its phase at a given pressure. Water, for example, on heating changes its phase from liquid to vapor at 100°C when the pressure is held constant at 1 atm or 101.325 kPa. However, 100°C is not the saturation temperature at other pressures. Saturation temperature increases with an increase in pressure. The saturation temperature for water at 200 kPa is 120.88°C.

Saturation pressure, p_{sat}, is defined as the pressure at which a pure substance changes its phase at a given temperature. For water, at a temperature of 100°C, p_{sat} is 101.42 kPa.

The temperature at which any two phases of a pure substance coexist in equilibrium depends on the pressure for solid–liquid, solid–vapor, and liquid–vapor equilibrium states. For most of the substances, there is a pressure below which solid and vapor phases coexist in equilibrium and the substance cannot exist as a liquid below this pressure. For each substance, there is a fixed relation between pressure and temperature for solid–vapor saturation states. The phase transformation from a solid to vapor is called sublimation and the solid–vapor saturation line on the p-T diagram (discussed later) is called the sublimation line. The well-known example of sublimation is the phase transformation of solid carbon dioxide (dry ice) to a gas instead of a liquid.

The amount of heat required to cause a phase change is called *latent heat.* Latent heat is also defined as the amount of energy absorbed or released during a phase change process. The amount of heat absorbed during the melting or the amount of heat released during freezing is called the *latent heat of fusion.* The amount of heat absorbed during vaporization or the amount of heat released during condensation is called the *latent heat of vaporization.* The magnitudes of latent heat depend on temperature or pressure at which phase change takes place. For water, at 1 atm pressure, the latent heat of fusion is 333.7 kJ/kg and the latent heat of vaporization is 2256.5 kJ/kg. From this, it is stated that at 1 atm pressure, solid ice absorbs 333.7 kJ/kg of energy to melt into liquid water; similarly, water releases the same amount, i.e., 333.7 kJ/kg, to freeze into ice.

The state of a pure substance at rest is specified entirely by the values of two independent intensive properties in the absence of electrical, magnetic, and surface tension effects. For example, for air, if the pressure and temperature are specified, then the values of other properties such as viscosity and density are fixed; similarly, if pressure and density are specified, then the other properties will have the fixed values. However, the two intensive properties chosen to specify the state should be independent; this is possible when either one can be varied throughout a range of values while the other remains constant.

If a substance exists in a single phase, its pressure and temperature can be varied independently over a wide range of values and therefore, these are independent properties. However, this is not the case when the two phases of a pure substance coexist in equilibrium. There exists a fixed relationship between the pressure and temperature; that is, at any given pressure, there is one temperature at which two phases of the substance coexist in equilibrium. On the other hand, two phases of a pure substance coexist in equilibrium at a given temperature only at a particular pressure. The pressure and temperature in this case are not independent properties; therefore, specifying these two properties will not fix the state of the substance and hence one more property is required.

4.3 p-v DIAGRAM OF A PURE SUBSTANCE

The heating of a unit mass of ice at –20°C in a piston-cylinder device at a constant pressure of 1 atm is considered as shown in Figure 4.1a. The ice is heated slowly keeping the pressure constant and the state changes are plotted on the p-v diagram as shown in Figure 4.1b. When heat is added to the ice, its temperature

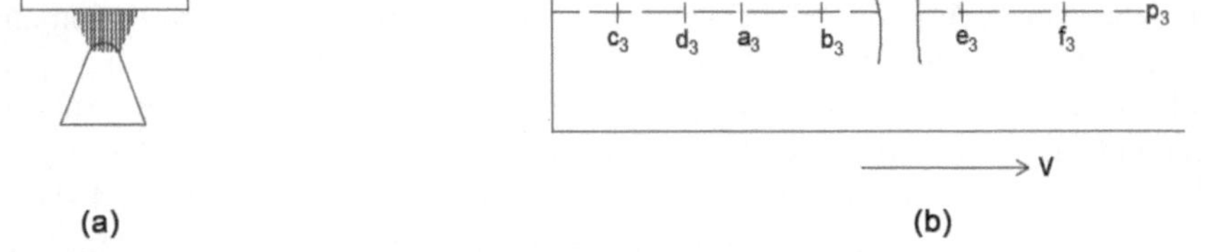

FIGURE 4.1 Changes in the volume of water during heating at a constant pressure. a. The heating of ice at –20°C at a constant pressure of 1 atm.
b. The state changes of water on p-v diagram

increases from −20°C to 0°C, the volume of ice increases as is the case for any solid upon heating. Ice melts into water at a constant temperature of 0°C, a phase change from solid to liquid, the volume of ice decreases which is a peculiar character of water. Upon further heating of water from 0°C to 100°C, the volume increases due to thermal expansion. Water starts boiling at a constant temperature of 100°C, a phase change from liquid to vapor. There is a large increase in volume. Heating of the vapor from 100°C to 200°C results in a further increase in volume.

In brief, the following things can be observed if the ice is heated from and at −20°C:

State 'a': initial condition of ice with a definite volume, the temperature is −20°C and the pressure is 1 atm.
Period 'a-b': sensible heat input, the temperature increases and the volume increases.
State 'b': saturated solid state.
Period 'b-c': latent heat input, the temperature remains constant and the volume decreases.
State 'c': the melting point of the substance.
Period 'c-d': sensible heat input, the temperature increases and the volume increases.
State 'd': saturated liquid state.
Period 'd-e': latent heat input, the temperature remains constant, while the volume increases.
State 'e': saturated vapor state.
Period 'e-f': superheated vapor state, the volume increases.

If the heating of ice at −20°C to steam at 200°C is carried at different constant pressures such as 2 and 3 atm, similar regimes of heating are obtained. The state changes of a pure substance (water) whose volume decreases on melting are plotted on the p-v diagram as shown in Figure 4.2 and a pure substance other than water whose volume increases on melting is shown in Figure 4.3. Referring to the Figures 4.2 and 4.3, all the saturated solid states 2 at different constant pressures can be joined by a line called saturated solid line; saturated liquid states 3 and 4 with respect to solidification and vaporization respectively can be joined by saturated liquid lines and similarly saturated vapor states 5 can be joined by a saturated vapor line. The region to the left of the saturated solid line is the solid region and the region between the saturated solid line and saturated liquid line (with respect to solidification) is the solid–liquid mixture region. The region between saturated liquid and saturated vapor is called the saturated liquid–vapor region. The region to the right of the saturated vapor line is called the vapor region. Figure 4.4 shows the general shape of the p-v diagram of a pure substance in which constant temperature lines have a downward trend.

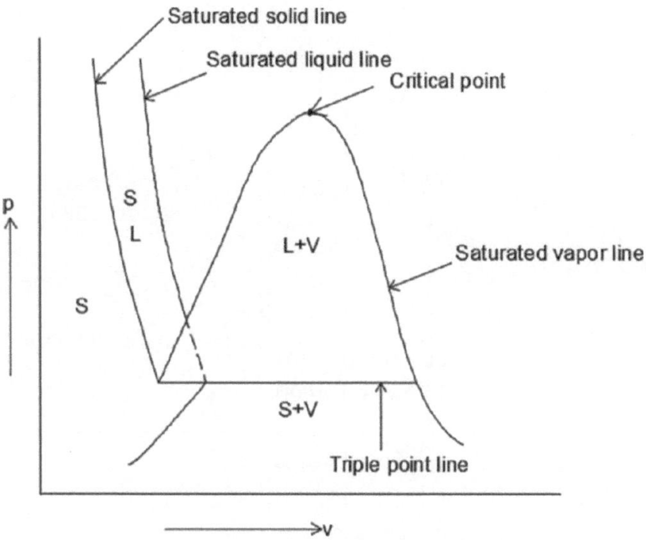

FIGURE 4.2 p-v diagram of a pure substance that contracts on freezing.

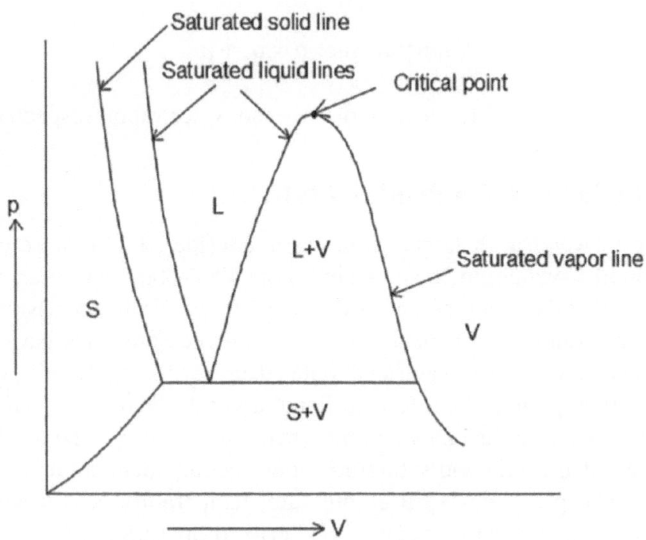

FIGURE 4.3 p-v diagram of a pure substance that expands on freezing.

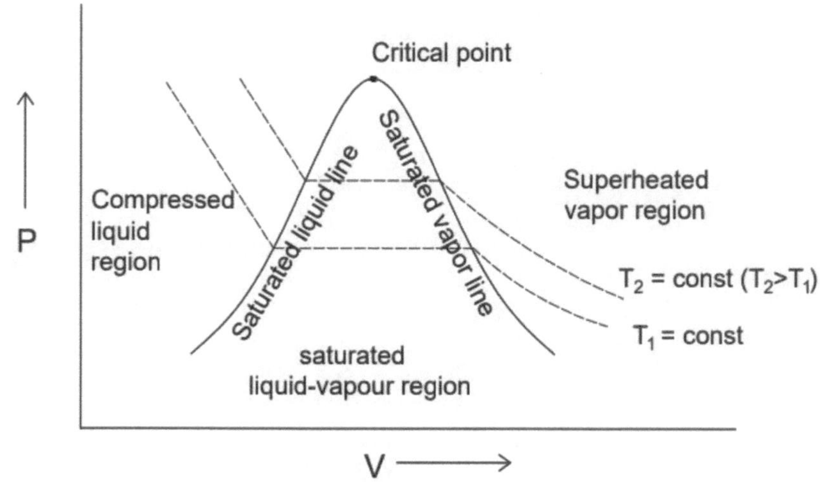

FIGURE 4.4 p-v diagram of a pure substance.

Vapor Pressure

Atmospheric air is regarded as a mixture of dry air and water vapor. The atmospheric air is the sum of partial pressures of dry air and water vapor called vapor pressure; it is the pressure that the vapor would exert if it existed alone at the temperature and volume of atmospheric air.

$$\text{Vapor pressure}, p = p_a + p_v \tag{4.1}$$

where p_a and p_v are partial pressures of dry air and water vapor, respectively.

4.4 T-v DIAGRAM OF A PURE SUBSTANCE

If the heating of water at different pressures is considered and phase change processes are plotted, T-v diagram as shown in Figure 4.5 is obtained. It can be observed from Figure 4.5 that the shape of the T-v diagram of a pure substance is much similar to that of the p-v diagram except that the constant pressure lines have an upward trend. The specific volume of saturated water decreases as the pressure increases. The saturation line, joining the saturated liquid and saturated vapor, narrows as the pressure increases and it becomes a point when the pressure is 220.06 bar. At this point, both saturated liquid and saturated vapor become identical and is called the *critical point*. The properties such as pressure, temperature, and specific volume of a substance at the critical point are called critical pressure, critical temperature, and critical specific volume, respectively. For water, these properties are 220.6 bar, 373.95°C, and 0.003106 m^3/kg respectively. Table 4.1 shows the critical pressure and critical temperature data of various substances.

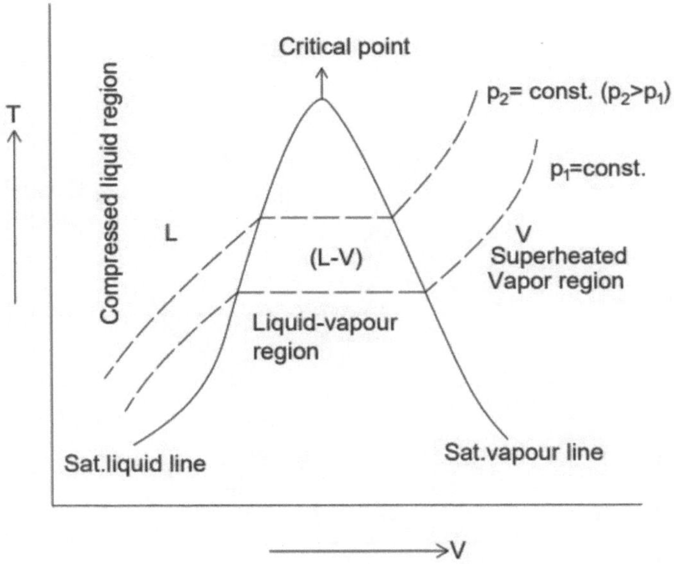

FIGURE 4.5 T-v diagram of a pure substance.

TABLE 4.1
Critical Point Data for Various Substances

Substance	Temperature	Pressure (MPa)
Ammonia, NH_3	405	11.3
Carbon dioxide, CO_2	304	7.39
Helium, He	5	0.23
Hydrogen, H_2	60	1.3
Mercury, Hg	1170	18.2
Nitrogen, N_2	126	3.4
Oxygen, O_2	154	5.0
Refrigerent, R-22	369	4.99
Refrigerent, R-134a	374	4.07
Water, H_2O	647	22.1
Ammonia, NH_3	405	11.3

4.5 p-T DIAGRAM OF A PURE SUBSTANCE

The pressure–temperature diagram is obtained for a substance, if either pressure or temperature or both are changed. Figure 4.6 shows how the three phases, solid, liquid, and vapor, exist together in equilibrium and the point at which this occurs is called the *triple point*. Table 4.2 shows the triple point pressure and triple point temperature data of various substances. Along the sublimation line, solid and vapor

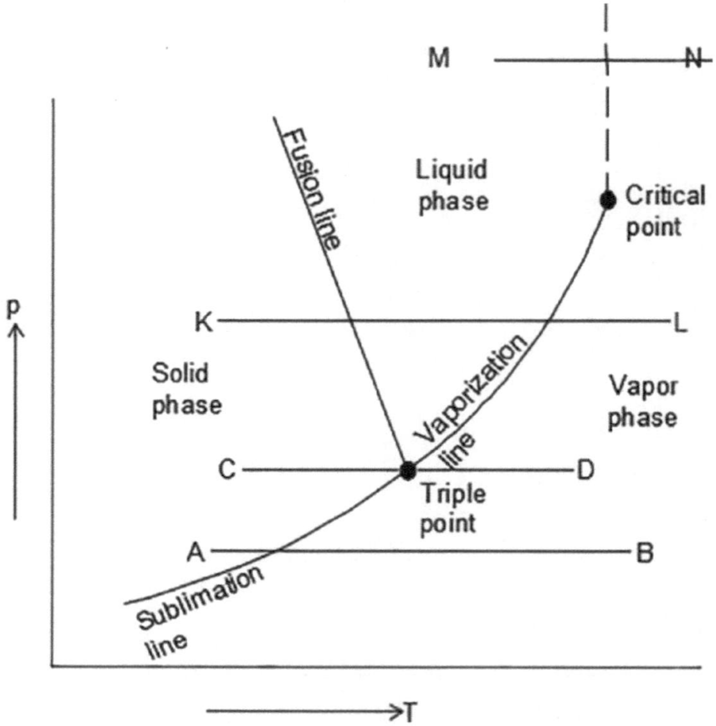

FIGURE 4.6 p-T diagram of a pure substance.

TABLE 4.2
Triple Point Data for Various Substances

Substance	Temperature (K)	Pressure (kPa)
Acetylene, C_2H_2	92.4	120
Ammonia, NH_3	195.40	6.076
Argon, Ar	83.81	68.9
Carbon (graphite), C	3900	10,100
Carbon dioxide, CO_2	216.55	517
Carbon monoxide, CO	68.10	15.37
Ethane, C_2H_6	89.89	8×10^{-4}
Helium, He	2.19	5.1
Hydrogen, H_2	13.84	7.04
Mercury, Hg	234.2	1.65×10^{-7}
Nitrogen, N_2	63.18	12.6
Oxygen, O_2	54.36	0.152
Water, H_2O	273.16	0.61

phases are in equilibrium; along the fusion line, solid and liquid phases are in equilibrium; and along the vaporization line, liquid and vapor phases are in equilibrium. The vaporization line ends at the critical point since there is no distinct change from the liquid phase to the vapor phase above the critical point.

Let us consider a solid substance at state A, as shown in Figure 4.6, whose temperature is raised keeping pressure (which is below the triple point pressure) constant; the substance passes directly from the solid phase to the vapor phase. Along the constant pressure line KL (which is above the triple point pressure), the substance passes directly from the solid phase to the liquid phase at one constant temperature and then from the liquid phase to the vapor phase at a higher temperature (below the critical point temperature). A constant pressure line CD passes through the triple point at which three phases exist together in equilibrium. At a pressure above the critical point pressure (line MN), there is no distinct transition from the liquid phase to the vapor phase. Though this discussion is pertinent to water, all the pure substances exhibit a similar behavior. However, it is important to notice that the triple point temperature and the critical point temperature vary considerably from substance to substance.

4.6 p-v-T SURFACE

Among the different properties, it is relatively easy to measure the pressure, specific volume, and temperature. From the experimental data, it can be understood that the pressure is the function of specific volume and temperature, and if the

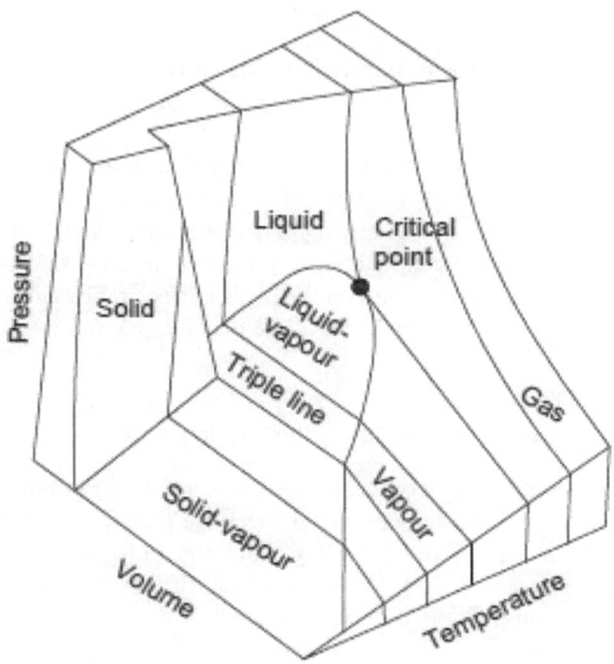

FIGURE 4.7 p-v-T surface of a substance that expands on freezing (water)

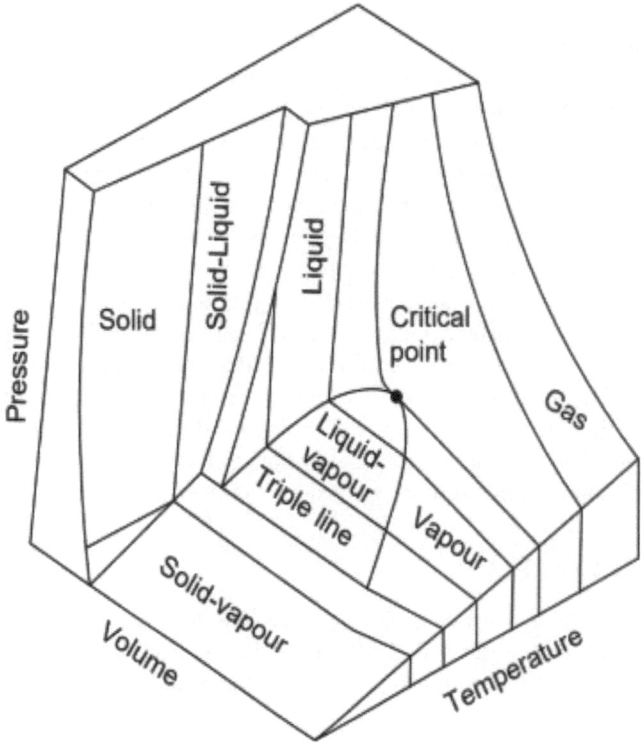

FIGURE 4.8 p-v-T surface of a substance that contracts on freezing.

values are plotted, the graph obtained is the p-v-T surface. Figure 4.7 shows the p-v-T surface of a substance that expands on freezing (water) while Figure 4.8 shows that of a substance that contracts on freezing. If any two of the properties are projected with different combinations, it results in p-v, T-v, and T-p diagrams of the concerned substance. Any point on the p-v-T surface represents an equilibrium state of the substance and the triple point line becomes a point when it is projected to the p-T plane.

4.7 T-s DIAGRAM OF A PURE SUBSTANCE

The heating of a unit mass of ice at a constant pressure of 1 atm from −20°C to steam at 200°C is considered and the entropy changes are plotted on the T-s diagram as shown in Figure 4.9. The curve 1-2-3-4-5-6 is the isobar of 1 atm pressure. Similarly, if the heating is carried out at different pressures, different regimes of similar type can be obtained. The phase equilibrium diagram of a pure substance as shown in Figure 4.10 can be obtained if the saturation states 2, 3, 4, and 5 at different pressures are joined. At a given pressure, if s_f and s_g are the specific entropies of saturated liquid and vapor respectively, then the entropy change of the system during the phase change from liquid to vapor at that pressure is

$$s_{fg} = s_g - s_f \tag{4.2}$$

The value of s_{fg} decreases as the pressure increases and becomes zero at a critical point. Liquid–vapor transformations are often encountered in most of the practical applications such as steam power plants and refrigeration and air-conditioning processes. Figure 4.11 shows the liquid, vapor, and transition zones.

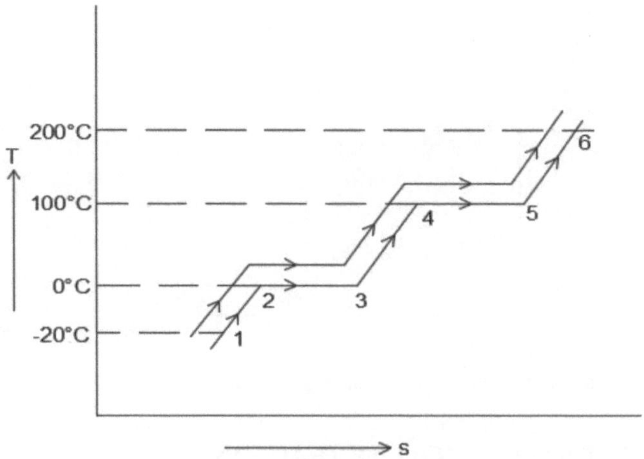

FIGURE 4.9 Isobars on the T-s diagram.

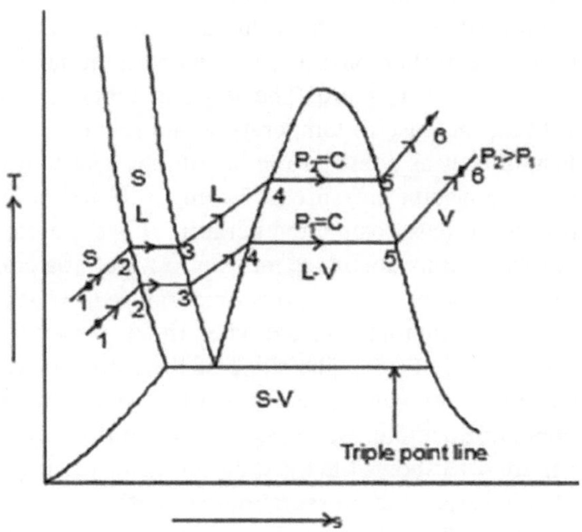

FIGURE 4.10 Phase equilibrium diagram of a pure substance on T-s coordinates.

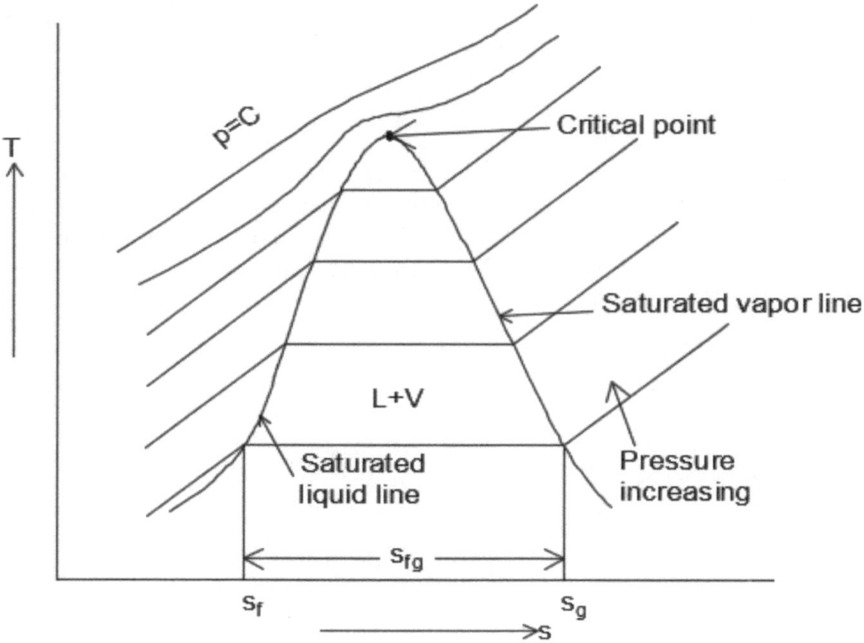

FIGURE 4.11 Saturation dome for water on T-s coordinates.

4.8 h-s DIAGRAM OR MOLLIER DIAGRAM

The thermodynamic property relation, Tds = dh − νdp, forms the basis for the establishment of an h-s diagram or Mollier chart that is useful for determining the properties of steam such as specific volume, enthalpy, internal energy, and entropy. The slope of the isobar on h-s coordinates is equal to the saturation temperature (absolute) at that pressure. The slope is dependent on temperature, which increases with an increase in temperature and remains constant when the temperature is constant. Let us consider the heating of a system of ice at −20°C to steam at 200°C at a constant pressure of 1 atm as shown in Figure 4.12. The slope of the isobar first increases as the temperature of ice increases from −20°C to 0°C. The slope remains constant as ice melts into water at a constant temperature of 0°C. The slope again increases as the temperature of water rises from 0°C to 100°C. The slope again remains constant when the water vaporizes into steam at a constant temperature of 100°C. The slope of the isobar increases further as the steam becomes superheated as the temperature of steam increases to 200°C. If we draw the isobars of different pressures on h-s coordinates, similar regimes can be obtained. Figure 4.13 shows the h-s diagram (Mollier diagram) indicating the liquid and vapor phases; as the pressure increases, the saturation temperature increases and the slope of the isobar also increases. Hence the constant pressure lines diverge from one another.

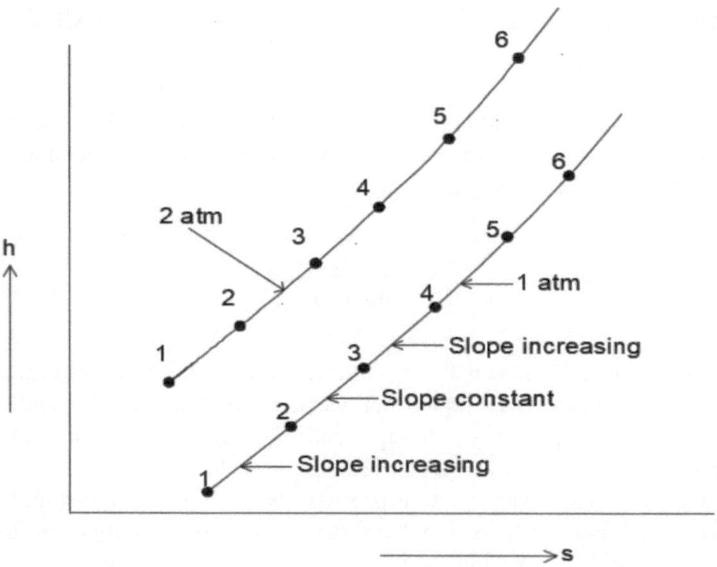

FIGURE 4.12 Phase equilibrium diagram of a pure substance on h-s coordinates.

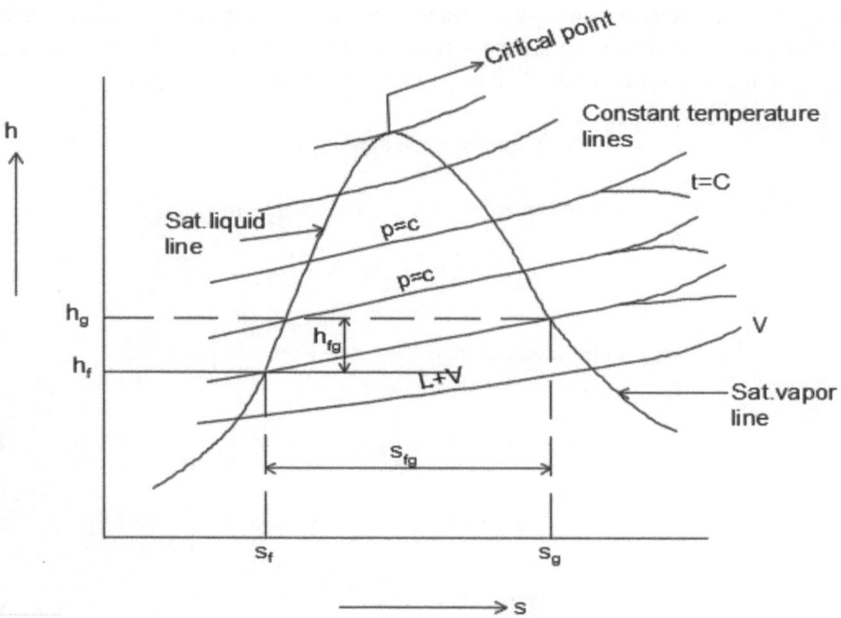

FIGURE 4.13 Enthalpy-entropy diagram of water.

4.9 QUALITY OR DRYNESS FRACTION—PROPERTY TABLES

4.9.1 Quality or Dryness Fraction

It is defined as the mass of vapor, m_v present in a mixture of liquid–vapor of mass $m_v + m_l$. In a liquid–vapor mixture of 1 kg, if x kg is the mass of vapor and (1 – x) kg is the mass of liquid, then x is quality, given as

$$x = \frac{m_v}{m_l + m_v} \tag{4.3}$$

The value of x varies between 0 for saturated water and 1 for saturated vapor. Saturated liquid is defined as a liquid that just starts boiling and contains no vapor. Saturated vapor is a vapor in which vaporization is just complete; it is also called dry saturated vapor.

A substance, during a vaporization process, exists in the liquid–vapor mixture region, that is, it exists partly as liquid and partly as vapor. Quality or dryness fraction serves as a useful tool to find the proportions of liquid and vapor so that we can analyze the mixture properly. A state of a substance in a saturated mixture region can be described with the quality. It is important to note that the properties of either saturated liquid or saturated vapor remain the same whether they exist alone or in a mixture.

To determine the amount of mass for each of the liquid and vapor phases, it is imagined that the two phases are mixed well. Let us consider a tank of volume V containing a saturated liquid–vapor mixture. The total volume $V = V_f + V_g$, where $V_f \rightarrow$ volume of saturated liquid and $V_g \rightarrow$ volume of saturated vapor.

$$\text{Now} \quad V = mv = m_f v_f + m_g v_g$$

$$\text{where} \quad m = m_f + m_g \Rightarrow m_f = m - m_g$$

$$\Rightarrow mv = \left(m - m_g\right) v_f + m_g v_g$$

$$\text{Dividing the above equation by m} \Rightarrow v = \left(1 - \frac{m_g}{m}\right) v_f + \frac{m_g}{m} v_g$$

$$v = (1 - x) v_f + x v_g \left(\text{since } x = \frac{m_g}{m}\right) \tag{4.4}$$

$$\text{or} \qquad v = v_f + x v_{fg} \left(v_{fg} = v_g - v_f\right)$$

$$\text{The quality,} \quad x = \frac{v - v_f}{v_{fg}} \tag{4.5}$$

$$\text{The internal energy,} \quad u = u_f + x u_{fg} \tag{4.6}$$

$$\text{Enthalpy} \quad h = h_f + xh_{fg} \tag{4.7}$$

$$\text{Entropy} \quad s = s_f + xs_{fg} \tag{4.8}$$

4.9.2 Compressed Liquid or Subcooled Liquid

It is defined as a liquid whose temperature is less than the saturation temperature at the given pressure. The pressure and temperature are independent properties for a compressed liquid. The properties of compressed liquid vary very little with pressure; however, they vary strongly with temperature. Thus the data for compressed liquid is not commonly available and therefore compressed liquid is treated as a saturated liquid at the given temperature. This is valid for all properties except enthalpy (h). When a liquid is cooled below its saturation temperature at a certain pressure, it is termed subcooled liquid. The difference between the saturation temperature and actual liquid temperature is referred to as the *degree of subcooling.*

4.9.3 Superheated Vapor

A vapor is said to be a superheated vapor, when its temperature is higher than the saturation temperature corresponding to the given pressure. The degree of superheat or simply superheat is the difference between the temperature of superheated vapor and saturation temperature at that pressure. A substance exists as a superheated vapor in the region to the right of the saturated vapor line and above the critical point temperature. In the superheated region, unlike in the saturated liquid–vapor region, the temperature and pressure are two independent properties. Figure 4.14 shows subcooling and superheat on the T-s diagram.

The superheated vapor is characterized by higher enthalpy, higher internal energy, and higher specific volume than the saturated vapor. The opposite is true for compressed liquid.

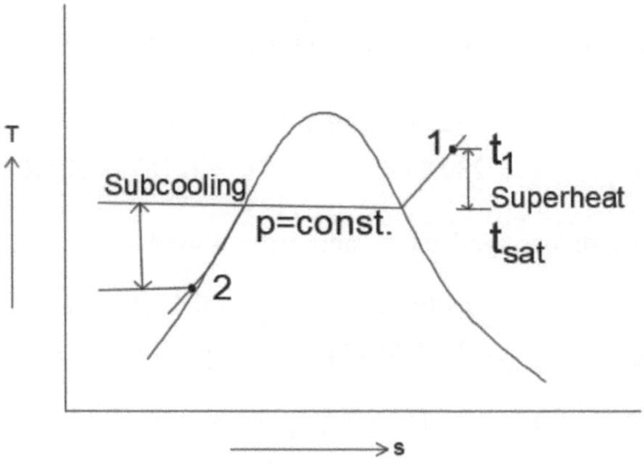

FIGURE 4.14 Subcooling and superheat.

EXAMPLE PROBLEMS

Example 4.1 A vessel of volume 50 L contains 5 kg of a mixture of saturated liquid and vapor at a pressure of 150 kPa. Determine (i) specific volume, (ii) temperature, (iii) enthalpy, (iv) entropy, and (v) internal energy.

Solution Specific volume, $\nu = \dfrac{0.050}{5} = 0.01\,\text{m}^3/\text{kg}\left(\nu = \dfrac{V}{m}\right)$ Ans.

From steam tables, at 150 kPa $\Rightarrow \nu_f = 0.00109$ m³/kg and $\nu_g = 0.3749$ m³/kg

Since $\nu_f < \nu < \nu_g$, the water is in the saturated mixture region; the temperature must be the saturated temperature at this pressure $t = t_{sat@150kpa} = 151.86°C$ Ans.

$$\text{Quality,}\quad x = \frac{\nu - \nu_f}{\nu_{fg}}$$

$$= \frac{0.01 - 0.00109}{0.3738}$$

$x = 0.023$ or 2.3%

At 500 kPa, $h_f = 646.23$ kJ/kg and $h_{fg} = 2108.5$ kJ/kg

$s_f = 1.8607$ kJ/kg K and $s_{fg} = 6.821$ kJ/kg K

$h = h_f + xh_{fg} = 640.23 + 0.023 \times 2108.5 = 688.73$ kJ/kg Ans.

$s = s_f + xs_{fg} = 1.8607 + 0.023 \times 6.821 = 2.017$ Ans.

$h = u + p\nu \Rightarrow u = h - p\nu$

$u = 688.73 - 150 \times 0.01 = 687.23$ kJ/kg Ans.

Example 4.2 15 kg of water at 55°C is heated at a constant pressure of 16 bar so that it becomes superheated vapor at 350°C. Determine the changes in (i) enthalpy, (ii) entropy, (iii) internal energy, and (iv) volume.

Solution At state 1, water is a saturated water, from saturated steam tables,

$$\left.\begin{array}{l}\text{At}\\ 55°\text{C}\end{array}\right\}\quad \begin{array}{ll} h_1 = h_f = 230.26\,\text{kJ/kg} & \nu_f = 0.00101\,\text{m}^3/\text{kg},\\ u_f = 230.24\,\text{kJ/kg} & s_f = 0.7679\,\text{kJ/kgK}\end{array}$$

At state 2, water becomes superheated vapor, from superheated steam tables,

$$\left.\begin{array}{l}16\text{ bar}\\ 350°\text{C}\end{array}\right\}\quad \begin{array}{ll} h_2 = 3146.0\,\text{kJ/kg} & \nu_2 = 0.1745\,\text{m}^3/\text{kg}\\ u_2 = 2866.6\,\text{kJ/kg} & s_2 = 7.0713\,\text{kJ/kgK}\end{array}$$

Change in enthalpy $= m(h_2 - h_1) = 15(3146.0 - 230.26) = 43{,}736.1$ kJ Ans.

Change in entropy $= m(s_2 - s_1) = 15(7.0713 - 0.7679) = 94.55$ kJ/K Ans.

Change in internal energy $\Delta U = m(u_2 - u_1) = 15(2866.6 - 230.24) = 39{,}545.4$ kJ Ans.

Change in volume $= m(\nu_2 - \nu_1) = 15(0.1745 - 0.00101) = 2.602$ m³ Ans.

Example 4.3 A 5 kg/s of steam, initially at 20 bar and 350°C, expands isentropically in a turbine to 2 bar. Compute the ideal power of the turbine.

Solution

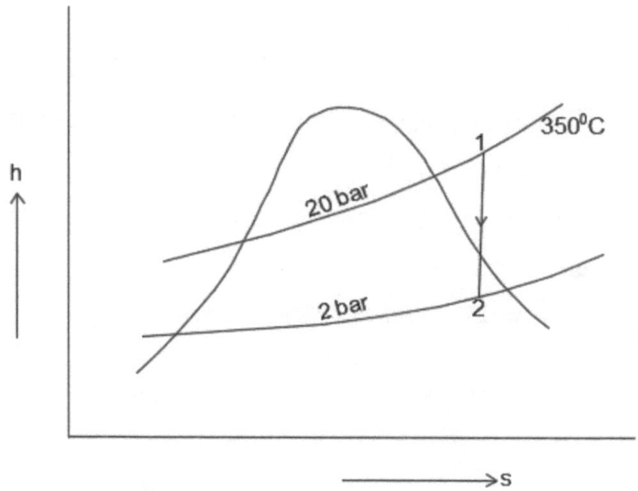

FIGURE EX. 4.3

Figure Ex. 4.3 shows the h-s diagram of the isentropic expansion of steam in the turbine.

From steam tables, $t_{sat@20bar} = 212.42°C$

The initial temperature (t) of steam is 350°C, ($t > t_{sat}$), therefore it is superheated.

From superheated steam tables, $h_1 = 3137.0$ kJ/kg.

Since the expansion is isentropic, the expansion on the h-s diagram is shown as a vertical line from 20 to 2 bar pressure line, so from the h-s diagram,

$$h_2 = 2620 \text{ kJ/kg}$$

The steady flow energy equation for a steam turbine is

$$\dot{m}(h_1) = \dot{m}(h_2) + \frac{dW_x}{dt} \left(\text{when heat loss and changes in KE and PE are negligible}\right)$$

$$\text{Power of the turbine}, \frac{dW_x}{dt} = \dot{m}(h_1 - h_2)$$

$$\frac{dW_x}{dt} = \dot{m}(h_1 - h_2) = 5(3137.0 - 2620) = 2585 \text{ kJ/s or kW} \qquad \text{Ans.}$$

Example 4.4 Steam at a rate of 2 kg/s flows through a turbine entering at 20 bar and 250°C and leaves at 0.2 bar with 5% moisture. The steam enters at 100 m/s and leaves at 50 m/s and the inlet is 3 m above the exit. Determine (i) the power of the turbine when heat loss is at a rate of 10 kJ/s and (ii) the diameters at inlet and exit. Neglect changes in kinetic and potential energies.

Solution From steam tables, saturation temperature, $t_{sat@20bar} = 212.42°C$.

The initial temperature of steam (t) is 250°C, ($t > t_{sat}$), therefore it is superheated.

From superheated steam tables,

State 1: at 20 bar and 250°C, $h_1 = 2902.5$ kJ/kg

$s_1 = 6.5453$ kJ/kg K $\quad\quad \nu_1 = 0.11144$ m³/kg

State 2: at 0.2 bar and 5% moisture, $h_{f2} = 251.40$ kJ/kg $\quad\quad h_{fg2} = 2358.3$ kJ/kg

$\nu_{f2} = 0.00101$ m³/kg $\quad\quad s_{f2} = 7.64$ kJ/kg K

Quality or dryness fraction, $x = 0.95$ (5% moisture)

$\therefore$ Enthalpy at state 2, $h_2 = h_{f2} + xh_{fg2}$

$$= 251.40 + 0.95 \times 2358.3 = 2491.78 \text{ kJ/kg}$$

$$\text{SFEE for steam turbine is} = \dot{m}\left[h_1 + \frac{V_1^2}{2} + z_1 g\right] + \frac{dQ}{dt} = \dot{m}\left[h_2 + \frac{V_2^2}{2} + z_2 g\right] + \frac{dW_x}{dt}$$

$$\text{or } \dot{m}\left[(h_1 - h_2) + \left(\frac{V_1^2 - V_2^2}{2}\right) + (z_1 - z_2)g\right] + \frac{dQ}{dt} = \frac{dW_x}{dt}$$

$$= 2\left[(2902.5 - 2491.78) + \frac{100^2 - 50^2}{2 \times 1000} + \frac{3 \times 9.8}{1000}\right] + (-10)$$

$$\text{Power of the turbine, } \frac{dW_x}{dt} = 2\left[(410.715) + 3.75 + 0.0294\right] - 10 = 818.98 \text{ kW}$$

Ans.

$$\text{Again } \dot{m} = \frac{A_1 V_1}{v_1} \Rightarrow A_1 = \frac{\dot{m} v_1}{V_1} = \frac{2 \times 0.1114}{100} = 2.22 \times 10^{-3} \text{ m}^2$$

$$A_2 = \frac{\dot{m} v_2}{V_2} \Rightarrow A_2 = \frac{2 \times 7.26}{50} = 0.2904 \text{ m}^2$$

$$A_1 = \frac{\pi}{4} d_1^2 \rightarrow d_1 = 5.31 \text{cm}$$ Ans.

$$A_2 = \frac{\pi}{4} d_2^2 \rightarrow d_2 = 60 \text{cm}$$ Ans.

Example 4.5 The feed water enters at 45°C into the steam boiler of 20 bar working pressure. Calculate the amount of heat required to produce 1 kg of steam under the following conditions (i) 0.9 dryness, (ii) dry saturated, and (iii) superheated with 30°C of superheat. Take $c_{pw} = 4.187$ and $c_{p,steam} = 2.1$ kJ/kg K.

Solution Temperature of feed water $t = 45°C$.

To calculate the net heat required to produce steam, sensible heat of feed water is to be subtracted from total heat supplied in the boiler.

Sensible heat of feed water, $h_f = 4.187 \times (45 - 0) = 188.415$ kJ/kg

Now from steam tables, at 20 bar, $h_{f1} = 908.79$ kJ/kg

$$h_{fg1} = 1890.7 \text{ kJ/kg}$$

$$h_{g1} = 2799.5 \text{ kJ/kg}$$

i. ∴ Heat required to produce 1 kg of steam with $x_1 = 0.9$

$$q_1 = (h_1 - h_f) = (h_{f1} + x_1 \times h_{fg1}) - h_f$$

$$q_1 = (908.79 + 0.9 \times 1890.7) - 188.415 = 2422.01 \text{ kJ/kg} \quad \text{Ans.}$$

ii. Heat required to produce 1 kg of dry saturated steam, $q_2 = (h_1 - h_f) = h_{g1} - h_f$

$$q_2 = 2799.5 - 188.415 = 2611.08 \text{ kJ/kg} \quad \text{Ans.}$$

iii. Heat required to produce 1 kg of superheated steam, $q_3 = (h_1 - h_f)$

where $h_1 = h_{g1} + c_{p,steam}(t_{sup} - t_s)$

From steam tables, at 20 bar, $t_{sat} = 212.42°C$, $t_{sup} = 212.42 + 30 = 242.42°C$

$$h_1 = 2799.5 + 2.1(242.42 - 212.42) = 2862.5 \text{ kJ/kg}$$

$$q_3 = 2862.5 - 188.415 = 2674.10 \text{ kJ/kg} \quad \text{Ans.}$$

Example 4.6 A throttling calorimeter has steam entering to it at 10 MPa and coming out of it at 0.05 MPa and 100°C. Determine (i) dryness fraction of steam and (ii) the maximum moisture that can be determined with this set up if at least 5°C of superheat is required after throttling.

Solution

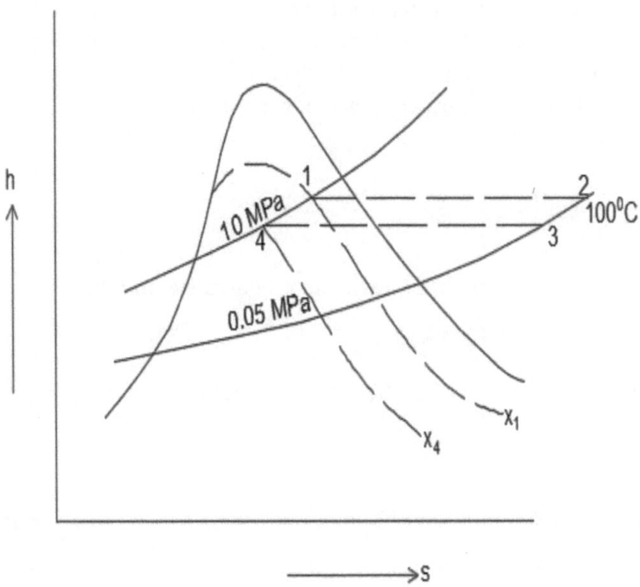

FIGURE EX. 4.6

Figure Ex. 4.6 shows the h-s diagram of the throttling process.

When $p_2 = 0.05$ MPa and $t = 100°C$, from superheated steam tables,

$$h_2 = 2682 \text{ kJ/kg}$$

At p_1(10 MPa), $h_{f1} = 1407$ kJ/kg and $h_{fg1} = 1317$ kJ/kg

For the throttling process, $h_1 = h_2$ (where $h_1 = h_{f1} + x_1 h_{fg1}$)

$h_2 = 2682 = 1407 + x_1 \times 1317$

Then $x_1 = 0.9624$

Quality (dryness fraction) of steam = 96.24% Ans.

5°C superheat after throttling

At p_2@0.05 MPa, $t_{sat} = 81°C$

$t = t_{sat} + 5 = 86°C$

From superheated steam tables,

$h_3 = 2660$ kJ/kg, $(h_3 = h_4)$

$h_4 = 2660 = h_{f4} + x_4 \times h_{fg4}$

$2660 = 1407 + x_4 \times 1317$

$x_4 = 0.9514$

Therefore, maximum moisture allowed is 4.86% Ans.

Example 4.7 Steam flows from a steam main at a pressure of 12 bar into a separating and throttling calorimeter. The condition of steam after throttling is pressure 2 bar and temperature 150°C. The moisture collected in the separator is 0.2 L at temperature 80°C and the mass of steam condensed after throttling is 2 kg. Determine the quality of steam in the main.

Solution

Figure Ex. 4.7

At 2 bar and 150°C ($t_{sat} = 120.23°C$), from superheated steam tables,

$$h_3 = 2768.8 \text{ kJ/kg}$$

During the throttling process, $h_2 = h_3$ (enthalpy before throttling = enthalpy after throttling)

where $h_2 = h_{f2} + x_2 h_{fg2}$

Now from saturated steam tables, at 12 bar, $h_{f2} = 798.65$ kJ/kg and $h_{fg2} = 1986.2$ kJ/kg

$\therefore h_3 = h_2 = 2768.8 = 798.65 + x_2 \times 1986.2 \Rightarrow x_2 = 0.991$

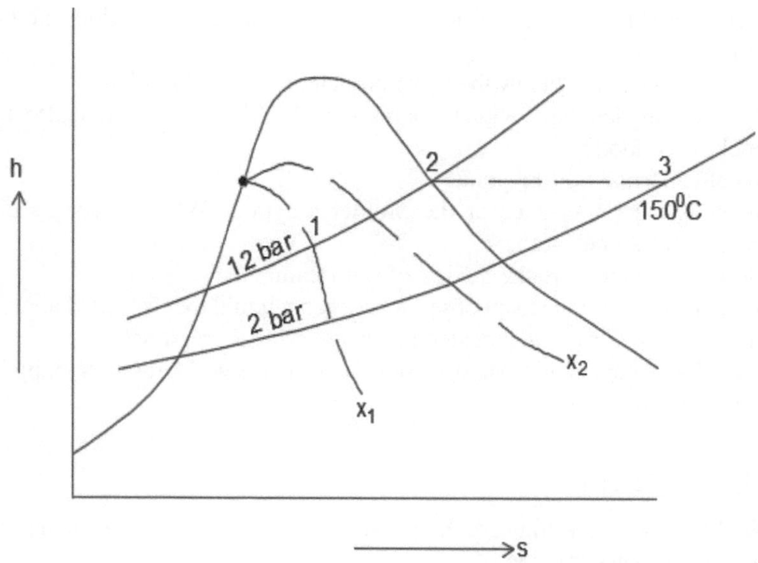

FIGURE EX. 4.7

Let x_1 and x_2 be the mass of moisture collected in the separator and steam condensed after throttling respectively, then

$$x_1 = \frac{x_2\, m_2}{m_1 + m_2} \text{ where } m_1 = \frac{V}{v}$$

$$V = 0.2\ L = 0.2 \times 10^{-3}\ m^3$$

From saturated steam tables at 80°C, $\nu = \nu_f = 0.001029\ m^3/kg$

$$m_1 = \frac{0.2 \times 10^{-3}}{0.001029} = 0.194\,kg, \quad m_2 = 2\,kg$$

$$x_1 = \frac{0.991 \times 2}{0.194 + 2} = 0.903 \qquad \text{Ans.}$$

REVIEW QUESTIONS

4.1 What is a pure substance? Give some examples.
4.2 Distinguish between saturated liquid and compressed liquid.
4.3 What is the difference between saturated vapor and superheated vapor?
4.4 Distinguish between the critical point and the triple point.
4.5 At higher pressures, water boils at higher temperatures. Explain.
4.6 Define saturated pressure and saturated temperature.
4.7 Temperature and pressure are dependent properties in the saturated mixture region. Explain.

4.8 Define quality or dryness fraction of steam. What are the various methods to measure it?
4.9 Does a vapor exist below the triple point temperature? Explain.
4.10 Air pressure decreases with an increase in altitude, how does that affect the cooking of food?
4.11 What is normal boiling point?
4.12 What is the importance of the Mollier diagram? Why do isobars on it diverge from one another?
4.13 How do you measure the quality of wet steam.
4.14 What is the difference you observe between a liquid that is heated above its critical pressure and that heated below the critical pressure?
4.15 How does latent heat of vaporization (h_{fg}) change with pressure? Explain.
4.16 Define vapor pressure.

EXERCISE PROBLEMS

4.1 Find the specific volume, enthalpy, and internal energy of wet steam at 19 bar and dryness fraction 0.9.
4.2 Find the dryness fraction, specific volume, and internal energy of steam at 6 bar and enthalpy 2950 kJ/kg.
4.3 Steam at 120 bar has a specific volume of 0.0198 m^3/kg, find the temperature, enthalpy, and internal energy.
4.4 Steam at 150 bar has an enthalpy of 309 kJ/kg, find the temperature, specific volume, and internal energy.
4.5 Steam at 19 bar is throttled to 1 bar and the temperature after throttling is found to be 175°C. Determine the initial dryness fraction of the steam.
4.6 Find the internal energy of 1 kg of steam at 14 bar under the following conditions: (i) when the steam is 0.85 dry; (ii) when the steam is dry saturated; and (iii) when the temperature of the steam is 285°C. Take c_{ps} = 2.35 kJ/kg K.
4.7 Calculate the internal energy of 0.3 m^3 of steam at 4 bar and 0.95 dryness. If this steam is superheated at a constant pressure through 36°C, determine the heat added and change in internal energy.
4.8 Water is supplied to the boiler at 12 bar and 90°C and steam is generated at the same pressure at 0.85 dryness. Estimate the heat supplied to the steam in passing through the boiler and change in entropy.
4.9 A cylindrical vessel of 7.5 m^3 capacity contains wet steam at 1 bar. The volumes of vapor and liquid in the vessel are 5.06 and 0.07 m^3 respectively. Heat is transferred to the vessel until the vessel is filled with saturated vapor. Estimate the heat transfer during the process.
4.10 A pressure cooker contains 1.5 kg of steam at 5 bar and 0.88 dryness when the gas was switched off. Determine the quantity of heat rejected by the pressure cooker when the pressure in the cooker falls to 1 bar.
4.11 A vessel of spherical shape having a capacity of 0.65 m^3 contains steam at 10 bar and 0.95 dryness. Steam is blown off until the pressure drops to

6 bar. The valve is then closed and the steam is allowed to cool until the pressure falls to 5 bar. Assuming that the enthalpy of steam in the vessel remains constant during blowing off periods, determine (i) the mass of steam blown-off, (ii) the dryness fraction of steam in the vessel after cooling, and (iii) the heat lost by steam per kg during cooling.

4.12 Two boilers one with a superheater and the other without a superheater are delivering equal quantities of steam into a common main. The pressure in the boilers and the main is 14 bar. The temperature of the steam from a boiler with a superheater is 320°C and the temperature of the steam in the main is 260°C. Determine the quality of steam supplied by the other boiler.

4.13 A tank of capacity 0.53 m^3 is connected to a steam pipe through a valve that carries steam at 16 bar and 350°C. The tank initially contains steam at 2.5 bar and at the saturated condition. The valve in the line connecting the tank is opened and the steam is allowed to pass into the tank until the pressure in the tank becomes 16 bar. Find the mass of steam that entered into the tank.

4.14 A 1.8-m^3 rigid tank contains steam at 220°C. One-third of the volume is in the liquid phase and the rest is in the vapor form. Determine (i) the pressure of the steam, (ii) the quality of the saturated mixture, and (iii) the density of the mixture.

4.15 Saturated steam coming off the turbine of a steam power plant at 30°C condenses on the outside of a 3-cm outer-diameter, 35-m-long tube at a rate of 45 kg/h. Determine the rate of heat transfer from the steam to the cooling water flowing through the pipe.

4.16 Water in a 45-mm deep pan is observed to boil at 100°C. At what temperatures will the water in a 400-mm-deep pan boil? Assume both pans are full of water.

4.17 A cooking pan whose inner diameter is 180 mm is filled with water and covered with a 3.8-kg lid. If the local atmospheric pressure is 100 kPa, determine the temperature at which the water starts boiling when it is heated.

4.18 A person cooks a meal in a 30-cm-diameter pot that is covered with a well-fitting lid and lets the food cool to the room temperature of 20°C. The total mass of the food and the pot is 8 kg. Now the person tries to open the pan by lifting the lid up. Assuming no air has leaked into the pan during cooling, determine if the lid will open or the pan will move up together with the lid.

4.19 Water is being heated in a vertical piston-cylinder device. The piston has a mass of 20 kg and a cross-sectional area of 100 cm^2. If the local atmospheric pressure is 100 kPa, determine the temperature at which the water starts boiling.

4.20 A rigid tank with a volume of 2.5 m^3 contains 15 kg of saturated liquid–vapor mixture of water at 75°C. Now the water is slowly heated. Determine the temperature at which the liquid in the tank is completely vaporized. Also, show the process on a *T-v* diagram with respect to saturation lines.

4.21 A 0.5-m^3 vessel contains 8.5 kg of refrigerant-134a at 15°C. Determine (i) the pressure, (ii) the total internal energy, and (iii) the volume occupied by the liquid phase.

4.22 A 1.2-m^3 rigid tank contains a saturated mixture of refrigerant-134a at 300 kPa. If the saturated liquid occupies 20% of the volume, determine the quality and the total mass of the refrigerant in the tank.

4.23 A rigid tank contains water vapor at 200°C and an unknown pressure. When the tank is cooled to 100°C, the vapor starts condensing. Estimate the initial pressure in the tank.

4.24 The air in an automobile tyre with a volume of 0.53 m^3 is at 30°C and 20 psi. Determine the amount of air that must be added to raise the pressure to the recommended value of 30 psi. Assume the atmospheric pressure to be 14.6 psi and the temperature and the volume to remain constant.

DESIGN AND EXPERIMENT PROBLEMS

4.25 A solid normally absorbs heat as it melts, but there is a known exception at temperatures close to absolute zero. Find out which solid it is and give a physical explanation for it.

5 First Law Analysis of Control Volumes

LEARNING OUTCOMES

After learning this chapter, students should be able to

- Formulate the first law of thermodynamics for closed systems and arrange the change in energy in the closed systems via heat and work transfer.
- Apply the first law of thermodynamics for closed systems and construct conservation of mass and energy equations.
- Apply the first law of thermodynamics to the open systems.
- Describe a steady-flow open system.
- Apply the first law of thermodynamics to the nozzles, diffusers, turbines, compressors, throttling valves, mixing chambers, and heat exchangers and construct energy and mass balance for unsteady-flow processes.
- Develop the general energy balance applied to closed systems.
- Solve energy balance problems for closed (fixed mass) systems that involve heat and work interactions for general pure substances, ideal gases, and incompressible substances.

5.1 CONTROL VOLUME

The first law analysis of closed systems or non-flow processes was presented in the previous chapter. In this chapter, the first law application to open systems or control volumes is considered. An open system involves energy as well as mass transfer across its boundary. Most of the engineering devices such as turbines, compressors, and nozzles are open systems. To analyze the open system with the conservation of mass and energy, the control volume technique is considered. This type of analysis can be carried out basically by studying a certain quantity of matter as it passes through the device. A control volume is a certain region within a prescribed boundary on which the attention is focused on the analysis of the system. The surface of a control volume is termed as a control surface that always consists of a closed surface. The size and shape of the control volume are so chosen that they best suit the analysis to be made. The surface may be fixed or moving. Figure 5.1 shows the automobile engine considered as a control volume. The dotted line describes the control volume surrounding the engine. It can be observed that both mass (air and fuel) and energy in the form of exhaust cross the system boundary.

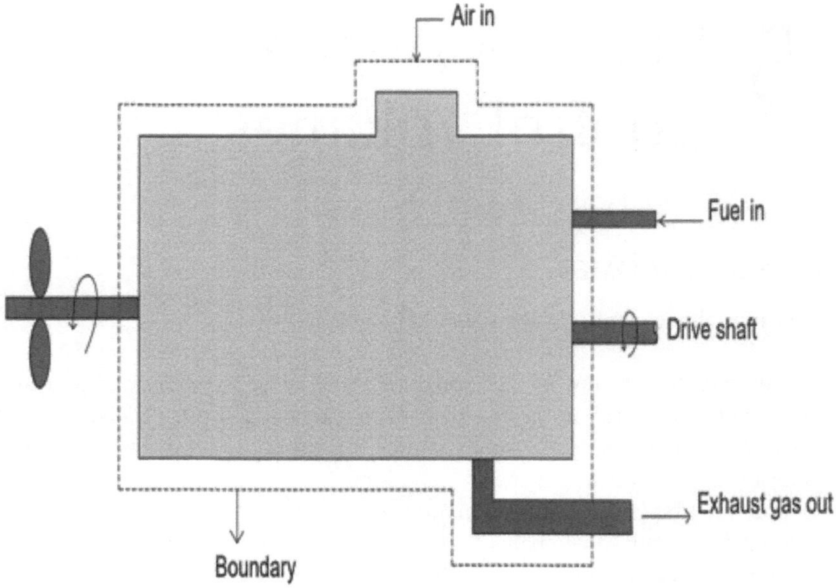

FIGURE 5.1 Control volume.

5.2 MASS BALANCE

If we consider the conservation of mass as it is related to the control volume, *the mass flow rate entering must equal the mass flow rate leaving the control volume, if there is no accumulation of mass within the system.*

$$\dot{m}_1 = \dot{m}_2$$

$$\rho_1 A_1 V_1 = \rho_2 A_2 V_2 \text{ or } \frac{A_1 V_1}{v_1} = \frac{A_2 V_2}{v_2} \tag{5.1}$$

where $\dot{m}_1$ and $\dot{m}_2$ are the mass flow rates entering and leaving the control volume. Equation 5.1 is known as the continuity equation.

When several mass flow quantities are involved, the continuity equation can be written as

$$\frac{dm_{c.v}}{dt} = \sum \dot{m}_i - \dot{m}_e \tag{5.2}$$

Equation 5.2 shows that the mass inside the control volume changes only when some mass is added or removed and it doesn't change by other means.

5.3 FLOW WORK

Closed systems do not involve any mass transfer across their boundaries while control volumes involve mass flow, and some work, known as the flow work, is required to push the mass either into or out of the system and for maintaining a continuous flow through a control volume. To obtain a relation for flow work, consider a fluid element of volume V. The fluid immediately forces upstream its fluid element to enter the control volume; thus, it can be regarded as an imaginary piston. The fluid element can be chosen to be sufficiently small so that it has uniform properties throughout. If the fluid pressure is P and the cross-sectional area of the fluid element is A, then the force applied on the fluid element by the imaginary piston is

$$F = PA \tag{5.3}$$

This force F must act through a distance L to push the fluid element into the control volume. Therefore, the work done in pushing the fluid element across the boundary (i.e., the flow work) is

$$W_{flow} = FL = PAL = PV\,(kJ) \tag{5.4}$$

The flow work per unit mass is obtained by dividing both sides of this equation by the mass of the fluid element:

$$W_{flow} = P\nu\,(kJ/kg) \tag{5.5}$$

The flow work relation remains the same irrespective of the fluid flow either into or out of the control volume. The flow work, unlike other work quantities, is expressed in terms of properties. It is the product of two properties of the fluid. For this reason, it can be viewed as a combination property (like enthalpy) and referred to as flow energy, convected energy, or transport energy instead of flow work. It can also be viewed that the product $P\nu$ represents energy for flowing fluids only and does not represent any form of energy for non-flow (closed) systems. Therefore, it should be treated as work. The flow energy is considered to be part of the energy of a flowing fluid, since this greatly simplifies the energy analysis of control volumes.

5.4 STEADY-FLOW PROCESSES

Usually, when a fluid passes through a control volume, its thermodynamic properties, intensive or extensive properties, vary along the space co-ordinates and with time. If the flow rates of mass and energy change with the time through the control surface, then mass and energy change with the time in the control volume also. Steady flow refers to the condition in which the rates of flow of mass and energy through the control surface are constant, that is, the total mass or energy entering the control volume must equal the total mass or energy leaving. The fluid properties within a control volume may vary along the space co-ordinates but remain unchanged with the time. In most of the engineering devices such as turbines, nozzles, and compressors, the flow occurs constantly under the same operating conditions for long periods of time once the flow is stabilized, i.e., the unsteady state is completed. They are classified as steady-flow devices. The steady-flow process can be suitably applied to such devices to represent the flow process. It is a process in which the fluid properties will not vary with time within the control volume, i.e., they remain constant at the inlet and exit. Moreover, heat and work interactions within the system and its surroundings also remain the same and will not vary with time. The devices such as turbines, compressors, and nozzles operate under steady-flow conditions and they are called steady-flow devices.

5.5 FIRST LAW ANALYSIS OF STEADY-FLOW PROCESSES

This section presents the energy analysis of steady-flow systems considering the mass and energy conservation equations.

Steady-Flow Energy Equation (SFEE)

Figure 5.2 shows a steady-flow device in which a single stream of mass enters and a single stream of mass leaves the control volume.

$\dot{m}_1$, V_1, z_1, h_1 and $\dot{m}_2$, V_2, z_2, h_2 are respectively mass flow rate, velocity, datum head, and enthalpy at inlet and outlet.

The total energy of a simple compressible system is a combination of three parts: internal energy, kinetic energy, and potential energy. That is $E = U + KE + PE$

and on a unit mass basis, $e = u + \frac{V^2}{2} + gz$

where V is the velocity and z is the elevation about a standard reference point. The flowing fluid possesses the flow energy PV, which is the work done in pushing the fluid element either into or out of the system. The force F acting on the fluid element by the piston is $F = PA$,

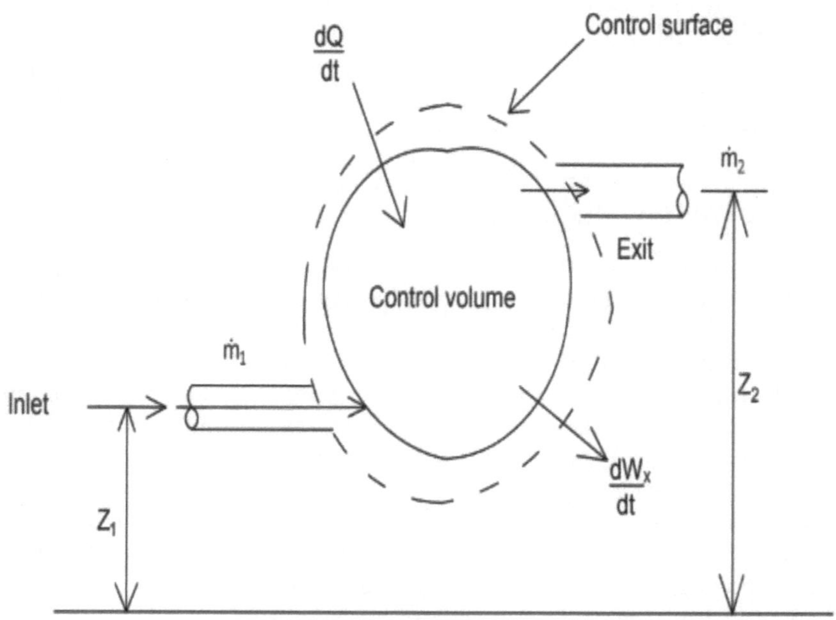

FIGURE 5.2 A steady-flow device.

where P is the fluid pressure and A is the cross-sectional area of the fluid. The work done in pushing the fluid is equal to the force acting through a distance, ds, thus flow work,

$$W_{flow} = F \cdot ds = PAds = pV(kJ)\ (V = A \cdot ds)$$

Flow work per unit mass $W_{flow} = p\nu$ kJ/kg

Energy Transfer by Mass

Therefore, the total energy of a flowing fluid per unit mass is

$$e + pv = u + ke + pe + pv = h + ke + pe(\text{since}\, h = u + pv) \tag{5.6}$$

To derive the SFEE for a control volume, the mass balance and energy balance are taken into consideration.

The mass balance equation for a general steady-flow system is given by

$$\Sigma_{in}\, \dot{m} = \Sigma_{out}\, \dot{m} \tag{5.7}$$

The mass balance equation for a steady-flow system in which a single stream is entering and leaving the control volume is given by

$$\dot{m}_1 = \dot{m}_2 = \rho_1 A_1 V_1 = \rho_2 A_2 V_2 \tag{5.8}$$

where the subscripts 1 and 2 represent the inlet and exit states of the control volume considered,

ρ is the density of the fluid, V is the velocity of flow, and A is the area normal to the direction of flow.

The general energy balance equation, considering the total energy transfer caused by heat, work, and mass, which causes the change of total energy of a system,

$$\underbrace{\dot{E}_{in} - \dot{E}_{out}}_{\text{Rate of net energy transfer}} = \underbrace{dE_{system}/dt}_{\text{Rate of change in total energy}} \tag{5.9}$$

During a steady-flow process, the rate of total energy transfer, caused by heat, work, and mass, entering into a control volume must equal the total energy leaving, therefore the change in total energy content $\Delta E = 0$. The energy balance equation in the rate form for a steady flow is

$$\underbrace{\dot{E}_{in}}_{\substack{\text{Rate of net}\\ \text{energy transfer in}}} = \underbrace{\dot{E}_{out}}_{\substack{\text{Rate of net}\\ \text{energy transfer out}}} \tag{5.10}$$

Eq. 5.10 can be expanded to include mass, heat, and work interactions at inlet and exit, and it is given by

$$\dot{Q}_{in} + \dot{W}_{in} + \Sigma_{in}\, \dot{m}e = \dot{Q}_{out} + \dot{W}_{out} + \Sigma_{out}\, \dot{m}e \tag{5.11}$$

$$\text{Where} \quad e = h + \frac{V^2}{2} + zg$$

$$\frac{\delta Q}{dt} - \frac{\delta W}{dt} = \dot{m}_2\left[h_2 + \frac{V_2^2}{2} + z_2 g\right] - \dot{m}_1\left[h_1 + \frac{V_1^2}{2} + z_1 g\right]$$

For a steady flow, the mass flow rates are constant; therefore,

$$\dot{m}_1 = \dot{m}_2 = \dot{m}$$

$$\dot{m}\left[(h_1 - h_2) + \left(\frac{V_1^2 - V_2^2}{2}\right) + (z_1 - z_2)g\right] + \frac{\delta Q}{dt} = \frac{\delta W}{dt}$$

where the subscripts 1 and 2 represent the inlet and exit states of the control volume and $\frac{\delta Q}{dt}$ and $\frac{\delta W}{dt}$ are the rates of heat input and work output respectively. To apply the SFEE to a practical problem, the magnitudes and directions of both heat and work transfers are to be known. Generally, it is assumed that the heat transferred into the system (heat input) and work done by a system (work output) are taken to be positive. It is to be noted here that if a negative quantity is obtained for the rate of heat input or work output, it means that the assumed direction is wrong and it is to be reversed.

$$\dot{m}\left[h_1 + \frac{V_1^2}{2} + z_1 g\right] + \frac{\delta Q}{dt} = \dot{m}\left[h_2 + \frac{V_2^2}{2} + z_2 g\right] + \frac{\delta W}{dt} (\text{J/s}) \tag{5.12}$$

Dividing the above equation by $\dot{m}$ or $\frac{\delta m}{dt}$

$$h_1 + \frac{V_1^2}{2} + z_1 g + \frac{\delta Q}{dm} = h_2 + \frac{V_2^2}{2} + z_2 g + \frac{\delta W}{dm} (\text{J/kg}) \tag{5.13}$$

where $\delta Q/dt$ and $\delta Q/dm$ are heat transfer rate and heat transfer per unit mass respectively and $\delta W/dt$ and $\delta W/dm$ are work transfer rate and work transfer per unit mass respectively. Equations 5.12 and 5.13 are SFEEs for a single stream of fluid flowing through the control volume, whereas Eq. 5.13 is the rate form of energy flow, i.e., energy flow per unit time, and Eq. 5.12 is on the unit mass of fluid.

5.6 STEADY-FLOW ENERGY EQUATION NEEDS

Most of the engineering devices and technical applications of energy conversion and energy transfer involve the flow of a fluid. Active devices such as turbines and pumps involve work, while heat exchangers involve the flow of heat either into or out of the flowing fluid. These devices are considered steady-flow devices as they operate under the same operating conditions for quite a long period of time. The components of some industrial plants such as turbines, compressors, and pumps operate continuously for some months; therefore, these devices can be conveniently considered as steady-flow devices and analyzed by SFEE. In this section, some common steady-flow devices are described including their purpose and the thermodynamic aspects of the flow through them are analyzed. The SFEE, developed from the conservation of mass and the conservation of energy principles, is applied to these devices.

5.7 STEADY-FLOW DEVICES

The various steady-flow devices are presented in the following sections. SFEE is derived for each of the devices.

5.7.1 Turbines and Compressors

Turbines are most widely used as prime movers in power generation and propulsion of aircraft. Certain mass of fluid strikes the buckets of the turbine and work is done on them to rotate the shaft at high speeds. As a result, mechanical work is produced, which drives the generator or so. Turbines and engines are termed as work-producing devices. Compressors and pumps, on the other hand, are used to raise the pressure of the fluid at the cost of work input given to them and they are termed work-consuming devices.

Though, the fluid velocities in most of the turbines are high enough to cause considerable changes in kinetic energies, when compared to the changes in enthalpy, it is very small and can be neglected. The changes in potential energies are neglected for all these devices. The general SFEE is given as

$$h_1 + \frac{V_1^2}{2} + z_1 g + \frac{\delta Q}{dm} = h_2 + \frac{V_2^2}{2} + z_2 g + \frac{\delta W}{dm}$$

For a well-insulated turbine as shown in Figure 5.3, with the changes in kinetic and potential energies being negligible, SFEE is reduced to

$$h_1 = h_2 + \frac{\delta W}{dt}$$

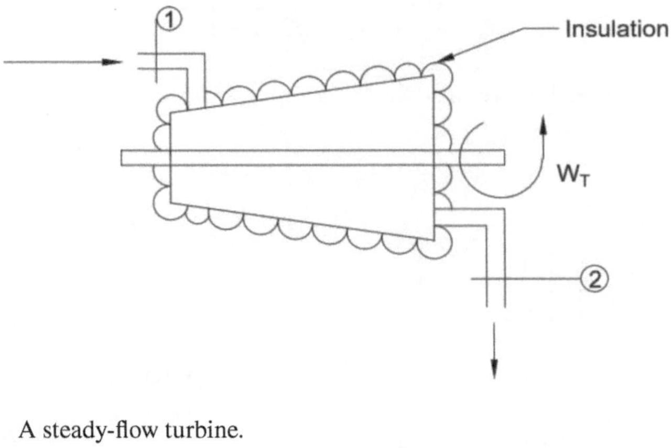

FIGURE 5.3 A steady-flow turbine.

Then the rate of work done or the power of the turbine,

$$\frac{\delta W}{dt} = h_1 - h_2 \tag{5.14}$$

From Eq. 5.14, it can be seen that a turbine produces net work at the expense of enthalpy drop of the steam that expands.

Similarly, for compressors and pumps, SFEE can be derived

$$h_1 = h_2 - \frac{\delta W}{dt}$$

Then the power required for the compressor is

$$\frac{\delta W}{dt} = h_2 - h_1 \tag{5.15}$$

5.7.2 Nozzles and Diffusers

A nozzle is a device used for accelerating the fluid at the expense of its pressure, i.e., the pressure energy is converted to kinetic energy, while the diffuser converts the kinetic energy into pressure energy. Both the nozzles and diffusers are used in jet engines and rocket engines and they perform opposite tasks. Figure 5.4 shows a nozzle and a diffuser. An insulated nozzle is assumed as an adiabatic nozzle for which $\frac{\delta Q}{dm} = 0$, since it doesn't involve any work transfer, $\frac{\delta W}{dm} = 0$, and change in potential energy is also zero. Then SFEE is reduced to

$$h_1 + \frac{V_1^2}{2} = h_2 + \frac{V_2^2}{2} \tag{5.16}$$

Equation 5.16 is also applicable for a diffuser.

If the velocity of approach (inlet velocity) is small compared to the velocity at the exit, then Eq. 5.16 becomes

$$h_1 = h_2 + \frac{V_2^2}{2} \text{ or } V_2 = \sqrt{2(h_1 - h_2)}\,\text{m/s}$$

where enthalpies h_1 and h_2 are in J/kg.

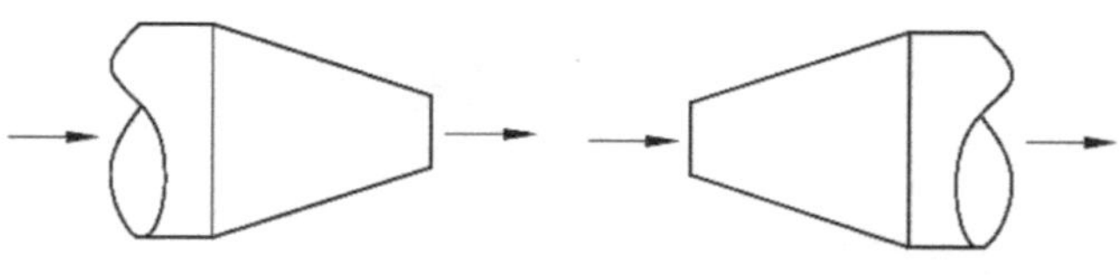

FIGURE 5.4 (a) Nozzle and (b) diffuser.

5.7.3 Throttling

If a fluid passes through a restricted passage or a narrow cross-section, it experiences a pressure drop. Throttling valves, capillary tubes, and porous plugs function based on this principle to produce a pressure drop accompanied by a large drop in temperature without involving any work. Figure 5.5 shows the throttling process in which the fluid passes through a partially opened valve. These devices are used in refrigeration and air-conditioning applications. Throttling devices are so small that there is neither enough time nor sufficient area for heat transfer to take place, and therefore, heat transfer is negligible. They also don't involve work and the changes in kinetic and potential energies are negligible.

$$\frac{\delta Q}{dm} = 0 \text{ and } \frac{\delta W}{dm} = 0$$

$$h_1 = h_2 \tag{5.17}$$

Therefore in a throttling process, the enthalpy of fluid before throttling is equal to the enthalpy of fluid after throttling; thus, throttling valves are also known as isenthalpic valves.

5.7.4 Heat Transfer

A heat exchanger is a device in which two moving fluids (hot and cold) exchange heat without mixing as shown in Figure 5.6. There are different designs of heat exchangers such as cross-flow, counter-flow, and shell-and-tube heat exchangers. Steam condensers and evaporators are typical heat exchangers used in industries for transferring heat from one fluid to the other. The SFEE for a heat exchanger is

$$\dot{m}_h h_1 + \dot{m}_c h_2 = \dot{m}_h h_3 + \dot{m}_c h_4 \tag{5.18}$$

where $\dot{m}_h$ and $\dot{m}_c$ are the mass flow rates of hot and cold fluids respectively.

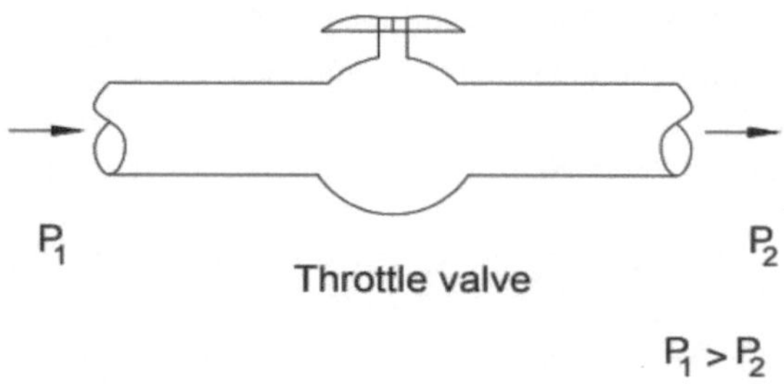

FIGURE 5.5 Fluid flow through a valve (a throttling device).

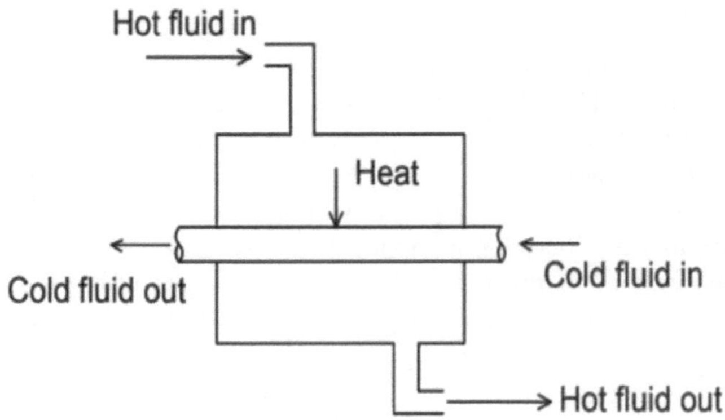

FIGURE 5.6 Heat exchanger.

5.8 FIRST LAW ANALYSIS OF UNSTEADY-FLOW PROCESSES

During a steady-flow process, flow in a control volume will not vary with time. However, in many processes, flow within a control volume varies with time. This type of process is called *unsteady-flow or transient flow process.* Filling and evacuating gas cylinders and charging of rigid vessels from supply lines are some of the notable cases of unsteady-flow processes. Figure 5.7 shows the charging and discharging of a rigid vessel. Unsteady-flow processes are analyzed by considering the mass and energy contents of the control volume and energy interactions across its boundary. The unsteady-flow process will continue for a definite time period as opposed to that of a steady flow, which continues indefinitely. Thus for analyzing an unsteady flow, the time interval 'Δt' is so chosen that the changes that occur over this time interval are considered.

The rate of a mass of fluid accumulated within the control volume is equal to the net rate of flow of mass within the control surface, that is

$$\dot{m} = \dot{m}_1 - \dot{m}_2 \tag{5.19}$$

where $\dot{m}$ is the mass of fluid within the control volume at any instant.

Over any finite period of time, $\Delta\dot{m} = \Delta\dot{m}_1 - \Delta\dot{m}_2$

The rate of energy of fluid accumulated within the control volume is equal to the net rate of flow of energy within the control surface; if $\dot{E}$ or $\frac{dE}{dt}$ is the rate of energy of fluid within the control volume at any instant, then

$$\text{Rate of energy increase} = \text{Rate of energy entering} - \text{Rate of energy leaving}$$

$$\left(\dot{Q}_{in} + \dot{W}_{in} + \Sigma_{in}\,\dot{m}e\right) - \left(\dot{Q}_{out} + \dot{W}_{out} + \Sigma_{out}\,\dot{m}e\right) = \left(\dot{m}_2\theta_2 - \dot{m}_1\theta_1\right)_{system} \tag{5.20}$$

where θ is the energy of the non-flowing fluid within the control volume per unit mass,

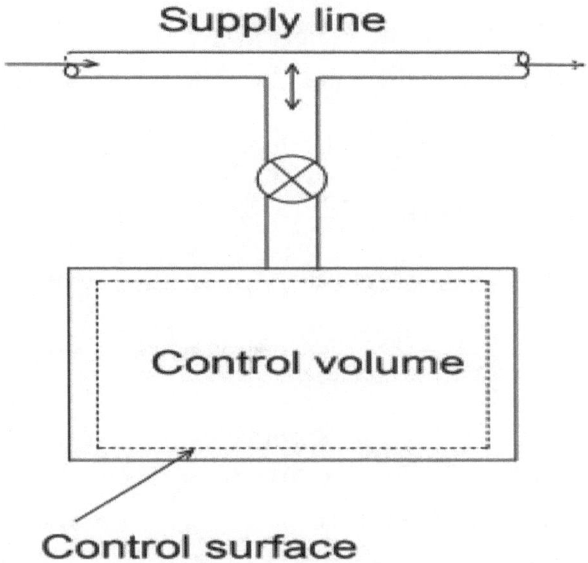

FIGURE 5.7 Charging and discharging a rigid vessel.

$$\theta = u + ke + pe$$

$$\dot{m}\left(u + \frac{V^2}{2} + zg\right)_{system} = \frac{\delta Q}{dt} + \dot{m}_1\left[h_1 + \frac{V_1^2}{2} + z_1 g\right] - \dot{m}_2\left[h_2 + \frac{V_2^2}{2} + z_2 g\right] - \frac{\delta W}{dt}$$

When kinetic and potential energies are neglected, the above equation is reduced to

$$\frac{\delta Q}{dt} - \frac{\delta W}{dt} = \dot{m}_2[h_2] - \dot{m}_1[h_1] + (\dot{m}_2 u_2 - \dot{m}_1 u_1)_{system} \tag{5.21}$$

EXAMPLE PROBLEMS

Example 5.1 Gas enters a gas turbine nozzle with an enthalpy of 2670 kJ/kg and a velocity of 72 m/s. At the exit of the nozzle, the enthalpy is 2580 kJ/kg. For a horizontal nozzle with negligible heat loss, find (i) the velocity at the exit of the nozzle, (ii) the mass flow rate of gas, if area $A_1 = 0.098\ m^2$ and $v_1 = 0.179\ m^3/kg$, and (iii) if $v_2 = 0.503\ m^3/kg$, find the area at the exit of the nozzle.

Solution At the inlet to the gas turbine nozzle, $h_1 = 2670$ kJ/kg, $V_1 = 72$ m/s.

$A_1 = 0.098\ m^2 \quad \nu_1 = 0.179\ m^3/kg$

At the exit of the gas turbine nozzle, $h_2 = 2580$ kJ/kg, $\nu_2 = 0.503\ m^3/kg$

$$\frac{\delta Q}{dt} = 0 \left(\text{Since heat loss is negligible}\right)$$

$$\frac{\delta W}{dt} = 0 \left(\text{work done is zero for nozzle}\right)$$

SFEE for nozzle, when there is no heat loss reduces to

$$h_1 + \frac{V_1^2}{2} = h_2 + \frac{V_2^2}{2} \Rightarrow h_1 - h_2 = \frac{V_2^2 - V_1^2}{2}$$

$$= 2670 - 2580 = \frac{V_2^2 - 5184}{2 \times 1000}$$

$\Rightarrow V_2 = 430.33$ m/s (velocity at exit) Ans.

Mass flow rate $\dot{m} = \frac{A_1 V_1}{v_1} = \frac{0.098 \times 72}{0.179} = 39.42\,\text{kg/s}$ Ans.

And also $\dot{m} = \frac{A_2 V_2}{v_2} \Rightarrow 39.42 = \frac{A_2 \times 430.33}{0.503}$

Area at the exit of nozzle, $A_2 = 0.046\ m^2$ Ans.

Example 5.2 A turbine is supplied with the following conditions, p = 110 bar, t = 200°C, h = 2900 kJ/kg, V = 36 m/s, and elevation z = 3.3 m. The steam leaves the turbine at 30 kPa, h = 2490 kJ/kg, V = 110 m/s, and elevation z = 0. The heat loss to the surroundings is at a rate of 0.32 kJ/s and the steam flow rate is 0.393 kg/s. Estimate the power output for steady-flow conditions.

Solution At turbine inlet, $p_1 = 110$ bar, $h_1 = 2900$ $V_1 = 36$ m/s

At turbine outlet, $p_2 = 30$ kpa $h_2 = 2490$ $V_2 = 110$ m/s

$z_1 = 3.3$ m, $z_2 = 0$

The rate of heat loss, $\frac{\delta Q}{dt} = -0.32$ kJ/s

Steam flow rate, $\dot{m} = 0.393\,\text{kg/s}$

$$\text{SFEE for turbine} = \dot{m}\left[h_1 + \frac{V_1^2}{2} + z_1 g\right] + \frac{\delta Q}{dt} = \dot{m}\left[h_2 + \frac{V_2^2}{2} + z_2 g\right] + \frac{\delta W}{dt}$$

$$\text{After rearranging} = \dot{m}\left[(h_1 - h_2) + \left(\frac{V_1^2 - V_2^2}{2}\right) + (z_1 - z_2)g\right] + \frac{\delta Q}{dt} = \frac{\delta W}{dt}$$

$$= 0.393\left[(2900 - 2490) + \left(\frac{36^2 - 110^2}{2 \times 1000}\right) + \frac{(3.3 - 0) \times 9.81}{1000}\right] - 0.32$$

Power output of turbine, $\frac{\delta W}{dt} = 158.7\,\text{kW}$ Ans.

Example 5.3 Air flows steadily through an air compressor at a rate of 30 kg/min, entering at 8 m/s with a pressure of 1.5 bar and a specific volume of 0.9 m³/kg and leaving at 3.5 m/s with a pressure of 8 bar and a specific volume of 0.145 m³/kg. The internal energy of air leaving is 100 kJ/kg greater than that of air entering. Cooling water circulated absorbs heat from air at a rate of 55 W. Calculate (i) the power required to drive the compressor and (ii) inlet and outlet cross-sectional areas.

Solution $p_1 = 1.5$ bar, $\nu_1 = 0.9\ m^3/kg$ $V_1 = 8$ m/s

$p_2 = 8$ bar $\nu_2 = 0.145\ m^3/kg$ $V_2 = 3.5$ m/s

$$z_1 - z_2 = 0 \qquad \frac{\delta Q}{dt} = -55\,kW$$

$u_1 - u_2 = 100$ kJ/kg

$$\dot{m}\left[h_1 + \frac{V_1^2}{2} + z_1 g\right] + \frac{\delta Q}{dt} = \dot{m}\left[h_2 + \frac{V_2^2}{2} + z_2 g\right] + \frac{\delta W}{dt}$$

$$\dot{m}\left[(h_1 - h_2) + \left(\frac{V_1^2 - V_2^2}{2}\right) + (z_1 - z_2)g\right] + \frac{\delta Q}{dt} = \frac{\delta W}{dt}$$

$$\text{Since}\, h = u + pv \Rightarrow h_1 - h_2 = (u_1 - u_2) + (p_1 v_1 - p_2 v_2)$$

Substituting the above in SFEE,

$$\dot{m}\left[(u_1 - u_2) + (p_1 v_1 - p_2 v_2) + \left(\frac{V_1^2 - V_2^2}{2}\right) + (z_1 - z_2)g\right] + \frac{\delta Q}{dt} = \frac{\delta W}{dt}$$

$$\frac{\delta W}{dt} = 0.5\left[-(100) + \frac{(1.5 \times 0.9 - 8 \times 0.145)10^5}{1000} + \left(\frac{8^2 - 3.5^2}{2 \times 1000}\right)\right] + (-55)$$

$$\frac{\delta W}{dt} = -95.48\,kW$$

Power required to drive the compressor is 95.48 kW Ans.

$$\dot{m} = \frac{A_1 V_1}{v_1} \Rightarrow A_1 = \frac{\dot{m} v_1}{V_1}$$

$$A_1 = \frac{0.5 \times 0.9}{8} = 0.056\,m^2$$ Ans.

$$\dot{m} = \frac{A_2 V_2}{v_2} \Rightarrow A_2 = \frac{\dot{m} v_2}{V_2}$$

$$\frac{0.5 \times 0.145}{3.5} = 0.0206\,m^2$$ Ans.

Example 5.4 Air enters a reciprocating air compressor at 1 bar, 25°C and is compressed to 2 bar, 120°C; it then enters the aftercooler where it is cooled to 27°C at a constant pressure. The work input required for the compressor is 5 kW. Estimate, for an airflow rate of 2.5 m³/min, the heat transfer rate in (i) the compressor and (ii) the cooler.

Solution $p_1 = 1$ bar, $t_1 = 25°C$

Volume flow rate of air, $\dot{V}_1 = 2.5\,m^3/min$

$p_2 = 2$ bar $t_2 = 120°C$

$$\frac{\delta W}{dt} = -5\,kW \quad t_3 = 27°C$$

Mass flow rate $\dot{m} = \dfrac{\dot{V}_1}{v_1}$

$$v_1 = \frac{RT_1}{P_1} = \frac{0.287 \times 298}{100} = 0.855\,m^3/kg$$

$$\dot{m} = \frac{2.5}{0.855} = 2.923\,kg/min = 0.0487\,kg/s$$

$$\dot{m}\left[h_1 + \frac{V_1^2}{2} + z_1 g\right] + \frac{\delta Q}{dt} = \dot{m}\left[h_2 + \frac{V_2^2}{2} + z_2 g\right] + \frac{\delta W}{dt}$$

For the compressor, after omitting the terms which are not necessary, SFEE becomes

$$\frac{\delta W}{dt} = \dot{m}\left[(h_1 - h_2)\right] + \frac{\delta Q}{dt} \text{ where } h_1 - h_2 = c_p(T_1 - T_2)$$

$$0.0487[1.005(298 - 393)] - \frac{\delta Q}{dt} = -5$$

$$4.65 - \frac{\delta Q}{dt} = -5 \Rightarrow \frac{\delta Q}{dt} = -0.35\,kJ/s \qquad \text{Ans.}$$

For the heat exchanger (cooler), SFEE becomes

$$\frac{\delta W}{dt} = \dot{m}\left[(h_2 - h_3)\right] + \frac{\delta Q}{dt} \quad \left(\frac{\delta W}{dt} = 0, \text{since no work is involved in cooler}\right)$$

$$0 = 0.0487[1.005(120 - 27)] + \frac{dQ}{dt}$$

$$0 = 4.55 + \frac{\delta Q}{dt} \Rightarrow \frac{\delta Q}{dt} = -4.55\,kJ/s \qquad \text{Ans.}$$

Example 5.5 Two streams of air, one at 1.2 bar, 30°C, and a velocity of 25 m/s and the other at 6 bar, 230°C, and a velocity of 45 m/s velocity, mix in equal proportions in a mixing chamber. Heat is removed from the chamber at a rate of 80 W. The mixture is then allowed to flow through an insulated nozzle. Estimate the velocity of air stream leaving the nozzle at a temperature of 30°C. Take c_p of air as 1.005 kJ/kg K.

Solution $p_1 = 1.2$ bar, $t_1 = 30°C$ $V_1 = 25$ m/s

$p_2 = 6$ bar $t_2 = 230°C$ $V_2 = 45$ m/s

$$\frac{\delta Q}{dt} = -80\,kJ/kg \quad t_4 = 30°C$$

The final temperature after mixing, $t_3 = \frac{t_1 + t_2}{2} = \frac{30 + 230}{2} = 130°C$

For mixing chamber, SFEE becomes

$$\dot{m}_1\left[h_1 + \frac{V_1^2}{2}\right] + \dot{m}_2\left[h_2 + \frac{V_2^2}{2}\right] = \dot{m}_3\left[h_3 + \frac{V_3^2}{2}\right] + \frac{\delta Q}{dt}$$

It is assumed that air streams mix in equal proportions on a mass basis, that is $\dot{m}_1 = \dot{m}_2$.

From mass balance, $\dot{m}_1 + \dot{m}_2 = \dot{m}_3$

From tables of ideal gas properties of air, $h_{1@303K} = 303$ kJ/kg, $h_{1@503K} = 503$ kJ/kg, and $h_{1@403K} = 403$ kJ/kg

$$= \dot{m}\left[303 + \frac{25^2}{2 \times 1000} + 503 + \frac{45^2}{2 \times 1000}\right] = 2\dot{m}\left[403 + \frac{V_3^2}{2}\right] - 0.8$$

$$= [303 + 0.3125 + 503 + 1.0125] + 0.8 = 2\left[403 + \frac{V_3^2}{2 \times 1000}\right]$$

Solving $\Rightarrow V_3 = 41.62$ m/s

For nozzle, SFEE becomes $\dot{m}_3\left[h_3 + \frac{V_3^2}{2}\right] = \dot{m}_3\left[h_4 + \frac{V_4^2}{2}\right]$

(mass flow rate remains constant)

$$\left[403 + \frac{41.62^2}{2 \times 1000}\right] = \left[303 + \frac{V_4^2}{2 \times 1000}\right]$$

Solving $\Rightarrow V_4 = 449.146$ m/s Ans.

Example 5.6 Air enters the diffuser of a turbojet engine steadily at 15°C and 90 kPa with a velocity of 215 m/s. The inlet area of the diffuser is 0.38 m². The velocity of air leaving the diffuser is very small compared with that at the inlet. Determine (i) the mass flow rate of the air and (ii) the temperature of the air at the exit of the diffuser.

Solution $T_1 = 273 + 15 = 298$ K $p_1 = 90$ kPa

$V_1 = 215$ m/s

$$\dot{m} = \frac{A_1 V_1}{v_1}$$

$$v_1 = \frac{RT_1}{p_1} = \frac{0.287 \times 288}{90} = 0.9184\,\text{m}^3/\text{kg}$$

$$\dot{m} = \frac{0.38 \times 215}{0.9184} = 88.96\,\text{kg/s}$$

For diffuser, SFEE is $h_1 + \frac{V_1^2}{2} = h_2 + \frac{V_2^2}{2}$ (mass flow rate remains constant in steady flow)

From tables of ideal gas properties of air, $h_{1@288K} = 288$ kJ/kg

$$288 + \frac{215^2}{2 \times 1000} = h_2 + 0 \Rightarrow h_2 = 312.55\,\text{kJ/kg}$$

The temperature corresponding to the enthalpy value of 312.55 kJ/kg from tables is

312 K = 39°C **Ans.**

Example 5.7 Hot air is cooled by the circulation of cooling water in a heat exchanger. Air enters the heat exchanger steadily at 100°C and leaves at 40°C, while water enters at 30°C and leaves at 40°C. Determine the mass flow rate of the cooling water required for cooling each kg/s of air.

Solution For a heat exchanger, SFEE is $\dot{m}_1 h_1 + \dot{m}_2 h_2 = \dot{m}_3 h_3 + \dot{m}_4 h_4$ (for a steady flow, mass flow rates are equal).

$$\dot{m}_{air}(h_1 - h_3) = \dot{m}_{water}(h_4 - h_2)$$

From tables of ideal gas properties of air, $h_{1@373K} = 373$ kJ/kg $h_{3@313K} = 313$ kJ/kg

From saturated water temperature tables, $h_{2@30°C} = h_f = 125.79$ kJ/kg

and $h_{4@40°C} = h_f = 167.53$ kJ/kg

Substituting the values $= 1(373 - 313) = \dot{m}_{water}(167.53 - 125.79)$

$\dot{m}_{water} = 1.44\,\text{kg/s}$ **Ans.**

REVIEW QUESTIONS

5.1 What are the different mechanisms for transferring energy to or from a control volume?
5.2 What is flow energy? Do fluids at rest possess any flow energy?
5.3 How do you compare the energies of a flowing fluid and a fluid at rest?
5.4 What is the difference between internal energy and enthalpy?
5.5 Define a steady-flow process.
5.6 What is conservation of mass?
5.7 How do you relate Bernoulli's equation with SFEE?
5.8 What are the various steady-flow devices?
5.9 What is a throttling process?
5.10 How does the pressure of a fluid vary during a throttling process?
5.11 How does the temperature of a fluid vary during a throttling process?
5.12 Write the SFEE for an adiabatic nozzle and explain each term.
5.13 Write the SFEE for a compressor and explain each term.
5.14 Explain how a nozzle increases the velocity of a fluid.
5.15 What is the function of a diffuser?
5.16 Why does the temperature of air increase as it is compressed in an air compressor?
5.17 What are the applications of the SFEE?
5.18 What is an unsteady-flow or a transient flow process?

EXERCISE PROBLEMS

5.1 A steam turbine has an inlet of 4.2 kg/s water at 1.3 bar and 375°C with a velocity of 20 m/s. The exit is at 1.03 bar, 170°C, and a very low velocity. Estimate the specific work and the power produced.

5.2 A steady-flow turbine receives 90 kg/s of steam. The steam enters the turbine at a velocity of 388 m/s, an elevation of 4.6 m, and a specific enthalpy of 2685 kJ/kg. It leaves the turbine at a velocity of 120 m/s, an elevation of 1.2 m, and a specific enthalpy of 2190 kJ/kg. Heat losses from the turbine to the surroundings amount to 4.8 kJ/s. Determine the power output of the turbine.

5.3 A stream of gases at 8 bar, 750°C, and 165 m/s is passed through a turbine of an aircraft. The stream comes out of the turbine at 2.3 bar, 589°C, and 325 m/s. The gas enthalpies at the entry and exit of the turbine are 1000 and 750 kJ/kg respectively. Determine the capacity of the turbine if the gas flow rate is 5.2 kg/s assuming the process is adiabatic.

5.4 At the inlet to a certain nozzle, the enthalpy of the fluid is 2895 kJ/kg and the velocity is 68 m/s. At the exit from the nozzle, the enthalpy is 2720 kJ/kg. The nozzle is horizontal and there is negligible heat loss from it. Find (i) the velocity at the nozzle exit, (ii) if the inlet area is 0.13 m^2 and the specific volume at the inlet is 0.186 m^3/kg, find the rate of flow of fluid, and (iii) if the specific volume at the nozzle exit is 0.48 m^3/kg, find the exit area of the nozzle.

5.5 Air enters a reciprocating air compressor with a flow rate of 2.5 m^3/s at 1 bar and 25°C and is compressed to 2 bar and 120°C and delivered to an aftercooler where it is cooled to its initial temperature at a constant pressure. The compressor takes 5 kW of power. Estimate the heat transfer in the (i) compressor and (ii) cooler.

5.6 Air enters a compressor at a flow rate of 3 kg/min, a velocity of 300 m/min, a pressure of 1 bar, and a specific volume of 0.785 m^3/kg. Air leaves at a pressure of 8 bar and a specific volume of 0.178 m^3/kg. The internal energy of air leaving is 80 kJ/kg greater than that of air entering. Cooling water circulated absorbs 70 W of heat from the air. Estimate (i) the power required to drive the compressor and (ii) the inlet and outlet cross-sectional areas.

5.7 A centrifugal air compressor used in gas turbine receives air at 110 kPa and 310 K and it discharges air at 380 kPa and 520 K. The velocity of air leaving the compressor is 125 m/s. Neglecting the velocity at the entry of the compressor, determine the power required to drive the compressor if the mass flow rate is 20 kg/s. Take c_p (air) = 1.03 kJ/kg K and assume that there is no heat transfer from the compressor to the surroundings.

5.8 The airspeed of a turbojet engine in flight is 300 m/s. The ambient air temperature is −20°C. The gas temperature outlet of the nozzle is 580°C. Corresponding enthalpy values for air and gas are respectively 275 and 930 kJ/kg. The fuel–air ratio is 0.0210. The chemical energy of the fuel is 43,500 kJ/kg. Owing to incomplete combustion, 5% of the chemical energy is not released in the reaction. Heat loss from the engine is 18 kJ/kg of air. Calculate the velocity of the exhaust jet.

5.9 The compressor of a large gas turbine receives air from the surroundings at 100 kPa, 25°C with a low velocity. At the exit of the compressor, air leaves at 1.45 MPa, 400°C with a velocity of 85 m/s. The power input required for the compressor is 4.5 MW. Estimate the mass flow rate of air through the unit.

5.10 An adiabatic diffuser receives 0.09 kg/s of steam at 0.45 MPa, 325°C. The steam exits are at 1 MPa, 400°C with negligible kinetic energy. Find the diffuser inlet velocity and the inlet area.

5.11 Compressed air enters at 5 bar, 55°C into a small, high-speed turbine producing a power output of 105 W. The exit state is 1.7 bar and –20°C. Estimate the mass flow rate of air required through the turbine, assuming the velocities to be low and the process to be adiabatic.

5.12 A compressor in an industrial air-conditioning application compresses saturated ammonia vapor from a state of 1.5 bar to a pressure of 12.5 bar. At the compressor exit, the temperature is found to be 95°C and the mass flow rate is 0.45 kg/s. Calculate the motor size required in kW for this compressor.

5.13 Steam enters a long horizontal pipe of an inlet diameter of 150 mm at 15 bar and 320°C with a velocity of 3.5 m/s. The conditions farther downstream are 10 bar and 270°C, and the diameter is 110 mm. Evaluate (i) the mass flow rate of the steam and (ii) the rate of heat transfer.

5.14 A shell-and-tube heat exchanger is used to heat the water in the tubes from 25°C to 90°C at a rate of 5 kg/s. Heat is supplied by hot oil that enters the shell side at 180°C at a rate of 12 kg/s. Heat loss from the heat exchanger is estimated to be at a rate of 10 kW. Determine the rate of heat transfer in the heat exchanger and the exit temperature of the oil. Take c_p of water and oil as 4.18 and 2.25 kJ/kg·°C respectively.

5.15 A double-pipe counter flow heat exchanger is used to cool the oil from 170°C to 30°C at a rate of 2.5 kg/s by water entering at 25°C at a rate of 2 kg/s. Determine the rate of heat transfer in the heat exchanger and the exit temperature of water. Take c_p of water and oil as 4.18 and 2.30 kJ/kg·°C respectively.

5.16 A water pump receives liquid water at 30°C, 1 bar and delivers it to a same-diameter short pipe having a nozzle with an exit diameter of 25 mm to the atmosphere at 1 bar. Determine the exit velocity and the mass flow rate if the pump draws 1.2 kW of power, neglecting the kinetic energy in the pipes and assuming constant u for the water.

5.17 Refrigerant-134a enters an insulated compressor operating steadily as saturated vapor at –17°C with a mass flow rate of 600 kg/min. The refrigerant exits at 5 bar, 60°C. Neglecting the changes in kinetic and potential energies from the inlet to exit, determine (i) the volumetric flow rates at the inlet and exit, each in m^3/s, and (ii) the power input to the compressor, in kW.

5.18 Refrigerant-134a at 5 bar, 60°C, and 0.16 kg/s is cooled by water in a condenser until it exists as a saturated liquid at the same pressure. The cooling water enters the condenser at 3 bar and 20°C and leaves at 27°C at the same pressure. Determine the mass flow rate of the cooling water required to cool the refrigerant.

5.19 Steam enters a heat exchanger operating steadily at 3 bar and a quality of 95% and exits as a saturated liquid at the same pressure. A separate stream of oil with a mass flow rate of 200 kg/min enters at 25°C and exits at 100°C with no significant change in pressure. The specific heat of the oil is 2.0 kJ/kg K. Kinetic and potential energy effects are negligible. If heat transfer from the heat exchanger to its surroundings is 10% of the energy required to increase the temperature of the oil, determine the steam mass flow rate.

5.20 A tank whose volume is 0.022 m^3 is initially evacuated. A pinhole develops in the wall, and air from the surroundings at 25°C, 103 kPa enters until the pressure in the tank is 103 kPa. If the final temperature of the air in the tank is 25°C, determine (i) the final mass in the tank, in g, and (ii) the heat transfer between the tank contents and the surroundings, in kJ.

5.21 In a counter-flow heat exchanger, oil enters at 170°C with a mass flow rate of 500 kg/min and exits at 250°C. A separate stream of liquid water enters at 25°C, 4.5 bar. Each stream experiences no significant change in pressure. Stray heat transfer with the surroundings of the heat exchanger and kinetic and potential energy effects can be ignored. The specific heat of the oil is constant at 2 kJ/kg K. Determine the allowed range of mass flow rates for the water if the designer wants to ensure no water vapor is present in the exiting water stream.

5.22 Air enters the evaporator section of a window air conditioner at 1 bar and 25°C with a volume flow rate of 25 m^3/min. Refrigerant-134a at 1.5 bar with a quality of 35% enters the evaporator at a rate of 2.5 kg/min and leaves as saturated vapor at the same pressure. Determine (i) the exit temperature of the air and (ii) the rate of heat transfer from the air.

5.23 A hot-water stream at 70°C enters a mixing chamber with a mass flow rate of 25 kg/min where it is mixed with a stream of cold water at 25°C. If it is desired that the mixture leaves the chamber at 45°C, determine the mass flow rate of the cold-water stream. Assume all the streams are at a pressure of 2 bar.

DESIGN PROBLEMS

5.24 Design an experiment with complete instrumentation to prove the SFEE for an air compressor of your thermal engineering laboratory. Discuss the instrumentation required and measurements need to be taken.

5.25 Design a 1200-W electric hair dryer such that the air temperature and velocity in the dryer will not exceed 50°C and 3 m/s, respectively.

5.26 Consider a thermal power plant in your locale for performing the energy audit of boiler feed pumps. Evaluate the operating performance of the feedwater pumps using suction and discharge pressure, discharge flow rate, and motor power consumption. Compare the operating efficiency for flow rate, total head developed, and rated motor power with design efficiency at operating conditions. Based on your findings, recommend the changes as warranted.

6 Second Law of Thermodynamics

LEARNING OUTCOMES

After learning this chapter, students should be able to

- Demonstrate understanding of key concepts related to the second law of thermodynamics, including alternative statements of the second law, the internally reversible process.
- Explain various irreversibilities that make the process irreversible and distinguish it from the reversible process.
- Assess the performance of power cycles, refrigeration, and heat pump cycles using, as appropriate, the corollaries.
- Discuss the concepts of perpetual-motion machines.
- Apply the second law of thermodynamics to cycles and cyclic devices.
- Apply the second law to develop the absolute thermodynamic temperature scale.
- Analyze the Carnot principles and distinguish between the idealized Carnot heat engines, refrigerators, and heat pumps and actual ones.

6.1 LIMITATIONS OF THE FIRST LAW OF THERMODYNAMICS

The first law of thermodynamics states that energy is neither created nor destroyed, it gets transformed from one form to another form. The first law is a valuable tool that expresses the relation between heat and work and defines stored energy. Joule proved experimentally that in a cyclic process, the net heat transfer will be equal to the net work transfer and as per Joule's assertion, both heat and work are equal. However, one cannot explain the energy conversion phenomenon using the first law alone. The first law alone is insufficient to predict the extent of energy conversion and whether the energy conversion process is possible. For example, if two metallic bars at different temperatures are placed in an insulated box, energy, in the form of heat, transfers from a metal bar of higher temperature to that of lower temperature. In this case, energy lost by one bar is equal to that gained by the other, if there is nothing else in the box. This process satisfies the first law. The first law would also be satisfied with the reverse process, i.e., transfer of heat from a lower-temperature bar to a higher-temperature one, but it never occurs. Satisfying the first law doesn't ensure that the process can occur.

One type of desirable energy conversion, that is, heat to work, cannot be carried out completely. It can be explained with the help of an example that a heat engine, even a best efficient one, requires about 250 kJ of energy input for producing 100 kJ of work.

The remaining 150 kJ of energy is rejected to the surroundings in another form but not as work. It was concluded in the first law that energy doesn't degrade quantity wise; however, there is no mention of the degradation of energy quality wise. Furthermore, the first law doesn't put a limitation on the direction of this energy transformation. It is the second law, which brings a certain direction to this energy transformation. All the spontaneous processes in nature take place in a certain direction only and the reverse process cannot take place spontaneously. If we think of the reverse process in the above example, i.e., heat transfer from the low-temperature bar to the high-temperature one, it cannot be possible. Hence, a process always proceeds in a certain direction only, that is, from high potential to low potential. Heat transfer always takes place from a body of high temperature to that of low temperature. The reverse process never takes place spontaneously. Thus, a process must satisfy both first and second laws to occur; satisfying the first law alone doesn't mean that the process has taken place. A process will occur only if both the first and second laws of thermodynamics are satisfied. The second law not only brings the direction to the energy transformations, but also brings about the qualitative distinction between heat and work. In a cyclic process, work can be completely converted to heat and only a part of heat can be converted to work. Work is considered to be high-grade energy and heat a low-grade energy.

6.2 SECOND LAW STATEMENTS

The second law of thermodynamics originates form the contributions of Nicolas Leonard Sadi Carnot (1796–1832), a French military engineer, who first discovered the theory of maximum efficiency of heat engines. He made two important conclusions:

First, it is impossible to make a water wheel that can convert all the energy available at the inlet to wheel into shaft work (output). There must be some outflow of water from the wheel with some accompanied energy.

Second, the maximum efficiency of a heat engine is solely dependent on the temperatures of two thermal energy reservoirs (high and low temperature) with which the heat engine exchanges and is independent of the working fluid of the heat engine. The contributions of Sadi Carnot laid the foundation for the development of a new concept called thermodynamics and were later used by Rudolf Clausius and Lord Kelvin to formalize the second law of thermodynamics.

The second law and inferences from it have so many applications as given below.

1. To predict the direction of a process.
2. To establish the conditions for equilibrium.
3. To evaluate the best theoretical performance of the cycles and other devices.
4. To assess quantitatively the factors that prevent the accomplishment of best theoretical performance.
5. To define the temperature scale that is independent of the properties of a thermometric substance.

There are two statements in the second law, namely Kelvin–Planck statement and Clausius statement.

6.2.1 Kelvin–Planck Statement

It has been mentioned in Section 6.1 that work can be completely converted into heat and complete conversion of heat into work in a cyclic process is impossible. A special device known as heat engine is required for this purpose. A heat engine is a device which produces useful work by converting a part of heat received from a high-temperature source and rejects the waste heat to the sink. The first law of thermodynamics states that during any cycle that a system undergoes, the cyclic integral of the heat is equal to the cyclic integral of the work. However, it places no restrictions on the direction of flow of heat and work. A cycle in which a given amount of heat is transferred from the system and an equal amount of work is done on the system satisfies the first law just as well as a cycle in which the flows of heat and work are reversed.

The second law of thermodynamics places a limit on energy transformation; according to it, heat always flows naturally from a high-temperature reservoir to another at a lower temperature, but the reverse process is not possible without any external means. A heat engine produces net work output by operating between two energy reservoirs at different temperatures. Moreover, the first law of thermodynamics establishes equivalence between the heat used and work but does not state the conditions under which conversion of heat into work is possible, neither the direction in which heat transfer can take place. It is the second law of thermodynamics that bridges the gap by bringing about the direction to the energy transfer.

Based on the discussions above, the second law of thermodynamics can be stated. The second law is based on two classical statements: the Kelvin–Planck statement and the Clausius statement.

The *Kelvin–Planck statement*: *It is impossible to construct a device that operates in a cycle and produces net work output by transferring energy with a single reservoir.*

It can also be expressed as *it is impossible to build a heat engine that has a thermal efficiency of 100%.* This statement was formulated by William Thomson (Lord Kelvin, 1824–1907) and later modified by Max Planck (1858–1947), hence it is named the Kelvin–Planck statement.

This implies that a heat engine operating in a cycle cannot perform an equal amount of work as it receives heat from a high-temperature source. Some heat has to be rejected from the working fluid at a higher temperature to a low-temperature body. Work can be done by the transfer of heat between two temperature levels, and heat is transferred from the high-temperature body to the heat engine and then from the heat engine to the low-temperature body.

6.2.2 Clausius Statement of the Second Law

The *Clausius statement*: *It is impossible to construct a device that operates in a cycle and produces no effect other than the transfer of heat from a cold body to a hot body.* This statement was formulated by Rudolph Clausius (1822–1888).

Heat engines and refrigerators can be considered as energy conversion devices, as the given input energy is conserved and comes out in a different form. The heat engine transforms the heat input at a high temperature to a work output and a heat output at a lower temperature, whereas the refrigerator and heat pump convert the input work and heat to an output heat at an elevated temperature.

6.2.3 Equivalence of Kelvin–Planck and Clausius Statements

The two statements of the second law of thermodynamics, Kelvin–Planck and Clausius statements, are fundamentally identical though they appear to be different and either of the statements can be used to express the second law of thermodynamics. Each statement is based on an irreversible process. The first one is based on the transformation of heat into work, while the second one is based on the transformation of heat between two thermal reservoirs. The equivalence of two statements can be proved if a device that violates the Kelvin–Planck statement also violates the Clausius statement and vice versa. This can be established with the help of the following case.

Proof of violation of the Clausius statement resulting in the violation of the Kelvin–Planck statement is shown in Figure 6.1a, which shows that a cyclic heat pump transfers heat in the amount of Q_l from a low-temperature body to a high-temperature body without any external means, i.e., without any work input, in violation of Clausius statement of the second law. The heat supplied by the heat pump, Q_l, is fed to a high-temperature reservoir. Now let us assume that a heat engine operates between the same thermal reservoirs and produces net work output $W_{net,out}$. The rate of heat input required for a heat engine is the same as that supplied by the heat pump since both heat pump and heat engine operate between the same temperature reservoirs. Then the hot reservoir seems to be unnecessary as the heat input supplied by the heat pump can be directly fed to the heat engine and the hot reservoir can be eliminated. The heat engine now produces net work output by exchanging heat with a single reservoir in violation of the Kelvin–Planck statement of the second law. Therefore, a violation of the Clausius statement results in the violation of the Kelvin–Planck statement.

It can also be shown in a similar manner that a violation of the Kelvin–Planck statement leads to the violation of the Clausius statement (Figure 6.1b).

Let us consider a heat engine driving the refrigerator operating between the same two reservoirs. Let us assume that the heat engine has a thermal efficiency of 100%, so that it converts all the heat Q_l it receives to work W, which is a violation of the Kelvin–Planck statement. The work developed by the heat engine is now supplied to a refrigerator that removes heat Q_2 from the low-temperature reservoir and rejects heat in the amount of Q_2+Q_l to the high-temperature reservoir. The high-temperature reservoir thus receives a net amount of heat Q_2, which is the difference between Q_l and Q_2+Q_l. The combined heat engine and refrigerator can be viewed as a refrigerator transferring heat from a low-temperature reservoir to a high-temperature reservoir without any external input violating the Clausius statement of the second law. The violation of the Kelvin–Planck statement results in the violation of the Clausius statement.

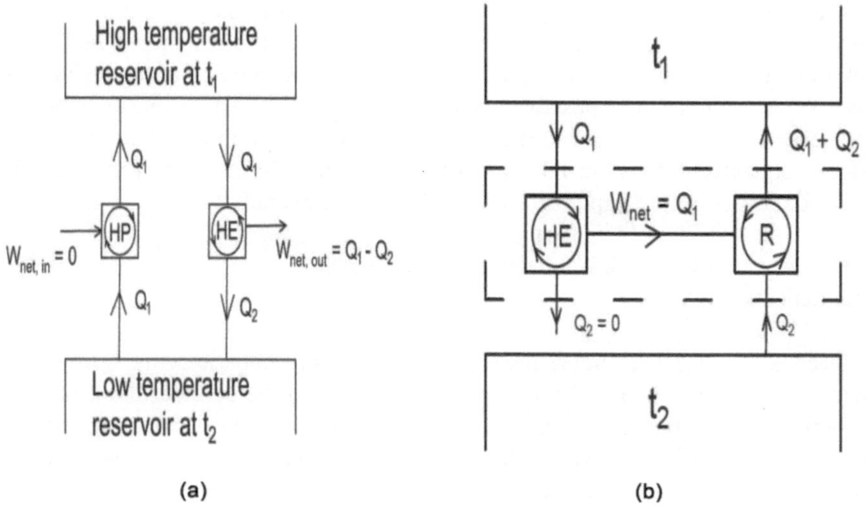

FIGURE 6.1 The systems used to demonstrate the equivalence of two statements. (a) Violation of Clausius statement. (b) Violation of Kelvin–Planck statement.

Therefore, the Clausius and the Kelvin–Planck statements are two equivalent statements of the second law of thermodynamics.

6.3 REVERSIBLE AND IRREVERSIBLE PROCESSES

One of the important uses of the second law of thermodynamics in engineering is to determine the best theoretical performance of the systems. By comparing actual performance with the best theoretical performance, an insight can be gained into the potential for improvement. Reversible processes are essentially important as they provide maximum possible performance (work output) from work-producing devices and minimum work input to the work-consuming devices. The best performance is evaluated in terms of idealized processes. In this section, such idealized processes are introduced and distinguished from actual processes that invariably involve irreversibilities.

6.3.1 Reversible Process

It is observed from the second law of thermodynamics that no heat engine is 100% efficient. If that is true, what could be the highest efficiency a heat engine can have. This can be addressed by defining the ideal process based on hypothetical assumptions. The spontaneous processes in nature occur in a certain direction only, the reverse process is not possible spontaneously, and the system cannot be restored to the initial state. Thus, they are called irreversible processes.

A reversible process is defined as the one which can be reversed without leaving any trace on the surroundings and at the end of the process, both the system and surroundings are returned to their initial states. If the net heat and net work exchange between system and surroundings for the combined process is zero, then the process is reversible. In both reversible and irreversible processes, the system can be brought back to its initial state at the completion of a process. Whereas for a reversible process, the restoration can be possible without leaving any net change on surroundings, in the case of an irreversible process, the surroundings do some work on the system and consequently it does not return to the initial state. Some of the processes that tend to be reversible are explained with the following examples:

1. a very slow frictionless adiabatic process and
2. deflection of a spring.

Let us consider a system comprising a gas trapped in an adiabatic cylinder fitted with a frictionless piston. If the piston is pushed slowly into the cylinder, both pressure and temperature of the gas increase uniformly throughout the system. If the external force on the piston is reduced slightly, the gas tends to expand, and if this expansion is very slow, the gas pressure decreases uniformly. The pressure during the expansion is the same as that during the compression for each position of the piston. As a result, the work done on the gas during the compression will be equal to that done by the gas during the expansion. The net work will be zero when the gas reaches its initial volume. Since the system is adiabatic, there is no heat transfer. Thus both the system and surroundings are returned to their initial states. Therefore, the very slow frictionless adiabatic process is reversible.

Deflection of a spring may also be considered as a reversible process. Let us consider that a very small load is applied slowly to a tension spring. The spring will elongate a very small distance because of the applied load. If the load is reduced after the elongation, the spring contracts to its original position. In this case, the work performed by the spring as it contracts is approximately equal to that needed to stretch it. If the force is applied and reduced incrementally, the process approaches a reversible one. If it is not so, then the presence of unconstrained vibrations and other effects will make the process irreversible.

6.3.2 Irreversible Process

A process is called irreversible, if both the system and its surroundings cannot be accurately restored to their respective initial states at the completion of the process. A system that has undergone an irreversible process is not necessarily excluded from being restored to its initial state. However, were the system restored to its initial state, it would not be possible also to return the surroundings to the state they were in initially. It is the second law that determines whether the system and surroundings can be returned to their initial states after a process has occurred and also whether a given process is reversible or irreversible. It might be apparent from the discussion of the Clausius statement of the second law that any process involving spontaneous heat transfer from a hotter body to a cooler body is irreversible. Otherwise, it would be possible to return this energy from the cooler body to the hotter body with no other effects within the two bodies or their surroundings. However, this possibility is denied by the Clausius statement. Processes involving other kinds of spontaneous events, such as an unrestrained expansion of a gas or liquid, are also irreversible. Friction, electrical resistance, hysteresis, and inelastic deformation are examples of additional effects whose presence during a process makes it irreversible. Some of the above factors, heat transfer through a finite temperature difference, friction, unrestrained expansion, and mixing of two different substances, are described in detail below.

Heat Transfer through a Finite Temperature Difference

This is a form of irreversibility. Let us consider that a hot metal bar is left in a cold room. The metal bar eventually cools after some time as a result of heat transfer due to the temperature difference between the bar and surroundings. To make the hot bar regain its initial temperature, heat is to be supplied in the form of electrical energy by a heater. At the end of the reverse process, the hot bar is restored to its original state and the surroundings will never be. It is known that the gain in the internal energy of the surroundings will be equal to the electrical energy supplied to the resistance heater. The surroundings can be restored to the initial state by converting this excess internal energy to electrical energy, which is impossible as it violates the second law. Since surroundings cannot be restored to the initial state, heat transfer through a finite temperature difference is irreversible.

Friction

When a body is in motion, or two bodies in contact are forced to move relative to each other, friction is present. For example, when an automobile is in motion, its tyre moves relative to the road, and friction is generated between the tyre surface and road surface in contact with each other. The friction is the opposing force developed at the contact surface and some work must be expended to overcome this friction. The energy supplied as work is converted to heat during this and is transferred to the surface of contact. The heat transfer is evident by the rise of temperature at the surface. If we consider the reverse process, that is, changing the direction of motion, the bodies are restored to the original position, the heat cannot be converted back to work, and even more of the work is converted to heat to overcome the friction in (opposite) reverse motion. In this case, both the system and surroundings are not returned to their original states, and thus the process is irreversible.

Mixing

It is possible to mix the gases in different conditions without any heat or work interactions; however, if we think of the reverse process, that is, the separation of the gases into each constituent gas (after mixing), it is impossible without any help from outside. Thus, the mixing process is always treated as an irreversible process since both the system and surroundings cannot be restored to their initial states.

From the above discussions, it is concluded that all actual processes are irreversible since every process involves effects such as those listed, whether the process is a naturally occurring one or the one which is devised for a particular purpose, ranging from the simplest mechanism to the largest industrial devices. The term irreversibility is used to identify any of these effects. The above list comprises a few of the irreversibilities that are commonly encountered.

Internal and External Irreversibilities

Irreversibilities are found to take place within a system and its surroundings when a system undergoes a process whether the system and surroundings are located in one place or other. There are two types of irreversibilities: (i) internal irreversibility and (ii) external irreversibility. Internal irreversibilities occur within the system caused by internal dissipative effects such as friction, electrical resistance, and turbulence. External irreversibilities occur within the surroundings, i.e., at the system boundary. A system exchanging heat with the surroundings when there is a temperature gradient between the system and surroundings causes external irreversibility. Since any natural process involves irreversibilities, it is essentially important to identify these irreversibilities present in the system and evaluate their effect and find the solutions to reduce them in order to improve the efficiency of the system. However, it becomes expensive in some of the cases to reduce the irreversibilities since it involves the change of design and operation.

6.4 SECOND LAW APPLICATION TO POWER CYCLES

In this section, cyclic devices such as power and refrigeration cycles, which are exchanging with two thermal energy reservoirs, are presented and analyzed based on the Kelvin–Planck statement and Clausius statement of the second law of thermodynamics respectively.

6.4.1 Thermal Efficiency of Power Cycles

A thermal energy reservoir (TER) is a body of infinite heat capacity that is capable of either absorbing or rejecting a huge quantity of heat energy with no changes in its thermodynamic co-ordinates. The heat transfer processes either into or out of the large body cause very slow and minute changes and all the processes within it are quasi-static. Two energy reservoirs at different temperatures are required for a heat engine to produce net work output by operating in a cyclic process. One is the high-temperature reservoir, known as source, from which heat is transferred to the system and the other is the low-temperature reservoir, known as a sink, to which heat is rejected from the system. A constant temperature furnace is considered as a source and a river is considered as a sink. Figure 6.2 shows the heat engine cycle developing net work output, $W_{net,out}$, by exchanging heat with a high-temperature source and a low-temperature sink.

The heat engine operates in a thermodynamic cycle as there is a net heat transfer to the system and net work transfer from the system and hence it is called a heat engine cycle. The heat engine cycle, as shown in Figures 6.3 and 6.4, uses a working substance, which receives heat energy from the source and rejects waste heat to sink. The heat engine is a broader term used to specify devices such as internal combustion engines, steam turbines, and gas turbines that produce work by converting heat into work. Figure 6.4 shows a heat engine cycle (steam power plant) in which a steady flow system interacts with the surroundings. Water is the working fluid to which heat in the amount of Q_1 is transferred from a source (furnace); water gets converted into steam in the boiler, which then expands doing work on the turbine rotor. Mechanical work (W_T) is the work output generated because of the rotation of the turbine rotor

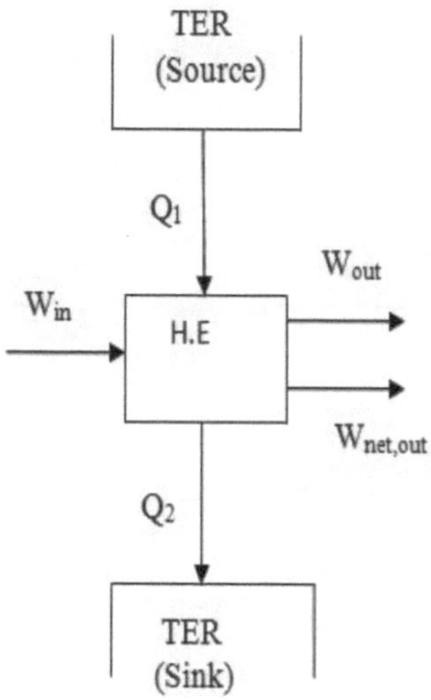

FIGURE 6.2 Heat engine with a source and a sink.

and is utilized to run the generator for the generation of electrical power. The steam then condenses in the condenser to water during which heat in the amount of Q_2 is rejected from the system to the surroundings. The water is circulated back into the boiler through a pump that takes work input (W_P) and the cycle repeats.

The net heat transfer in the heat engine cycle,

$$Q_{net} = Q_1 - Q_2 \tag{6.1}$$

The net work transfer in the heat engine cycle,

$$W_{net,out} = W_T - W_P \tag{6.2}$$

According to the first law of thermodynamics,

$$\delta Q_{cycle} = \delta W_{cycle} \text{ for a cyclic process.}$$

The thermal efficiency of a heat engine ($\eta_{thermal}$) is defined as the ratio of desired output to the required input, that is

$$\eta_{thermal} = \frac{\text{Net work output}}{\text{Total heat input}} \tag{6.3}$$

$$\eta_{thermal} = \frac{W_{net,out}}{Q_{in}} = \frac{Q_1 - Q_2}{Q_1} \quad \left(\because W_{net,out} = Q_1 - Q_2\right) \tag{6.4}$$

$$\eta_{thermal} = 1 - \frac{Q_2}{Q_1} \tag{6.5}$$

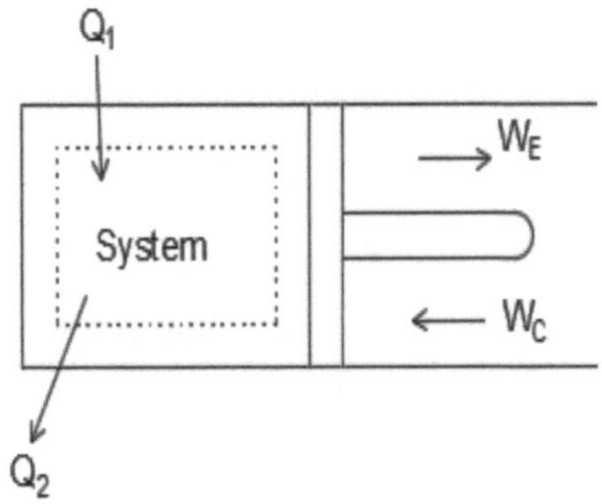

FIGURE 6.3 A closed system heat engine cycle.

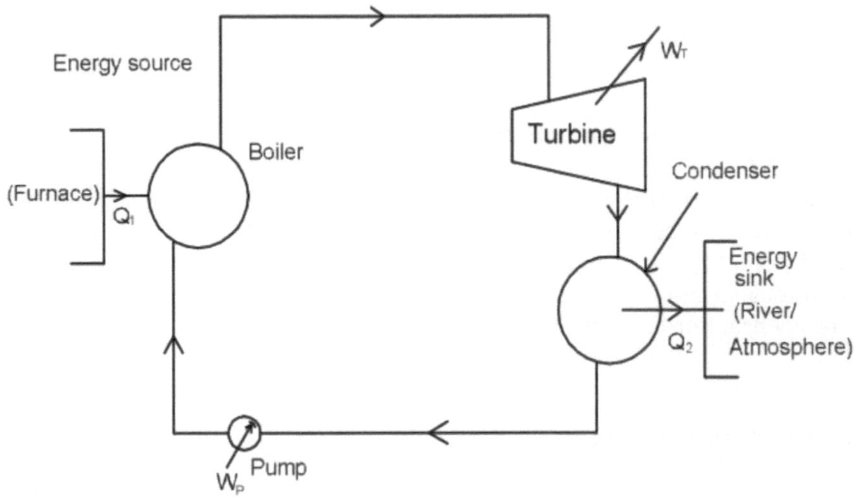

FIGURE 6.4 An open system heat engine cycle.

The thermal efficiency is a conversion efficiency for the process of going from the necessary input to the desired output. There are different heat engines that vary in size and shape; they are large steam engines used in marine applications, steam turbines used in power generation, gas turbine and jet engines used in aircraft propulsion, gasoline and diesel engines used in cars, and diesel engines used in trucks. The thermal efficiencies of actual engines are in the range of 35%–50% for large power plants, 30%–35% for gasoline engines, and 30%–40% for diesel engines. However, with the advents in technology, there are efforts to improve efficiency further in recent times.

6.4.2 Corollaries of the Second Law for Power Cycles

As mentioned in the previous section, the thermal efficiency of a power cycle cannot be 100%. A question then arises, "What is the maximum theoretical efficiency a power cycle can have?" The maximum theoretical efficiency of power cycles exchanging thermal energy with the same two thermal reservoirs at different temperatures can be developed by the two corollaries of the second law called the Carnot corollaries, as given below.

1. The thermal efficiency of all irreversible power cycles is often less than that of reversible cycles exchanging thermal energy with the same two thermal reservoirs at different temperatures.
2. All reversible power cycles exchanging thermal energy with the same two thermal reservoirs at different temperatures will have the same thermal efficiency.

A cycle is considered to be a reversible one when there are no irreversibilities within itself and there occurs the heat transfer between the system and surroundings in a reversible manner. It is in accordance with the second law that when the two heat engines operating between the same two thermal reservoirs, with each receiving the heat energy in the amount of Q_1, the heat engine that is operating in the reversible manner is more efficient than an irreversible heat engine.

The second Carnot corollary is concerned with reversible cycles only. Let us consider the two reversible cycles operating between the same two thermal reservoirs, receiving the heat energy in the amount of Q_1; if one cycle is capable of producing more work than the other, it can be possible by either a better working fluid or better choice of processes that make up the system. This corollary rules out either of such possibilities and makes it clear that both the cycles must have the same work output irrespective of other factors such as working fluid and the series of operations that make up the system.

Perpetual-Motion Machine of the Second Kind—PMM2

An imaginary machine that creates energy in violation of the first law of thermodynamics is called a perpetual-motion machine of the first kind (PMM1). Again, according to the Kelvin–Planck statement of the second law of thermodynamics, there can be no heat engine that has 100% efficiency. A fictitious machine that is capable of converting completely as much heat as it received from a source into another form of energy and thus becomes 100% efficient in violation of the Kelvin–Planck statement

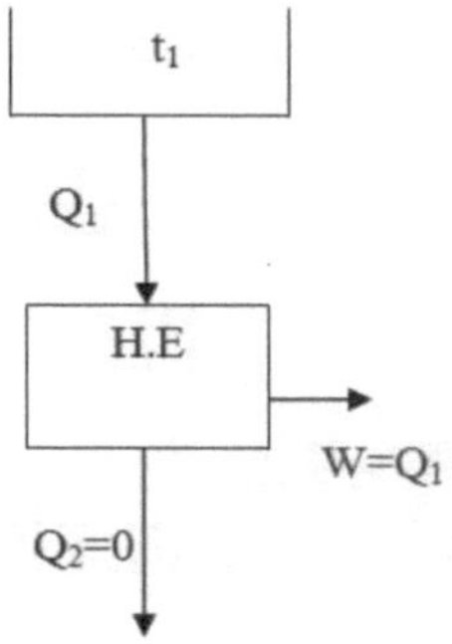

FIGURE 6.5 The PMM2.

of the second law is called a *perpetual-motion machine of the second kind (PMM2).* A PMM2 is thus impossible. Figure 6.5 shows the PMM2.

Perpetual-Motion Machine of the Third Kind—PMM3

A movable device in continual motion in the absence of friction is termed *PMM3.* Any movable device would experience some friction while in motion. It is possible with some means to minimize this friction but it cannot be eliminated completely. A PMM3 is also impossible.

6.5 REFRIGERATION AND HEAT PUMP CYCLES

The Clausius statement of the second law forms the basis for the development of refrigeration and heat pump cycles. The following sections present the operation of the refrigeration and heat pump cycles and their performance called the coefficient of performance (COP).

6.5.1 Refrigeration Cycles

The second law of thermodynamics places limits on the performance of refrigeration and heat pump cycles as it does in the case of power cycles. As shown in Figure 6.6, a system undergoes a cycle while communicating thermally with two thermal reservoirs, a hot and a cold reservoir, and thus operates on a thermodynamic cycle. The cycle discharges heat in the amount of Q_1 by heat transfer to the hot reservoir equal

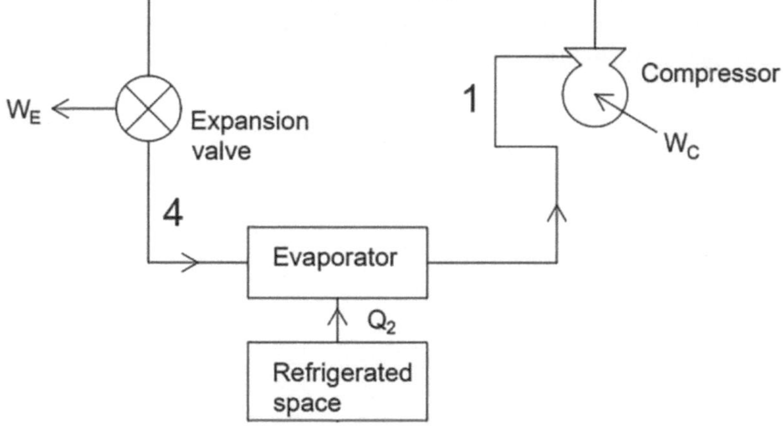

FIGURE 6.6 Schematic diagram of a refrigerator and a heat pump.

to the sum of the heat in the amount of Q_2 received by heat transfer from the cold reservoir and the net work input. This cycle can serve both as a refrigeration cycle or a heat pump cycle, depending on its purpose whether to remove heat Q_2 from the cold reservoir or deliver heat Q_1 to the hot reservoir.

Heat always flows in the direction of decreasing temperature, i.e., from the high-temperature body to low-temperature ones without requiring any devices. However, the reverse process, that is, transfer of heat from a low-temperature medium to a high-temperature cannot occur spontaneously; it requires special devices called refrigerators. Refrigerators are cyclic devices that use the refrigerant as a working fluid in the refrigeration cycle. The most commonly used refrigeration cycle is the vapor-compression refrigeration cycle. It comprises four main components: compressor, condenser, expansion valve, and evaporator.

The vapor refrigerant enters the compressor where it is compressed to the condenser pressure and leaves it at a comparatively high temperature. The refrigerant flows through the coils of the condenser, cools down, and condenses by rejecting heat to the circulating fluid. It then enters an expansion valve or capillary tube, where it is throttled to low pressure, so that the temperature drops appreciably. The low-temperature refrigerant then enters the evaporator, absorbs latent heat of vaporization from the refrigerated space, evaporates, and comes out as a vapor. The cycle repeats as the refrigerant leaves the evaporator and reenters the compressor. The amount of heat removed from the refrigerated space is Q_2 at temperature T_2, the amount of heat rejected to the warm environment is Q_1 at temperature T_1, and $W_{net,in}$ is the net work input to the refrigerator. Q_2 and Q_1 represent magnitudes and thus are positive quantities. Figure 6.7 shows the refrigerator removing heat from cooled space.

The COP of refrigeration and heat pump cycles is defined as

$$\text{COP} = \frac{\text{Desired effect}}{\text{Required input}} \tag{6.6}$$

The COP of a refrigeration cycle (COP_R) is

$$COP_R = \frac{Q_2}{W_{net,in}} = \frac{Q_2}{Q_1 - Q_2} \tag{6.7}$$

where $W_{net,in} = W_C - W_E$

6.5.2 Heat Pump Cycles

A heat pump, like a refrigerator, transfers heat from a low-temperature body to a high-temperature one. Figure 6.6 can also be considered for the heat pump cycle. Though refrigerators and heat pumps operate on the same cycle, their objectives are different. The objective of a refrigerator is to keep the refrigerated space comparatively at low temperature by continuous removal of heat from it and discharging this heat to a higher-temperature medium. When the refrigerated space is maintained at low temperatures, some heat will always leak into it by virtue of temperature difference, so this heat must be continuously removed to maintain the space at low temperatures. However, discharging heat to a higher-temperature medium is merely

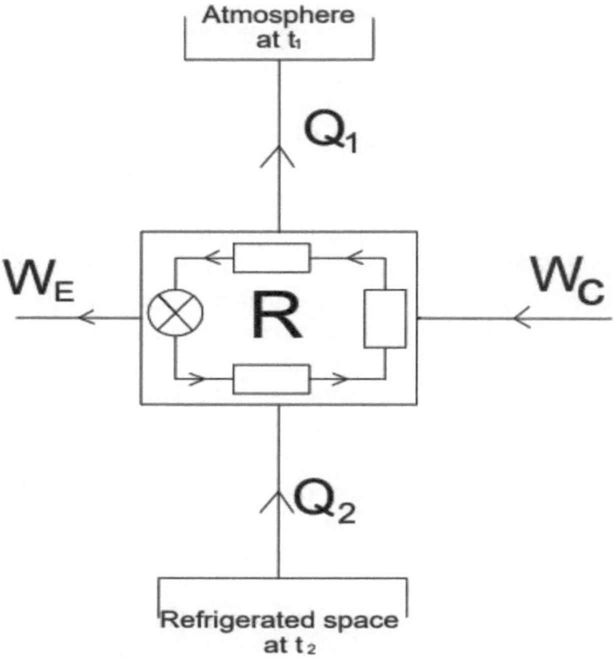

FIGURE 6.7 The refrigerator removing heat from cooled space.

a necessary part of the operation, not the purpose. On the other hand, the objective of a heat pump is to maintain a heated space at a high temperature than the surrounding environment. This is accomplished by absorbing heat from a low-temperature source, such as well water or cold outside air in winter, and supplying this heat to the high-temperature medium such as a house. Figure 6.8 shows the heat pump supplying heat to the warmer space. The same device can be operated as a refrigerator and heat pump if its direction can be changed. For example, in winter conditions, a refrigerator placed in the window of a house with its door open to the cold outside air tends to do as a heat pump, since it attempts to cool the outside by absorbing heat from it and discharging this heat into the house through the coils behind it.

For a heat pump cycle, the COP ($COP_{H.P}$) is

$$COP_{H.P} = \frac{Q_1}{W_{net,in}} = \frac{Q_1}{Q_1 - Q_2} \tag{6.8}$$

From the Eqs. 6.7 and 6.8, it can be observed that

$$COP_{H.P} = COP_R + 1 \tag{6.9}$$

for constant values of Q_2 and Q_1. It can be observed from Eq. 6.9 that Q_1 is greater than W, meaning that the heat supplied by a heat pump is essentially greater than the energy supplied to it, since COP_R is a positive quantity. Thus the COP of a heat

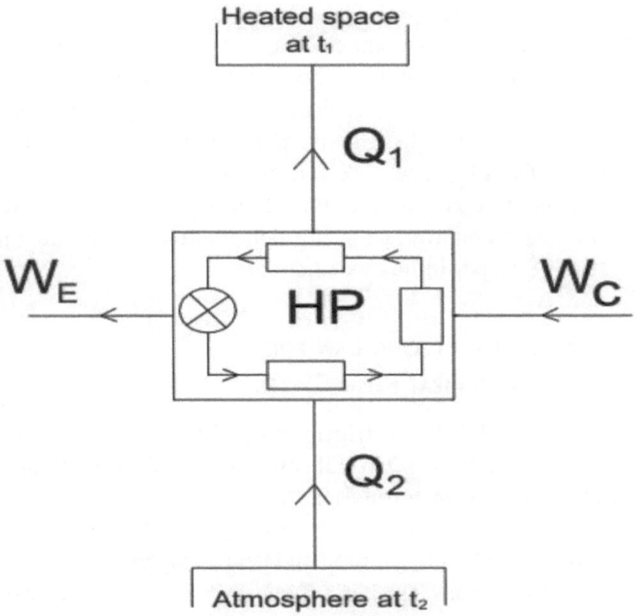

FIGURE 6.8 The heat pump supplying heat to the warmer space.

pump is always greater than unity. More specifically, a heat pump will perform, as a resistance heater, supplying as much energy to the house as it receives. However, in actual practice, there may be some losses in the form of heat lost to the surroundings through piping and other devices. Consequently, $COP_{H.P}$ tends to drop below unity in a cold environment. The average COP of most of the present heat pumps ranges from 2 to 3.

6.5.3 Energy Efficiency Ratio and Seasonal Energy Efficiency Ratio

The *energy efficiency ratio* (EER) is a measure of the performance of the heat pump and air-conditioning devices. It is defined as the system output in Btu/h per watt of electrical energy. It is the ratio of heat removal from the cooled space by the cooling equipment to the rate of electric power consumption. COP is the equivalent measure using SI units. A COP of 1.0 equates to an EER of 3.412. The units for EER is Btu/kWh.

$$EER = 3.412\ COP_R \tag{6.10}$$

From Eq. 6.10, it can be said that a cooling equipment with a COP of 1, i.e., which removes 1 Btu of heat from the cooled space for each kWh of electric power consumption, will have an EER of 3.412.

The seasonal EER is widely used in many countries to evaluate the energy efficiency of an air conditioner. *The seasonal energy efficiency ratio* (SEER) is defined as the ratio of total heat removed (in BTU) by an air conditioner or a heat pump during a normal cooling season to the total electric power consumption (in watt-hours). Since the temperature varies season-wise, the cooling and amount of energy required will vary. Accordingly, the Bureau of Energy and Efficiency (BEE) defined Indian seasonal energy efficiency ratio (ISEER) as a star rating for air-conditioning (AC) units in India effective from January 2018. It is the ratio of cooling seasonal total load to cooling seasonal energy consumption. Based on this, the energy efficiency of an AC can be evaluated based on average performance at the surrounding temperature between 24°C and 43°C as per Indian weather data.

6.5.4 Corollaries of the Second Law for Refrigeration and Heat Pump Cycles

The maximum theoretical COP of refrigeration and heat pump cycles exchanging thermal energy with the same two thermal reservoirs at different temperatures can be developed by the below-given corollaries.

1. The COP of irreversible refrigeration and heat pump cycles is less than that of reversible ones exchanging thermal energy with the same two thermal reservoirs.
2. All reversible refrigeration and heat pump cycles exchanging thermal energy with the same two thermal reservoirs will have the same COP.

As the net work input ($W_{net,in}$) to the cycle tends to zero, the COPs of both refrigeration and heat pump cycles may attain a value of infinity. If $W_{net,in}$ were zero, the cyclic process would extract energy Q_2 from the cold reservoir and discharge the same energy to the hot reservoir, in violation of the Clausius statement of the second law, and therefore, it is not possible. It follows that the COPs of both the devices must always be a finite value. This is attributed to another corollary of the second law.

6.6 THERMODYNAMIC TEMPERATURE SCALE

The temperature scale which measures the temperature, independent of the properties of the substances used for it, is called the thermodynamic temperature scale. This temperature scale is based on some reversible heat engines and it is more flexible in thermodynamic calculations. According to the Carnot theorem, the thermal efficiency of all reversible heat engines operating between the same two reservoirs is the same. The reversible cycle efficiency is independent of the working fluid used and its properties, the way the cycle is executed, or the type of reversible engine used. As the thermal energy reservoirs are described by their temperatures, the thermal efficiency of reversible heat engines solely depends on reservoir temperatures only. The thermal efficiency of any heat engine, with Q_1 and Q_2 as the heat addition and heat rejection, is

$$\eta_{thermal} = \frac{W_{net,out}}{Q_{in}} = \frac{Q_1 - Q_2}{Q_1} = 1 - \frac{Q_2}{Q_1}$$

For a reversible heat engine, when t_1 and t_2 are temperatures at which heat is transferred, since thermal efficiency is a function of temperatures only

$$\eta_{rev} = f(t_1, t_2) \tag{6.11}$$

Then $$1 - \frac{Q_2}{Q_1} = f(t_1, t_2)$$

By introducing another arbitrary function F,

$$\frac{Q_1}{Q_2} = F(t_1, t_2) \tag{6.12}$$

To define the temperature scale, some functional relationship is to be assigned to t_1, t_2 and Q_1/Q_2. This can be done with the help of three reversible heat engines as shown in Figure 6.9.

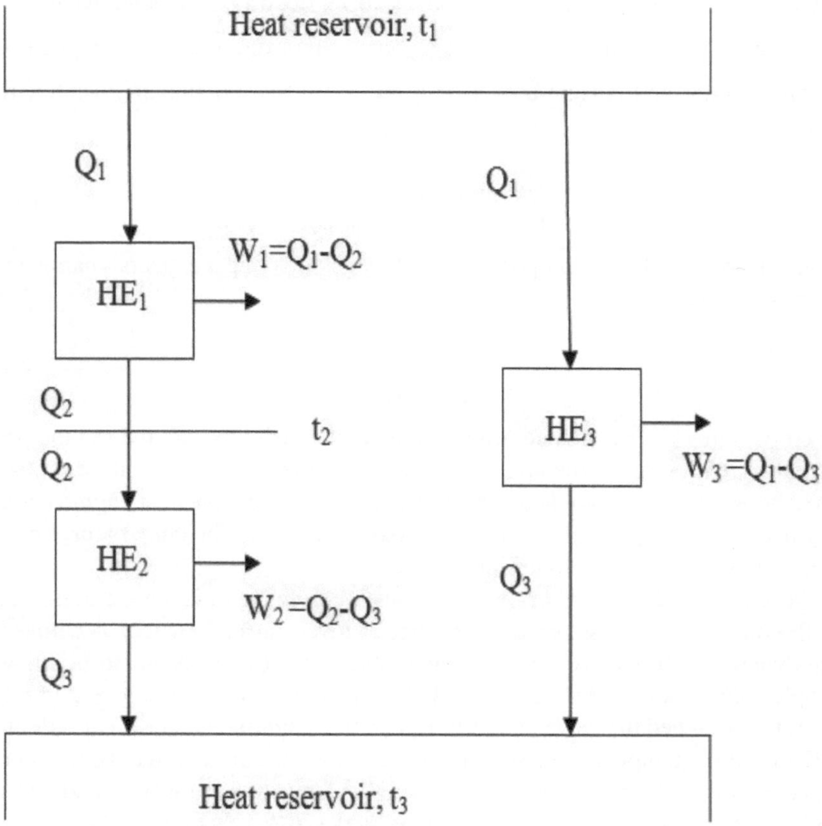

FIGURE 6.9 Three reversible heat engines.

Heat engine HE_1 receives heat in the amount of Q_1 from a source at t_1 and rejects heat in the amount of Q_2 at t_2 to heat engine HE_2, which in turn rejects heat in the amount of Q_3 at t_3 to the sink. Then

$$\frac{Q_1}{Q_2} = F(t_1, t_2) \text{and} \frac{Q_2}{Q_3} = F(t_2, t_3)$$

Another heat engine, HE_3 operates between temperatures t_1 and t_3, so that

$$\frac{Q_1}{Q_3} = F(t_1, t_3)$$

The ratio $\frac{Q_1}{Q_2}$ can be rearranged as $\frac{Q_1}{Q_2} = \frac{Q_1 / Q_3}{Q_2 / Q_3}$

Then

$$\frac{Q_1}{Q_2} = F(t_1, t_2) = \frac{F(t_1, t_3)}{F(t_2, t_3)} \tag{6.13}$$

For randomly chosen temperatures t_1, t_2, and t_3, from Eq. 6.19, it is found that the ratio $\frac{Q_1}{Q_2}$ depends on temperatures t_1, t_2 and is independent of t_3; therefore, t_3 can be taken out from the ratio on the right-hand side of Eq. 6.13. If we define another function ϕ, then

$$\frac{Q_1}{Q_2} = F(t_1, t_2) = \frac{\phi(t_1)}{\phi(t_2)} \tag{6.14}$$

Lord Kelvin was the first to propose taking $\phi(t) = T$ to define a thermodynamic temperature scale as

$$\frac{Q_1}{Q_2} = \frac{T_1}{T_2} \tag{6.15}$$

This temperature scale is named after Kelvin as the Kelvin scale and the temperatures that appear on this scale are absolute temperatures. The temperature ratios on this scale rely on the ratios of heat transfers between a reversible heat engine and the energy reservoirs with which the engine is communicating. The temperatures on the Kelvin scale vary from zero to infinity.

However, Eq. 6.15 doesn't fully define the thermodynamic temperature scale since the only ratio of absolute temperatures is given in that. Hence, to define the thermodynamic temperature scale, the magnitude of a Kelvin needs to be known. The triple point of water, the state at which three phases of water exist in equilibrium, is arbitrarily assigned the value of 273.16 K, and the magnitude of a Kelvin is defined as 1/273.16 of the temperature interval between absolute zero and the triple point of water (273.16 K). The magnitudes of temperature units of both Kelvin and Celsius scales are identical and temperatures on these scales differ by a constant 273.15.

$$T°(C) = T(K) - 273.15 \tag{6.16}$$

The thermodynamic temperature scale, defined with the help of reversible heat engines, may not determine the numerical values on an absolute temperature scale, since it is not practical to operate such an engine. However, to measure the absolute temperatures, there are other means such as a constant volume gas thermometer, which can effectively measure it.

6.7 CARNOT CYCLE

6.7.1 The Carnot Power Cycle

A heat engine is a cyclic device, in which the working fluid, after undergoing a series of processes returns to the initial state. During one cycle, the work is done by the working fluid in one part and on the working fluid in another part. The net work developed by the heat engine is the difference between the two, and the efficiency of the device depends on how each process of the cycle is executed. The net work and the cycle efficiency can be maximized by reversibly carrying the process so that the work required is minimized and the output maximized. The cycles that operate in this method are called reversible cycles. However, a reversible cycle cannot be a reality due to the reason that it cannot be possible to eliminate the irreversibilities related to the process, and hence they are the limiting cases for both engines and refrigerators and serve as ideal models for comparing the actual cycles.

The Carnot cycle can be executed in different types of systems. The system can be a gas, a liquid, an electric cell, and a steel wire. The cycle involves

1. a system,
2. two energy reservoirs one at a higher temperature T_1 and another at a lower temperatureT_2,
3. the arrangement for periodically insulating the system from one or both of the reservoirs, and
4. a part of the surroundings to absorb work and do work on the system.

Let us consider a closed system Carnot heat engine, as shown in Figure 6.10, comprising a gas in an adiabatic piston-cylinder arrangement. The insulation of the cylinder head can be removed so that the cylinder can be placed in contact with either of the reservoirs, maintained at constant temperatures T_1 and T_2 respectively. The cycle consists of four reversible processes.

Reversible Isothermal Expansion Process 1-2: The gas is initially at a temperature of T_1 and the system is brought in contact with the reservoir at T_1. The gas expands slowly, doing work on surroundings, and its temperature decreases. Owing to the temperature difference between the source and the system, heat transfers from the source to the system so that the gas temperature is maintained constant at T_1. This process is termed reversible heat transfer as the temperature difference between the source and the system is always an infinitesimal difference dT. Applying the first law, we can determine heat transfer Q_1 for this process,

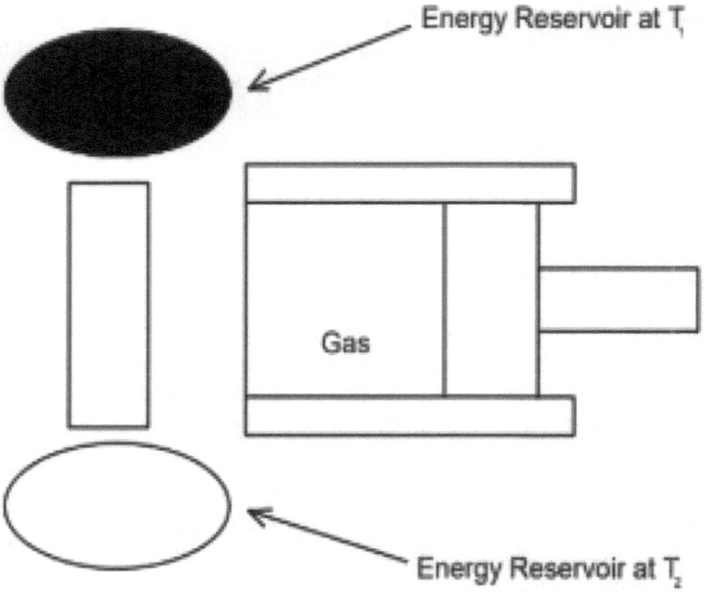

FIGURE 6.10 Carnot heat engine-closed system.

$$Q_1 = U_2 - U_1 + W_{1\text{-}2} \left(\text{For an ideal gas } U_1 = U_2\right) \tag{6.17}$$

Reversible Adiabatic Expansion Process 2-3: Now the system is placed on the insulating stand. The gas continues to expand adiabatically doing work on surroundings and its temperature decreases. The process is reversible and adiabatic as the piston is assumed to be frictionless and the process quasi-static. Applying the first law,

$$0 = U_3 - U_2 + W_{2\text{-}3} \tag{6.18}$$

Reversible Isothermal Compression Process 3-4: Again the system, after the insulation, is removed, placed in contact with the reservoir at T_2. The gas is compressed and work is done on the gas. The temperature of the gas decreases to T_2 due to the heat transfer from gas to the sink. This process is termed reversible heat transfer since the temperature difference between the gas and the sink always remains constant at dT (differential amount).

The heat transfer Q_2 for this process can be determined by the first law,

$$Q_2 = U_4 - U_3 - W_{3\text{-}4} \left(\text{For an ideal gas } U_3 = U_4\right) \tag{6.19}$$

Reversible Adiabatic Compression Process 4-1: The gas is compressed adiabatically to state 1. Work is done on the gas.

$$0 = U_1 - U_4 - W_{4\text{-}1} \tag{6.20}$$

The cycle constitutes two reversible adiabatic and two reversible isothermal processes.

Adding Eq. 6.16–6.19

$$Q_1 - Q_2 = (W_{1\text{-}2} + W_{2\text{-}3}) - (W_{3\text{-}4} + W_{4\text{-}1})$$

The above equation can also be written as $\sum_{cycle} Q_{net} = \sum_{cycle} W_{net}$

Carnot's Theorem

Carnot's theorem states that *of all the heat engines operating between the same temperature limits, none has a higher efficiency than a Carnot heat engine (reversible heat engine).*

It can also be stated as the *Carnot heat engine is more efficient than any other heat engine operating between the same temperature limits.*

The Carnot cycle is the most prominent reversible cycle, which is named after Nicolas Leonard Sadi Carnot, who first proposed it in 1824. The Carnot heat engine is a theoretical heat engine that functions on the Carnot cycle. Figure 6.11 shows the p-v and T-s diagrams of the Carnot cycle. As shown in the figure, the cycle constitutes two reversible adiabatic processes and two reversible isothermal processes. Heat Q_1 is added to the working fluid at constant temperature T_1 in the process 1-2. The working fluid then expands isentropically in the process 2-3. Heat Q_2 is rejected from the working fluid at constant temperature T_2 in the process 3-4 and it is then compressed isentropically in the process 4-1.

As shown earlier, the thermal efficiency of any heat engine, whether it is reversible or irreversible, can be expressed as

$$\eta_{thermal} = 1 - \frac{Q_2}{Q_1}$$

where Q_1 and Q_2 are respectively the heat transferred to the heat engine from a high-temperature source at T_1 and heat rejected to the low-temperature sink at T_2. The efficiency of the Carnot heat engine can be developed by replacing the heat transfer

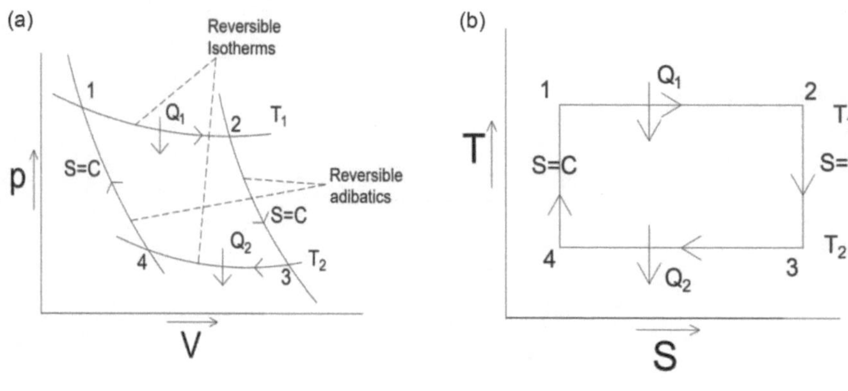

FIGURE 6.11 Carnot cycle (a) p-v diagram and (b) T-s diagram.

ratio with the ratio of absolute temperatures of two reservoirs with which the heat engine is communicating. Carnot efficiency is expressed as

$$\eta_{thermal} = \frac{Q_1 - Q_2}{Q_1} = \frac{T_1(S_2 - S_1) - T_2(S_3 - S_4)}{T_1(S_2 - S_1)}$$

$$= \frac{T_1 - T_2}{T_1} = 1 - \frac{T_2}{T_1} \quad (\text{Since } S_2 = S_3 \text{ and } S_1 = S_4)$$

$$\eta_{Carnot} = 1 - \frac{T_2}{T_1} \tag{6.21}$$

This expression is also known as the efficiency of a reversible heat engine and it is the highest efficiency a heat engine working between two reservoirs at temperatures T_1 and T_2 can have, since the Carnot heat engine is the famous ideal or reversible heat engine. T_1 and T_2 in Eq. 6.21 are the absolute temperatures. The efficiencies of all irreversible heat engines working between these two reservoirs are lower than those for reversible ones. It is possible to maximize the thermal efficiency of actual heat engines by adding the heat input to the engine at the maximum possible temperature (based on metallurgical conditions) and rejecting heat from it at the lowest possible

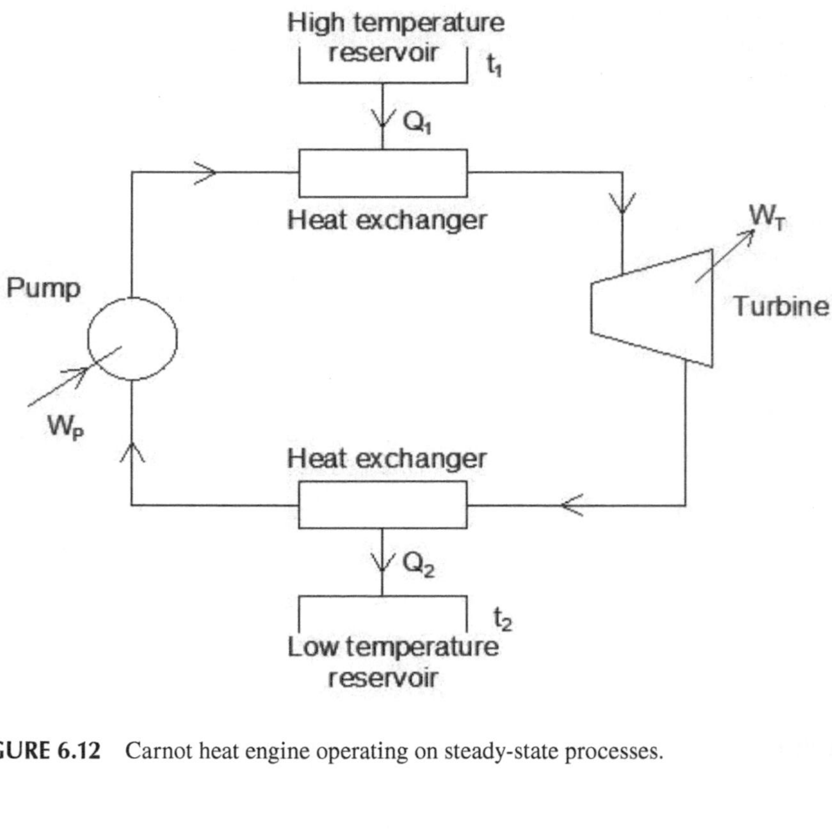

FIGURE 6.12 Carnot heat engine operating on steady-state processes.

temperature (based on the temperature of the cooling medium). The lowest possible temperature of heat rejection is the temperature of the surroundings.

Figure 6.12 shows the steady-flow Carnot heat engine cycle in which heat in the amount of Q_1 is added to the system in the heat exchanger-I reversibly and isothermally from a source at t_1, the work is done by the system reversibly and adiabatically in the turbine, heat is rejected reversibly and isothermally from the system in the heat exchanger-II to a sink at t_2, and work is done on the system reversibly and adiabatically by the pump.

6.7.2 The Carnot Refrigerator and Heat Pump Cycles

Since the Carnot cycle is reversible, all the processes constituted by the cycle can be reversed by reversing the directions of heat and work transfer interactions without changing the cycle, so that it becomes the reversed Carnot cycle. The cycle so developed can be an ideal cycle for both refrigerators and heat pumps and they are called the Carnot refrigeration cycle and Carnot heat pump cycle. Figure 6.13 shows a reversed Carnot heat engine. Figure 6.14 shows the p-v diagram of the reversed Carnot cycle. Heat Q_2 is absorbed from the low-temperature reservoir and Q_1 is rejected to a high-temperature reservoir with the help of the work input of $W_{net,in}$.

The COP of the refrigerator or heat pump is expressed as

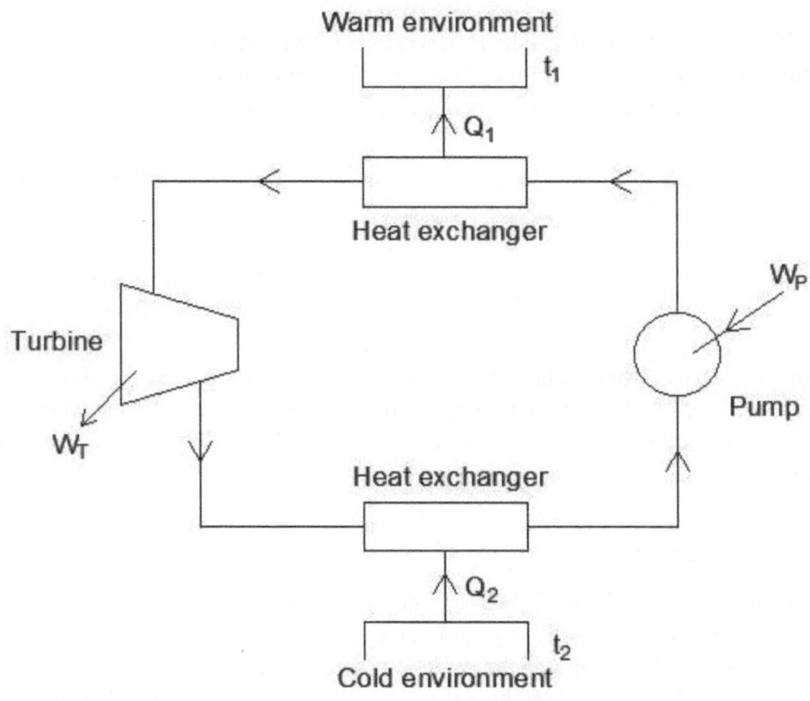

FIGURE 6.13 Reversed Carnot heat engine operating on steady-state processes.

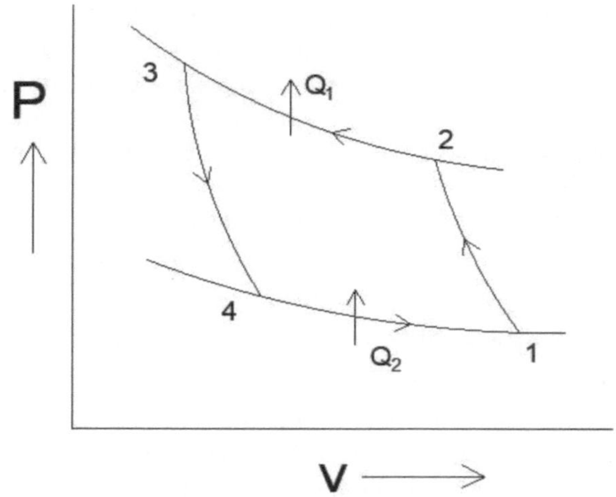

FIGURE 6.14 p-v diagram of the reversed Carnot cycle.

$$COP_R = \frac{Q_2}{W_{net,in}} = \frac{Q_2}{Q_1 - Q_2}$$

$$COP_{H.P} = \frac{Q_1}{W_{net,in}} = \frac{Q_1}{Q_1 - Q_2}$$

The COP of a reversible (Carnot) refrigerator or heat pump can be developed by replacing the heat transfer ratio with the ratio of absolute temperatures of two reservoirs with which the refrigerator or heat pump is communicating.

$$COP_{R,Rev} = \frac{T_2}{T_1 - T_2} \tag{6.22}$$

$$COP_{HP,Rev} = \frac{T_1}{T_1 - T_2} \tag{6.23}$$

A reversible refrigerator or a heat pump, like the reversible heat engine, functioning between two reservoirs at temperatures T_1 and T_2, can have the highest COP while the actual refrigerators or heat pumps functioning between the above temperatures have lower COPs.

EXAMPLE PROBLEMS

Example 6.1 A reversed Carnot heat engine cycle is used as a refrigeration plant for food, a storage which maintains the store at a temperature of –7°C. The heat transfer from the food storage unit to the cycle is at a rate of 6 kW. Determine the power required to drive the plant, if the heat transfer from the cycle to the atmosphere is at a temperature of 27°C.

Solution

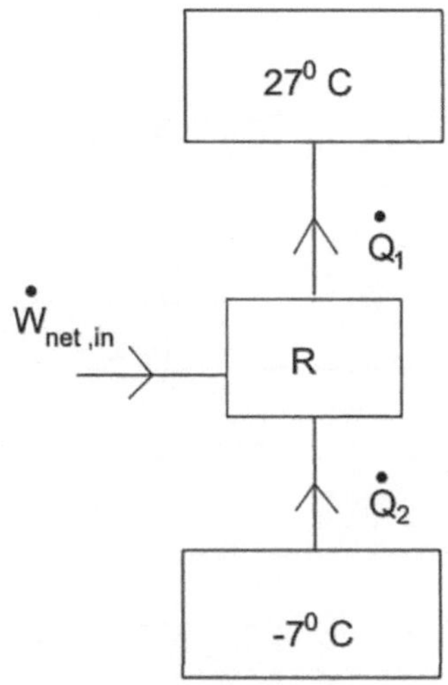

FIGURE EX. 6.1

Store temperature $T_2 = 273 + (-7) = 266$ K

Atmospheric temperature $T_1 = 273 + 27 = 300$ K

The rate of heat transfer (removed) from store, $\dot{Q}_2 = 6$ kW

As the cycle operates on a reversed Carnot cycle,

$$COP_{R,Rev} = \frac{T_2}{T_1 - T_2} = \frac{266}{300 - 266} = 7.8$$

$$\text{Again} \quad COP_{R,Rev} = \frac{\dot{Q}_2}{\dot{W}_{net,in}}$$

$$7.8 = \frac{6}{\dot{W}_{net,in}} \text{ then } \dot{W}_{net,in} = 0.77\,\text{kW}$$

Therefore, power required to drive the plant = 0.77 kW Ans.

Example 6.2 A domestic refrigerator, operating on 20% of ideal COP, keeps the items inside it at a temperature of 4°C. The door of the refrigerator is opened 12 times a day with warm material placed inside at each door opening, which introduces an average of 360 kJ of heat energy, without any considerable change in the temperature of the refrigerator. The cost of work is 25 paise/kW hour. Estimate the monthly bill of the refrigerator. Take surrounding temperature, t_{atm}, as 30°C.

Solution

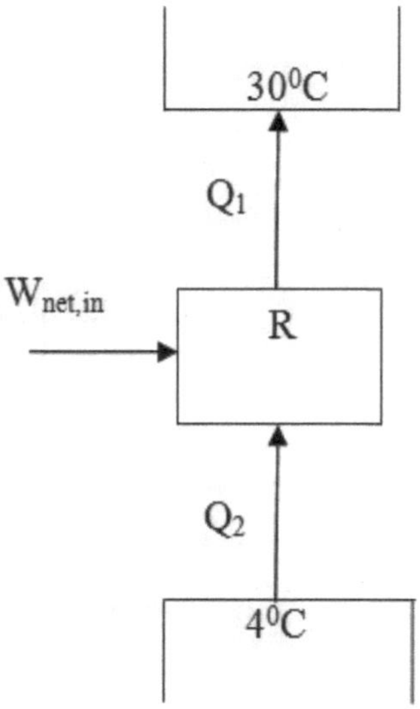

FIGURE EX. 6.2

Ideal COP of the refrigerator,

$$COP_{R,Rev} = \frac{T_2}{T_1 - T_2} = \frac{277}{303 - 277} = 10.65$$

Actual COP = 20% of ideal COP   Actual $COP_{act} = 0.20 \times 10.65 = 2.13$

Heat to be removed from the refrigerated space,

$$Q_2 = 360 \times 12 = 4320\,kJ$$

$$COP_R = \frac{Q_2}{W_{net,in}} \Rightarrow W_{net,in} = \frac{4320}{2.13} = 2028.16\,kJ$$

The cost of work is 25 paise/kW h

1 kW h = 3600 kJ

Then 2028.16 kJ = 0.563 kW h

∴ Cost of work = 0.563 kW h × 0.25 = Rs. 0.1407

Monthly bill of the refrigerator = 0.1407 × 30 = Rs. 4.23   Ans.

Example 6.3 A refrigerator is driven by a heat engine of 28% thermal efficiency. Determine the heat input required into the heat engine for each MJ of heat removed from the low-temperature medium by the refrigerator. The COP of the refrigerator is 4.5. Determine the heat that is available for heating if the above is used as a heat pump, for each MJ of heat input to the engine.

Solution

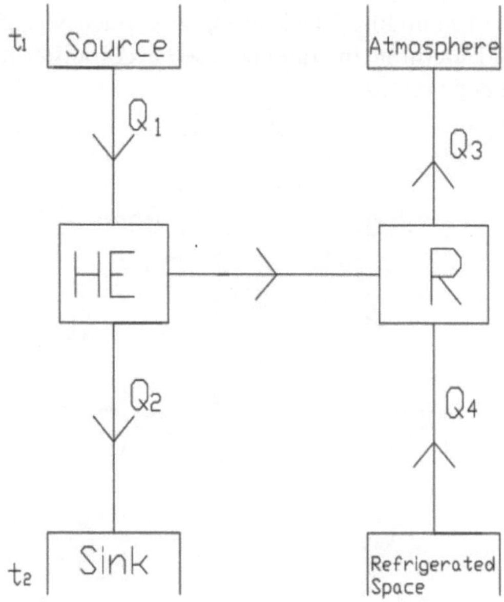

FIGURE EX. 6.3

For refrigerator, $COP_R = \frac{Q_4}{W_{net,in}} = 4.5$

but $Q_4 = 1\ MJ = 1000\ kJ$

$$\therefore W_{net,in} = \frac{1000}{4.5} = 222.22\,kJ$$

Again $\eta_{thermal} = \frac{W_{net,out}}{Q_1}$

$$Q_1 = \frac{222.22}{0.28} = 793.7\,kJ$$

$\therefore$ Heat input required to the engine for each MJ of heat removal is 793.7 kJ Ans.

Now for heat pump, $Q_1 = 1\ MJ = 1000\ kJ$ (Heat input to the engine)

$$\eta_{thermal} = \frac{W_{net,out}}{Q_1} \Rightarrow W_{net,out} = 0.28 \times 1000 = 280\,kJ$$

$$COP_{H.P} = 1 + COP_R = 1 + 4.5 = 5.5$$

$$COP_{H.P} = \frac{Q_3}{W_{net,in}} \Rightarrow Q_3 = 5.5 \times 280 = 1540\,kJ$$

$\therefore$ Heat available for heating by the heat pump is 1680 kJ = 1.68 MJ Ans.

Example 6.4 A heat pump is driven by a heat engine. The heat transfers from both the heat engine and heat pump are utilized in heating the water, which is circulated through the radiators of a building. The heat engine efficiency is 25% and the COP of the heat pump is 3.5. Determine the ratio between heat transfer to circulating water and heat input to the engine.

Solution

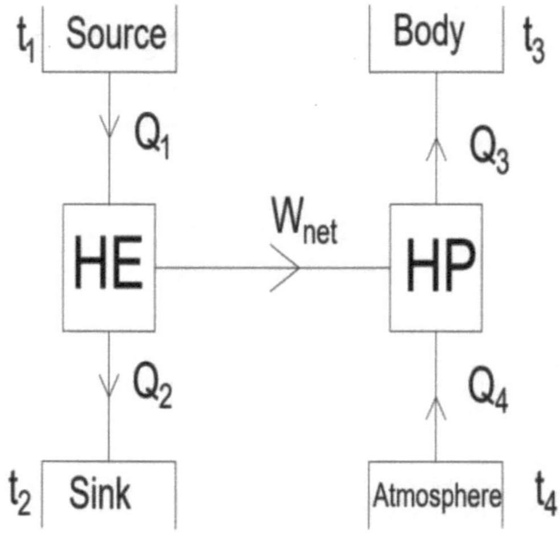

FIGURE EX. 6.4

For heat pump, $COP_{H.P} = \dfrac{Q_3}{W_{net}}$

$$W_{net} = \frac{Q_3}{3.5}$$

The thermal efficiency of heat engine $\eta_{thermal} = \dfrac{W_{net}}{Q_1} \Rightarrow W_{net} = 0.25 \times Q_1$

$$W_{net} = 0.25 \times Q_1 = \frac{Q_3}{3.5} \text{ then } \frac{Q_3}{Q_1} = 0.875 \tag{6.24}$$

Again $\eta_{thermal} = \dfrac{W_{net,out}}{Q_1} = 0.25$

The ratio,

$$\frac{Q_2}{Q_1} = 1 - 0.25 = 0.75 \tag{6.25}$$

Adding Equations 6.24 and 6.25 $\Rightarrow \dfrac{Q_3}{Q_1} + \dfrac{Q_2}{Q_1} = \dfrac{Q_3 + Q_2}{Q_1}$

$= 0.875 + 0.75 = 1.625$ Ans.

The ratio $\dfrac{Q_3 + Q_2}{Q_1}$ is the heat transfer to water to the heat input to the engine.

Example 6.5 A heat engine receives heat at a rate of 900 kJ/s from a source at a temperature of 800 K and rejects waste heat to the surroundings at 298 K. Work produced is at a rate of 400 kW. Based on both the actual and reversible operating conditions, estimate (i) the heat rejected to the surroundings and (ii) the engine efficiency.

Solution

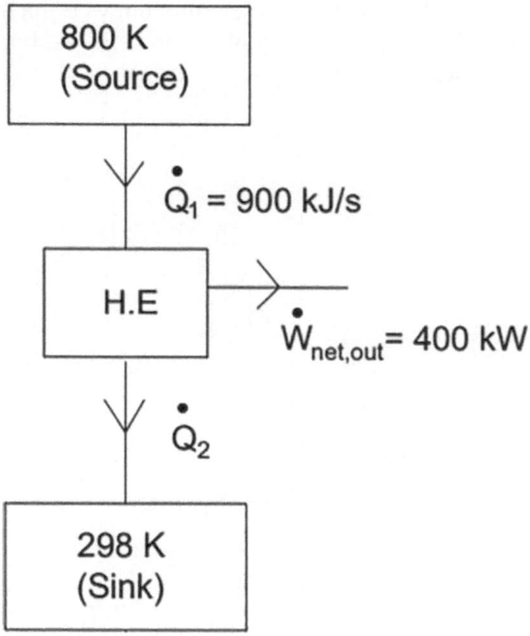

FIGURE EX. 6.5

The rate of heat input, $\dot{Q}_1 = 900$ kW and net power output, $\dot{W}_{net,out} = 400$ kW

The thermal efficiency of heat engine, $\eta_{thermal} = \dfrac{\dot{W}_{net,out}}{\dot{Q}_1}$

$$\frac{400}{900} = 0.444 = 44.4\%$$

$$\dot{Q}_2 = \dot{Q}_1 - \dot{W}_{net,out} = 900 - 400 = 500 \text{ kJ/s}$$

$$\eta_{Carnot} = 1 - \frac{T_2}{T_1}$$

$$= 1 - \frac{298}{800} = 0.6275 = 62.75\%$$

The power output and rate of heat rejection based on Carnot efficiency are

$$\dot{W}_{net,out} = \eta_{Carnot}\dot{Q}_1 = 0.6275 \times 900 = 564.75 \text{ kW}$$

$$\dot{Q}_2 = \dot{Q}_1 - \dot{W}_{net,out} = 900 - 564.75 = 335.25 \text{ kW}$$

Thus, the actual heat engine has a lower thermal efficiency than the Carnot one, and also the actual engine rejects a larger amount of heat energy to the surroundings compared with that of the Carnot heat engine.

Example 6.6 An ice producing plant produces ice at atmospheric pressure and 0°C from water. The temperature of the water which is used as a coolant in the condenser of the refrigeration unit is 20°C. Estimate the minimum work input required in kW hour to produce 1 ton of ice. Take enthalpy of fusion of ice at atmospheric pressure as 335 kJ/kg. Take c_p of water as 4.2 kJ/kg K.

Solution

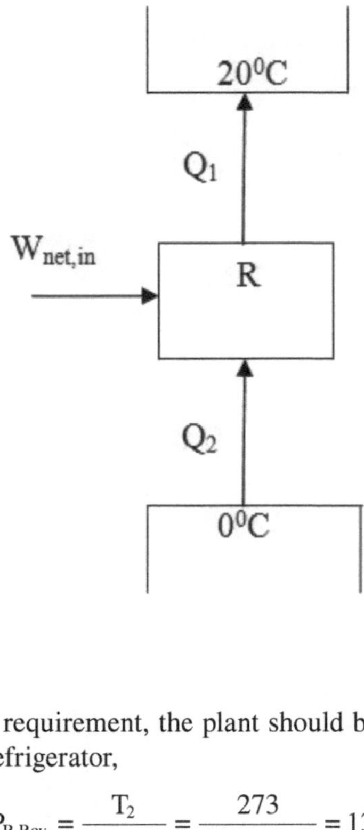

FIGURE EX. 6.6

For minimum work input requirement, the plant should be operated in a reversible manner. For a reversible refrigerator,

$$COP_{R,Rev} = \frac{T_2}{T_1 - T_2} = \frac{273}{293 - 273} = 13.65$$

where $T_1 = 20 + 273 = 293$ K and $T_2 = 0 + 273 = 273$ K

$$\text{COP of refrigerator, COP} = \frac{Q_2}{W_{net,in}} \Rightarrow W_{net,in} = \frac{Q_2}{COP}$$

where Q_2 is heat removed by the refrigerator from water to produce 1 ton of ice.

Q_2 (latent heat + sensible heat) = $1000 \times 335 + 1000 \times 4.2 \times 20 = 419{,}000$ kJ

$$W_{net,in} = \frac{419{,}000}{13.65} = 30{,}695.97\,\text{kJ}$$

Also 1 kW hour = 3600 kJ

$$\text{Minimum work input required in kWh} = \frac{30{,}695.97}{3600} = 8.53\,\text{kWh} \qquad \text{Ans.}$$

Example 6.7 150 kg of fish, initially at 10°C, is to be frozen at −5°C by a refrigerating machine. The specific heat c_p of fish above and below the freezing point is 3.2 and 1.699 kJ/kg K respectively. The freezing point of fish is −2°C and the latent heat of fusion is 232 kJ/kg. The refrigerator, operating at 40% of ideal COP, rejects heat to the surroundings at 35°C. Evaluate the power required to remove heat in 5 hours.

Solution $m = 150$ kg

Initial temperature $t_i = 10°C$ freezing point $t_f = -2°C$

Ambient temperature, $t_1 = 35°C$ or $T_1 = 35 + 273 = 308$ K

Temperature required, $t_2 = -5°C$ or $T_2 = -5 + 273 = 268$ K

Specific heat of fish above the freezing point, $c_{p1} = 3.2$ kJ/kg K

Specific heat of fish below the freezing point, $c_{p2} = 1.699$ kJ/kg K

Latent heat of fusion = 232 kJ/kg

$$COP_{R,Rev} = \frac{T_2}{T_1 - T_2} = \frac{268}{308 - 268} = 6.7$$

Actual COP = 0.4 × 6.7 = 2.68

$$\text{Now } COP = \frac{Q_2}{W_{net,in}}$$

$$\text{where } Q_2 = m\left[c_{p1}(t_i - t_f) + h_{fg} + c_{p2}(t_f - t_2)\right]$$

$$= 150[3.2(10 - (-2)) + 235 + 1.699(-2 - (-5))] = 41{,}774.55\,\text{kJ}$$

$$W_{net,in} = \frac{Q_2}{COP} = \frac{41{,}774.55}{2.68} = 15{,}587.52\,\text{kJ}$$

This heat has to be removed in 5 hours; therefore, the rate of work

$$\dot{W} = \frac{W_{net,in}}{\Delta t} = \frac{15{,}587.52}{18{,}000} = 0.866\,\text{kW} \qquad \text{Ans.}$$

REVIEW QUESTIONS

6.1 What is a thermal energy reservoir? Give some examples.

6.2 What is a perpetual-motion machine of the second kind (PMM2)? Does it exist?

6.3 What is an imaginary process that violates the second law of thermodynamics?

6.4 Is it possible for a heat engine to operate without a sink? Explain.

6.5 What are the characteristics of all heat engines?

6.6 Is it possible to operate a refrigerator as a heat pump? Describe.

6.7 What is the Kelvin–Planck expression of the second law of thermodynamics?

6.8 Is the Carnot heat engine 100% efficient? Explain.
6.9 What are the factors that make the process irreversible? Explain.
6.10 What is Carnot's theorem? What is it its significance?
6.11 How do you justify that both the Kelvin–Planck and Clausius statements are two parallel statements of the second law? Explain.
6.12 Distinguish between a refrigerator and a heat pump.
6.13 Distinguish between a refrigerator and an air conditioner.
6.14 What is the coefficient of performance of a refrigerator and a heat pump?
6.15 How can you convert heat into work in a cyclic process? Explain.
6.16 What is a perpetual-motion machine of the third kind (PMM3)? Does it exist?
6.17 What is energy efficiency ratio (EER)?
6.18 How is the energy efficiency ratio (EER) related to COP?
6.19 How do you compare the electrical resistance heater and heat pump for the same purpose of space heating?
6.20 How are COPs of a refrigerator and a heat pump related?
6.21 Derive an expression for thermal efficiency of a heat engine.
6.22 Define source and sink and give some examples.

EXERCISE PROBLEMS

6.1 A refrigerator removes heat from the food storage device at a rate of 150 kJ/min for each kW of power it consumes. Estimate the COP and the rate of heat rejection to the outside air.
6.2 A heat pump supplies energy to a house at a rate of 8000 kJ/h for each kW of electric power it draws. Estimate the COP and the rate of energy absorption from the outdoor air.
6.3 It is required by a refrigerator to remove heat from the cooled space at a rate of 10 kJ/s to maintain it at a temperature of –5°C when the surrounding temperature is 27°C. Estimate the minimum power input required for this refrigerator.
6.4 A heat pump used for heating a house in winter has to maintain it at 20°C continuously. When the outdoor temperature drops to 5°C, the heat losses from the house are estimated to be 20 kJ/s. Estimate the minimum power required to run the heat pump if heat is extracted from (i) the outdoor air at –5°C and (ii) the well water at 10°C.
6.5 A heat engine, used to drive a Carnot heat pump, operates between two reservoirs at 1000 and 300 K. One-half of the work output of the heat engine is used to drive the heat pump that removes heat from the cold surroundings at 3.5°C and transfers it to a house maintained at 27°C. If the house is losing heat at a rate of 55 MJ/h, determine the minimum rate of heat supply to the heat engine required to keep the house at 27°C.
6.6 A Carnot heat engine receives heat at 750 K and rejects the waste heat to the environment at 298 K. The entire work output of the heat engine is used to drive a Carnot refrigerator that removes heat from the cooled space at –12°C at a rate of 0.38 MJ/min and rejects it to the same environment at

298 K. Determine (i) the rate of heat supplied to the heat engine and (ii) the total rate of heat rejection to the environment.

6.7 Find the thermal efficiency of a steam power plant that generates steam at 1000 K in the boiler. The turbine work output is 580 kJ/s and pump work is 25 kJ/s. The heat removed in the condenser is 800 kJ/s at 320 K. What is the turbine power if the plant is running on a Carnot cycle assuming the same pump work and heat transfer to the boiler?

6.8 A power cycle operates between a lake's surface water at a temperature of 298 K and water at a depth whose temperature is 280 K. At steady state, the cycle develops a power output of 12 kW, while rejecting energy by heat transfer to the lower-temperature water at the rate of 200 kJ/s. Determine (i) the thermal efficiency of the power cycle and (ii) the maximum thermal efficiency for any such power cycle.

6.9 An inventor claims to have developed a refrigeration unit that maintains the cold space at −15°C while operating in a room at 25°C. The COP of the plant claimed is 7.8. How do you evaluate this?

6.10 A coal-burning steam power plant, operating with an overall thermal efficiency of 30%, produces a net power of 300 MW. The actual gravimetric air–fuel ratio in the furnace is calculated to be 13 kg air/kg fuel. The heating value of the coal is 30,000 kJ/kg. Find (i) the amount of coal consumed during a 24-hour period and (ii) the rate of air flowing through the furnace.

6.11 A heat engine is used to drive a heat pump, which extracts heat from the reservoir at 300°C at a rate 1.5 times that at which the engine rejects heat to it. The heat engine operates between two reservoirs at 700°C and 300°C. If the efficiency of the heat engine is 35% of ideal efficiency and the COP of the heat pump is 40% of ideal COP, determine the temperature of the reservoir to which the heat pump rejects the heat. What is the rate of heat rejection from the heat pump if the rate of heat supply to the engine is 65 kW.

6.12 To maintain the interior of a house at 22°C when the outside temperature is 5°C, 12 kW heat energy is required by burning a gas. Determine the power required if 12 kW heat flow was supplied by a reversible engine with the house as the high-temperature source and surroundings as the low-temperature sink so that the power was used to perform work needed to operate the engine.

6.13 Two reversible refrigeration cycles are operating in series, the first cycle receives energy by heat transfer from a cold reservoir at 298 K and rejects energy by heat transfer to a reservoir at an intermediate temperature T greater than 298 K. The second cycle receives energy by heat transfer from the reservoir at temperature T and rejects energy by heat transfer to a higher-temperature reservoir at 900 K. The two refrigeration cycles have the same COP. Evaluate (i) T, in K, and (ii) the value of each COP.

6.14 A reversible heat engine operates between source and sink temperatures of 550 and 250 K respectively receiving 125 kJ from the source. Estimate (i) the efficiency of reversible heat engine, (ii) the net work transfer, and (iii) heat rejected to the sink.

6.15 A heat pump receives energy from a source at 80°C and delivers it to a boiler that operates at 5 bar. The input to the boiler is saturated liquid water and the exit is saturated vapor, both at 5 bar. The heat pump is driven by a 2000 kW motor and operates with 50% of a Carnot COP. What is the maximum mass flow rate of water the system can deliver?

6.16 A heat engine has a solar collector receiving 0.2 kW/m^2 of energy, which is used to heat a transfer medium to 200°C. The collected energy of the solar collector powers a heat engine that rejects heat at 50°C. If the heat engine should deliver 2 kW, what is the minimum size (area) of the solar collector?

6.17 In a Carnot power cycle with a piston-cylinder assembly, water is the working fluid. During isothermal expansion, the water is heated from a saturated liquid at 40 bar until it is a saturated vapor. The vapor then expands adiabatically to a pressure of 2.5 bar while doing 350 kJ/kg of work. (i) Sketch the cycle on p-v coordinates, (ii) evaluate the heat transfer per unit mass and work per unit mass for each process, in kJ/kg, and (iii) evaluate the thermal efficiency.

6.18 A heat pump, operating with a COP of 3, supplies heat energy to a house in the winter season at a rate of 20 kJ/s. Estimate (i) the power required and (ii) the rate of heat absorption by the heat pump from the outside air.

6.19 The food compartment of a refrigerator is to be maintained at 6°C by heat removal from it at a rate of 5 kJ/s. The electrical power input required for the refrigerator is 2.5 kW. Determine the COP of the refrigerator and the rate of heat rejection to the room in which the refrigerator is located.

6.20 An inventor claims to have developed a resistance heater that supplies 1.2 kWh of energy to a room for each kWh of electricity it consumes. Is this a reasonable claim, or has the inventor developed a perpetual-motion machine? Explain.

6.21 The solar energy stored in solar ponds is being used to generate electricity. If such a solar power plant has an efficiency of 5% and a net power output of 325 kW, determine the average value of the required solar energy collection rate, in Btu/h.

6.22 The power generation involves the utilization of geothermal energy, the energy of hot water that exists naturally underground as the heat source. If a supply of hot water at 120°C is discovered at a location where the environmental temperature is 22°C, determine the maximum thermal efficiency a geothermal power plant built at that location can have.

6.23 Two reversible power cycles are arranged in series. The first cycle receives energy by heat transfer from a hot reservoir at temperature T_1 and rejects energy by heat transfer to a reservoir at an intermediate temperature $T < T_1$. The second cycle receives energy by heat transfer from the reservoir at temperature T and rejects energy by heat transfer to a cold reservoir at temperature $T_2 < T$. (i) Obtain an expression for the thermal efficiency of a single reversible power cycle operating between hot and cold reservoirs at T_1 and T_2, respectively, in terms of the thermal efficiencies of the two cycles. (ii) Obtain an expression for the intermediate temperature T in terms of T_1 and T_2 for the special case where the thermal efficiencies of the two cycles are equal.

6.24 Two Carnot heat engines are operating in series, with the first engine receiving heat from a hot reservoir at 1600 K and rejecting the waste heat to a cold reservoir at temperature T. The second engine receives this energy rejected by the first one, converts some of it into work, and rejects the rest to a reservoir at 300 K. Determine the temperature T, if the thermal efficiencies of both engines are the same.

DESIGN PROBLEMS

6.25 Design a humidifier with efficient refrigeration coils, compressors, and fans that has a humidification capacity of 220 ml/h with an effective area of 25 m^2 but uses about 25% less energy than conventional models.

6.26 Design a cost-effective dual-source heat pump (DSHP) for heating, cooling, and domestic hot water production. The DSHP should be capable of choosing the most favorable source/sink in such a way that it can work as an air-to-water heat pump using the air as a source/sink, or as a brine-to-water heat pump coupled to the ground. Compare this DSHP with a pure ground source heat pump (GSHP) system in terms of area requirements, cost, and efficiency.

7 Entropy

LEARNING OUTCOMES

After learning this chapter, students should be able to

- Interpret the Clausius inequality, which is the basis for the development of concepts such as entropy, entropy production, and entropy balance.
- Demonstrate an understanding of key concepts related to entropy and the second law including entropy transfer, entropy production, and the entropy principle.
- Evaluate entropy change during a process and analyze isentropic processes, using appropriate property tables.
- Apply entropy balances to closed systems and control volumes.
- Apply the entropy balance equation to develop the relations for entropy change of an ideal gas, pure substances, solids, and liquids.
- Evaluate isentropic efficiencies for turbines, nozzles, compressors, and pumps.

7.1 INEQUALITY OF CLAUSIUS

The first law of thermodynamics leads to the definition of a very useful property known as stored energy 'E', which is often used in everyday conversation. The second law leads to the definition of another useful property known as entropy 'S'. The corollaries of the second law introduced in the previous chapter are meant for systems undergoing cycles while exchanging thermal energy with one or two thermal energy reservoirs. In this section, another corollary of the second law called *Clausius inequality* is introduced. It is applicable to any cycle without regard for the body, or bodies, from which the cycle receives energy by heat transfer or to which the cycle rejects energy by heat transfer. The Clausius inequality forms the basis for the development of concepts such as entropy, entropy production, and entropy balance.

However, entropy, unlike energy, is not used in everyday conversation and hence it is an unfamiliar term. The second law of thermodynamics brings about certain inequality of expressions. The efficiency of an irreversible heat engine is less than that of a reversible one operating between the same temperature limits. The coefficient of performance of an irreversible refrigerator or heat pump is less than that of a reversible one. Another such inequality is Clausius inequality, which has influential implications in thermodynamics, which forms the basis for defining the property entropy. It is expressed as follows: the cyclic integral of $\frac{\delta Q}{T}$ is less than or equal to zero. It can be applicable for all cycles, reversible or irreversible. To prove the inequality of Clausius, let us consider a system connected to a thermal reservoir maintained at a temperature T_1 through a reversible cyclic device as shown in Figure 7.1.

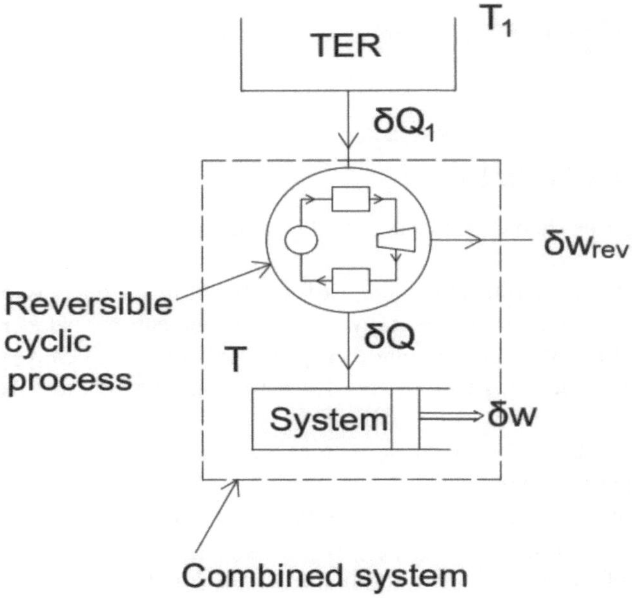

FIGURE 7.1 A system used for the proof of Clausius inequality.

The cyclic device receives heat δQ_1 from the thermal reservoir and produces work δW_{rev} while supplying heat δQ to the system maintained at temperature T. The δW_{sys} is the work produced by the system, which receives heat δQ. The energy balance according to the first law yields

$$\delta Q_1 = \delta W + dE$$

on rearranging,

$$\delta W = \delta Q_1 - dE \tag{7.1}$$

where δW is the total work $(\delta W_{rev} + \delta W_{sys})$ of the combined system, and dE is the change in the total energy of the system.

Since the cyclic device is a reversible cycle,

$$\frac{\delta Q_1}{T_1} = \frac{\delta Q}{T} \tag{7.2}$$

For a reversible cycle, $\frac{Q_1}{T_1} = \frac{Q_2}{T_2}$ (as proved in Section 6.6, Eq. 6.15)

From 7.1 and 7.2,

$$\delta W = T_1 \frac{\delta Q}{T} - dE \tag{7.3}$$

Let the system undergo a cycle, while the cyclic device undergoes an integral number of cycles. Then the above equation becomes

$$W = T_1 \oint \frac{\delta Q}{T} - dE \tag{7.4}$$

$W = T_1 \oint \frac{\delta Q}{T}$, since the cyclic integral of any property (energy) is zero. W is the cyclic integral of δW and is the net work of the combined cycle.

From the above discussion, it is concluded that the combined system is producing work by exchanging heat with a single reservoir. It is a violation of Kelvin Planck's statement of the second law of thermodynamics, so W cannot be a work output and hence cannot be a positive quantity. It is considered that T_1 is the thermodynamic temperature and hence a positive quantity, therefore $\oint \frac{\delta Q}{T} \leq 0$ is the Clausius inequality.

If $\oint \frac{\delta Q}{T} > 0$, work can be a positive quantity and it is not possible. This inequality is valid for all cycles reversible or irreversible.

if $\oint \frac{\delta Q}{T} < 0$ the cycle is irreversible,

if $\oint \frac{\delta Q}{T} = 0$ the cycle is reversible,

and if $\oint \frac{\delta Q}{T} > 0$ the cycle is impossible.

7.2 ENTROPY—A PROPERTY OF A SYSTEM

The second law of thermodynamics introduces a property of a system known as entropy based on the equation $\oint \frac{\delta Q}{T} \leq 0$. Let a system undergo a cycle, as shown in Figure 7.2, following a reversible process from state 1 to state 2 along a path P, and return to the initial state along path Q, which is also reversible.

$$\oint \frac{\delta Q}{T} = 0 = \int_1^2 \left(\frac{\delta Q}{T}\right)_P + \int_2^1 \left(\frac{\delta Q}{T}\right)_Q \tag{7.5}$$

Now consider another reversible cycle, which proceeds first along path R and returns to the initial state along path Q. For this cycle,

$$\oint \frac{\delta Q}{T} = 0 = \int_1^2 \left(\frac{\delta Q}{T}\right)_R + \int_2^1 \left(\frac{\delta Q}{T}\right)_Q \tag{7.6}$$

If we subtract the second equation from the first,

$$\int_1^2 \left(\frac{\delta Q}{T}\right)_P = \int_1^2 \left(\frac{\delta Q}{T}\right)_R \tag{7.7}$$

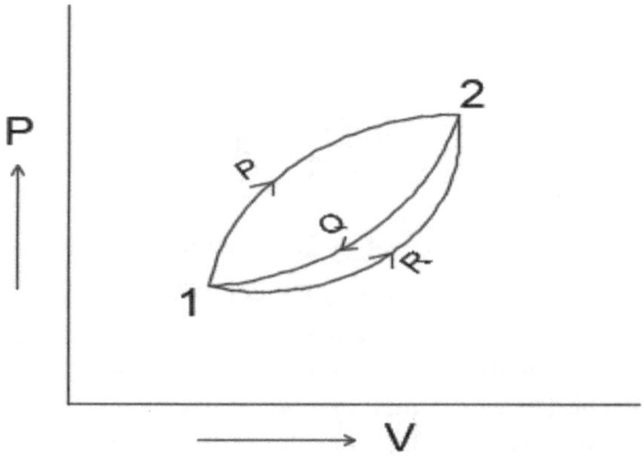

FIGURE 7.2 Two reversible cycles for demonstrating that entropy is a property of a system.

For all reversible paths between states 1 and 2, $\oint \frac{\delta Q}{T}$ is the same, thus it can be concluded that the quantity $\oint \frac{\delta Q}{T}$ is independent of the path the system follows in a change of state and is a function of initial and final states only. Therefore, it is a property and it is called entropy and is designated S. It follows that entropy may be defined as a property of a substance in accordance with the relation

$$dS = \left(\frac{\delta Q}{T}\right)_{rev} \tag{7.8}$$

Entropy is an extensive property, and the entropy per unit mass is designated s. It is important to note that entropy is defined in terms of a reversible process.

The change in the entropy of a system as it undergoes a change of state may be found by integrating Eq. 7.8. Thus,

$$S_2 - S_1 = \int_1^2 \left(\frac{\delta Q}{T}\right)_{rev} \tag{7.9}$$

Isothermal Heat Transfer Process

Isothermal heat transfer processes are internally reversible; the change in the entropy of a system during a reversible isothermal heat transfer process can be found by integrating Eq. 7.9.

$$\Delta S = \int_1^2 \left(\frac{\delta Q}{T}\right)_{rev} = \int_1^2 \left(\frac{\delta Q}{T_0}\right)_{rev} = \frac{1}{T_0}\int_1^2 (\delta Q)$$

where T_0 is the constant temperature of the system.

The above equation can become

$$\Delta S = \frac{Q}{T_0} (\text{kJ/K}) \tag{7.10}$$

Equation 7.10 is useful for determining the change in the entropy of thermal energy reservoirs, used in steam power plants and combustion-related applications, which supply or absorb an infinite amount of heat at a constant temperature.

Integration can be performed if we know the relation between T and Q. The change in the entropy of a substance in a change of state is the same for all processes, both reversible and irreversible between these two states, since entropy is a property. It is to be noted that Eq. 7.9 allows us to find the change in entropy only along a reversible path and this value is the magnitude of the entropy change for all processes between these two states. However, the absolute values of entropy cannot be evaluated with this expression. From the third law of thermodynamics, which is based on observations of low-temperature chemical reactions, it is concluded that the entropy of all pure substances (in the appropriate structural form) can be assigned the absolute value of zero at the absolute zero of temperature. It also follows from the subject of statistical thermodynamics that all pure substances in the (hypothetical) ideal-gas state at absolute zero temperature have zero entropy.

However, when the composition doesn't change, as the one which occurs in a chemical reaction, it is quite adequate to give values of entropy relative to some arbitrarily selected reference state. In each case, whatever reference value is chosen, it will cancel out when the change of property is calculated between any two states. Q is a path function, and therefore δQ is an inexact differential.

However, since $(\delta Q/T)_{rev}$ is a thermodynamic property, it is an exact differential. From a mathematical perspective, an integrating factor is required to convert an inexact differential to an exact differential. 1/T serves as the integrating factor in converting the inexact differential δQ to the exact differential δQ/T for a reversible process.

The infinitesimal change in entropy dS due to reversible heat transfer δQ at temperature T is given as

$$dS = \left(\frac{\delta Q}{T}\right)_{rev}$$

If the process is reversible and adiabatic, then $\delta Q_{rev} = 0$, $dS = 0$, and $S = \text{constant}$; thus, the reversible and adiabatic process is isentropic.

If the system is taken from i to f reversibly as shown in Figure 7.3,

$$\delta Q_{rev} = TdS \text{ then } Q_{rev} = \int_i^f TdS$$

Thus the area under the curve $\int_i^f TdS$ is equal to the heat transferred in the process.

For the reversible isothermal heat transfer process shown in Figure 7.4 ($T = \text{constant}$),

$$Q_{rev} = T\int_i^f dS = T(S_f - S_i)$$

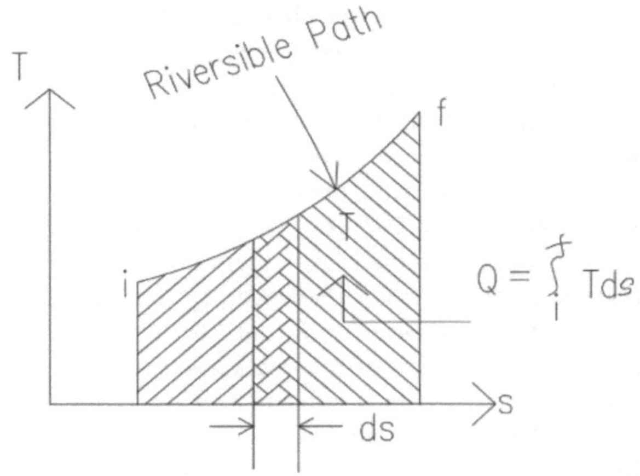

FIGURE 7.3 Area under a reversible path on the T-s diagram representing heat transfer.

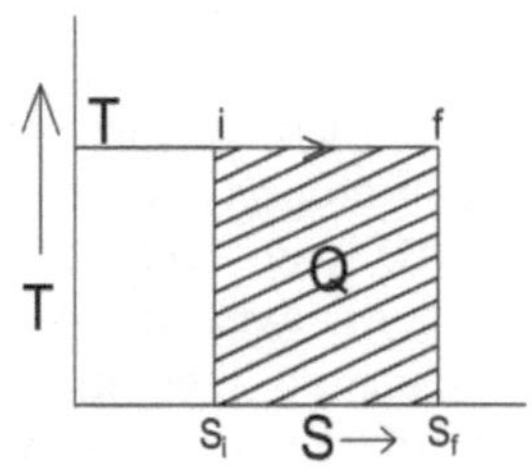

FIGURE 7.4 Reversible isothermal heat transfer.

7.3 PRINCIPLE OF ENTROPY

The property entropy provides a means to find whether a process is possible, and if so, we have to find what kind of process it is, whether reversible or irreversible. This is done with the principle of increase of entropy or entropy principle stated as *the entropy of an isolated system will never decrease*. For a cyclic process as shown in Figure 7.5, which involves two processes, one is an arbitrary (reversible or irreversible, 1-2) and the other is an internally reversible process (2-1), from Clausius inequality

$$\oint \frac{\delta Q}{T} \leq 0 \rightarrow \int_1^2 \frac{\delta Q}{T} + \int_2^1 \left(\frac{\delta Q}{T}\right)_{\text{int.rev}} \leq 0$$

$$s_1 - s_2 + \int_1^2 \frac{\delta Q}{T} \leq 0$$

$$\therefore \int_1^2 \frac{\delta Q}{T} \leq s_2 - s_1 \text{ or } s_2 - s_1 \geq \int_1^2 \frac{\delta Q}{T}$$

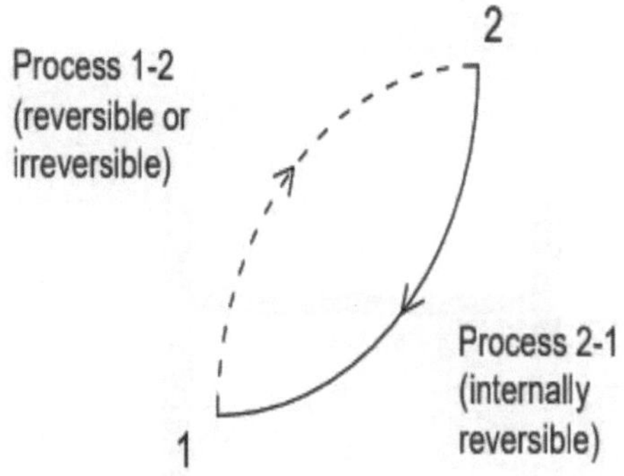

FIGURE 7.5 A cyclic process with a reversible and an irreversible process.

In differential form, $ds \geq \frac{\delta Q}{T}$

$\therefore ds = \frac{\delta Q}{T}$ for a reversible process and

$ds \geq \frac{\delta Q}{T}$ for the irreversible process

$\delta Q \rightarrow$ differential heat transfer between system and surroundings and T is the thermodynamic temperature at the boundary.

For a reversible process, the entropy change Δs or $(s_2 - s_1)$ is equal to entropy transfer (with heat).

For an irreversible process $\left(ds > \frac{\delta Q}{T}\right)$, entropy change is always greater than entropy transfer. This is due to the entropy generation within the system due to irreversibilities. It is indicated by s_{gen}.

$\therefore$ For an irreversible process, $ds = \int_1^2 \frac{\delta Q}{T} + s_{gen}$

Entropy generation is a positive quantity.

In the absence of entropy transfer, $ds = s_{gen}$

in the equation, $ds \geq \int_1^2 \frac{\delta Q}{T}, \frac{\delta Q}{T} = 0$ for an adiabatic closed process.

$$\therefore (ds)_{iso} \geq 0, (\Delta s)_{iso} \geq 0$$

Therefore, the entropy of an isolated system during a process does not decrease, but it increases (for a reversible process, it remains constant). This is known as the principle of entropy increase or the principle of entropy.

Figure 7.6 shows an isolated system formed by a system and surroundings which interact within a single boundary. Therefore, the system and its surroundings are considered as the subsystems of an isolated system. The combination of system, surroundings, and everything that is affected by the process is sometimes called the *universe.* The term universe refers to everything that is involved in a process

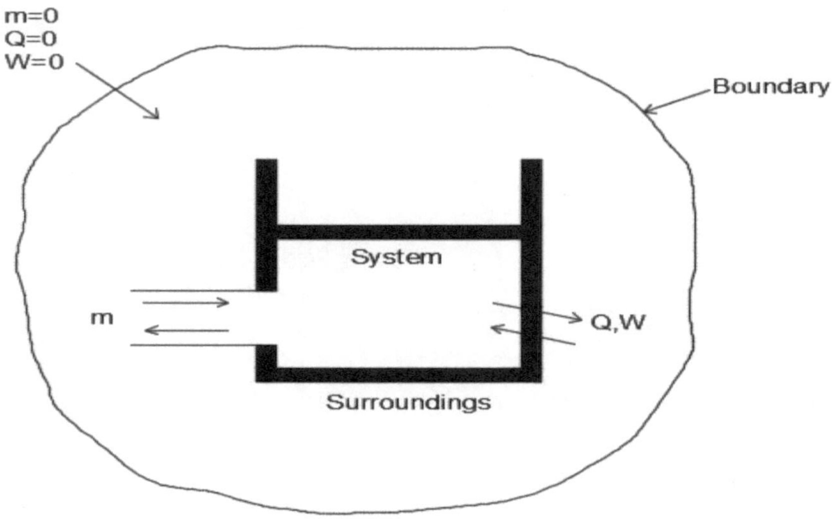

FIGURE 7.6 An isolated system with the system and its surroundings.

and doesn't consider the things that have no effect on the process. The principle of entropy increase is stated as

$$\Delta S_{universe} \geq 0 \tag{7.11}$$

where $\Delta S_{universe} = \Delta S_{sys} + \Delta S_{surr}$

For an isolated system, the entropy change is equal to entropy generation, since no entropy transfer takes place either into or out of the isolated system.

$$(\Delta S)_{Total} = S_{gen} = (\Delta S)_{system} + (\Delta S)_{surr} \geq 0 \tag{7.12}$$

where $S_{gen} = 0$ for a reversible process and $S_{gen} > 0$ for an irreversible process. Hence, it is concluded that the entropy of the universe is always increasing, since all the processes in nature are irreversible. The entropy generation increases with the irreversibility of the process. The entropy of an isolated system increases continuously. For a process, entropy change of a system may increase, decrease, or remain constant, but entropy generation cannot be. The entropy principle is useful for determining whether a process is reversible, irreversible, or impossible.

If $S_{gen} > 0 \rightarrow$ the process is irreversible,
if $S_{gen} = 0 \rightarrow$ the process is reversible, and
if $S_{gen} < 0 \rightarrow$ the process is impossible.

The entropy of an isolated system increases until it reaches a maximum value at the state of equilibrium. According to the principle of entropy, there is no change in the entropy of a system further after the equilibrium state is attained.

7.4 THE CONCEPT OF ENTROPY

The property entropy is introduced by the second law of thermodynamics and is useful in the second law analysis of engineering systems. It is considered as a measure of molecular disorder or randomness. In a disordered system, molecules move randomly in a disordered manner, and entropy increases. When compared to solids, gas molecules move randomly and their position is unpredictable at any instant. They frequently collide and change the direction and hence, entropy of a substance in the gaseous phase is high when compared to that in the solid phase. In the solid phase, the molecules of a substance oscillate continuously about their position. With the decrease in temperature, these oscillations (fade) reduce. At absolute zero, these molecules become motionless and this state is characterized by a complete molecular order.

Finally, we must note that the above formulation enables us to calculate only the changes in entropy. It does not give us a way to obtain the absolute value of entropy. In this formalism, entropy can be known only up to an additive constant. However, in 1906, Walther Nernst (1864–1941) formulated a law that stated that the entropy of all systems approaches zero as the temperature approaches zero The third Law enables us to give the absolute value for the entropy. Walther Nernst, in 1906, formulated the *third law of thermodynamics* also known as the *Nernst heat theorem*, which is stated as follows: the entropy of all systems approaches zero as the temperature approaches zero (0 K) as there is no molecular motion at this state.

$$S \to 0 \text{ as } T \to 0$$

This law is useful in thermodynamic analysis of chemical reactions, since it is possible to determine absolute entropy at a reference point provided by the third law of thermodynamics. The quantum theory demonstrates the behavior of matter at low temperatures, which forms the basis of defining the absolute values of entropy as the theory of relativity does with the absolute values of energy.

7.5 THE TdS EQUATIONS

There are two important thermodynamic property relations for a simple compressible substance.

To derive the first relation, let us consider the first law equation for a simple compressible system,

$$\delta Q = dU + \delta W \tag{7.13}$$

The heat transfer for a reversible process (according to the second law) $\delta Q = TdS$ and the first law equation for a closed non-flow system, which involves pdV work only, is $\delta W = pdV$

Substituting $\delta Q = TdS$ and $\delta W = pdV$ into Eq. 7.13

$$TdS = dU + pdV \text{ (kJ)} \tag{7.14}$$

$$\mathrm{Tds = du + pdv\,(kJ/kg)(on\,a\,unit\,mass\,basis)} \tag{7.15}$$

This is the first TdS equation or Gibbs equation.

The second equation can be derived from the relation, enthalpy $H = U + PV$.

Differentiating above $dH = dU + pdV + Vdp$

Substituting $TdS = dU + pdV$

$$dH = TdS + VdP \Rightarrow TdS = dH - Vdp$$

On a unit mass basis,

$$Tds = dh - vdp \tag{7.16}$$

This is the second TdS equation or Gibbs equation.

Equation 7.16 holds good for any process reversible or irreversible, undergone by a closed system involving pdV work only.

The equation $TdS = dH - Vdp$ also holds good for any process when pdV work is present.

7.6 ENTROPY CHANGE OF PURE SUBSTANCES

Entropy of a Pure Substance

The values of specific entropy, s, entropy per unit mass, are mentioned in thermodynamic property tables along with other properties such as specific enthalpy and specific internal energy. These values are given with respect to an arbitrary reference state. In steam tables, the entropy of saturated liquid at 0.01°C is assigned the value of zero, while in refrigeration tables, the entropy of saturated liquid is assigned zero

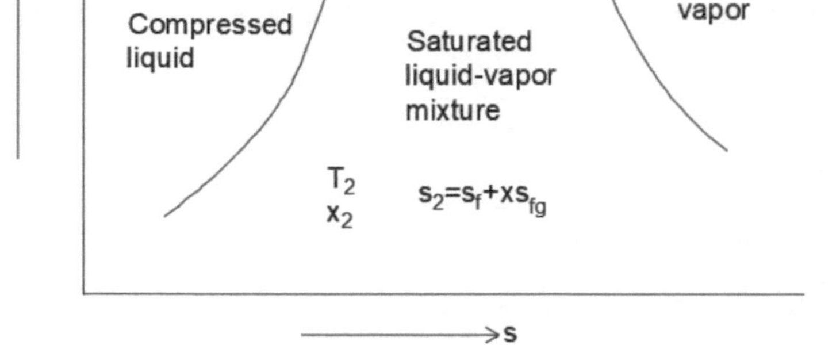

FIGURE 7.7 Determination of entropy of a pure substance.

at −40°C. Figure 7.7 shows how the entropy of a pure substance is evaluated from measurable property data.

In the saturation region, the relation for calculating the entropy is

$$s = (1 - x)s_f + xs_{fg}$$

$$\text{or } s = s_f + xs_{fg}, \text{where x is quality or dryness fraction}$$

(s_f and s_{fg} values are available in the tables)

Entropy is a property, it has a definite value at each state of a system.

The entropy change of a process is

$$\Delta S = m\Delta s = m(s_2 - s_1) \tag{7.17}$$

where m is the mass of the substance. The entropy change is the difference between entropy at final and initial states.

7.7 ENTROPY CHANGE OF AN IDEAL GAS

The relation for entropy change of an ideal gas can be developed by using thermodynamic property relations for ideal gases.

Differential changes in entropy can be obtained by solving Eqs. 7.15 and 7.16.

$$ds = \frac{du}{T} + \frac{Pdv}{T} \tag{7.18}$$

$$ds = \frac{dh}{T} - \frac{vdP}{T} \tag{7.19}$$

For ideal gas, $du = c_v dt$ and $pv = RT$, substituting these into Eq. 7.19

$$ds = \frac{c_v dt}{T} + R\frac{dv}{v} \tag{7.20}$$

The entropy change for a process is obtained by integrating Eq. 7.20 between initial and final states,

$$s_2 - s_1 = \int_1^2 c_v(T)\frac{dT}{T} + R\ln\frac{V_2}{V_1} \tag{7.21}$$

The other entropy change relation is obtained similarly by substituting $dh = cp\ dT$ and $pv = RT$ in Eq. 7.19

$$ds = \frac{c_p dT}{T} - R\frac{dP}{P} \tag{7.22}$$

Integrating Eq. 7.22, we get

$$s_2 - s_1 = \int_1^2 c_p(T)\frac{dT}{T} - R\ln\frac{P_2}{P_1} \tag{7.23}$$

However, the integration of Eqs. 7.21 and 7.23 cannot be performed unless the dependence of c_p and c_v on temperature is known, since the specific heats of ideal gases depend on temperature. Besides, it is not practical, calculating entropy change by performing integrations though $c_v(T)$ and $c_p(T)$ functions are available. In this case, there appear two possible choices, one is to assume constant specific heats and the other to evaluate integrals and tabulate results.

Constant Specific Heats: For ideal gases, specific heats vary nearly linearly with temperature and the assumption of constant specific heats is an approximation with some error. This error can be minimized by considering specific heat values evaluated at an average temperature.

With the above assumptions, the entropy change relations for ideal gases can be obtained by using $c_{v,avg}$ and $c_{p,avg}$ in place of $c_v(T)$ and $c_p(T)$ in Eqs. 7.21 and 7.23 respectively and integrating

$$s_2 - s_1 = c_{v,avg}\ln\frac{T_2}{T_1} + R\ln\frac{v_2}{v_1} \tag{7.24}$$

$$s_2 - s_1 = c_{p,avg}\ln\frac{T_2}{T_1} - R\ln\frac{P_2}{P_1} \tag{7.25}$$

Variable Specific Heats: The assumption of constant specific heats may lead to appreciable errors in entropy change calculations when the temperature change during a process is large and specific heats of ideal gas vary nonlinearly with temperature. In such cases, the specific heat variation with temperature is accounted for by using $c_v(T)$ and $c_p(T)$ relations. The entropy change during a process can then be obtained by substituting $c_v(T)$ and $c_p(T)$ relations in Eqs. 7.21 and 7.23 and integrating them.

However, performing the integrals each time for a new process is a laborious task and it can be eliminated by performing integrals once and tabulating the results. This can be done by choosing absolute zero as the reference temperature and defining a function s^0, which is a function of temperature only

$$s^0 = \int_0^T c_p(T)\frac{dT}{T} \tag{7.26}$$

The value of s^0 is zero at absolute zero temperature and its values at other temperatures are evaluated. The integral in Eq. 7.26 becomes

$$\int_1^2 c_p(T)\frac{dT}{T} = s_2^0 - s_1^0 \tag{7.27}$$

where s_1^0 and s_2^0 are the values of s^0 at T_1 and T_2 respectively. Therefore,

$$s_2 - s_1 = s_2^0 - s_1^0 - R\ln\frac{P_2}{P_1} \tag{7.28}$$

7.8 ENTROPY CHANGE OF SOLIDS AND LIQUIDS

Solids and liquids are almost incompressible substances as the specific volumes of both solids and liquids during a process are constant. Entropy change for a solid or liquid can be derived by using the thermodynamic property relation, Eq. 7.18. When specific volume is neglected, it becomes

$$ds \cong \frac{du}{T} \cong \frac{c}{T}dT \quad \left(\text{since } c_p = c_v = c \text{ for incompressible substances}\right) \tag{7.29}$$

Integration of Eq. 7.29 gives

$$s_2 - s_1 = c\ln\frac{T_2}{T_1} \quad \left(\text{specific heat is assumed as constant}\right) \tag{7.30}$$

In a reversible and adiabatic process (isentropic), the approximation of constant specific volume leads to a constant temperature, which demonstrates that the pumping of a liquid does not change the temperature.

The relation for the isentropic process of solids and liquids is obtained by setting the entropy change relation equal to zero.

$$s_2 - s_1 = c\ln\frac{T_2}{T_1} = 0 \rightarrow T_2 = T_1 \tag{7.31}$$

In an isentropic process, the temperature of an incompressible substance remains constant, and hence, the isentropic process of an incompressible substance is isothermal.

7.9 ENTROPY BALANCE

As per the second law of thermodynamics, entropy can be created but it can't be destroyed. The entropy balance equation is expressed in terms of entropy transfer with heat and mass, entropy generation, and change in entropy, that is,

$$\underbrace{S_{in} - S_{out}}_{\substack{\text{Entropy transfer with}\\ \text{heat and mass}}} + \underbrace{S_{gen}}_{\text{Entropy generation}} = \underbrace{(\Delta S)_{sys}}_{\text{Change in entropy}} \tag{7.32}$$

Eq. 7.32 can also be expressed as

$$\underbrace{\int_1^2 \frac{\delta Q}{T}}_{\text{Entropy transfer}} \le \underbrace{S_2 - S_1}_{\text{Entropy change}} \tag{7.33}$$

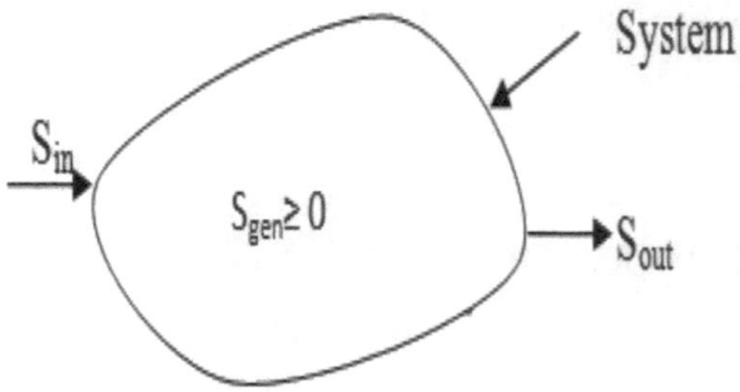

FIGURE 7.8 Entropy balance.

The second law of thermodynamics for a process, Eq. 7.33, states that algebraically the entropy transfer never exceeds the entropy change of a system during a process.

Figure 7.8 shows the entropy balance, which is stated as follows: the entropy change $(\Delta S)_{sys}$ of a system during a process is equal to net entropy transfer $(S_{in} - S_{out})$ through the boundaries of the system plus the entropy generation (S_{gen}) within the system.

It can be expressed as

$$dS = \int_1^2 \frac{\delta Q}{T} + S_{gen} \tag{7.34}$$

Entropy generation is

$$S_{gen} = S_2 - S_1 - \int_1^2 \frac{\delta Q}{T} \geq 0 \tag{7.35}$$

Therefore, to express the entropy balance equation for any process, three terms are taken into consideration: (i) entropy change, (ii) entropy transfer, and (iii) entropy generation.

7.9.1 Entropy Change of a System

The entropy change of a system during a process is determined by calculating the difference between the entropy at the beginning and at the end of the process.

$$(\Delta S)_{system} = S_2 - S_1 \tag{7.36}$$

where $(\Delta S)_{system}$ is the change of entropy of a system and S_1 and S_2 are the entropies at the initial and final states respectively. Entropy, because it is a property, does not change for a process unless the system undergoes a change of state. Therefore, the entropy change of steady-flow devices is zero during steady-flow operation. Some of the examples of steady-flow devices are turbines, pumps, and compressors.

7.9.2 Entropy Transfer by Heat and Mass Transfer

Entropy transfer does take place either into or out of a system by two means, one is by heat transfer and the other by mass transfer. Entropy transfer does not take place via work transfer. The entropy transfer of a closed system is due to heat transfer only.

Heat Transfer

Heat is a disorganized form of energy, transfer of which increases randomness (a disorder of the system). When heat is transferred to a system, entropy of the system increases, and when heat is removed, it decreases. For example, when a gas is heated, its molecules will move in a disorganized manner as the molecular interaction increases with an increase in temperature (molecules collide frequently).

The entropy transfer due to heat transfer is the heat transfer at a location divided by the absolute temperature at that location, that is $S = \frac{Q}{T}$ where S is the entropy transfer due to heat transfer, Q is the heat transfer, and T is the absolute temperature (T = constant).

$\frac{Q}{T}$ is positive when heat is added and negative when heat is rejected. When the temperature is changing, entropy transfer is determined by integrating

$$S = \int_{T_1}^{T_2} \frac{\delta Q}{T} \tag{7.37}$$

where T_1 and T_2 are the temperatures at initial and final states of a process.

7.9.3 Entropy Generation—Closed System and Control Volume

Unlike energy that cannot be created, entropy is generated within the system due to the irreversibilities such as friction, inelasticity, heat transfer through a finite temperature difference, mixing, and so on. Entropy generated is quantified by the entropy created by the above effects. A reversible process doesn't involve any irreversibilities; therefore, entropy generation is zero and hence entropy change is equal to the entropy transfer.

For a reversible process, the entropy balance equation is

$$S_{in} - S_{out} = (\Delta S)_{sys} \tag{7.38}$$

and for an irreversible process, the entropy balance equation is

$$S_{in} - S_{out} + S_{gen} = (\Delta S)_{sys} \tag{7.39}$$

Entropy Generation in a Closed system

A closed system is a system of fixed mass that involves no mass flow either into or out of the system. Therefore, entropy transfer is due to heat transfer only. The entropy

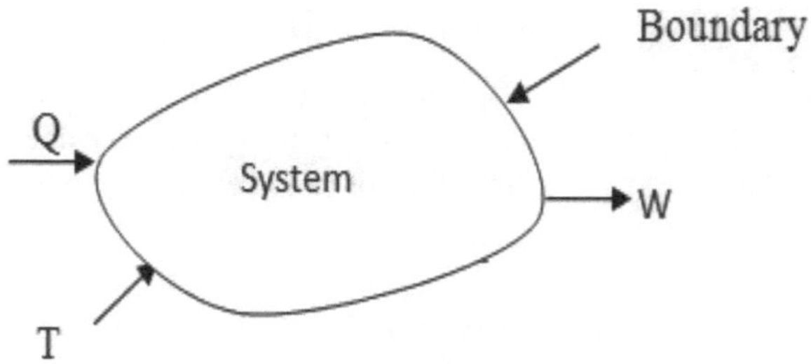

FIGURE 7.9 A closed system interacting with its surroundings.

change is the sum of entropy transfer with the heat transfer and entropy generation within the system boundaries. Figure 7.9 shows a closed system interacting with its surroundings.

$$(\Delta S)_{sys} = S_2 - S_1 = \frac{Q}{T} + S_{gen} \tag{7.40}$$

If the system involves more heat transfer quantities, then

$$(\Delta S)_{sys} = \sum \frac{Q_n}{T_n} + S_{gen} \quad \left(\text{Where n is the number of heat transfer processes}\right) \tag{7.41}$$

According to the second law $dS \geq \int_1^2 \frac{\delta Q}{T}$.

From the above equation, entropy change of the system exceeds entropy transfer, the difference is equal to entropy generation.

$$dS - \int_1^2 \frac{\delta Q}{T} = S_{gen} \tag{7.42}$$

$$S_{gen} \geq 0$$

(As the heat transfer to the system is taken to be positive).

Therefore, any thermodynamic process is accompanied by the entropy generation.

Entropy Generation in an Open System

An open system, as shown in Figure 7.10, involves the transfer of mass and energy (heat and work). The entropy transfer for an open system (control volume) takes place because of heat and mass transfers. Therefore, the entropy balance equation of a control volume differs from that of a closed system in that it consists of an additional quantity of mass flow.

$$\sum \frac{Q_n}{T_n} + \sum m_i s_i - \sum m_e s_e + s_{gen} = s_2 - s_1 \tag{7.43}$$

or in the rate form

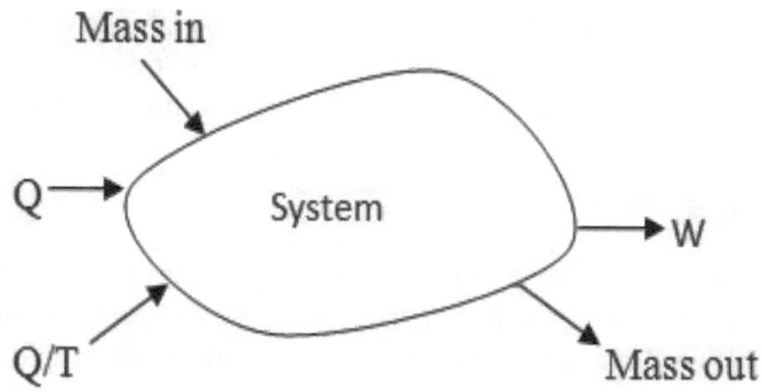

FIGURE 7.10 An open system interacting with its surroundings.

$$\sum \frac{Q_n}{T_n} + \sum m_i s_i - \sum m_e s_e + s_{gen} = \frac{ds}{dt} \tag{7.44}$$

The above equations are entropy balance equations for a control volume involving energy and mass interactions. The entropy change of a control volume (open system) during a process is the sum of the rate of entropy transfer due to heat transfer through the system, net entropy transfer due to mass flow, and the rate of entropy generation within the boundary of the system due to irreversibilities. For steady-flow devices such as turbines, nozzles, compressors, and pumps (control volumes), the rate of change of entropy is zero as they operate steadily and hence

$$s_{gen} = \sum m_e s_e - \sum m_i s_i - \sum \frac{Q_n}{T_n} \tag{7.45}$$

$$s_{gen} = m(s_e - s_i) - \sum \frac{Q_n}{T_n} \tag{7.46}$$

For the adiabatic process, $\left(\text{since } \frac{Q}{T} = 0\right) \quad s_{gen} = m(s_e - s_i)$

If the process is reversible and adiabatic, then entropy remains constant, since $s_{gen} = 0$.

7.10 ISENTROPIC PROCESS

The isentropic process is a process in which entropy is constant. It has been observed that the entropy of a closed system can be varied by heat transfer and irreversibilities. If the process is reversible and adiabatic, the entropy remains constant and it is called an isentropic process. The entropy of a substance will be the same at the end of the process and the beginning, if the process is carried out isentropically. The term efficiency is frequently used by engineers in many engineering applications. Generally, it is defined as the ratio of output to the input. In this section, the isentropic efficiencies for steady-flow devices such as nozzles, turbines, and compressors are presented.

Isentropic efficiency is derived by comparing the actual performance of a device and the performance under idealized conditions for the same inlet and the exit conditions. It is defined differently for work-producing and work-consuming devices.

7.11 ISENTROPIC EFFICIENCY

The performance of an actual process is highest when it approaches that of an ideal process. Isentropic efficiency is a comparison of the actual performance of a device to that obtained under idealized conditions when the inlet state and exit pressure are the same. In the following sections, the isentropic efficiency of turbines, compressors, and nozzles are presented.

7.11.1 Isentropic Efficiency of a Turbine

The isentropic efficiency of a turbine is derived from the assumptions that there is negligible heat transfer between the turbine and its surroundings and also are kinetic and potential energy effects. With the above assumptions, the mass and energy balance equations reduce, at a steady state, to give the work developed per unit of mass flowing through the turbine. The state of the matter entering the turbine and the exit pressure are fixed. The ideal process for the turbine is the isentropic process between the inlet state and exit pressure. Figure 7.11 shows the actual and isentropic processes of an adiabatic turbine.

$$\eta_T = \frac{\text{Actual work output of the turbine}}{\text{Isentropic work output of the turbine}} = \frac{W_a}{W_i} \quad (7.47)$$

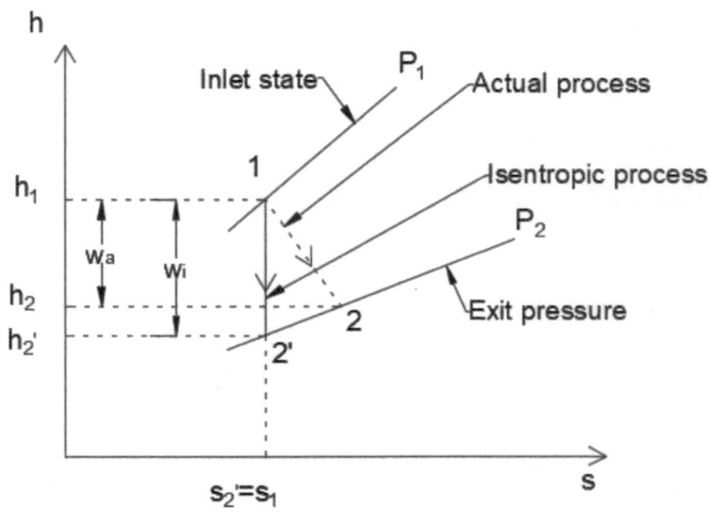

FIGURE 7.11 The h-s diagram for an adiabatic turbine.

where W_a is the actual work output and W_i is the isentropic work output. The value of the work depends on the specific enthalpy h_2 only, since specific enthalpy h_1 is fixed. Work increases as h_2 is reduced. The maximum value for the turbine work corresponds to the smallest allowed value for the specific enthalpy at the turbine exit.

$$\eta_T = \frac{h_1 - h_2}{h_1 - h_2^1} \tag{7.48}$$

7.11.2 Isentropic Efficiency of a Compressor and a Pump

The isentropic efficiency of a compressor can be defined as the ratio of the work input required to raise the pressure of a gas under isentropic conditions to the actual work input.

$$\eta_C = \frac{\text{Isentropic work input of compressor}}{\text{Actual work input of compressor}} = \frac{W_i}{W_a} \tag{7.49}$$

It is to be noted that W_i is smaller than W_a, and hence the isentropic compressor efficiency is always indicated with the isentropic work input in the numerator instead of in the denominator. This definition implies that η_c cannot be greater than 100%, which would falsely imply that the actual compressors performed better than the isentropic ones. Also notice that the inlet conditions and the exit pressure of the gas are the same for both the actual and the isentropic compressors. When the kinetic and potential energy changes of the gas being compressed are negligible, then the work input to an adiabatic compressor becomes equal to the change in enthalpy, and the compressor efficiency becomes

$$\eta_C = \frac{h_2^1 - h_1}{h_2 - h_1} \tag{7.50}$$

where h_2 and h_2^1 are the enthalpy values at the exit state for actual and isentropic compression processes, respectively, as shown in Figure 7.12.

The value of compressor efficiency greatly depends on the design of the compressor, and with better designs, it is possible to have isentropic efficiencies in the range of 75%–85%. When the gas is not cooled after its compression, then the actual compression process approaches the adiabatic process. To reduce the work input requirements, the compressors are cooled deliberately with the help of fins or a water jacket placed around the casing.

We can derive the isothermal efficiency of a pump in a similar way as that of a compressor. The pump usually handles incompressible fluid (liquid water) whose density or specific volume undergoes a negligible change when pressure increases. For an isentropic compression, using the thermodynamic property relation

$$Tds = dh - vdp \text{ and } ds = 0$$

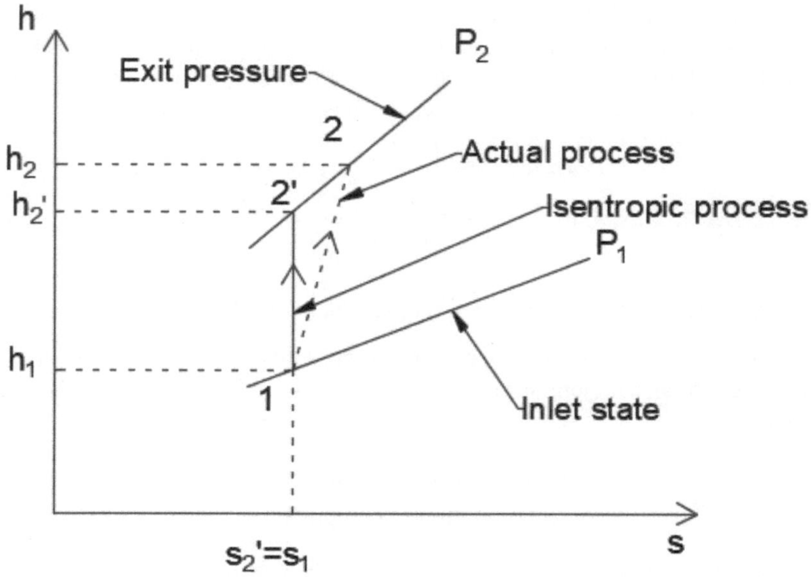

FIGURE 7.12 The h-s diagram for an adiabatic compressor.

Then $dh = \nu dp$ or $\Delta h = \nu \Delta p$ (since change in specific volume is negligible). Then the isothermal efficiency of a pump becomes

$$\eta_P = \frac{W_i}{W_a} = \frac{v(p_2 - p_1)}{h_2 - h_1} \tag{7.51}$$

where ν is the specific volume and p_1 and p_2 are the pressures before and after the compression respectively.

7.11.3 Isentropic Efficiency of a Nozzle

Nozzles are used to accelerate a fluid and they are adiabatic devices. Therefore, the isentropic process serves as a suitable model for nozzles. The isentropic efficiency of a nozzle can be defined as the ratio of the actual kinetic energy of the fluid at the nozzle exit to the kinetic energy at the exit of a nozzle under isentropic conditions for the same inlet state and exit pressure. That is,

$$\eta_N = \frac{\text{Actual kinetic energy at nozzle exit}}{\text{Isentropic kinetic energy at nozzle exit}} = \frac{V_2^2}{V_{2^1}^2} \tag{7.52}$$

Though the exit pressure is the same for both the actual and isentropic processes, the exit state is different. Usually, nozzles involve no work interactions, and the change in potential energy of the fluid is almost negligible as it flows through the device. If the inlet velocity of the fluid is small when compared to the exit velocity, then the energy balance relation for this steady-flow device reduces to

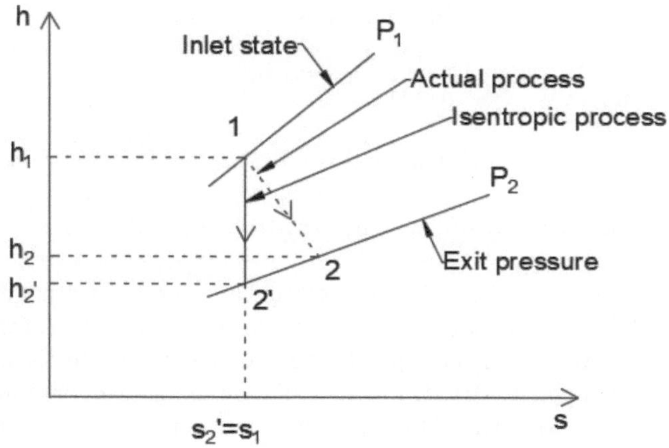

FIGURE 7.13 The h-s diagram for an adiabatic nozzle.

$$h_1 = h_2 + \frac{V_2^2}{2}$$

Then the isentropic efficiency of the nozzle can be expressed in terms of enthalpies as

$$\eta_N = \frac{h_1 - h_2}{h_1 - h_{2^1}} \qquad (7.53)$$

where h_2 and h_2^1 are the enthalpy values at the nozzle exit for the actual and isentropic processes, respectively. Figure 7.13 shows the isentropic nozzle.

Isentropic efficiencies of nozzles are typically above 90%, and nozzle efficiencies above 95% are not rare.

EXAMPLE PROBLEMS

Example 7.1 Determine the increase in entropy of water due to adiabatic mixing of 1.5 kg at 75°C with 4 kg at 40°C in a constant pressure process of 1 atm (take $c_{pw} = 4.2$ kJ/kg K).

Solution According to the first law of thermodynamics, total energy before mixing = total energy after mixing.

$1.5 \times c_{pw} (75 - 0) + 4 \times c_{pw} (40 - 0) = 5.5 \times c_{pw} (t_f - 0)$

$= 112.5\ c_{pw} + 160\ c_{pw} = 5.5\ c_{pw}\ t_m$

$t_m = 49.5$°C (t_f is the final equilibrium temperature of water)

The entropy change ΔS can be calculated by the equation

$$\Delta S = S_2 - S_1 = \int_{T_1}^{T_2} mc_p \frac{dT}{T}$$

Initial entropy of the system, S_1 (before mixing) =

$$1.5\times 4.2\times \ln\left(\frac{348}{273}\right)+4\times 4.2\times \ln\left(\frac{313}{273}\right)=3.83\,\text{kJ/K}$$

Final entropy of the system, S_2 (after mixing) $= 5.5\times 4.2\times \ln\left(\frac{322.5}{273}\right)=3.85\,\text{kJ/K}$

Therefore, increase in entropy of the system $(\Delta S)=S_2-S_1=0.02$ kJ/K Ans.

Example 7.2 Heat is supplied at 600 K and rejected at 320 K in a Carnot cycle. Water is the working fluid in the cycle, which receives heat and evaporates from water at 600 K to steam at 600 K. The entropy change associated with this is 1.5 kJ/kg K. Determine (i) heat supplied, (ii) work done if the cycle operates on a unit mass of water, and (iii) steam flow rate, if the cycle operates in a steady flow with a power output of 18.3 kW.

Solution

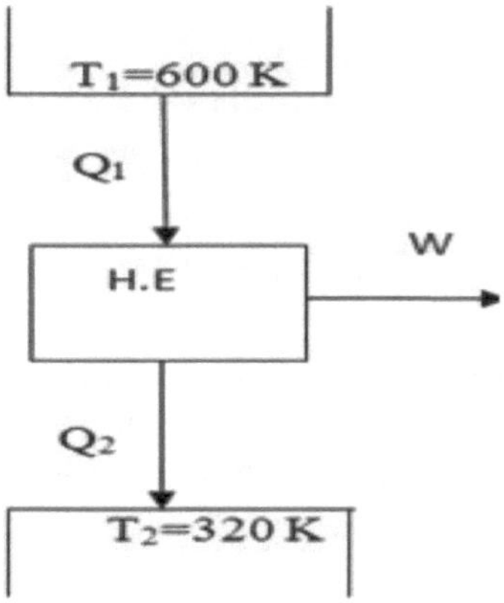

FIGURE EX. 7.2

i. Heat supplied per unit mass q_1 is found from entropy change of source maintained at 600 K.

$$\Delta s=\frac{q_1}{T_1}=1.5\,\text{kJ/kg K}$$

For a reversible cycle, $\frac{q_1}{T_1}=\frac{q_2}{T_2}$

$$\frac{q_1}{600}=1.5\rightarrow q_1=900\,\text{kJ/kg}$$ Ans.

ii. For a heat engine cycle, $w=q_1-q_2\rightarrow q_2=q_1-w$

then $\frac{q_2}{T_2}=\frac{q_1-w}{T_2}$

$$\frac{900 - w}{320} = 1.5 \rightarrow w = 420\,\text{kJ/kg} \qquad \text{Ans.}$$

$$\text{Alternatively}, \eta = 1 - \frac{T_2}{T_1} = 1 - \frac{320}{600} = 46.6\%$$

$$\eta = \frac{w}{q_1} \rightarrow w = \eta \times q_1$$

$$w = 900 \times 0.466 = 420 \text{ kJ/kg}$$

iii. Power output = mass flow rate × work done/kg

$$18.3 = \dot{m} \times 420$$

$$\dot{m} = 0.043\,\text{kg/s} \qquad \text{Ans.}$$

Example 7.3 250 g of water at 25°C is converted at constant pressure into ice at –5°C. Estimate the entropy change of the system. The latent heat of fusion of ice at 0°C is 335 J/g. Take specific heat c_p of water as 4.182 J/g K and that of ice to be half of this value.

Solution There are three stages: (i) the system (water) is converted to ice from 25°C to 0°C, (ii) water freezing at 0°C, and (iii) ice formation from 0°C to –5°C.

Entropy change of the system (water) when it is converted to ice from 25°C to 0°C

$$(\Delta S)_{sys\text{-}I} = mc_p \ln\left(\frac{T_2}{T_1}\right) \quad \text{where } T_1 = 298\,\text{K and } T_2 = 273\,\text{K}$$

$$(\Delta S)_{sys\text{-}I} = 250 \times 4.2 \times \ln\left(\frac{273}{293}\right) = -74.23\,\text{J/K}$$

Entropy change of the system (water) when water freezes at 0°C,

$$(\Delta S)_{sys\text{-}II} = \frac{Q}{T} \quad \text{where } Q = m \times L = 250 \times 335\,\text{J}$$

$$\therefore \frac{Q}{T} = \frac{250 \times 335}{273} = -306.78\,\text{J/K}$$

Entropy change of the system (water) when ice formation takes place from 0°C to –5°C,

$$(\Delta S)_{sys\text{-}III} = mc_p \ln\left(\frac{T_2}{T_1}\right) \text{where } T_1 = 273\,\text{K and } T_2 = 268\,\text{K}$$

$$250 \times 2.1 \times \ln\left(\frac{268}{273}\right) = -9.704\,\text{J/K}$$

$$\text{Total entropy change of the system}, (\Delta S)_{total} = (\Delta S)_{sys\text{-}I} + (\Delta S)_{sys\text{-}II} + (\Delta S)_{sys\text{-}III}$$

$$-92.002 + (-306.78) + (-9.704) = -408.486\,\text{J/K} \qquad \text{Ans.}$$

(– sign is due to that water freezes into ice releasing the latent heat of fusion)

Example 7.4 An iron block of 0.5 kg initially at 120°C is placed in a lake at 10°C; the same block at 10°C is dropped from 150 m height into a lake and two such blocks at 120°C and 10°C are joined together. Estimate the entropy change of the universe as a result of the above. Take c_p of iron as 0.45 kJ/kg K.

Solution Case I: Iron block is placed in a lake at 10°C.

$$(\Delta S)_{system}\text{(Iron block loses entropy)} = mc_p \ln\frac{T_2}{T_1} \quad \text{where } T_1 = 393\,K \text{ and } T_2 = 283\,K$$

$$\Rightarrow 0.5 \times 0.45 \times \ln\left(\frac{283}{393}\right) = -0.073\,kJ/K$$

$$(\Delta s)_{surr} = \frac{Q}{T}\left(\text{Entropy of lake increases}\right)$$

$$\text{Where } Q = mc_p\Delta t = 0.5 \times 0.45 \times (120 - 10) = 24.75\,kJ$$

$$(\Delta S)_{surr} = \frac{Q}{T} = \frac{24.75}{283} = 0.087\,kJ/K$$

$$\therefore (\Delta S)_1 = (\Delta S)_{sys} + (\Delta S)_{surr}$$

$$= -0.073 + 0.087 = 0.0142\,kJ/K$$

Case II: Iron block is dropped in lake from 150 m height

$$Q = mgh = 0.5 \times 9.8 \times 150 = 730\text{ kJ}$$

$$(\Delta S)_2 = \frac{Q}{T} = \frac{735}{283} = 2.597\,kJ/K$$

Case III: Two blocks are joined together, the final temperature $T_f = \dfrac{120 + 10}{2} = 65°C$

$$(\Delta S)_{block\text{-}I} = mc_p \ln\frac{T_2}{T_1} \text{ where } T_1 = 393\,K \text{ and } T_2 = 338\,K$$

$$= 0.5 \times 0.45 \times \ln\left(\frac{338}{393}\right) = -0.033\,kJ/K$$

$$(\Delta S)_{block\text{-}II} = mc_p \ln\frac{T_2}{T_1} \text{ where } T_1 = 283\,K \text{ and } T_2 = 338\,K$$

$$= 0.5 \times 0.45 \ln\left(\frac{338}{283}\right) = 0.039\,kJ/K$$

$$(\Delta S)_3 = (\Delta S)_{block\text{-}I} + (\Delta S)_{block\text{-}II}$$

$$= -0.033 + 0.039 = 0.006\,kJ/K \qquad \text{Ans.}$$

Example 7.5 A steam power plant operating at 1000 K loses 2.4 MJ of heat to a river at (i) 550 K and (ii) 800 K. Find which of the heat transfer processes is more irreversible.

Solution Entropy change for each heat transfer process can be found using Eq. 7.10 since each process is an isothermal process.

Case I: $(\Delta S)_{source} = \frac{Q}{T} = \frac{-2400}{1000} = -2.4\,kJ/K$

$$(\Delta S)_{sink} = \frac{Q}{T} = \frac{2400}{550} = 4.37\,kJ/K$$

$$(\Delta S)_{total} = (\Delta S)_{source} + (\Delta S)_{sink}$$

$$-2.4 + 4.37 = 1.96\ kJ/K$$

Case II: $(\Delta S)_{source} = -2.4\,kJ/K$

$$(\Delta S)_{sink} = \frac{Q}{T} = \frac{2400}{800} = 3\,kJ/K$$

$$(\Delta S)_{total} = (\Delta S)_{source} + (\Delta S)_{sink}$$

$$-2.4 + 3 = 0.6\ kJ/K$$

The total entropy change for the heat transfer process in case II is small and therefore it is less irreversible.

Example 7.6 2.3 kg/s of steam flows isentropically through a turbine, entering at 100 bar and 500°C, and leaves at a pressure of 10 bar. Determine the power of the turbine.

Solution Steady-flow energy equation for the turbine in this case reduces to

$$\dot{m}[h_1] = \dot{m}[h_2] + \frac{\delta W}{dt}$$

$$\therefore \frac{\delta W}{dt} = \dot{m}[h_1 - h_2]$$

Steam flow rate = 2.3 kg/s

From the Mollier diagram, at 100 bar and 500°C,

$h_1 = 3375.1$ kJ/kg and $s_1 = 6.6$ kJ/kg K

and $s_1 = s_2 = 6.6$ kJ/kg K, since steam flows isentropically.

At 10 bar and $s_1 = s_2 = 6.6$ kJ/kg K, $h_2 = 2570$ kJ/kg

$$\therefore \text{Power of the turbine}, \frac{\delta W}{dt} = \dot{m}[h_1 - h_2] = 2.3(3375.1 - 2570)$$

$$= 1851.73\ kW \qquad \text{Ans.}$$

Example 7.7 2.5 kg of saturated water vapor at 120°C is condensed to a saturated liquid at 120°C in a constant pressure process of 200 kPa by heat transfer to surrounding air, which is at 30°C. Determine the entropy increase of the universe.

Solution Entropy increase of the universe, $(\Delta S)_{univ} = (\Delta S)_{sys} + (\Delta S)_{surr}$

From steam tables, for saturated water vapor at 200 kPa, $s_{fg} = 5.597$ kJ/kg K

$$(\Delta S)_{sys} = (\Delta S)_{water} = -m \times s_{fg} = -2.5 \times 5.597 = -13.99\ kJ/K$$

(– sign is due to that the latent of condensation is released when water vapor is condensed to saturated liquid)

$$\therefore (\Delta S)_{surr} = \frac{Q}{T_o} \text{ where } Q = m \times h_{fg}$$

From steam tables, $h_{fg@120°C} = 2202$ kJ/kg

$$Q = 2.5 \times 2202 = 5505 \text{ kJ}$$

$$(\Delta S)_{surr} = \frac{5505}{303} = 18.168 \text{ kJ/K}$$

$$\therefore (\Delta S)_{univ} = (\Delta S)_{sys} + (\Delta S)_{surr}$$

$$= -13.99 + 18.168 = 4.178 \text{ kJ/K}$$ Ans.

Example 7.8 Steam enters an adiabatic turbine at 50 bar, 600°C, and 100 m/s and leaves at 0.5 bar, 150°C, and 180 m/s. The turbine develops a power output of 5 MW. Find (i) the isentropic efficiency of the turbine and (ii) the mass flow rate of steam flowing through the turbine.

Solution

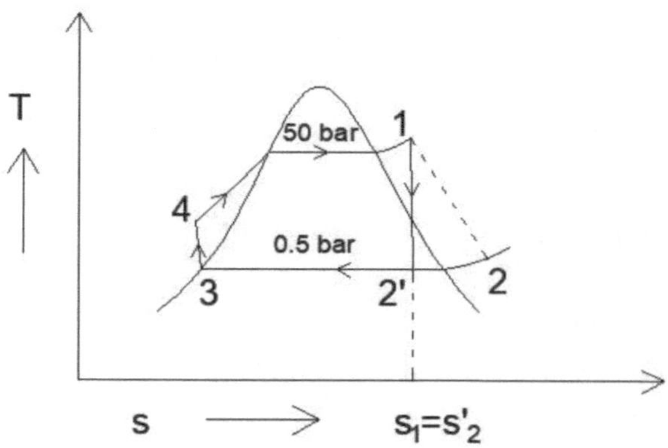

FIGURE EX. 7.8

The T-s diagram (Figure Ex. 7.8) shows the expansion of steam in the turbine.

It is assumed that changes in kinetic and potential energies are ignorable.

Enthalpy values can be found from superheated steam tables,

State 1: At $\left\{\begin{array}{l} p_1 = 50 \text{ bar} \\ T_1 = 600°C \end{array}\right\} h_1 = 3666.5 \text{ kJ/kg}, s_1 = 7.259 \text{ kJ/kgK}$

State 2: At $\left\{\begin{array}{l} p_2 = 5 \text{ bar} \\ T_2 = 150°C \end{array}\right\} h_2 = 2780.2 \text{ kJ/kg}, s_2 = 7.9413 \text{ kJ/kgK}$

For the isentropic process, entropy remains constant, i.e., $s_1 = s_2^1$.

The enthalpy at the exit for the isentropic process h_2^1 can be found by determining the quality or dryness fraction since steam exits as a saturated mixture at the end of the isentropic case.

From saturated steam tables,

At $p_2 = 0.5$ bar → $s_f = 1.091$ kJ/kg K, $s_g = 7.593$ kJ/kg K, $s_{fg} = 6.502$ kJ/kg K

$$h_f = 340.5 \text{ kJ/kg}, \; h_{fg} = 2304.7 \text{ kJ/kg}$$

$$x_2^1 = \frac{s_2^1 - s_f}{s_{fg}} = \frac{7.259 - 1.091}{6.502} = 0.94$$

$$h_2^1 = h_f + x_2^1 \cdot h_{fg} = 340.5 + 0.94 \times 2304.7 = 2506.92 \text{ kJ/kg}$$

i. Isentropic efficiency of turbine, $\eta_T = \dfrac{h_1 - h_2}{h_1 - h_2^1} = \dfrac{3666.5 - 2780.2}{3666.5 - 2506.92} = 76.43\%$

ii. The mass flow rate of steam $\dot{m}$ can be found from the energy balance equation,

$$\dot{m}(h_1) = \dot{m}(h_2) + \dot{W}_{out} \text{ or } \dot{W}_{out} = \dot{m}(h_1 - h_2)$$

$$5 \times 1000 = \dot{m}(3666.5 - 2780.2) \rightarrow \dot{m} = 5.64 \text{ kg/s}$$

Example 7.9 Air enters steadily into a compressor at a rate of 0.25 kg/s at 17°C and 100 kPa and is compressed to a pressure of 1000 kPa and 610 K. Find (i) the isentropic efficiency of the compressor and (ii) the power required to drive the compressor.

Solution

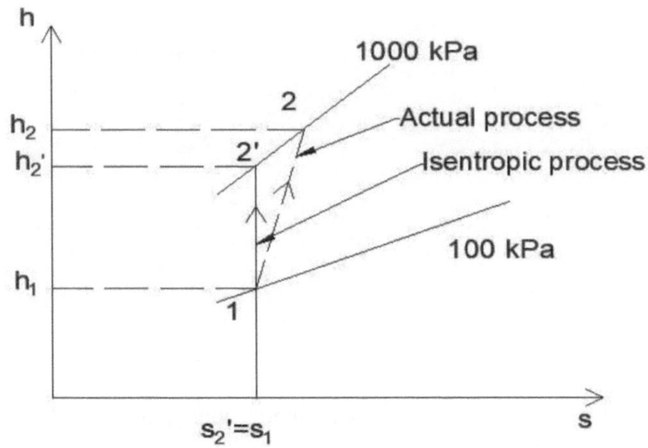

FIGURE EX. 7.9

It is assumed that air is an ideal gas and changes in kinetic and potential energies are ignorable.

From tables of ideal gas properties of air,

At $T_1 = 290\ K \rightarrow h_1 = 290.16\ kJ/kg,\ P_{r1} = 1.2311$

At $T_2 = 610\ K \rightarrow h_2 = 617.53\ kJ/kg$

The enthalpy at the end of the isentropic compression process is determined from the isentropic relation of an ideal gas

$$\frac{P_{r2}}{P_{r1}} = \frac{P_2}{P_1} \rightarrow P_{r2} = 1.2311\left(\frac{1000}{100}\right) = 12.311$$

$$\text{At}\ P_{r2} = 12.311 \rightarrow h_2^1 = 559.23\ \text{kJ/kg}$$

i. Isentropic efficiency of compressor, $\eta_c = \dfrac{h_2^1 - h_1}{h_2 - h_1} = \dfrac{559.23 - 290.16}{617.53 - 290.16} = 82.19\%$ Ans.

ii. The power required to run the compressor is determined from the energy balance equation

$$\dot{m}h_1 + \dot{W} = \dot{m}h_2$$

$$\text{Then}\ \dot{W} = \dot{m}(h_2 - h_1) = 0.25(617.53 - 290.16) = 81.84\ \text{kW}$$ Ans.

Example 7.10 Air enters an adiabatic nozzle at 250 kPa, 880 K with negligible velocity and is discharged at 90 kPa and 690 K. Assuming constant specific heats of air, find (i) the isentropic efficiency of nozzle (ii) the exit temperature, and (iii) the actual exit velocity of air.

Solution

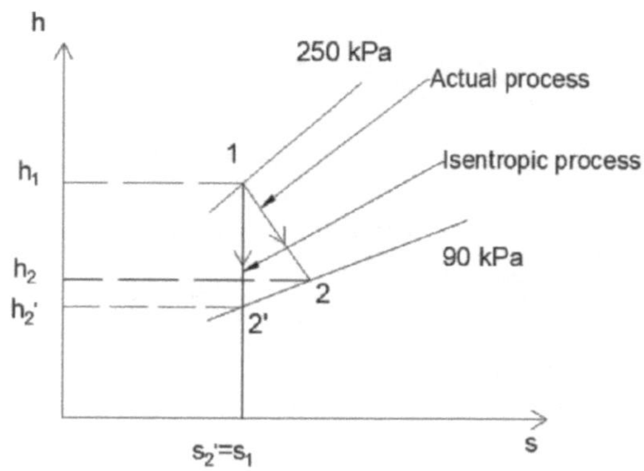

FIGURE EX. 7.10

It is assumed that air is an ideal gas and kinetic energy at the inlet is ignorable. Assume constant temperature of air as 800 K.

From tables of ideal gas specific heats,

Average specific heat, $c_{p,ave} = 1.099\ kJ/kg\ K$ and $k = 1.354$

Isentropic efficiency of nozzle, $\eta_N = \dfrac{h_1 - h_2}{h_1 - h_{2^1}} = \dfrac{c_{p,ave}(T_1 - T_2)}{c_{p,ave}(T_1 - T_2^1)} = \dfrac{880 - 690}{880 - 674} = 92\%$ Ans.

Exit temperature is based on isentropic conditions

$$\frac{T_2^1}{T_1} = \left(\frac{P_2}{P_1}\right)^{\frac{\gamma-1}{\gamma}} \rightarrow T_2^1 = 880\left(\frac{90}{250}\right)^{\frac{0.354}{1.354}} = 674\,\text{K} \qquad \text{Ans.}$$

So the average temperature will be 777 K which is close to 800 K (guessed value).

The isentropic exit velocity of the nozzle can be found from the steady-flow energy equation

$$h_1 + \frac{V_1^2}{2} = h_2^1 + \frac{V_{2^1}^2}{2} \text{ or}$$

$$V_{2^1} = \sqrt{2\left(h_1 - h_2^1\right)} = \sqrt{2 \cdot c_{p,ave}\left(T_1 - T_2^1\right)}$$

$$= \sqrt{2(1.099)(880 - 674) \times 1000} = 673\,\text{m/s}$$

The actual exit velocity of air can be found from

$$\eta_N = \frac{V_2^2}{V_{2^1}^2}$$

$$V_2 = \sqrt{\eta_N \times V_{2^1}^2} = \sqrt{0.92 \times (673)^2} = 645\,\text{m/s} \qquad \text{Ans.}$$

REVIEW QUESTIONS

7.1 What is Clausius inequality?

7.2 Does a cycle for which $\oint \delta Q > 0$ violate the Clausius inequality? Why?

7.3 What is the principle of entropy?

7.4 What is the property which is introduced by the second law of thermodynamics?

7.5 What is the difference between energy and entropy?

7.6 What are the different mechanisms that cause the entropy of a control volume to change?

7.7 How do you compare the entropy change of a reversible and an irreversible process?

7.8 The entropy of a hot metallic bar decreases when it is dropped into a lake. How do you apply an increase of entropy principle to this? Explain.

7.9 Is the value of the integral dQ/T equal to ds for all the processes? Explain.

7.10 How does the entropy change for an adiabatic mixing of two fluids?

7.11 What is the third law of thermodynamics?

7.12 Is entropy transfer associated with work transfer? Explain.

7.13 Explain how entropy increases for an adiabatic process.

7.14 What is the difference between a reversible adiabatic and an adiabatic process?

7.15 How do you compare the entropy of a relaxed person to that of a disturbed one?

7.16 What is the isentropic efficiency of a turbine?

7.17 What is the isentropic efficiency of a compressor?
7.18 What is the isentropic efficiency of a nozzle?

EXERCISE PROBLEMS

7.1 In a nozzle, air enters steadily at 0.3 MPa and 82°C with a velocity of 45 m/s and leaves at 0.08 MPa and 350 m/s. The heat losses from the nozzle to the surroundings, which is at 20°C, are estimated to be 2.8 kJ/kg. Estimate (i) the temperature at the exit and (ii) the total entropy change for this process.

7.2 The nitrogen gas of 1.2 kg mass is compressed slowly in a piston-cylinder device in a polytropic process during which $PV^{1.4}$ = constant. The gas is initially at 0.02 MPa and 30°C. The process ends when the volume is halved. Estimate the entropy change of nitrogen during this process.

7.3 Helium gas is compressed in a piston-cylinder device from an initial state of 0.8 m^3/kg and 300 K to a final state of 0.21 m^3/kg and 588 K. Estimate the entropy change of the helium during this process assuming constant specific heats.

7.4 In a Carnot cycle, 900 kJ of heat is added to the working fluid isothermally from a source at 400°C. Estimate (i) the entropy change of the working fluid, (ii) the entropy change of the source, and (iii) the total entropy change for the process.

7.5 Steam enters an adiabatic turbine at 120 bar, 450°C, and 80 m/s and exits at 40 kPa, 135°C, and 130 m/s. If the power output of the turbine is 4600 kW, estimate (i) the mass flow rate of the steam flowing through the turbine and (ii) the isentropic efficiency of the turbine.

7.6 Air enters a compressor, operating at steady state, at 300 K and 103 kPa and is compressed adiabatically to an end state of 395 K, 342 kPa. For the compressor, estimate (i) the rate of entropy production in kJ/K kg of air flowing and (ii) the isentropic compressor efficiency. Assume the air as an ideal gas, and kinetic and potential energy effects are negligible.

7.7 Carbon dioxide (CO_2) at 298 K, 102 kPa enters a compressor operating at steady state and is compressed adiabatically to an end state of 535 K, 1200 kPa. For the compressor, estimate (i) the work input per kg of CO_2 flowing, (ii) the rate of entropy production, in kJ/K kg of CO_2 flowing, and (iii) the isentropic efficiency of the compressor. Assume the air as an ideal gas, and kinetic and potential energy effects are negligible.

7.8 Nitrogen initially occupying 0.08 m^3 at 500 kPa, 520 K undergoes an internally reversible expansion process during which $pV^{1.2}$ = constant to an end state where the temperature is 320 K. Assuming the ideal gas model, determine (i) the pressure at the final state, (ii) the work and heat transfer, and (iii) the entropy change, in kJ/K.

7.9 Air is contained in a rigid, insulated tank of the volume of 2.03 m^3 fitted with a paddle wheel, initially at 100 kPa, 300 K. It receives an energy transfer by work from the paddle wheel in an amount of 450 kJ. Assuming the ideal gas model for the air, work out (i) the final temperature, in K, (ii) the final pressure, and (iii) the amount of entropy produced, in kJ/K neglecting the changes in kinetic and potential energy.

7.10 Water vapor with a volumetric flow rate of 0.596 m^3/s enters a turbine, operating at steady state, at 480 kPa, 600 K and expands adiabatically to an end state of 100 kPa, 450 K. Estimate for the turbine (i) the power developed, (ii) the rate of entropy production, and (iii) the isentropic turbine efficiency. Neglect kinetic and potential energy effects.

7.11 Air is contained in a rigid, insulated tank of the volume of 2.5 m^3 fitted with a paddle wheel initially at 295 K, 200 kPa. It is stirred until its temperature is 450 K. Determine (i) the final pressure, (ii) the work, in kJ, and (iii) the amount of entropy produced, in kJ/K. Assume the ideal gas model for the air and neglect kinetic and potential energy changes.

7.12 Liquid water at a pressure of 110 kPa enters a 10-kW pump where its pressure is raised to 4.5 MPa. The elevation difference between the exit and the inlet levels is 12 m. Evaluate the highest mass flow rate of liquid water this pump can handle, neglecting the kinetic energy change of water. Take the specific volume of water to be 0.001 m^3/kg.

7.13 A heat engine receives 390 kJ/cycle of heat reversibly from a source at 700 K and rejects it reversibly to the atmosphere at 298 K. Calculate the cyclic integral of $\frac{dQ}{dT}$ for the heat rejections: (i) 200 kJ/cycle, (ii) 120 kJ/cycle, and (iii) 300 kJ/cycle. Discuss your results.

7.14 The water initially at a temperature of 10°C is heated at a constant pressure of 1 atm and converted to steam. The boiling point is 170°C and the latent heat of vaporization is 2100 kJ/kg. Estimate the entropy increase in water.

7.15 A current of 12 amp is allowed to flow through an electrical resistor of 25 Ω for 1.5 s. The resistor is maintained at a constant temperature of 30°C. Estimate the entropy change of the resistor and universe.

7.16 A rigid tank is divided into two equal parts by a partition with one part of the tank containing 2 kg of compressed liquid water at 4 bar and 70°C and the other part being evacuated. The water is now allowed to expand to fill the entire tank by removing the partition. Determine the entropy change of water during this process, if the final pressure in the tank is 20 kPa.

7.17 1 kg of water contained in a piston-cylinder assembly, initially at 145°C, 1.8 bar, undergoes an isothermal compression process to saturated liquid. For the process, W = −450 kJ. Determine for the process, (i) the heat transfer, in kJ and (ii) the change in entropy, in kJ/K. Show the process on a T-s diagram.

7.18 2 kg of air is heated from 30°C to 300°C while pressure drops from 5 to 2.5 bar. Estimate the change in entropy assuming constant specific heat and variable specific heat.

7.19 In a mixing chamber, water entering at 2 bar and 25°C at a rate of 100 kg/s mixes steadily with steam entering at 2 bar and 120°C. The mixture then leaves the chamber at 2 bar and 75°C, and heat is lost to the surrounding air at 35°C at a rate of 3 kJ/s. Estimate the rate of entropy generation during this process, neglecting the changes in kinetic and potential energies.

7.20 A gas within a piston-cylinder assembly undergoes an isothermal process at 425 K during which the entropy change is −0.42 kJ/K. Assuming the ideal gas model for the gas and negligible kinetic and potential energy effects, evaluate the work, in kJ.

7.21 A 4.5 kW compressor compresses air steadily from 1 bar and 20°C to 7 bar and 180°C at a rate of 1.45 kg/min. Heat transfer takes place between the compressor and the surrounding medium at 20°C during this process. Evaluate the rate of entropy change of air during this process.

7.22 In a Carnot cycle, 850 kJ of heat is added isothermally to the working fluid from a source at 700 K. Estimate (i) the entropy change of the working fluid, (ii) the entropy change of the source, and (iii) the total entropy change for the process.

7.23 Find the entropy increase when the air inside a rigid tank is heated from 30°C to 80°C. If the tank is heated from 1000°C to 1060°C, what is the entropy increase?

7.24 Heat is transferred to the refrigerant R-134a, which is initially at 1.8 bar and 35% quality in a rigid tank of 0.48 m^3, from a source at 320 K until the pressure rises to 4.5 bar. Estimate (i) the entropy change of the refrigerant, (ii) the entropy change of the heat source, and (iii) the total entropy change for this process.

7.25 Saturated steam of 4.5 kg at 100°C is contained in a rigid tank. The steam is cooled to the surrounding temperature of 27°C. Estimate (i) the entropy change of the steam and (ii) for the steam and its surroundings, the total entropy change or S_{gen} associated with this process.

7.26 A 25 kg iron block, initially at 100°C, is dropped into an insulated tank that contains 100 l of water at 30°C. Estimate (i) the final equilibrium temperature and (ii) the total entropy change for this process.

7.27 Steam enters an adiabatic turbine at 6 MPa, 500°C, and 75 m/s and leaves at 45 kPa, 120°C, and 150 m/s. If the power output of the turbine is 5 MW, find (i) the mass flow rate of the steam flowing through the turbine and (ii) the isentropic efficiency of the turbine.

7.28 Air enters an adiabatic compressor steadily at 100 kPa and 12°C at a rate of 0.24 kg/s, and it exits at 245°C. The compressor has an isentropic efficiency of 80%. Neglecting the changes in kinetic and potential energies, find (i) the exit pressure of air and (ii) the power required to drive the compressor.

7.29 Air is compressed by an adiabatic compressor from 90 kPa and 25°C to 500 kPa and 250°C. Assuming variable specific heats and neglecting the changes in kinetic and potential energies, find (i) the isentropic efficiency of the compressor and (ii) the exit temperature of air if the process were reversible.

7.30 Hot gases from the combustor of a turbojet engine enter the nozzle at 250 kPa, 750°C, and 90 m/s and exit at a pressure of 80 kPa. Assuming an isentropic efficiency of 90% and treating the combustion gases like air, find (i) the exit velocity, (ii) the maximum possible exit velocity, and (iii) the exit temperature.

DESIGN AND EXPERIMENT PROBLEMS

7.31 Select an industry that uses compressed air for the operation of power tools such as drill bits and brakes and air cylinders. Assuming the compressors to operate at full load during one-third of the time on average and the average motor efficiency to be 80%, determine how much energy and money will be saved per year if the energy consumed by compressors is reduced by at least 5% as a result of implementing some conservation measures. Take the unit cost of electricity to be Rs. 0.50/kWh.

7.32 What is the difference you observe between a plant that operates at the sea level (where the pressure is taken as 1 atm) and one that operates above the sea level at an altitude of 100 m? Determine the pressure at this location. The compressor in both the plants is compressing the air to 800 kPa. Estimate the energy requirements and cost of the energy for both the plants if the compressor operates 3000 hours/year driven by a motor that has an efficiency of 85% and operates at 80% of the rated load. Take the price of electricity as 0.5/kWh.

7.33 A piston-cylinder device contains 0.5 kg of water at room temperature, 25°C, 101.32 kPa, and at constant pressure. An electric heater of 450 W heats the water to 450°C. Sketch the temperature and total accumulated entropy generation as a function of time assuming no heat losses to the surroundings. Analyze the first part of the process that is bringing the water to the boiling point, by measuring it in your kitchen and knowing the rate of power added.

7.34 Design an experimental set-up with complete instrumentation to obtain property data required to evaluate the entropy change of gases such as hydrogen, helium, and nitrogen undergoing a process. Compare the experimentally determined entropy change with a value obtained from published engineering data. Discuss the full description of the data needed and sample calculations.

8 Properties of Gases and Gas Mixtures

LEARNING OUTCOMES

After learning this chapter, students should be able to

- Explain the quantitative relationship between T, v, and P as described by the kinetic theory of gases and ideal gas models.
- Interpret the relationship between partial pressures and the total pressure as described in Dalton's law of partial pressure.
- Determine the mole fractions of gases within and gas mixture and relate mole fraction to the partial pressure of a gas within a gas mixture.
- Explain the relationship between kinetic energy and temperature of a gas; between temperature and the velocity of a gas; and between molar mass and the velocity of a gas.
- Explain the deviation of ideal gas models with the behavior of real gases observed in nature.
- Explain the general principles of the hard-sphere model and the Van der Waals model of gas.

8.1 IDEAL GAS EQUATION OF STATE

Equation of State

It is defined as the functional relationship among the properties pressure p, specific volume ν, and temperature T, expressed as $f(p,\nu,T)=0$. The equation of state is useful for finding the properties of a gas; that is, if two of the properties are known, the third can be found. There are numerous equations of state, including simple and complex ones. The ideal gas equation of state is comparatively simple and it can accurately predict the behavior of substances in the gas phase.

Robert Boyle proved experimentally that the pressure of gases is inversely proportional to their volume. Charles and Gay Lussac showed that the volume of a gas is directly proportional to its temperature at low pressures, which is expressed as

$$p\nu = RT \tag{8.1}$$

which is an ideal gas equation of state. R is a characteristic gas constant; its value is different for different gases.

Also specific volume, $\nu = \frac{V}{m}(m^3/kg)$

then

$$pV = mRT \tag{8.2}$$

$$pV = m\frac{\bar{R}}{\mu}T \text{ where } R = \frac{\bar{R}}{\mu}(\text{kJ/kg K})$$

A mole of a substance has a mass numerically equal to its molecular weight, that is, 1 g mol of nitrogen has a mass of 28 g and 1 kg mol of hydrogen has a mass of 2 kg.

$$\text{Number of moles of a gas, } n = \frac{\text{mass of gas}}{\text{molecular weight of gas}} = \frac{m}{\mu} \tag{8.3}$$

$$pV = n\bar{R}T \tag{8.4}$$

where $\bar{R}$ is the universal gas constant and μ is the molar mass or molecular weight of the gas.

$$\bar{R} = 8.314 \text{ kJ/kmol K or } 8314 \text{ Nm/kmol K}$$

For example, for nitrogen, characteristic gas constant, $R_{N_2} = \frac{8.314}{28} = 0.296 \text{ kJ/kg K}$

An ideal gas is a gas in which the molecules are spaced far apart in such a way that at low densities, the behavior of a molecule is not affected by the presence of other molecules. An ideal gas obeys the ideal gas equation of state. However, real gases approximate the ideal gas behavior at low pressure or high temperature relative to their critical point values. An ideal gas is similar to a perfect gas, as both obey the ideal gas equation $p\nu = RT$. However, they differ in some aspects, i.e., the specific heats of an ideal gas vary with temperature, whereas for a perfect gas, specific heats are assumed to be constant with temperature.

In reality, there seems to be no gas that behaves like an ideal gas or a perfect gas. The real gases such as oxygen, hydrogen, nitrogen, and air are assumed to behave like a perfect gas at very low pressures or very high temperatures.

8.2 OTHER EQUATIONS OF STATE

Van der Waals Equation of State

The kinetic theory of gases, proposed by Clerk Maxwell, forms the basis for establishing the ideal gas equation of state. The ideal gas equation is developed based on certain assumptions as given below:

1. The molecules of a gas are spaced apart that there is little or no attraction between them.
2. The volume occupied by the molecules themselves is quite low compared to the volume of gas.

3. Molecules are in random motion, which follows Newton's law of motion.
4. The kinetic energy of the molecules and their momentum are conserved as molecules and walls of the container are perfectly elastic.

The real gases do not obey the assumptions made in the kinetic theory of gases. At very low pressures or very high temperatures, real gas obeys the ideal gas equation as intermolecular attraction and volume occupied by molecules compared to the total volume are not considered at this state. When the pressure increases, intermolecular forces increase and the volume of molecules becomes considerable when compared to that of gas. Thus real gases deviate from the ideal gas equation of state appreciably with an increase in pressure. Van der Waals introduced two correction factors 'a' and 'b' in the ideal gas equation, first one to account for intermolecular attraction and the second one to account for volume of molecules. The Van der Waals equation is

$$\left(p+\frac{a}{v^2}\right)(v-b)=RT \tag{8.5}$$

the term $\frac{a}{v^2}$ is called the force of cohesion and b the co-volume. At high pressures, all real gases obey the Van der Waals equation of state; however, it is not true at all ranges of pressures and temperatures. Table 8.1 shows the Van der Waals constants for some commonly used real gases. There are other equations of the state apart from Van der Waals equation of state. Those are Berthelot, Dieterici, and Redlich–Kwong equations of state.

Berthelot equation of state

$$p=\frac{RT}{v-b}-\frac{a}{Tv^2} \tag{8.6}$$

Dieterici equation of state

$$p=\frac{RT}{v-b}\cdot e^{-a/RTv} \tag{8.7}$$

TABLE 8.1
Van der Waals Constants

Gas	a [kN m^4/(kg mol)^2]	b (m^3/kg mol)
Oxygen (O_2)	138.0	0.0318
Hydrogen (H_2)	2510.5	0.2262
Helium (He)	3417.62	0.0228
Nitrogen (N_2)	136.7	0.0386
Carbon dioxide (CO_2)	362.9	0.0314
Air	135.5	0.0362

Redlich–Kwong equation of state

$$p = \frac{RT}{\nu - b} - \frac{a}{T^{\frac{1}{2}}\nu(\nu + b)} \tag{8.8}$$

8.3 COMPRESSIBILITY FACTOR—THE DEVIATION OF REAL GASES FROM THE IDEAL GAS BEHAVIOR

Since no gas is a perfect gas, all gases (real) deviate from the ideal gas equation of state at states close to the saturation region and critical point. This deviation of real gases from the ideal gas equation of state can suitably be accounted for by the introduction of a factor known as compressibility factor, indicated by Z. The ideal gas equation of state with the compressibility factor Z is given as

$$p\nu = ZRT$$

$$Z = \frac{p\nu}{RT} \tag{8.9}$$

$Z = 1$ for ideal gases, while for real gases, Z is either greater than or less than unity. For real gases, the p-ν-T behavior is expressed by $p\nu = ZRT$. The large deviation of real gases from the ideal gas equation of state implies that the Z is farther away from unity. Z can also be given by

$$Z = \frac{\nu_{actual}}{\nu_{ideal}} \text{ where } \nu_{ideal} = \frac{RT}{p} \tag{8.10}$$

The specific volumes of different gases are different at the same temperature and pressure. However, the reduced volumes of different gases are almost the same at the same reduced temperature and reduced pressure. Thus, the gases behave differently at a particular temperature and pressure; however, they behave much the same at the same temperatures and pressures normalized with respect to their critical temperatures and pressures. This normalization leads to what are known as reduced properties; the reduced pressure is the ratio of actual pressure to critical pressure of a substance and similarly, we can define the reduced temperature and reduced volume.

$$p_r = \frac{p}{p_c}, T_r = \frac{T}{T_c} \text{ and } \nu_r = \frac{\nu}{\nu_c} \tag{8.11}$$

The p_r, T_r, and ν_r are reduced pressure, reduced temperature, and reduced volume respectively, whereas p_c, T_c, and ν_c are critical pressure, critical temperature, and critical volume respectively. The Z value is almost the same for gases at the same reduced pressure and reduced temperature, this is known as the law of corresponding states. Compressibility charts, prepared with the experimental values of Z against p_r and T_r for different gases, are useful to predict the properties of gases for which more precise data are not available. Figure 8.1 shows the compressibility chart.

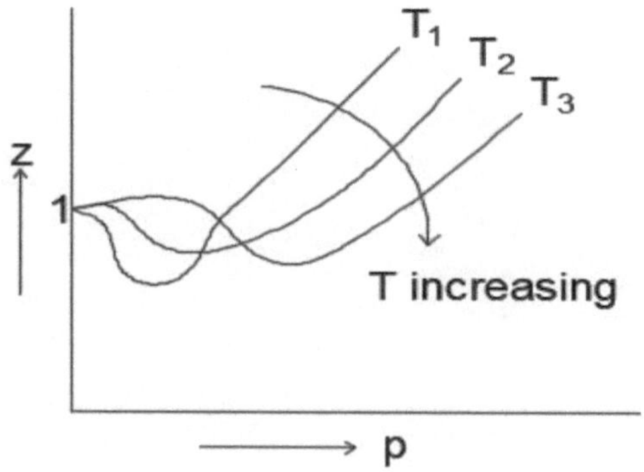

FIGURE 8.1 Compressibility chart.

At very low pressures, gases follow the ideal gas equation of state irrespective of temperature, and at very high temperatures, gases follow the ideal gas equation of state irrespective of pressure. There is a greater deviation of gases from the ideal gas behavior in the surrounding area of the critical point.

8.4 GAS COMPRESSION—REDUCING THE WORK OF COMPRESSION

Gas Compression

A compressor raises the pressure of air or gas at the expense of work. The density of gas also increases considerably with compression. For a steady-flow device, compressor, the work input, neglecting kinetic energy and potential energy changes

$$W = -\int_1^2 v\,dp \tag{8.12}$$

If the compression process is reversible polytrophic with the relation $p\nu^n = c$, then

$$W = \frac{n}{n-1} p_1 v_1 \left[\left(\frac{p_2}{p_1} \right)^{\frac{n-1}{n}} - 1 \right] \tag{8.13}$$

If the compression is reversible adiabatic compression, n is replaced by γ

$$W = \frac{\gamma}{\gamma-1} p_1 v_1 \left[\left(\frac{p_2}{p_1} \right)^{\frac{\gamma-1}{\gamma}} - 1 \right] \tag{8.14}$$

The work of compression for a reversible isothermal process is given by

$$W = p_1 v_1 \ln \frac{p_1}{p_2} \tag{8.15}$$

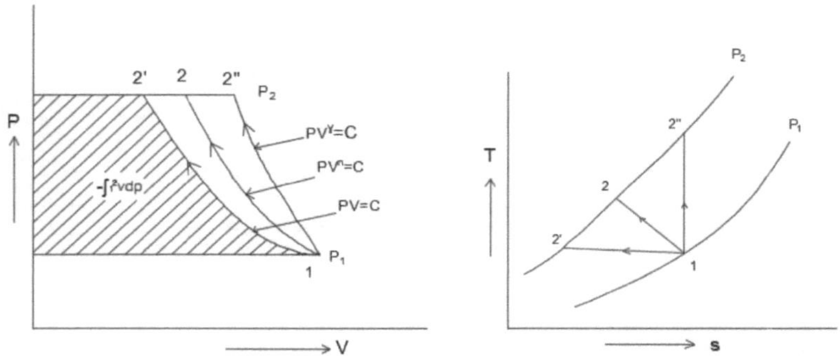

(a) p-V diagram (b) T-s diagram

FIGURE 8.2 Comparison of three reversible compression processes on (a) p-V and (b) T-s diagrams.

The three reversible compression processes, namely polytropic, adiabatic, and isothermal processes, are shown in Figure 8.2 with the help of p-V and T-s diagrams.

From Figure 8.2, it can be observed that isothermal compression requires minimum work input as it involves maximum cooling. The isentropic process requires maximum work input as it involves no cooling. The work input in the polytropic case lies in between these two, as polytropic compression involves some cooling. It is important that all the three processes are carried out between the same pressure limits. The work input requirement of polytropic compression can be reduced further by increasing the heat rejection during compression. If the polytropic index n is decreased, it approaches unity and the process tends to be isothermal.

Reducing the Work of Compression

The work input requirement is minimized by cooling the gas during the compression. However, it is not always a viable solution to maintain cooling jackets through the casing of a compressor. Multi-stage compression with intercooling is an alternative method for minimizing the work input requirements. In this, the gas is compressed in stages instead of a single stage and cooled in between each stage by being passed through a heat exchanger, known as an intercooler. Multi-stage with intercooling is advantageous in the case of higher pressure requirements.

Figure 8.3 shows the two-stage compression with intercooling on p-V and T-s diagrams. The gas is compressed in the first stage from initial pressure p_1 to intermediate pressure p_i, cooled at this constant pressure (p_i) to initial temperature $T_{1,}$ and compressed further to final pressure p_2 (desired pressure). The compression process is assumed to follow a polytrophic process $\left(pv^n = C\right)$ in which n value varies between γ and 1. The shaded area in Figure 8.3a represents the work saved due to two-stage compression with intercooling. The amount of work saved depends on the conditions in which the compression is carried out. However, the intermediate pressure p_i plays an important role in reducing the work input by varying the size of the shaded area.

For the two-stage compression, the work input is the sum of the work inputs of each stage, that is

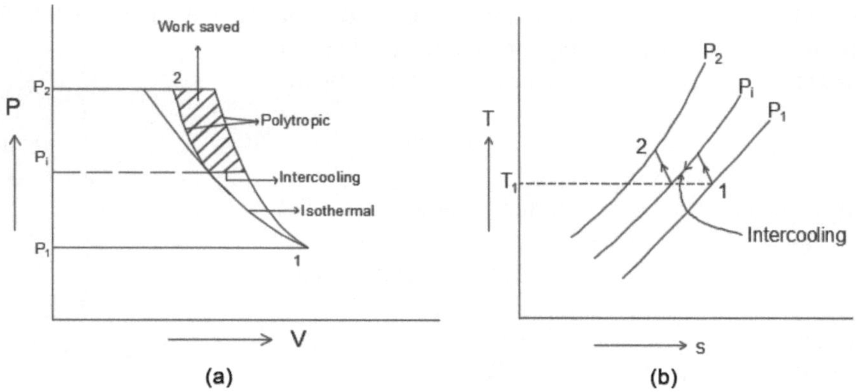

FIGURE 8.3 Two-stage compression with intercooling. (a) p-V diagram (b) T-s diagram.

$$W = W_1 + W_2 = \frac{n}{n-1}\left[\left(\frac{p_i}{p_1}\right)^{\frac{n-1}{n}} - 1\right] + \frac{n}{n-1}\left[\left(\frac{p_2}{p_i}\right)^{\frac{n-1}{n}} - 1\right] \tag{8.16}$$

where W_1 and W_2 represent the work inputs in stages 1 and 2 respectively.

The minimum work required is dependent on p_i in the above equation.

The value of p_i, which minimizes the work input, can be obtained by differentiating the above expression with respect to p_i and equating it to zero, that is,

$$\frac{dW}{dP_i} = \frac{d}{dP_i}\left\{\frac{n}{n-1}\left[\left(\frac{p_i}{p_1}\right)^{\frac{n-1}{n}} - 1\right] + \frac{n}{n-1}\left[\left(\frac{p_2}{p_i}\right)^{\frac{n-1}{n}} - 1\right]\right\} = 0$$

Differentiation yields $p_i^{\frac{2(n-1)}{n}} = (p_1 p_2)^{\frac{n-1}{n}}$

$$p_i = \sqrt{p_1 p_2} \quad \text{or} \quad \frac{p_i}{p_1} = \frac{p_2}{p_i} \tag{8.17}$$

It can be concluded from Eq. 8.17 that the pressure ratio should be the same in each stage to minimize the work of compression.

Specific Heat Relations

Specific heat at constant volume,

$$c_v = \left(\frac{\partial u}{\partial T}\right)_v \tag{8.18}$$

and specific heat at constant pressure,

$$c_p = \left(\frac{\partial h}{\partial T}\right)_p \tag{8.19}$$

Then $du = c_v \Delta t$ and this equation applies for an ideal gas for any process while for other substances, it applies only for a constant volume process.

u doesn't change when ν changes at constant temperature,

$$u = f(T) \tag{8.20}$$

This is true for ideal gases only and is known as Joule's law.

$$\Delta u = c_v \Delta T \quad (c_v \text{ is constant})$$

The enthalpy of any substance $h = u + p\nu$

For ideal gas, $p\nu = RT \Rightarrow h = u + RT$

Therefore for an ideal gas, enthalpy is a function of absolute temperature only, that is,

$$h = f(T) \tag{8.21}$$

$$dh = du + RdT$$

$$\Delta h = \Delta u + R\Delta T$$

$$= c_v \Delta T + R\Delta T \Rightarrow \Delta T(c_v + R)$$

The equation $c_p = \left(\frac{\partial h}{\partial T}\right)_p \Rightarrow dh = c_p dT \text{ or } \Delta h = c_p \Delta T$ applies to ideal gas at all pressures while for other substances, it is applicable for constant pressure only.

$$c_p \Delta T = \Delta T(c_v + R)$$

$$c_p - c_v = R \tag{8.22}$$

The ratio of specific heats of an ideal gas is indicated by γ and is called the adiabatic index

$$\gamma = \frac{c_p}{c_v} \Rightarrow c_p = \gamma c_v$$

c_p and c_v can be expressed as

$$c_p = \frac{\gamma R}{\gamma - 1} \text{ and } c_v = \frac{R}{\gamma - 1} \text{ kJ/kgK} \tag{8.23}$$

Since $c_p > c_v$, it follows that γ is always greater than unity.

$$\gamma = 1.67 \text{ for mono-atomic gases}$$

$$\gamma = 1.4 \text{ for di-atomic gases}$$

$$R = \frac{\bar{R}}{\mu}$$

Reversible Adiabatic Process

Thermodynamic property relations are

$$TdS = du + pdv \tag{8.24}$$

$$TdS = dh - vdp \tag{8.25}$$

Now for a reversible adiabatic process, entropy remains constant, ds = 0.

Since $du = c_v dT$ for an ideal gas,

Equation 8.24 becomes $c_v dT + pd\nu = 0$

$$c_v dT = -pdv \tag{8.26}$$

Also $dh = c_p dT$ for an ideal gas

Equation 8.25 becomes

$$c_p dT = vdp \tag{8.27}$$

Specific heat relation $\Upsilon = \frac{c_p}{c_v}$

Dividing Eq. 8.26 by 8.27,

$$\Upsilon = \frac{-vdp}{pdv}\left(\text{since } \Upsilon = \frac{c_p}{c_v}\right)$$

$$\frac{\Upsilon dv}{v} = \frac{-dp}{p}$$

$$\frac{\Upsilon dv}{v} + \frac{dp}{p} = 0$$

on integration, $d(\ln p) + \Upsilon d(\ln v) = d(\ln c)$

$$\ln(p) + \Upsilon \ln(v) = \ln(c)$$

$$pv^{\Upsilon} = c \tag{8.28}$$

Equation 8.28 is the p-ν relation for an ideal gas for a reversible adiabatic process.

$$p_1 v_1^{\Upsilon} = p_2 v_2^{\Upsilon} = c$$

Again $p\nu = RT$ for an ideal gas.
From Eq. 8.28 $p = c\nu^{-\gamma}$

$$c\nu^{-\gamma}\nu = RT$$

On rearranging $c\nu^{1-\gamma} = RT$

$$T\nu^{\gamma-1} = c \quad (8.29)$$

Equation 8.29 is the T-ν relation for an ideal gas for a reversible adiabatic process.

$$\frac{T_2}{T_1} = \left(\frac{\nu_1}{\nu_2}\right)^{\gamma-1}$$

From Eq. 8.28, $\nu = \left(\frac{c}{p}\right)^{\frac{1}{\gamma}}$ substituting this in the equation in $p\nu = RT$

$$pc^1p^{-1/\gamma} = RT \Rightarrow c^1p^{1-\frac{1}{\gamma}} = RT$$

$$Tp^{\frac{1-\gamma}{\gamma}} = c \quad (8.30)$$

Equation 8.30 is the p-T relation for an ideal gas for a reversible adiabatic process

$$\frac{T_2}{T_1} = \left(\frac{p_2}{p_1}\right)^{\frac{\gamma-1}{\gamma}}$$

Internal Energy Change of an Ideal Gas

$$du = Tds - pd\nu \quad (Tds = 0 \text{ for a reversible adiabatic process})$$

Integrating the above equation between states 1 and 2

$$\int_1^2 du = -\int_1^2 pd\nu = -\int_1^2 c\nu^{-\gamma}\, d\nu$$

$$u_2 - u_1 = \frac{c\nu_2^{1-\gamma} - \nu_1^{1-\gamma}}{\gamma - 1}$$

$$u_2 - u_1 = \frac{p_2\nu_2 - p_1\nu_1}{\gamma - 1}\left(p_1\nu_1^{\gamma} = p_2\nu_2^{\gamma}\right)$$

Again

$$pv = RT \Rightarrow u_2 - u_1 = \frac{R(T_2 - T_1)}{\gamma - 1} = \frac{RT_1}{\gamma - 1}\left[\left(\frac{p_2}{p_1}\right)^{\frac{\gamma-1}{\gamma}} - 1\right] \tag{8.31}$$

Enthalpy change of an ideal gas $dh = Tds + \nu dp$ ($Tds = 0$ for a reversible adiabatic process)

Integrating the above equation between states 1 and 2

$$\int_1^2 dh = \int_1^2 v\,dp = \int_1^2 c^{\frac{1}{\gamma}} p^{\frac{-1}{\gamma}}\,dp$$

$$h_2 - h_1 = \frac{\gamma}{\gamma - 1} c^{\frac{1}{\gamma}}\left[\left(p_2^{\frac{\gamma-1}{\gamma}} - p_1^{\frac{\gamma-1}{\gamma}}\right)\right]$$

$$= \frac{\gamma}{\gamma - 1}\left(p_1 v_1^{\gamma}\right)^{\frac{1}{\gamma}}\left[\left(p_2^{\frac{\gamma-1}{\gamma}} - p_1^{\frac{\gamma-1}{\gamma}}\right)\right]$$

$$= \frac{\gamma}{\gamma - 1}\left(p_1 v_1^{\gamma}\right)^{\frac{1}{\gamma}} p_1^{\frac{\gamma-1}{\gamma}}\left[\left(\frac{p_2}{p_1}\right)^{\frac{\gamma-1}{\gamma}} - 1\right]$$

$$= \frac{\gamma}{\gamma - 1} p_1 v_1\left[\left(\frac{p_2}{p_1}\right)^{\frac{\gamma-1}{\gamma}} - 1\right] \tag{8.32}$$

Work Done by an Ideal Gas in a Reversible Adiabatic Process

According to the first law, $\delta Q = dU + \delta W$

$$\delta W = -dU (\text{Since } \delta Q = 0 \text{ for a reversible adiabatic})$$

$$W_{1\text{-}2} = -(U_2 - U_1) = U_1 - U_2 = m(u_1 - u_2)$$

$$= \frac{m(p_1 v_1 - p_2 v_2)}{\gamma - 1}$$

$$W_{1\text{-}2} = \frac{mR(T_1 - T_2)}{\gamma - 1} = \frac{mRT_1}{\gamma - 1}\left[1 - \left(\frac{p_2}{p_1}\right)^{\frac{\gamma-1}{\gamma}}\right] \tag{8.33}$$

Reversible Isothermal Process

The work done in a reversible isothermal process is given by

$$W_{1\text{-}2} = \int_1^2 p\,dv$$

$$W_{1\text{-}2} = \int_1^2 \frac{mRT}{v}\,dv$$

$$W_{1\text{-}2} = mRT\ln\frac{v_2}{v_1} \tag{8.34}$$

It can also be written as $W_{1\text{-}2} = mRT\ln\frac{p_1}{p_2}$

Now $\delta Q = dU + \delta W$ or $Q_{1\text{-}2} = U_2 - U_1 + W_{1\text{-}2}$

$$Q_{1\text{-}2} = mRT\ln\frac{v_1}{v_2}\left(\text{since } U_{1\text{-}2} = 0 \text{ for isothermal process}\right) \tag{8.35}$$

Reversible Polytropic Process

Heat and Work in a Polytropic Process

For a polytropic process, $pv^n = c$ (where 'n' is the polytropic index)

And also $Tv^{n-1} = c$ or $\frac{T_2}{T_1} = \left(\frac{v_1}{v_2}\right)^{n-1}$

$$Tp^{\frac{1-n}{n}} = c \text{ or } \frac{T_2}{T_1} = \left(\frac{p_2}{p_1}\right)^{\frac{n-1}{n}}$$

$$\delta Q - \delta W = U_2 - U_1 = m(u_2 - u_1)$$

For a unit mass of a substance, $\delta Q - \delta W = (u_2 - u_1)$

$$= c_v(T_2 - T_1)$$

$$= \frac{R(T_2 - T_1)}{\Upsilon - 1}\left(\text{where } c_v = \frac{R}{\Upsilon - 1}\right)$$

$$= \frac{p_2v_2 - p_1v_1}{\Upsilon - 1}$$

For a polytropic process, work done

$$\delta Q - \delta W = \frac{p_1 v_1}{\Upsilon - 1}\left[\left(\frac{p_2}{p_1}\right)^{\frac{n-1}{n}} - 1\right] = \frac{p_1 v_1}{\Upsilon - 1}\left[\left(\frac{v_1}{v_2}\right)^{n-1} - 1\right] \quad (8.36)$$

8.5 PROPERTIES OF GAS MIXTURES

In most of the practical applications, the working substance happens to be a mixture of gases; for example, in an open cycle gas turbine power plant, the products of combustion are a mixture of various gases such as CO_2, CO, O_2, and N_2 in varying proportions. The p-V-T behavior of gas mixtures can be well predicted by Dalton's law of partial pressures and Amagat's law of partial volumes. However, these two laws hold approximately for real-gas mixtures while they hold exactly for ideal-gas mixtures.

The properties of a gas mixture depend on the properties of individual constituent gases and their mass or volume.

Gravimetric Analysis

This analysis is based on the mass of individual gases. Let the mass of individual gases be m_1, m_2, m_3 … m_i (subscript 1, 2, 3,…i refer to the individual gases in the mixture)

$$m = m_1 + m_2 + m_3 + \cdots m_i$$

Dividing both sides of the above equation by 'm' gives

$$1 = \frac{m_1}{m} + \frac{m_2}{m} + \frac{m_3}{m} + \cdots \frac{m_i}{m}$$

Each term on the right-hand side is termed the mass fraction.

$\frac{m_1}{m}$ is the mass fraction of gas 1 in the mixture.

Let $\frac{m_1}{m} = y_1, \frac{m_2}{m} = y_2 \ldots \Rightarrow y_1 + y_2 + \cdots y_i = 1$

$\varepsilon y_n = 1$ where $n = 1, 2, 3 \ldots i$

Therefore, the sum of the mass fractions of the individual constituents in a mixture is equal to 1.

Molar analysis: This analysis is based on volume.

Let N be the total number of molecules of a mixture which contains 1, 2, 3…i number of individual molecules, then

$$N = N_1 + N_2 + N_3 + \cdots N_i$$

Dividing both sides by the Avogadro number

$$\frac{N}{N_A} = \frac{N_1}{N_A} + \frac{N_2}{N_A} + \cdots \frac{N_i}{N_A}$$ $N_1, N_2, \ldots N_i$ and N are proportional to corresponding masses

$$= \frac{m}{M} = \frac{m_1}{M_1} + \frac{m_2}{M_2} + \cdots \frac{m_i}{M_i}$$

$$n = n_1 + n_2 + \cdots n_i$$

Dividing by n both sides

$$\frac{n_1}{n} + \frac{n_2}{n} + \cdots \frac{n_i}{n} = 1$$

$$x_1 + x_2 + x_3 + \cdots x_i = 1 \text{ or } \varepsilon x_n = 1$$

where $x_1, x_2 \ldots x_i$ are mole fractions of individual gases.

Dalton's Law of Partial Pressures

Dalton's law of partial pressures states that the total pressure of a mixture of ideal gases is equal to the sum of partial pressures of all the constituent gases of the mixture, and the partial pressure of an individual constituent gas is the pressure it exerts, if it were to exist alone at the mixture temperature and volume. If we consider a homogeneous mixture of inert ideal gases at temperature T, pressure p, and volume V, and if there are n_1 moles of gas k_1, n_2 moles of gas k_2, and up to n_i moles of gas k_i, then according to the ideal gas equation of state in the absence of any chemical reaction,

$$pV = (n_1 + n_2 + n_3 + \cdots + n_i)\bar{R}T \quad (\bar{R} = 8.314 \text{ kJ/kg mol K})$$

$$p = \frac{n_1 \bar{R}T}{V} + \cdots \frac{n_i \bar{R}T}{V}$$

In the above expression, $\frac{n_i \bar{R}T}{V}$ is the pressure, the ith gas would exert if it occupied the total volume V at temperature T. This pressure is called the partial pressure of ith gas, that is, p_i,

$$p_1 = \frac{n_1 \bar{R}T}{V}, p_i = \frac{n_i \bar{R}T}{V}$$

$$p = p_1 + p_2 + p_3 + \cdots p_i \tag{8.37}$$

This is known as Dalton's law of partial pressures.

$p_1, p_2, p_3 \ldots p_i$ are partial pressures of constituents.

$p_i = n_i RT$. Dividing and multiplying the right-hand side of the equation by the number of moles (n) of the mixture,

$$p_i = n_i \left(\frac{n}{n}\right)\frac{RT}{V} = \left(\frac{ni}{n}\right)\frac{nRT}{V}$$

The term in the parenthesis on the right-hand side is the mole fraction of the ith component and other term, i.e., $\frac{nRT}{V}$ is equal to mixture pressure

$$\therefore p_i = x_i \cdot p$$

The partial pressure of an individual constituent in a mixture of ideal gas is equal to the product of the mole fraction of the constituent and the pressure of the mixture.

From the above expressions, $V = (n_1 + n_2 + n_3 + \cdots + n_i)\frac{\overline{R}T}{p} = \sum n_k \frac{\overline{R}T}{p}$

The partial pressure of kth gas is $p_k = \frac{n_k \overline{R}T}{V} = \frac{n_k p}{\sum n_k}$

where $\sum n_k = n_1 + n_2 + n_3 + \cdots + n_i$

The ratio $\frac{n_k}{\sum n_k}$ is termed the mole fraction of kth gas, x_k

The mole fractions of the constituents are $x_1 = \frac{n_1}{\sum n_k}, x_i = \frac{n_i}{\sum n_i}$

Then the partial pressure $p_1 = x_1 p \ldots p_i = x_i p$

$$p_k = x_k \cdot p$$

$$x_1 + x_2 + \cdots + x_i = 1$$

Gas Constant (R_m) of a Mixture

The ideal gas equation of state is $pV = mRT$. For constituent gases,

$$p_1 V = m_1 R_1 T, p_2 V = m_2 R_2 T \ldots p_i V = m_i R_i T$$

According to Dalton's law, $pV = (m_1 R_1 + m_2 R_2 + m_3 R_3 + \cdots m_i R_i)T$.

If R_m is assumed as the gas constant of the mixture, then $pV = (m_1 + m_2 + m_3 + \cdots m_i) R_m T$

Then,

$$R_m = \frac{m_1 R_1 + m_2 R_2 + m_3 R_3 + \cdots m_i R_i}{m_1 + m_2 + m_3 + \cdots m_i} \tag{8.38}$$

From the above equations, it is found that $m = m_1 + m_2 + m_3 + \cdots m_i$, where m is the mass of the mixture.

Amagat's Law of Partial Volumes

Amagat's law of partial volumes states that the volume of an ideal gas mixture is the sum of partial volumes of the constituent gases of the mixture, and the partial volume of an individual constituent gas is the volume it alone occupies at the mixture temperature and pressure. The partial volume of a component of a mixture is the volume the component would occupy at the pressure and temperature of the mixture.

For an ideal gas, $pV_1 = m_1R_1T \ldots pV_i = m_iR_iT$

$$p(V_1 + V_2 + V_3 + \cdots + V_i) = (m_1R_1 + m_2R_2 + m_3R_3 + \cdots m_iR_i)T$$

Then,

$$V = V_1 + V_2 + V_3 + \cdots + V_i \tag{8.39}$$

Eq. 8.39 is known as Amagat's law of partial volumes.

$$v_i = \left(\frac{n_iRT}{P}\right)$$

Dividing and multiplying RHS of the equation by n

$$v_i = \left(\frac{n_i}{n}\right)\frac{nRT}{P}$$

The partial volume of an individual constituent gas in a mixture of ideal gas is equal to the product of its mole fraction and the volume of the mixture.

8.6 INTERNAL ENERGY, ENTHALPY, AND SPECIFIC HEATS OF GAS MIXTURES

If the gases of equal pressures and temperatures are mixed in an adiabatic constant volume container, involving no work, the internal energy of the mixture remains constant. Moreover, in the absence of heat and work interactions, the temperature also remains constant. Thus, the total internal energy of the mixture of gases is equal to the sum of internal energy of all the constituents of the mixture, each considered at temperature and volume of the mixture. This statement equally applies to other thermodynamic properties such as enthalpy and specific heats as well. Equations for the average specific internal energy, average specific enthalpy, and average specific heats of the gas mixtures on a unit mass basis are as follows:

$$mu_m = m_1u_1 + m_2u_2 + m_3u_3 + \cdots m_iu_i$$

$$u_m = \frac{m_1u_1 + m_2u_2 + m_3u_3 + \cdots m_iu_i}{m_1 + m_2 + m_3 + \cdots m_i} \tag{8.40}$$

$$mh_m = m_1h_1 + m_2h_2 + m_3h_3 + \cdots m_ih_i$$

$$h_m = \frac{m_1h_1 + m_2h_2 + m_3h_3 + \cdots m_ih_i}{m_1 + m_2 + m_3 + \cdots m_i} \tag{8.41}$$

$$c_{Pm} = \frac{m_1c_{P1} + m_2c_{P2} + m_3c_{P3} + \cdots m_ic_{Pi}}{m_1 + m_2 + m_3 + \cdots m_i} \tag{8.42}$$

$$c_{v_m} = \frac{m_1c_{v_1} + m_2c_{v_2} + m_3c_{v_3} + \cdots m_ic_{v_i}}{m_1 + m_2 + m_3 + \cdots m_i} \tag{8.43}$$

8.7 ENTROPY OF GAS MIXTURES

Entropy of Gas Mixtures

According to Gibb's theorem, the total entropy of a mixture of gases is the sum of the partial entropies of the constituent gases. Partial entropy of the component is the entropy, the gas would occupy the total volume alone at the same temperature. If a number of inert ideal gases are separated from each other by partitions at the same temperature and pressure, then

$$s = n_1s_1 + n_2s_2 + \cdots n_is_i = \sum n_ks_k$$

Thermodynamic properties relation $Tds = dh - \nu dp$

$$\text{and}\, dh = c_p \cdot dT$$

$$Tds = c_p dT - \nu dp \Rightarrow ds = c_p \frac{dT}{T} - \frac{\nu}{T}dp$$

$$ds = c_p \frac{dT}{T} - \overline{R}\frac{dp}{p}$$

On integrating the above equation to obtain entropy of 1 mol of kth gas at T and P,

$$\overline{s}_k = \int c_{p_K} \frac{dT}{T} - \overline{R}\ln p + \overline{s}_{ck}$$

where s_{ck} is the constant of integration.

Then, $S_i = \overline{R}\Sigma n_K \left[\frac{1}{\overline{R}}\int c_{p_K}\frac{dT}{T} + \frac{\overline{s}_{ck}}{\overline{R}} - \ln p\right]$

If we assume $\alpha_K = \frac{1}{\overline{R}}\int c_{p_K}\frac{dT}{T} + \frac{\overline{s}_{ck}}{\overline{R}}$

then $S_j = \overline{R}\Sigma n_K (\alpha_K - \ln p)$

The gases diffuse into one another when the partitions are removed at the same pressure and temperature. The entropy of the mixture s_j (final) is the sum of partial entropies according to Gibb's theorem.

Again $p_K = x_K \cdot p \Rightarrow S_j = \overline{R}\Sigma n_K(\alpha_K - \ln x_K - \ln p)$

The change in entropy due to diffusion is

$$\begin{aligned} S_j - S_i &= -\overline{R}\Sigma n_K \ln x_K \\ &= -\overline{R}[n_1\ln x_1 + n_2\ln x_2 + \cdots n_i \ln x_i] \end{aligned} \tag{8.44}$$

Now $x_1 = \frac{p_1}{p}, x_2 = \frac{p_2}{p} \cdots x_i = \frac{p_i}{p}$

$$S_j - S_i = -\bar{R}\left[n_1 \ln\frac{p_1}{p} + n_2 \ln\frac{p_2}{p} + \cdots n_i \ln\frac{p_i}{p}\right] \quad (8.45)$$

Equation 8.45 implies that each gas diffuses, a free expansion from the total pressure p to its partial pressure at a constant temperature.

On a mass basis, the entropy change due to diffusion is

$$S_j - S_i = -\Sigma m_k R_k \ln\frac{p_k}{p} = -\left[m_1 R_1 \ln\frac{p_1}{p} + m_2 R_2 \ln\frac{p_2}{p} + \cdots m_i R_i \ln\frac{p_i}{p}\right] \quad (8.46)$$

EXAMPLE PROBLEMS

Example 8.1 A vessel of 8 m^3 volume contains 10 kg of N_2 gas at 25°C. Determine the pressure exerted by the N_2 gas (i) when the gas obeys the ideal gas equation and (ii) when the gas follows Van der Waals equation. Constants a and b are 136.7 kN m^4/(kg mol)^2 and 0.0386 m^3/kg mol respectively.

Solution $V = 8\ m^3 \qquad m = 10\ kg \qquad T = 298\ K$

Gas constant of $N_2, R_{N_2} = \frac{\bar{R}}{\mu_{N_2}}$ where μ_{N_2} is the molecular weight of nitrogen

$$= \frac{8.314}{28} = 0.296\ kJ/kg\,K$$

When the gas obeys ideal gas equation, $p = \frac{mRT}{V} = \frac{10 \times 0.296 \times 298}{8} = 110.60\ kPa$

Ans.

When the gas follows Vander Waals equation, $\left(p + \frac{a}{v^2}\right)(v - b) = RT$

Molar volume, $v = \frac{V}{n}\left(\text{where n is number of moles, } n = \frac{m}{\mu}\right)$

Therefore $v = \frac{V}{m} \times \mu = \frac{8}{10} \times 28 = 22.4\ m^3/kg\,mol$

$$\left(p + \frac{136.7}{22.4^2}\right)(22.4 - 0.0386) = 8.314 \times 298$$

Then p = 110.53 kPa Ans.

Example 8.2 A gas mixture consists of 5 kg of O_2, 8 kg of N_2, and 10 kg of CO_2. Evaluate (i) mole fraction of each gas, (ii) mass fraction of each gas, and (iii) average gas constant of the mixture.

Solution Total mass of the mixture, $m_m = 5 + 8 + 10 = 23$ kg.

Mole fraction of a component can be found by finding the mole number of that component

$$n_{O_2} = \frac{\text{mass of } O_2}{\text{molecular weight of } O_2} = \frac{5}{32} = 0.156\,\text{kmol}$$

$$n_{N_2} = \frac{\text{mass of } N_2}{\text{molecular weight of } N_2} = \frac{8}{28} = 0.285\,\text{kmol}$$

$$n_{CO_2} = \frac{\text{mass of } CO_2}{\text{molecular weight of } CO_2} = \frac{10}{44} = 0.227\,\text{kmol}$$

Total number of moles, $n_m = 0.15 + 0.285 + 0.227 = 0.668$ kmol

$$\text{Mole fraction} = \frac{\text{Mole number of a component in the mixture}}{\text{Total mole numbers of the mixture}}$$

$$\text{Mole fraction of } O_2 = \frac{0.156}{0.668} = 0.233$$ Ans.

$$\text{Mole fraction of } N_2 = \frac{0.285}{0.668} = 0.426$$ Ans.

$$\text{Mole fraction of } CO_2 = \frac{0.227}{0.668} = 0.339$$ Ans.

$$\text{Mass fraction} = \frac{\text{Mass of a component in the mixture}}{\text{Total mass of the mixture}}$$

$$\text{Mass fraction of } O_2 = \frac{5}{23} = 0.217$$

$$\text{Mass fraction of } N_2 = \frac{8}{23} = 0.347$$

$$\text{Mass fraction of } CO_2 = \frac{10}{23} = 0.434$$

Mass $m = \mu\, n$ where $n \rightarrow$ number of moles, $\mu \rightarrow$ molar mass

$$\text{Then average molar mass of mixture, } \mu_m = \frac{\text{Total mass of the mixture}}{\text{Total mole numbers of the mixture}}$$

$$\frac{m_m}{n_m} = \frac{23}{0.668} = 34.43\,\text{kg / kmol}$$

$$\text{Average gas constant of mixture, } R = \frac{\overline{R}}{\mu_m} = \frac{8.314}{34.43} = 0.2414\,\text{kJ / kg K}$$ Ans.

Example 8.3 1.5 kg of air at a pressure 6 bar occupies a volume of 0.2 m^3. If this air is expanded to a volume of 1.1 m^3, find the work done and heat absorbed or rejected by the air for each of the methods, (i) isothermal and (ii) adiabatic.

Solution $p_1 = 6$ bar $= 600$ kPa $V_1 = 0.2\ m^3$, $V_2 = 1.1\ m^3$

$m = 1.5$ kg

i. Work done in isothermal process, $W_{1\text{-}2} = mRT\ln\left(\frac{V_2}{V_1}\right)$

or $W_{1\text{-}2} = p_1V_1\ln\left(\frac{V_2}{V_1}\right) = 600\times 0.2\ln\left(\frac{1.1}{0.2}\right)$

Work done $= 204.56$ kJ **Ans.**

Heat transfer, $Q_{1\text{-}2} = U_2 - U_1 + W_{1\text{-}2}$ $(U_2 - U_1 = 0)$

$Q_{1\text{-}2} = 204.56$ kJ (since internal energy change is zero) **Ans.**

ii. Work done in adiabatic process, $W_{1\text{-}2} = \frac{p_1V_1 - p_2V_2}{\gamma - 1}$

$$pV = C \text{ but} \Rightarrow \frac{p_2}{p_1} = \left(\frac{V_1}{V_2}\right)^{\gamma} \Rightarrow p_2 = p_1\times\left(\frac{V_1}{V_2}\right)^{\gamma} = 600\times\left(\frac{0.2}{1.1}\right)^{1.4}$$

$p_2 = 55.16$ kPa or 0.551 bar

$$\text{Work done} = \frac{600\times 0.2 - 55.16\times 1.1}{1.4-1} = 148.31\ \text{kJ}$$ **Ans.**

Heat transfer $= 0$ (since it is adiabatic)

Example 8.4 2 kg of air expands isentropically in a closed stationary system from an initial pressure of 600 kPa and a temperature of 20°C to 300 kPa. Compute (i) the change in enthalpy, (ii) the work done, (iii) the heat transferred, and (iv) the final temperature.

Solution For an ideal gas, for a reversible adiabatic (isentropic) process,

$$\frac{T_2}{T_1} = \left(\frac{p_2}{p_1}\right)^{\frac{\gamma-1}{\gamma}} \Rightarrow T_2 = 293\left(\frac{300}{600}\right)^{\frac{1.4-1}{1.4}} \Rightarrow 240.47\,\text{K (final temp)}$$

$$\text{Enthalpy change} = \frac{\gamma mRT_1}{\gamma-1}\left[\left(\frac{p_2}{p_1}\right)^{\frac{\gamma-1}{\gamma}} - 1\right]$$

$$\frac{1.4\times 2\times 0.284\times 293}{1.4-1}\left[\left(\frac{300}{600}\right)^{\frac{0.4}{1.4}} - 1\right] = -104.26\,\text{kJ}$$ **Ans.**

Heat transferred $= 0$ (Since the process is reversible adiabatic)

$$Q_{1-2} = U_2 - U_1 + W_{1-2} \Rightarrow (-U_{1-2}) = W_{1-2}$$

$$W_{1\text{-}2} = U_1 - U_2 = m(u_1 - u_2) = \frac{mRT_1}{\gamma-1}\left[1 - \left(\frac{P_2}{P_1}\right)^{\frac{\gamma-1}{\gamma}}\right]$$

$$W_{1\text{-}2} = \frac{2\times 0.284\times 293}{1.4-1}\left[1-\left(\frac{300}{600}\right)^{\frac{0.4}{1.4}}\right] = 74.58\,\text{kJ} \quad \text{Ans.}$$

Example 8.5 Heat is transferred at a constant pressure to a 5 kg gas which occupies 0.3 m^3 volume in a chamber. The temperature of the gas rises from 10°C to 120°C. Calculate (i) the final volume, (ii) the heat transferred, (iii) the work done, (iv) the changes in enthalpy and entropy, (v) the gas constant, and (vi) the molecular weight. For gas, take $c_p = 1.88$, $c_v = 1.43$ kJ/kg K.

Solution $V_1 = 0.3$ m^3 $\quad m = 5$ kg

$t_1 = 10°C \qquad t_2 = 120°C$

Gas constant, $R = c_p - c_v = 1.88 - 1.43 = 0.45$ kJ/kg K

For constant pressure process, $Q_{1\text{-}2} = mc_p(t_2 - t_1) = 5\times 1.88(120-10) = 1034$ kJ

$$p_1V_1 = mRT_1 \Rightarrow p_1 = \frac{5\times 0.45\times 283}{0.3} = 2122.5\,\text{kPa}$$

$$p_2V_2 = mRT_2 \Rightarrow V_2 = \frac{5\times 0.45\times 393}{2122.5} = 0.416\,\text{m}^3$$

$$W_{1\text{-}2} = \int_1^2 p\,dV = p_1(V_2 - V_1) = 2122.5(0.416 - 0.3) = 246.21\,\text{kJ}$$

Change in internal energy, ΔU or $U_{1\text{-}2}$

$Q_{1\text{-}2} = U_{1\text{-}2} + W_{1\text{-}2} \Rightarrow U_{1\text{-}2} = 1034 - 246.21 = 787.79\,\text{kJ}$ Ans.

Change in enthalpy, $\Delta H = mc_p(t_2 - t_1) = 5\times 1.88\times 110 = 1034\,\text{kJ}$ Ans.

Change in entropy, $S_2 - S_1 = m\left(c_p \ln\frac{T_2}{T_1} - R\ln\frac{p_2}{p_1}\right), \left(R\ln\frac{p_2}{p_1} = 0\right.$, since it is constant pressure process$\left.\right)$

$$\therefore S_2 - S_1 = mc_p \ln\frac{T_2}{T_1} = 5\times 1.88\times \ln\left(\frac{393}{283}\right) = 3.0866\,\text{kJ/K} \quad \text{Ans.}$$

Gas constant, $R = c_p - c_v = 1.88 - 1.43 = 0.45\,\text{kJ/kg K}$ Ans.

Molecular weight, $\mu = \frac{\bar{R}}{R} = \frac{8.314}{0.45} = 18.47\,\text{kg/kg mol}$ Ans.

Example 8.6 Determine the specific volume of superheated vapor at 15 MPa and 350°C using (i) the ideal gas equation, (ii) the generalized compressibility chart, and (iii) steam tables. Determine the error involved in the first two cases.

Solution From superheated steam tables, at 15 MPa, 350°C,

Specific volume $\upsilon = 0.011481$ m^3/kg Ans.

This is the most accurate value.

From data of water vapor, $R = 0.4615$ kJ/kg K

$$p_c = 22.06 \text{ MPa}, \ T_c = 647.1 \text{ K}$$

i. Ideal gas equation $p\upsilon = RT \rightarrow \upsilon = \frac{0.4615 \times 623}{15 \times 10^3} = 0.01916 \text{m}^3/\text{kg}$ Ans.

ii. Generalized compressibility chart

$$p_r = \frac{p}{p_c} = \frac{15}{22.06} = 0.68$$

$$T_r = \frac{T}{T_c} = \frac{623}{647.1} = 0.962$$

From the compressibility chart, at $p_r = 0.68$ and $T_r = 0.962$, $Z = 0.65$.

Again $Z = \frac{\nu_{actual}}{\nu_{ideal}} \Rightarrow \upsilon_{actual} = 0.65 \times 0.01916 = 0.01245 \text{m}^3/\text{kg}$ Ans.

$$\text{Error in case I} = \frac{0.01916 - 0.011481}{0.011481} \times 100 = 67\%$$ Ans.

$$\text{Error in case II} = \frac{0.01246 - 0.011481}{0.011481} \times 100 = 8.5\%$$ Ans.

Example 8.7 An ideal gas of 2.5 kg mass occupies a volume of 1.8 m^3 at 103 kPa and 100°C. The gas is compressed in a polytropic process ($pv^{1.3} = c$) to a pressure of 600 kPa. The gas then expands at a constant pressure to the initial volume. Determine (i) the volume and temperature at the end of compression, (ii) total work transfer, (iii) total heat transfer, and (iv) total change of entropy.

Solution $V_1 = 1.8$ m^3 $p_1 = 103$ kPa

$T_1 = 373$ K $p_2 = 600$ kPa

$$p_1 V_1 = mRT_1 \Rightarrow R = \frac{103 \times 1.8}{2.5 \times 373} = 0.2 \text{kJ / kgK}$$

$$p_1 v_1^{1.3} = p_2 v_2^{1.3} \Rightarrow \frac{v_2}{v_1} = \left(\frac{p_1}{p_2}\right)^{1/1.3} = \frac{V_2}{V_1}$$

$$V_2 = 1.8\left(\frac{103}{600}\right)^{0.763} = 0.464 \text{m}^3$$ Ans.

$$p_2 V_2 = mRT_2$$

$$\Rightarrow T_2 = \frac{600 \times 0.464}{2.5 \times 0.2} = 556.8 \text{K} = 283.8°\text{C}$$ Ans.

$$W_{1\text{-}2} = \frac{p_1 V_1 - p_2 V_2}{n - 1} = \frac{103 \times 1.8 - 600 \times 0.464}{0.3} = -310 \text{kJ}$$

$$W_{2\text{-}3} = \int_2^3 p_2 \, dV = p_2(V_3 - V_2) = 600(1.8 - 0.464) = 801.6 \text{kJ} \quad (\because v_1 = v_3)$$

Total work transfer $= W_{1\text{-}2} + W_{2\text{-}3} = -310 + 801.6 = 491.6$ kJ Ans.

$$Q_{1\text{-}2} = U_{1\text{-}2} + W_{1\text{-}2} = mc_v(T_2 - T_1) + W_{1\text{-}2}$$

$$= 2.5 \times 0.783(556.8 - 373) + (-310) = 49.78\,\text{kJ}$$

$$Q_{2\text{-}3} = mc_p\,(T_3 - T_2)$$

$$\frac{p_2V_2}{T_2} = \frac{p_3V_3}{T_3} \Rightarrow T_3 = \frac{T_2V_1}{V_2}$$

$$\Rightarrow T_3 = \frac{1.8}{0.464} \times 388.013 = 1505.22\,\text{K}$$

$$\therefore Q_{2\text{-}3} = 2.5 \times 1.005(1505.22 - 388.013) = 2806.98\,\text{kJ}$$

Total heat transferred $= -280.61 + 2806.98 = 2526.37$ kJ Ans.

$$S_2 - S_1 = m\left[c_v \ln\frac{T_2}{T_1} + R\ln\frac{V_2}{V_1}\right]$$

$$= 2.5\left[0.783\ln\frac{556.8}{373} + 0.287\ln\frac{0.464}{1.8}\right] = -0.1884\,\text{kJ/K}$$

$$S_3 - S_2 = m\left(c_v \ln\frac{T_3}{T_2} + R\ln\frac{V_3}{V_2}\right) = 2.5\left[0.783\ln\frac{1505.22}{556.8} + 0.287\ln\frac{1.8}{0.464}\right] = 2.9124\,\text{kJ/K}$$

Total entropy change $= -0.1884 + 2.9124 = 2.731$ kJ/K Ans.

Example 8.8 2 kg of H_2 at 30°C is mixed with 5 kg of O_2 at the same temperature. The initial pressure of both the gases is 101 kPa and remains the same after mixing. Estimate the increase in entropy due to mixing.

Solution $m_{O_2} = 5\,\text{kg}$ $m_{H_2} = 2\,\text{kg}$

$T_1 = 303$ K

$$\text{Number of moles of } O_2 = \frac{\text{Mass}}{\text{Molecular weight}} = \frac{5}{32} = 0.156$$

$$\text{Number of moles of } H_2 = \frac{2}{2} = 1$$

$$\text{Mole fraction of } O_2 = \frac{0.156}{0.156 + 1} = 0.135$$

Then mole fraction of $H_2 = 1 - 0.135 = 0.865$

The partial pressure of each constituent gas is equal to its mole fraction

$$\Delta S = S_2 - S_1 = -m_{O_2} \times R_{O_2} \ln\left(\frac{p_{O_2}}{p}\right) - m_{H_2} \times R_{H_2} \ln\left(\frac{p_{H_2}}{p}\right)$$

$$R_{O_2} = \frac{\bar{R}}{\mu_{O_2}} = \frac{8.314}{32} = 0.259\,\text{kJ/kg K}$$

$$R_{H_2} = \frac{\bar{R}}{\mu_{H_2}} = \frac{8.314}{2} = 4.157\,\text{kJ/kg K}$$

$$\Delta S = -5 \times 0.259\ln(0.135) - 2 \times 4.157\ln(0.865) = 3.798\,\text{kJ/K}$$ Ans.

Example 8.9 A perfect gas mixture contains 5 kg of O_2 and 2 kg of CO_2 at a temperature of 25°C and a pressure of 250 kPa. Compute the (i) specific heats and (ii) change in internal energy, enthalpy, and entropy of the mixture, if the mixture is heated to 50°C at a constant pressure. Take c_p of $O_2 = 0.9094$, c_p of $CO_2 = 0.662$, c_V of $O_2 = 0.649$, and c_V of $CO_2 = 0.472$ kJ/kg K.

Solution $m_{O_2} = 5\,kg$ $m_{CO_2} = 2\,kg$

$T_1 = 298$ K $T_2 = 323$ K

Specific heat at constant pressure of mixture, $m_{O_2} \times c_{pO_2} + m_{CO_2} \times c_{pCO_2} = (m_{O_2} + m_{CO_2})c_p$

$$= 5 \times 0.9094 + 2 \times 0.662 = (5+2)\ c_p \Rightarrow c_p = 0.838\,kJ/kg\,K$$

Specific heat at constant volume of mixture, $m_{O_2} \times c_{vO_2} + m_{CO_2} \times c_{vCO_2} = (m_{O_2} + m_{CO_2})c_v$

$$= 5 \times 0.649 + 2 \times 0.472 = (5+2)\ c_\upsilon \Rightarrow c_\upsilon = 0.598\,kJ/kg\,K$$

Change in internal energy, $\Delta U = m\ c_\upsilon\ (T_2 - T_1) = 7 \times 0.598 \times (323 - 298) = 104.65$ kJ

Change in enthalpy, $\Delta H = m\ c_p\ (T_2 - T_1) = 7 \times 0.838 \times (323 - 298) = 146.45$ kJ

$$\text{Change in entropy}, \Delta S = m\left[c_p \ln\left(\frac{T_2}{T_1}\right) - R\ln\left(\frac{P_2}{P_1}\right)\right]$$

The term $R\ln\left(\frac{P_2}{P_1}\right) = 0$, since the gas is heated at constant pressure.

$$= 7 \times 0.838 \times \ln\left(\frac{323}{298}\right) = 0.47\,kJ/K$$ Ans.

Example 8.10 A gas mixture, consisting of 3 kg of N_2 and 5 kg of CO_2 at a temperature of 25°C and a pressure of 103 kPa, is compressed isentropically to 500 kPa. Compute (i) the final temperature, (ii) the final partial pressure of components, and (iii) the change in internal energy of the mixture during the process. Take Υ of N_2 and CO_2 as 1.4 and 1.286 respectively.

Solution $m_{N_2} = 3\,kg$ $m_{CO_2} = 5\,kg$

$P_1 = 103$ kPa $P_2 = 500$ kPa

$$\text{Gas constant of } N_2, R_{N_2} = \frac{\bar{R}}{\mu_{N_2}} \text{ where } \mu_{N_2} \text{ is the molecular weight of nitrogen}$$

$$R_{N_2} = \frac{8.314}{28} = 0.296\,kJ/kg\,K$$

$$\text{Gas constant of } CO_2, R_{CO_2} = \frac{\bar{R}}{\mu_{CO_2}} = \frac{8.314}{44} = 0.188\,kJ/kg\,K$$

$$c_{vN_2} = \frac{R_{N_2}}{\gamma - 1} = \frac{0.296}{1.4 - 1} = 0.74$$

$$c_{pN_2} = 1.4 \times 0.74 = 1.036\,kJ/kg\,K \quad \text{where } \gamma = \frac{c_p}{c_v}$$

$$c_{vCO_2} = \frac{R_{CO_2}}{\gamma - 1} = \frac{0.188}{1.4 - 1} = 0.47\,kJ/kg\,K$$

$$c_{pCO_2} = 1.286 \times 0.47 = 0.604\,kJ/kg\,K$$

For the mixture, $m_{N_2} \times c_{pN_2} + m_{CO_2} \times c_{pCO_2} = \left(m_{N_2} + m_{CO_2}\right)c_p$

Then $$c_p = \frac{3 \times 1.036 + 5 \times 0.604}{3 + 5} = 0.766\,kJ/kg\,K$$

$$m_{N_2} \times c_{vN_2} + m_{CO_2} \times c_{vCO_2} = \left(m_{N_2} + m_{CO_2}\right)c_v$$

$$c_v = \frac{3 \times 0.76 + 5 \times 0.47}{3 + 5} = 0.578\,kJ/kg\,K$$

Final temperature of the mixture $\frac{T_2}{T_1} = \left(\frac{P_2}{P_1}\right)^{\frac{\gamma-1}{\gamma}} \Rightarrow T_2 = 298\left(\frac{500}{103}\right)^{\frac{1.4-1}{1.4}} = 467\,K$

The final partial pressure of component:

$$\text{Mole fraction of } N_2, \chi_{N_2} = \frac{\frac{3}{28}}{\frac{3}{28} + \frac{5}{44}} = 0.485$$

$$\text{Mole fraction of } CO_2, \chi_{CO_2} = \frac{\frac{5}{44}}{\frac{3}{28} + \frac{5}{44}} = 0.514$$

$$\text{Final partial pressure of } N_2, p_{N_2} = \chi_{N_2} \times p_2 = 0.485 \times 500 = 242.5\,kPa$$

$$\text{Final partial pressure of } CO_2, p_{CO_2} = \chi_{CO_2} \times p_2 = 0.514 \times 500 = 257.45\,kPa$$

Change in internal energy of the mixture, $\Delta U = mc_\upsilon\,(T_2 - T_1)$

$= 7 \times 0.578 \times (467 - 298) = 683.77$ kJ Ans.

REVIEW QUESTIONS

8.1 What is the significance of the compressibility factor Z?

8.2 Write the Van der Waal's equation of state.

8.3 What is the principle of corresponding states?

8.4 What is the difference between characteristic gas constant and universal gas constant? How are these two related?

8.5 Under what conditions do the real gases obey the ideal-gas assumption?

8.6 Define reduced pressure and reduced temperature.

8.7 What is the difference between molal mass and molar mass? How are these two related?

8.8 What is the difference between component pressure and partial pressure? When are these two equivalent?

8.9 What is the difference between component volume and partial volume? When are these two equivalent?
8.10 What is the purpose of multi-stage compression of gases?
8.11 Why is intercooling required in the case of multi-stage compression of gases?
8.12 What is Dalton's law of partial pressures?

EXERCISE PROBLEMS

8.1 A piston-cylinder arrangement has air at a pressure of 90 kPa and temperature 25°C with a mass of 0.2 kg. Air is compressed reversibly according to the law $p\nu^{1.4}=$constant. The final volume is 1/6th of the initial volume. Find the heat and work transfer associated with the process.
8.2 An ideal gas, initially at a temperature of 20°C, is heated reversibly at a constant volume until its pressure becomes double, and again heated at a constant pressure until the volume is tripled. For total path, find the heat transfer, work transfer, and entropy change. Take the mass of ideal gas as 0.8 kg.
8.3 From experiments, it is found that the specific heat ratio of CH_4 is 1.6. Determine the two specific heats.
8.4 Two chambers X and Y, each of volume 2 m^3, are connected by a tube with a negligible volume. Chamber X contains air at 5 bar and 100°C, while Y contains air at 2.5 bar and 220°C. Find the entropy change when X is connected with Y. Assume chambers are insulated from surroundings.
8.5 Air enters at a rate of 1000 Kg/h into a reciprocating compressor with a water cooling jacket approximating a quasi-static compression following a path $p\nu^{1.4}=$constant. Initially, air is at 25°C and 1 atm and is compressed to 5 bar. Determine the discharge temperature of air, power required to drive the compressor, and heat transfer associated with the process.
8.6 A 3.27-m^3 tank contains 100 kg of nitrogen at 175 K. Determine the pressure in the tank using (i) the ideal-gas equation, (ii) the Van der Waals equation, and (iii) the Beattie–Bridgeman equation. Compare your results with the actual value of 1505 kPa.
8.7 1 kg of water vapor and 3 kg of air are held in a cylinder of volume 1.23 m^3. If the temperature of the mixture is 400 K, estimate the mixture pressure. Also estimate the pressure to which the mixture could be compressed isothermally before it ceases to be a pure substance.
8.8 A constant volume chamber of 0.42 m^3 contains 1 kg of air at 7°C. Heat is transferred to the air until its temperature becomes 85°C. Estimate (i) the work done, (ii) the heat transferred, and (iii) the changes in internal energy, enthalpy, and entropy.
8.9 A gas mixture, consisting of 3 kg of N_2 and 5 kg of CO_2 at a temperature of 25°C and a pressure of 103 kPa, is heated to a temperature of 125°C. Compute the changes in internal energy, enthalpy, and entropy (i) at constant volume and (ii) at constant pressure.
8.10 A gas mixture consists of 2 kg of CO_2 and 1 kg of CH_4 at a pressure of 150 kPa and a temperature of 302 K. Determine the partial pressure of each gas and the apparent molar and mass fractions of the gas mixture.

8.11 1 kg of oxygen and 1 kg of nitrogen are mixed at 25°C and a total pressure of 103 kPa. Determine (i) the volume of the mixture, (ii) partial volumes of the components, (iii) partial pressures of the components, (iv) specific heats of the mixture, and (v) gas constant of the mixture.

8.12 A cylinder fitted with a frictionless piston contains 0.45 m^3 of air at 1.3 bar, 25°C. The air is then compressed reversibly according to the law pv^n = constant until the final pressure is 5.5 bar, at which point the temperature is 130°C. Compute (i) the polytropic index n, (ii) the final volume of air, (iii) the work done on the air and the heat transfer, and (iv) the net change in entropy.

8.13 A closed system allows nitrogen to expand reversibly from a volume of 0.3–0.8 m^3 along the path $pv^{1.3}$ = const. The original pressure of the gas is 3 bar and its initial temperature is 90°C. Compute (i) the final temperature and the final pressure of the gas, (ii) the work done and heat transferred, and (iii) the entropy change of nitrogen. Draw the p-v and T-s diagrams.

8.14 A piston-cylinder device contains a mixture of 0.75 kg of helium and 1.5 kg of nitrogen at 101.3 kPa and 30°C. Heat is now transferred to the mixture at a constant pressure until the volume is doubled. Compute (i) the heat transfer and (ii) the entropy change of the mixture assuming constant specific heats at the average temperature.

8.15 A process requires a mixture that is 21% oxygen, 78% nitrogen, and 1% argon by volume. All three gases are supplied from separate tanks to an adiabatic, constant-pressure mixing chamber at 1.8 bar but at different temperatures. The oxygen enters at 20°C, the nitrogen at 50°C, and the argon at 150°C. Compute total entropy change for the mixing process per unit mass of the mixture.

8.16 A volume of 0.286 m^3 of O_2 at 180 K and 7.5 MPa is mixed with 0.54 m^3 of N_2 at the same temperature and pressure, forming a mixture at 180 K and 7.5 MPa. Determine the volume of the mixture, using (i) the ideal-gas equation of state and (ii) the compressibility chart.

8.17 A gas mixture at 25°C, 105 kPa is 60% N_2, 20% H_2O, and 20% O_2 on a mole basis. Compute the mass fractions, mixture gas constant, and volume for 5 kg of the mixture.

8.18 A 3-kg mixture of 50% N_2, 25% O_2, and 25% CO_2 by mass is at 125 kPa and 298 K. Find the mixture gas constant and the total volume.

8.19 A steady flow of 0.15 kg/s carbon dioxide at 800 K in one line is mixed with 0.25 kg/s nitrogen at 500 K from another line, both at 105 kPa. The mixing chamber is insulated and has a constant pressure of 105 kPa. Determine the mixing chamber's exit temperature, assuming constant heat capacity.

8.20 A rigid insulated vessel contains 10 kg of oxygen at 200 kPa, 300 K separated by a membrane from 20 kg carbon dioxide at 370 kPa, 350 K. The membrane is removed and the mixture comes to a uniform state. Find the final temperature and pressure of the mixture.

DESIGN AND EXPERIMENT PROBLEMS

8.21 Hot combustion gases from the combustor of a turbojet engine enter the nozzle at high pressure where they expand so that a high-velocity jet is produced, which provides the thrust for propelling the aircraft. Plot the graph between the mass flow rate of gases entering and their velocity leaving the nozzle. Determine the optimal mass flow rate of gases that produce the highest thrust.

9 Concept of Available Energy (Exergy)

LEARNING OUTCOMES

After learning this chapter, students should be able to

- Define exergy and reversible work.
- Evaluate the performance of engineering devices in light of the second law of thermodynamics.
- Apply the second-law efficiency to engineering systems for evaluating their best possible performance.
- Develop the exergy relation to closed systems and control volumes.
- Apply exergy balance to closed systems and control volumes to assess the useful and wasted work potential of engineering devices.

9.1 AVAILABLE ENERGY (EXERGY)

In previous chapters, we have discussed the concept of reversible process and reversible efficiency that are used to compare the performance of actual efficiency of the systems. In this chapter, we will develop the concept of reversible work that is essential to evaluate the work potential of a system or a process. As far as the first law of thermodynamics is concerned, energy doesn't degrade quantity wise; however, as per the second law, it degrades quality wise. Thus, the second law is also called the law of degradation of energy. For example, a gas at higher pressure can do more work than the same gas at lower pressure though they both have the same stored energy. One kilogram of a gas at 1 atm pressure and 50°C temperature has the same stored energy as that of the same mass of gas at 2 atm pressure and the same temperature.

It is important here to define the system, surroundings, and atmosphere that are used in the analysis of maximum work, useful work, and available energy. A system is defined as a region in space within the prescribed boundaries. Surroundings are everything external to the system boundaries and for most of the thermodynamic systems, atmosphere with uniform temperature and pressure is one part of the surroundings. Although the atmospheric temperature and pressure are not influenced by any process of the system as the atmosphere is very large in comparison with the system, the atmosphere can influence the activities of the system. Figure 9.1 shows the system and surroundings with the atmosphere in which the system is the gas in the cylinder and surroundings are the atmosphere, piston, cylinder, energy reservoir, and coil spring. The system exchanges heat with the energy reservoir and with the atmosphere. If the system changes its volume, then it exchanges work with the coil spring and atmosphere. Figure 9.1 shows a system and surroundings with the atmosphere.

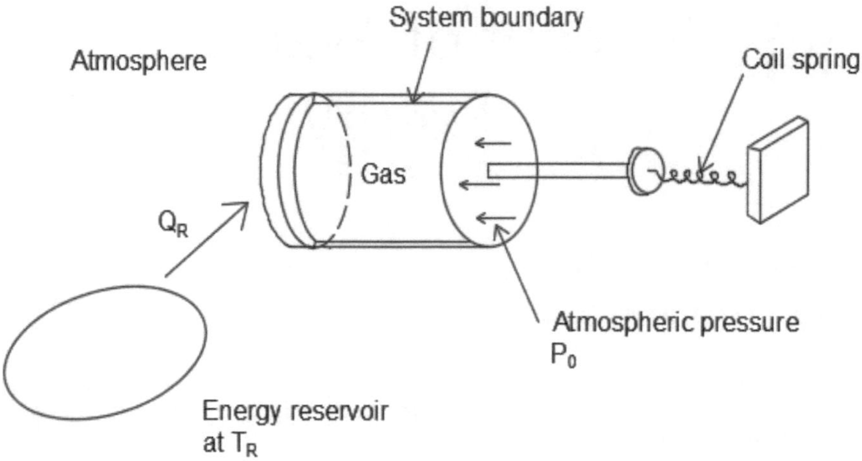

FIGURE 9.1 System and surroundings with the atmosphere.

9.2 REVERSIBLE WORK AND IRREVERSIBILITY

9.2.1 Useful Work

Useful work is the work done by a system excluding the work done on the atmosphere. The *available energy or exergy* is defined as the maximum useful work that can be obtained from the system and surroundings as the system undergoes a process from the initial state to the dead state while exchanging heat with the surroundings only. It is the work potential of the energy source. The amount of energy that is discarded as waste energy is called unavailable energy or anergy. The work potential is the maximum useful work that can be obtained from a system at a specified state. The work extracted from a source can be maximum when the final state approaches the dead state, the state of surroundings that is characterized by temperature T_0 and pressure p_0. Thus the work done during a process depends on the initial state, final state, and the path the system follows during the process. It can be concluded from the above discussion that if the process is carried out in a reversible manner, its work output will be maximum.

Dead state is defined as the state which is in thermodynamic equilibrium with its surroundings, that is, the state of zero exergy. At the dead state, the system is at a state of environment and its properties are characterized by a subscript zero. The temperature at dead state T_0 is 25°C and pressure p_0 is 101.325 kPa. All other properties such as specific enthalpy (h), specific entropy (s), and specific internal energy (u) are also specified by a subscript zero at the dead state. To maximize the work output, the final state must be the dead state. For example, if the steam in a turbine is expanded to the pressure above the atmospheric still, there is scope for obtaining work output by expanding it further to the atmospheric pressure. For this reason, a vacuum is created in steam condensers to take the advantage of large expansion and increased work output.

The work potential of a system can be best assessed by comparing its relative performance with other systems operating between the same initial and final states. Exergy is a property that forms the basis for finding out the quality of energy by comparing the relative performance of the systems. However, in evaluating the engineering devices, it is often assumed that the final state is a dead state, which is the state of the environment. For most of the engineering devices, the final state may not be the dead state and thus, it may not be sufficient to determine the energy quality with the exergy alone. Reversible work and irreversibility serve as important tools in evaluating the thermodynamic systems. The useful work output of a work-producing device is always less than that of actual work. The reason for this is some work must be expended to overcome the surroundings, which cannot be recovered to perform any useful purpose, and is called surroundings work W_{surr}. This can be demonstrated with the help of an example in which a gas expands doing work in a piston-cylinder arrangement. However, a part of this work must be expended to push the atmospheric air out of the piston and it represents a loss. The surroundings work is given by

$$W_{surr} = p_0 \left(V_2 - V_1 \right) \tag{9.1}$$

where p_0 is the atmospheric pressure and V_1 and V_2 are initial and final volumes.

Now the useful work is the difference between actual work and surroundings work, given as

$$W_{useful} = W - W_{surr} \tag{9.2}$$

In contrast to the above, when a gas is compressed, the atmospheric pressure helps the compression process and, therefore, W_{surr} is a gain.

9.2.2 Reversible Work

Reversible work can be defined as the maximum amount of useful work that can be produced or the minimum work that has to be supplied when a system undergoes a process within the specified inlet and exit states. It can also be stated as the useful work output that is obtained or input that is expended when a process is carried out in a fully reversible manner within the specified inlet and exit states. For work-producing devices, reversible work represents the maximum work that can be produced in a process and for work-consuming devices, it is the minimum work that is required for performing the process. Any natural process is always accompanied by irreversibilities, and owing to these irreversibilities, there occurs some difference between the useful work and the reversible work, that is called irreversibility and it can be shown as

$$I = W_{max,out} - W_{useful,out} \left(\text{for work producing devices} \right) \tag{9.3}$$

$$I = W_{useful,in} - W_{min,in} \left(\text{for work consuming devices} \right) \tag{9.4}$$

9.3 EXERGY CHANGE OF A SYSTEM

The exergy is a property and the exergy change of a system can be positive, negative, or zero. Exergy change of open and closed systems is presented in the following sections.

9.3.1 Exergy of a Flow Stream (Open System) Exchanging Heat Only with Surroundings

An open system, as shown in Figure 9.2, exchanging energy with surroundings (T_0,p_0) is considered for deriving the exergy of a flow stream. A mass dm_1 enters and dm_2 leaves the system at states 1 and 2, respectively. Heat in the amount of δQ is absorbed by the system, δW is the work delivered by the system, and dE is the energy change of the system. Applying the first law,

$$\delta Q - \delta W = dE$$

$$\delta W = \delta Q + dm_1\left(h_1 + \frac{v_1^2}{2} + z_1 g\right) - dm_2\left(h_2 + \frac{v_2^2}{2} + z_2 g\right) - dE$$

$$\text{where } dE = d\left(U + \frac{mv^2}{2} + mgz\right) \tag{9.5}$$

But for maximum work, the process must be reversible. There is a temperature difference between the system and surroundings, and a heat engine operates in a reversible manner between the two. Heat δQ_0 is supplied to the heat engine from the source at T_0. Heat engine performs a work output of δW_c and δQ is the amount of heat rejected to the control volume at temperature T. For a reversible heat engine,

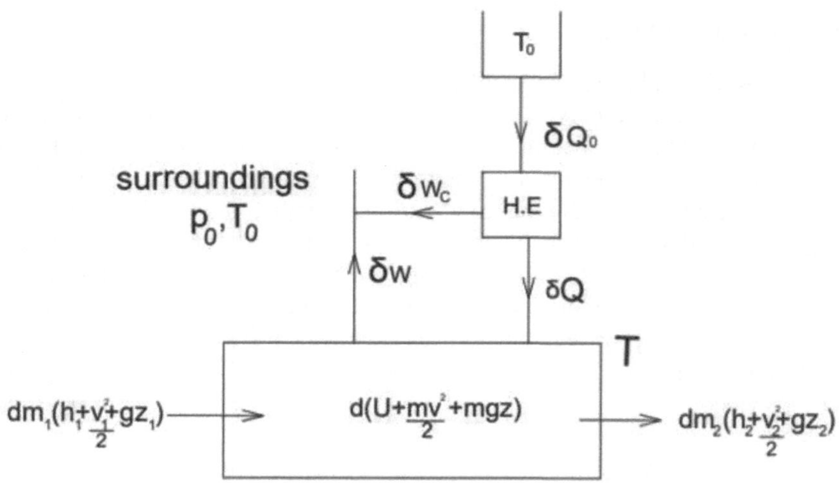

FIGURE 9.2 Maximum work done by a steady-flow system.

$$\frac{\delta Q_0}{T_0} = \frac{\delta Q}{T} \tag{9.6}$$

The work output of the engine, $\delta W_c = \delta Q_0 - \delta Q = \delta Q\frac{T_0}{T} - \delta Q$.

On rearranging the above equation,

$$\delta W_c = \delta Q\left(\frac{T_0}{T} - 1\right)$$

For a reversible process, the entropy balance equation is $dS = \frac{\delta Q}{T} + dm_1 s_1 - dm_2 s_2$

The maximum work output will be the sum of the work outputs of both the system and heat engine, that is $\delta W_{max} = \delta W_{rev} = \delta W + \delta W_c$

$$\delta W_{max} = \delta Q + dm_1\left(h_1 + \frac{v_1^2}{2} + z_1 g\right) - dm_2\left(h_2 + \frac{v_2^2}{2} + z_2 g\right) - d\left(U + \frac{mv^2}{2} + mgz\right) + \delta Q\left(\frac{T_0}{T} - 1\right)$$

$$\delta W_{max} = \delta Q + dm_1\left(h_1 + \frac{v_1^2}{2} + gz_1\right) - dm_2\left(h_2 + \frac{v_2^2}{2} + gz_2\right) - d\left(U + \frac{mv^2}{2} + mgz\right)$$
$$+ T_0\left(ds - dm_1 s_1 + dm_2 s_2\right) - \delta Q$$

where $\frac{\delta Q}{T} = dS - dm_1 s_1 + dm_2 s_2$

On rearranging the above equation,

$$\delta W_{max} = \delta Q + dm_1\left(h_1 + \frac{v_1^2}{2} + gz_1\right) - dm_2\left(h_2 + \frac{v_2^2}{2} + gz_2\right) - d\left(U + \frac{mv^2}{2} + mgz\right)$$

For a steady-flow process, $dm_1 = dm_2 = dm$

$$d\left[U - T_0 S + \frac{mv^2}{2} + mgz\right] = 0$$

$$\delta W_{max} = \delta Q + dm_1\left(h_1 + \frac{v_1^2}{2} + gz_1\right) - dm_2\left(h_2 + \frac{v_2^2}{2} + gz_2\right) - d\left(U + \frac{mv^2}{2} + mgz\right)$$

When the changes in kinetic energy and potential energy are neglected, then

$$\delta W_{max} = dm\left[(h_1 - h_2) - T_0(s_1 - s_2)\right]$$

$$w_{max} = (h_1 - h_2) - T_0(s_1 - s_2)\ (\text{on a unit mass basis}) \tag{9.7}$$

$$\text{or} \quad W_{max} = (H_1 - H_2) - T_0(S_1 - S_2)$$

9.3.2 Exergy of Non-Flowing Fluids (Closed Systems)

For a closed system, $dm_1 = dm_2 = 0$

$$\therefore \delta W_{max} = -d\left[U - T_0S + \frac{mv^2}{2} + mgz\right] \quad (9.8)$$

$$= -d[E - T_0S] \quad \text{where} \quad E = U + \frac{mv^2}{2} + mgz$$

$$W_{max} = E_1 - E_2 - T_0(S_1 - S_2)$$

$$\text{or} \quad W_{max} = (U_1 - U_2) - T_0(S_1 - S_2)$$

$$w_{max} = (u_1 - u_2) - T_0(s_1 - s_2) \text{(on a unit mass basis)}$$

$$= (u_1 - T_0s_1) - (u_2 - T_0s_2) \quad (9.9)$$

For an unsteady open or closed system, the volume of the system changes.

$$\therefore (\delta W_u)_{max} = \delta W_{max} - p_0 dv$$

$$\therefore \delta W_{max} = -d\left[U + p_0 v - T_0S + \frac{mv^2}{2} + mgz\right]$$

$$= d[E + p_0 v - T_0S]$$

$$W_{max} = (E_1 - E_2) + p_0(v_1 - v_2) - T_0(S_1 - S_2)$$

$$W_{max} = (U_1 + p_0v_1 - T_0S_1) - (U_2 + p_0v_2 - T_0S_2)$$

$$\phi = U + p_0 v - T_0S$$

$$W_{max} = \phi_1 - \phi_2 = (U_1 - U_2) + p_0(v_1 - v_2) - T_0(S_1 - S_2) \rightarrow \text{Unsteady closed system} \quad (9.10)$$

$$W_{max} = \psi_1 - \psi_2 = (H_1 - H_2) + p_0(v_1 - v_2) - T_0(S_1 - S_2) \rightarrow \text{Unsteady open system} \quad (9.11)$$

where ϕ is the exergy of the non-flowing stream or closed system and ψ is the exergy of the flowing stream or open system.

9.4 EXERGY TRANSFER BY HEAT, WORK, AND MASS

Exergy transfer accompanies energy transfer as the transfer of either of them is by heat, mass, and work. Exergy transfer is identified, as it crosses the system boundaries and quantifies as exergy gain or loss by a system during a process. Both energy and exergy share common units; however, they are essentially different concepts. Energy is related to the first law of thermodynamics, whereas exergy is related to the second law of thermodynamics; the former is a conserved quantity and the latter is destroyed by the irreversibilities such as friction, viscosity, and so on. Exergy transfer associated with a closed system is by heat transfer and work since a closed system does not involve any mass interactions.

Exergy Transfer by Heat

Heat and work are two means of energy transfer, according to the second law, heat is a low-grade energy and work a high-grade energy. Complete conversion of heat into work in a cyclic process is not possible, only a part of it can be converted to work. This can be explained by the fact that when heat is supplied to a heat engine, it produces work output and rejects waste heat to surroundings. Thus, heat transfer results in useful work output and involves exergy transfer. We can express the exergy transfer of a system associated with heat transfer Q at temperature T when the temperature of the environment is T_0 as

$$X_q = \left(1 - \frac{T_0}{T}\right)Q \tag{9.12}$$

In Eq. 9.12, when $T > T_0$, heat transfers into the system and exergy of the system increases; when $T < T_0$, heat transfers from the system and exergy of the system decreases. The exergy transfer is zero when $T = T_0$ since heat transfer is zero. It is important to note that when $T > T_0$, the directions of both exergy transfer and heat transfer are one and the same; when $T < T_0$, they are opposite to each other. In the former case, as the heat is transferred to a system, both energy and exergy of the system increase. In the latter case, when heat is transferred to the system, its energy increases but its exergy decreases. This happens when the heat engine to which heat is supplied is at a lower temperature than the source (surroundings) temperature.

For the case of variable temperature at which heat transfer takes place, the exergy transfer can be obtained by integrating Eq. 9.12.

$$X_q = \int \left(1 - \frac{T_0}{T}\right)\delta Q \tag{9.13}$$

Exergy Transfer by Work

Exergy transfer associated with work in the case of boundary work is given as

$$X_w = W - W_{surr} \tag{9.14}$$

where surroundings work, $W_{surr} = p_0(V_2 - V_1)$, in which p_0 is the atmospheric pressure and V_1 and V_2 are initial and final volumes of the system. For the work quantities other than boundary work, exergy transfer is

$$X_w = W \tag{9.15}$$

A piston-cylinder device involves the boundary work; in this case, the work done against the atmospheric air cannot be available as a useful work and should be subtracted. The exergy transfer for the devices that don't involve boundary work is equal to W.

If we consider a piston-cylinder device containing a gas at atmospheric pressure, when heat is transferred to the device, the gas expands raising the piston. The work done in this case is boundary work and it is just sufficient to push the atmospheric air out of the way of the piston and thus it is not available for any useful purpose. Similarly, when the work is done by the atmospheric air (when the gas is cooled), it is also not available for any useful purpose.

Exergy Transfer by Mass

It was mentioned in previous chapters that both the energy and entropy transfer are associated with mass transfer. Exergy transfer, like energy and entropy transfer, is also associated with mass transfer. Mass transfer is a means of transfer of energy, entropy, and exergy either into or out of the system. However, since closed systems don't involve any mass transfer across their boundaries, no exergy transfer due to mass takes place for the closed systems. If the exergy of a mass of m is mψ, then the associated exergy transfer is given as

$$X_m = m\psi \tag{9.16}$$

where $\psi = (h - h_0) - T_0(s - s_0) + \frac{V^2}{2} + zg$

In integral form, Eq. 9.16 can be written as

$$X_m = \int \psi \, \delta m \tag{9.17}$$

Eq. 9.17 can be used for evaluating the exergy transfer of fluid streams in which fluid properties vary.

Therefore, the exergy of a system increases by mψ when the mass in the amount of m enters, and decreases by the same amount when the same amount of mass at the same state leaves the system.

9.5 SECOND-LAW EFFICIENCY

As far as the first law is concerned, it always speaks about the quantity of energy and disregards its quality and the forms in which energy exists. The first-law efficiency of devices, such as thermal efficiency of heat engines and coefficient of performance (COP) of refrigerators and heat pumps, doesn't consider the best possible performance and therefore is not a realistic measure of the performance of engineering devices. This argument leads to the development of another performance measure that is the second-law efficiency, which provides a means of allocating a quality index to energy. The second-law efficiency is defined as the ratio of minimum available energy (exergy) that needs to be supplied to do a task and actual available energy consumed to perform the same task.

$$\eta_{II} = \frac{\text{Minimum exergy required to do a task}}{\text{Actual exergy required to do the same task}} \tag{9.18}$$

It can also be expressed as

$$\eta_{II} = \frac{\text{Exergy recovered}}{\text{Exergy supplied}} \text{ or } \eta_{II} = 1 - \frac{\text{Exergy destroyed}}{\text{Exergy supplied}} \tag{9.19}$$

For work-producing devices, it is defined as the ratio of useful work output and maximum possible (reversible) work output:

$$\eta_{II} = \frac{W_{useful}}{W_{max}} \tag{9.20}$$

where W_{useful} and W_{max} are useful and maximum work outputs, respectively.

For heat engines, it can be expressed as the ratio of actual thermal efficiency and reversible thermal efficiency:

$$\eta_{II} = \frac{\eta_{thermal}}{\eta_{rev.thermal}} \tag{9.21}$$

For work-consuming devices such as pumps and compressors, it is defined as the ratio of minimum (reversible) work input and actual work input:

$$\eta_{II} = \frac{W_{min}}{W_{useful}} \left(\text{where } W_{min} \text{ is the minimum work input}\right) \tag{9.22}$$

For refrigerators and heat pumps, it can be expressed as the ratio of actual COP and COP on a reversible basis.

$$\eta_{II} = \frac{COP}{COP_{rev}} \tag{9.23}$$

If the second-law efficiencies (the ratio of actual thermal efficiency and reversible thermal efficiency) of the two heat engines are 50% and 70%, then the first heat engine is converting 50% of the available work potential to useful work while it is 70% for the second one. Thus, the second heat engine is performing better than the first one. The second-law efficiency will be zero when the useful work obtained from the exergy supplied is zero.

9.6 EXERGY DESTRUCTION

Unlike energy, exergy is destroyed. Entropy is generated within the system due to irreversibilities such as friction, inelasticity, heat transfer through a finite temperature difference, and mixing process. The entropy generated is essentially accompanied by a loss of available energy known as exergy destruction. Entropy generated quantifies the exergy destruction

$$\dot{X}_{des} = T_0\dot{S}_{gen} \geq 0 \tag{9.24}$$

In accordance with the second law, the exergy destruction is positive when irreversibilities are present within the system during the process and vanishes in the limiting case where there are no irreversibilities. It is to be noted that exergy destruction is a positive quantity when irreversibilities are present within the system according to the second law (for any natural process) and becomes zero for a reversible process. So any natural process is accompanied by an exergy decrease. The principle of decrease of exergy is stated as exergy destroyed cannot be negative and is given below:

$$\dot{X}_{des}\begin{cases} > 0 \text{ irreversible process} \\ = 0 \text{ reversible process} \end{cases} \tag{9.25}$$

9.7 EXERGY BALANCE

Exergy is destroyed but it cannot be generated. It is not a conserved quantity like entropy. The exergy balance equation can be derived considering the exergy transfer and exergy destroyed during a process, the difference being the exergy change of a system. The exergy transfer of a system is due to heat, work, and mass. For any process and for any system, exergy balance can be

$$X_{in} - X_{out} - X_{des} = \Delta X_{system} \tag{9.26}$$

$$\underbrace{X_{in} - X_{out}}_{\substack{\text{Net exergy transfer by} \\ \text{heat, work and mass}}} - \underbrace{X_{des}}_{\text{Exergy destruction}} = \underbrace{\Delta X_{system}}_{\text{Change in exergy}} \tag{9.27}$$

A closed system, however, doesn't involve any mass transfer across its boundaries; exergy transfer in that case is due to heat and work.

EXAMPLE PROBLEMS

Example 9.1 A refrigeration plant for food storage maintains the store at a temperature of −5°C when the heat transfer from cycle to the atmosphere is at a temperature of 27°C. The heat transfer from the food storage unit to the cycle is at a rate of 7.8 kJ/s. The power required to drive the plant is 2.2 kJ/s. Determine (i) the reversible power input required to drive the plant and (ii) irreversibility for the process.

Solution $\dot{Q}_2 = 7.8\,\text{kJ/s}$ and $\dot{W}_{in} = 2.2\,\text{kJ/s}$

$$\text{COP of the reversible refrigeration plant, } COP_{R,Rev} = \frac{T_2}{T_1 - T_2} = \frac{268}{300 - 268} = 8.37$$

$$COP_{R,Rev} = \frac{\dot{Q}_2}{\dot{W}_{in,rev}} \Rightarrow \dot{W}_{in,rev} = \frac{7.8}{8.37} = 0.932\,\text{kW}$$

This is the minimum power required to drive the plant when it is run on reversible conditions.

Now the irreversibility for the process is the difference between useful power and power required on reversible conditions, that is

$$I = \dot{W}_{useful,in} - \dot{W}_{min,in} = 2.2 - 0.932 = 1.268\,kW$$ Ans.

That is, the wasted work potential of the refrigeration plant is 1.268 kW.

Example 9.2 Air enters the compressor in a steady-flow process at 140 kPa, 17°C, and 70 m/s and leaves it at 350 kPa, 127°C, and 110 m/s. The environment is at 100 kPa and 7°C. Calculate per kg of air (i) the actual amount of work required, (ii) the minimum work required, and (iii) the irreversibility of the process.

Solution For a steady-flow process, from thermodynamic property relations,

$$Tds = dh - vdp$$

$$Tds = dh - vdp \Rightarrow \int_1^2 ds = \int_1^2 \frac{dh}{T} - R\int_1^2 \frac{dp}{p}$$

$$s_2 - s_1 = c_p \ln\frac{T_2}{T_1} - R\ln\frac{p_2}{p_1}$$

$$\text{Change in availability, } \varphi_1 - \varphi_2 = (h_1 - T_0 s_1) - (h_2 - T_0 s_2) + \frac{V_1^2 - V_2^2}{2}$$

$$= (h_1 - h_2) - T_0(s_1 - s_2) + \frac{V_1^2 - V_2^2}{2}$$

$$= c_p(T_1 - T_2) - T_0\left[R\ln\left(\frac{p_2}{p_1}\right) - C_p\ln\left(\frac{T_2}{T_1}\right)\right] + \frac{V_1^2 - V_2^2}{2}$$

$$= 1.005(290 - 400) - 280(0.263 - 0.323) + \frac{4900 - 12{,}100}{2 \times 1000} = 97.3\,kJ/kg$$

The minimum work required is equal to change in availability, that is, the amount of work input required when the compressor is run on reversible conditions.

$$\therefore w_{min,in} = \psi_1 - \psi_2 = 97.3\,kJ/kg$$

The actual work input required for the compressor can be determined from

$$h_1 + \frac{V_1^2}{2} = h_2 + \frac{V_2^2}{2} - w_{useful,in}\ (\text{where } w_{useful,in} \text{ is the useful work input})$$

$$w_{useful,in} = (h_2 - h_1) + \frac{V_2^2 - V_1^2}{2} = c_p(T_2 - T_1) + \frac{V_2^2 - V_1^2}{2}$$

$$= 1.005(400 - 290) + \frac{12{,}100 - 4900}{2 \times 1000} = 114.15\,kJ/kg$$ Ans.

$$\text{Irreversibility of process, } I = w_{useful,in} - w_{min,in} = 114.15 - 97.3 = 16.85\,kJ/kg$$ Ans.

When the compressor is run on reversible conditions, 16.85 kJ/kg of work input could have been reduced.

Example 9.3 Air expands in a turbine adiabatically from 500 kPa, 400 K, and 150 m/s to 100 kPa, 300 K, and 70 m/s. The environment is at 100 kPa and 17°C. Calculate per kg of air (i) the availability at states 1 and 2, (ii) irreversibility or exergy destruction, and (iii) the second-law efficiency.

Solution Actual work output of turbine, $w_{useful,out} = (h_1 - h_2) + \dfrac{V_1^2 - V_2^2}{2}$

$$= C_p(T_1 - T_2) + \frac{V_1^2 - V_2^2}{2 \times 1000}$$

$$= 1.005(400 - 300) + \frac{150^2 - 70^2}{2 \times 1000} = 109\,\text{kJ/kg}$$

The maximum work output is equal to the change in availability, that is

$$w_{max,out} = (h_1 - h_2) - T_0(s_1 - s_2) + \frac{V_1^2 - V_2^2}{2} = \psi_1 - \psi_2$$

$$s_2 - s_1 = c_p \ln\frac{T_2}{T_1} - R\ln\frac{p_2}{p_1} \Rightarrow s_1 - s_2 = R\ln\frac{p_2}{p_1} - c_p\ln\frac{T_2}{T_1}$$

The availability at state 1, $\psi_1 = c_p(T_1 - T_0) - T_0(s_1 - s_0) + \dfrac{V_1^2}{2}$

$$\psi_1 = c_p(T_1 - T_0) - T_0\left[R\ln\frac{p_0}{p_1} - c_p\ln\frac{p_0}{p_1}\right] + \frac{V_1^2}{2}$$

$$\psi_1 = 1.005(400 - 290) - 290\left[0.287\ln\frac{100}{500} - 1.005\ln\frac{290}{400}\right] + \frac{150^2}{2 \times 1000} = 161\,\text{kJ}$$

Ans.

The availability at state 2, $\psi_2 = c_p(T_2 - T_0) - T_0(s_2 - s_0) + \dfrac{V_2^2}{2}$

$$= 1.005(300 - 290) - 290\left[0.287\ln\frac{100}{100} - 1.005\ln\frac{290}{300}\right] + \frac{70^2}{2 \times 1000} = 2.64\,\text{kJ/kg}$$ Ans.

$$w_{max} = \psi_1 - \psi_2 = 161 - 2.64 = 158.36\,\text{kJ/kg}$$

Irreversibility, $I = w_{max,out} - w_{useful,out} = 158.36 - 109 = 49.36\,\text{kJ/kg}$ Ans.

The second-law efficiency is the ratio of actual work output and reversible work output, $\eta_{II} = \dfrac{w_{useful,out}}{w_{max,out}} = \dfrac{109}{158.36} = 68.83\%$ Ans.

The work potential of the steam is 68.83% or 31.17% of the work potential of the steam is wasted.

Example 9.4 In a piston-cylinder device, steam at 12 bar and 350°C expands to a state of 3 bar and 200°C, doing work. Heat lost to the surroundings is estimated to be 2.2 kJ. The mass of steam is 0.052 kg and the surroundings are $T_o = 25°C$ and $p_o = 1$ bar. Determine (i) the exergy of steam at the initial and final states, (ii) exergy destruction, and (iii) the second-law efficiency.

Solution Given data $\left.\begin{array}{l} p_1 = 12 \text{ bar} \\ T_1 = 350°C \end{array}\right\}$ $u_1 = 2872.7 \text{ kJ/kg}$

$\nu_1 = 0.2345 \text{ m}^3/\text{kg}$

$s_1 = 7.2139 \text{ kJ/kg K}$

$\left.\begin{array}{l} p_2 = 3 \text{ bar} \\ T_2 = 200°C \end{array}\right\}$ $u_2 = 2651 \text{ kJ/kg}$

$\nu_2 = 0.7164 \text{ m}^3/\text{kg}$

$s_2 = 7.313 \text{ kJ/kg K}$

$u_o = 104.8 \text{ kJ/kg}$

$\left.\begin{array}{l} p_o = 1 \text{ bar} \\ T_o = 25°C \end{array}\right\}$ $v_o = 0.00103 \text{ m}^3/\text{kg}$

$s_o = 0.3672 \text{ kJ/kg K}$

$$\text{Exergy of steam at state 1}, X_1 = m\left[(u_1 - u_0) - T_0(s_1 - s_0) + p_0(v_1 - v_0)\right]$$

$$= 0.052[(2872.7 - 104.83) - 298(7.2139 - 0.3672) + 100(0.2345 - 0.00103)]$$

$$= 39.046 \text{ kJ}$$ Ans.

$$\text{Exergy of steam at state 2}, X_2 = m\left[(u_2 - u_0) - T_0(s_2 - s_0) + p_0(v_2 - v_0)\right]$$

$$= 0.052[(2651 - 104.83) - 298(7.313 - 0.3672) + 100(0.7164 - 0.00103)]$$

$$= 28.48 \text{ kJ}$$ Ans.

$$\text{Change of exergy}, \Delta X = X_1 - X_2 = 39.046 - 28.48 = 10.56 \text{ kJ}$$

$$\text{Exergy destroyed}, X_{destroyed} = T_0 S_{gen} = T_0\left[m(s_2 - s_1) + \frac{Q}{T_0}\right]$$

where $S_{gen} = (\Delta S)_{sys} + (\Delta S)_{surr}$

$$(\Delta S)_{sys} = m(s_2 - s_1) = 0.052(7.313 - 7.2139) = 0.00515 \text{ kJ/K}$$

$$(\Delta S)_{surr} = \frac{Q}{T_0} = \frac{2.2}{298} = 0.00738 \text{ kJ/K}$$

$$X_{destroyed} = 298[0.00515 + 0.00738] = 3.734 \text{ kJ}$$ Ans.

The wasted work potential of the steam is 3.734 kJ.

$$\text{Exergy expended} = \Delta X = X_1 - X_2 = 39.046 - 28.48 = 10.56 \text{ kJ}$$

$$\eta_{II} = \frac{\text{Exergy recovered}}{\text{Exergy expended}} = 1 - \frac{\text{Exergy destroyed}}{\text{Exergy expended}}$$

$$\therefore \eta_{II} = 1 - \frac{3.734}{10.56} = 64.64\%$$ Ans.

Therefore, the work potential of the steam is 64.64%.

Example 9.5 Air expands in a turbine adiabatically in a closed system from 500 kPa, 400 K and 0.23 m³/kg to 100 kPa, 300 K and 0.86 m³/kg. The environment is at 100 kPa and 17°C. Calculate per kg of air, (i) W_{max}, (ii) W_{act}, (iii) change in availability (iv) irreversibility, and (v) second-law efficiency, η_{II}

Solution $w_{useful,out} = q - \Delta U \Rightarrow -\Delta U = c_v(T_1 - T_2)$

$$= 0.718 \times 100 = 71.8\,\text{kJ/kg} \qquad \text{Ans.}$$

For a closed system, maximum work, $w_{max} = (u_1 - u_2) - T_0(s_1 - s_2)$

$$\text{Where } s_2 - s_1 = c_v \ln\frac{T_2}{T_1} + R\ln\frac{v_2}{v_1}$$

$$\therefore w_{max} = c_v(T_1 - T_2) + T_0\left(c_v \ln\frac{T_2}{T_1} + R\ln\frac{v_2}{v_1}\right)$$

$$= 0.718 \times 100 + 290\left[0.718\ln\left(\frac{300}{400}\right) + 0.287\ln\left(\frac{0.86}{0.23}\right)\right]$$

$$= 121.8\ \text{kJ/kg} \qquad \text{Ans.}$$

For a closed system, change in availability is
$\Rightarrow (w_{useful})_{max} = w_{max} - p_0(dv) = w_{max} - p_0(v_2 - v_1)$
(since work done against atmosphere is to be subtracted)
$121.8 - 100(0.63) = 58.8$ kJ/kg

$$I = w_{max,out} - w_{useful,act} = 121.8 - 71.8 = 50\,\text{kJ/kg} \qquad \text{Ans.}$$

$$\text{Alternatively } I = T_0(s_{gen}) = T_0(\Delta s_{sys} + \Delta s_{surr})$$

$$(\Delta s)_{surr} = 0$$

$$I = T_0 \cdot \Delta s_{sys} = T_0\left(c_v \ln\frac{T_2}{T_1} + R\ln\frac{v_2}{v_1}\right)$$

$$= 290 \times \left(0.718\ln\left(\frac{300}{400}\right) + 0.287\ln\left(\frac{0.86}{0.23}\right)\right) = 50\,\text{kJ/kg}$$

$$\eta_{II} = \frac{w_{useful,out}}{w_{max,out}} = \frac{71.8}{121.8} = 59\% \qquad \text{Ans.}$$

The best possible efficiency of the turbine is 59%.

Example 9.6 In a counter-flow heat exchanger, oil (c_p = 2.1 kJ/kg K) is cooled from 440 to 320 K, while water (c_p = 4.2 kJ/kg K) is heated from 290 K to temperature, T. The pressure drop, kinetic energy and potential energy effects, and heat loss are negligible. Determine (i) temperature T and (ii) rate of exergy destruction. Take T_0 = 17°C, p_0 = 1 atm, and mass flow rates of oil and water as 800 and 3200 kg/h, respectively.

Solution Taking the energy balance between oil and water,

$$m_h c_{p_h}(T_{h_1} - T_{h_2}) = m_c c_{p_c}(T_{c_2} - T_{c_1})$$

$$800 \times 2.1(440 - 320) = 3200 \times 4.2(T_{c_2} - 290)$$

∴ The final temperature of water after heating, T = 305 K Ans.

Oil is treated as system and water as the surroundings.

$$\text{Exergy destroyed}, X_{\text{destroyed}} = T_0 S_{\text{gen}} = T_0\left(S_{\text{sys}} + S_{\text{surr}}\right)$$

$$(\Delta S)_{\text{oil}} = \int_{T_1}^{T_2} \frac{dQ}{T} = \int_{T_1}^{T_2} \frac{mc_p dT}{T} = mc_p \ln\frac{T_2}{T_1}$$

$$(\Delta S)_{\text{oil}} = 800 \times 2.1 \times \ln\frac{320}{440} = -535\,\text{kJ/K}$$

$$(\Delta S)_{\text{water}} = \int_{T_1}^{T_2} \frac{mc_p dT}{T} = 3200 \times 4.2 \times \ln\frac{305}{290} = 677\,\text{kJ/K}$$

The total entropy change of the system and surroundings,

$$(\Delta S)_T = (\Delta S)_{\text{oil}} + (\Delta S)_{\text{water}} = -535 + 677 = 142\,\text{kJ/K}$$

$$\therefore X_{\text{destroyed}} = T_0 S_{\text{gen}} = 290 \times 142 = 41.18\,\text{MJ/h}$$ Ans.

Example 9.7 Water is evaporated in a steam generator at 300°C and combustion gas ($c_p = 1.2$ kJ/kg K) is cooled from 1500°C to 350°C. Estimate the increase in unavailable energy due to the above heat transfer per kg of water evaporated. The surroundings are at 25°C. Take latent heat of vaporization of water at 300°C is 1404 kJ/kg.

Solution Heat lost by combustion gases = heat gained by water

where heat gained by water is latent heat, L

$$m_g c_{p_g} (\Delta T)_g = m_w c_{pw} (\Delta T)_w = 1 \times 1404\,\text{kJ/kg}$$

$$m_g \times 1.2 \times (1500 - 350) = 1 \times 1404$$

Mass of gas = m_g = 1.017 kg/kg of water

Entropy increase of water due to the evaporation, $(\Delta S)_{\text{water}} = \dfrac{mL}{T} = \dfrac{1 \times 1404}{573} = 2.45\,\text{kJ/K}$

Entropy decrease of combustion gas as it is cooled, $(\Delta S)_{\text{gas}} = \displaystyle\int_{T_1}^{T_2} m_g c_{p_g} \frac{dT}{T} \Rightarrow m_g c_{p_g} \ln\left(\frac{T_2}{T_1}\right)$

$$= 1.017 \times 1.2 \ln\left(\frac{623}{1773}\right) = -1.276\,\text{kJ/K}$$

Total change of entropy,

$$(\Delta S)_{\text{Total}} = (\Delta S)_w + (\Delta S)_g = 2.45 - 1.276 = 1.173\,\text{kJ/K}$$

Loss in (available energy) exergy = $T_0(\Delta S)_{\text{Total}}$

= 298 × 1.173 = 349.73 kJ Ans.

Example 9.8 A 10 kg of iron block, initially at 300°C, is cooled in an insulated tank containing 80 kg of water at 27°C when the surroundings are at 25°C. Estimate the exergy destroyed or loss of available energy. Take c_p of the iron block as 0.55 kJ/kg K.

Solution Let the final equilibrium temperature be t_f.

According to the first law, $\delta Q = \delta W + \Delta U$

$\Delta U = 0$ (Since no heat and work interactions are involved, $\delta Q = \delta W = 0$)

$$m_I c_{pI}(t_f - t_{1,I}) + m_w c_{pw}(t_f - t_{1,w}) = 0$$

$$= 10 \times 0.55(t_f - 300) + 80 \times 4.18(t_f - 27) = 0 \Rightarrow t_f = 31.41°C$$

$$\text{Exergy loss of iron block} = (U_1 - U_2) - T_0(S_1 - S_2) = mc_v(T_{1,I} - T_f) - T_0 mc_v \ln\left(\frac{T_{1,I}}{T_f}\right)$$

$$= 10 \times 0.55\left[(573 - 304.4) - 298 \ln\left(\frac{673}{304.4}\right)\right] = 176.91\,\text{kJ}$$

$$\text{Exergy gain of water} = mc_v(T_{1,w} - T_f) - T_0 mc_v \ln\left(\frac{T_{1,w}}{T_f}\right)$$

$$= 80 \times 4.18\left[(300 - 304.4) - 298 \ln\left(\frac{300}{304.4}\right)\right] = 34.44\,\text{kJ}$$

Exergy loss of the system = 176.91 − 34.44 = 142.47 kJ Ans.

REVIEW QUESTIONS

9.1 Define available energy.
9.2 What is reversible work?
9.3 Define unavailable energy?
9.4 What is the law of degradation of energy?
9.5 Distinguish between energy and exergy.
9.6 What is a dead state? Give two examples.
9.7 Define surrounding work and maximum useful work.
9.8 Define irreversibility.
9.9 Define isentropic efficiency of a turbine and a compressor.
9.10 Define second-law efficiency.
9.11 What are the means of exergy transfer?
9.12 How do you compare the exergy of the same gas at a higher temperature and lower temperature?
9.13 Distinguish between first law and second-law efficiencies.
9.14 Energy doesn't degrade quantity wise but degrade quality wise, explain.
9.15 What is exergy destruction?
9.16 What is the difference between useful work and actual work?

EXERCISE PROBLEMS

9.1 A source is at 800 K and the system receives 800 kJ from 400 K of that source. The temperature of the sink is at 300 K. Assuming heat transfer is taking place isothermally between source and sink, find (i) the entropy generation during the process and (ii) the change in available energy.

9.2 In a power station, saturated steam is generated by transferring heat from the hot gases produced in the combustion chamber at 190°C. The hot gases are cooled from 1000°C to 450°C during transferring the heat for steam generation. Find (i) the change in total entropy of the combined system of hot gases and steam and (ii) the change in unavailable energy by considering 1 kg of steam generated. Assume water is saturated while it enters the boiler and while it leaves.

9.3 Air in the mixing chamber is heated from 30°C to 50°C adiabatically in a steady-flow process by the use of a quantity of air at 110°C. Neglect changes in kinetic and potential energy, calculate the ratio of the mass flow of air initially at 110°C to that initially at 30°C. Also calculate the effectiveness of the heating process, if the ambient temperature is 30°C.

9.4 In a constant pressure process, a liquid having a specific heat of 5.8 kJ/kg K is heated from 25°C to 90°C by passing it through tubes that are immersed in a furnace. The temperature of the furnace is 1800°C. Calculate the effectiveness of the heating process when the atmospheric temperature is 10°C.

9.5 Consider a 2 kg of gas initially at 2 bar and 350 K in a closed tank and 600 kJ of heat from an infinite source at 1200 K. Determine the change in available energy due to heat transfer. Take c_v (gas) = 0.75 kJ/kg K and surrounding temperature = 310 K.

9.6 In a parallel-flow heat exchanger, water enters at 55°C and leaves at 85°C while oil of specific gravity 0.75 enters at 280°C and leaves at 120°C. The specific heat of oil is 2.52 kJ/kg K and the surrounding temperature is 310 K. Determine the change in availability on the basis of 1 kg of oil flowing per second.

9.7 8 kg of water at 40°C is used to mix with 1 kg of ice at 0°C. Calculate the net change in the entropy and unavailable energy when the system reaches a common temperature. Assume that the surrounding temperature is 15°C. Take specific heat of water = 4.18 kJ/kg K; specific heat of ice = 1.9 kJ/kg K; and latent heat of ice = 336 kJ/kg.

9.8 The hot gases leave 800 kJ of heat at 1200°C from a firebox of a boiler and go to steam at 200°C. The ambient temperature is 28°C. Determine the energy into available and unavailable portions for leaving hot gases and entering steam.

9.9 In a certain process, an evaporator uses water, which takes the heat from a vapor which is at 350°C and starts to evaporate at 220°C. The obtained steam is used in a power cycle, which rejects heat at 40°C. Determine the fraction of the available energy in the heat transferred from the process vapor at 350°C that is lost due to the irreversible heat transfer process at 220°C.

9.10 In a steam boiler, heat transfer takes place between hot gases and water, which vaporizes at a constant temperature. The gases are cooled from 1000°C to 500°C while the water evaporates at 190°C. The specific heat of gases is 1.005 kJ/kg K, and the latent heat of water at 190°C is 1750.5 kJ/kg. All the heat transferred from the gases goes to the water. Determine the increase in total entropy of the combined system of gas and water as a result of irreversible heat transfer. Assume that 1 kg of water is evaporated and the temperature of the surroundings is 28°C. Find the increase in unavailable energy due to irreversible heat transfer.

9.11 Determine the unavailable energy in 30 kg of water at 80°C with respect to the ambient conditions which are at 8°C, the pressure being 1 atm.

9.12 Calculate the decrease in available energy when 25 kg of water at 95°C mixes with 35 kg of water at 35°C, the pressure is taken as constant and the temperature of the surroundings is 15°C. Take cp of water = 4.18 kJ/kg K.

9.13 In a closed system, 1.5 kg of air at 4 bar at 70°C expands adiabatically until its volume is doubled and its temperature becomes equal to that of the ambient which is at 1.5 bar, 10°C. Determine the following for the process: (i) the change in availability; (ii) the irreversibility; and (iii) maximum work. For air, take $c_v = 0.718$ kJ/kg K, $u = c_vT$, where c_v is constant, and $pV = mRT$, where p is in bar, V is the volume in m^3, m mass in kg, R is a constant equal to 0.287 kJ/kg K, and T is the temperature in K.

9.14 Certain air of mass in kg at a pressure p, and temperature 700 K is mixed with 1 kg of air at the same pressure and 400 K. Determine the change in availability if the ambient temperature is 270 K.

9.15 In an insulated tank, 8 kg of water is heated by a churning process from 280 to 330 K. Determine the change in availability for the process, if the surrounding is maintained at a temperature of 270 K.

9.16 In a closed system, air is cooled at constant pressure to ambient conditions which are 0.5 bar and 280 K and the initial conditions are 500 K and 4 bar. Assume 8 kg of air contains 10 kg of air at 600 K and 5 bar. Determine the exergy of the system.

9.17 In a turbine, air expansion takes place from 4.5 bar, 480°C to 0.5 bar, 280°C. During the expansion process, heat is lost to the surroundings of magnitude 9 kJ/kg which is at 0.85 bar, 25°C. Ignoring the potential and kinetic energy changes, determine per kg of air: (i) the change in availability; (ii) the irreversibility; and (iii) the maximum work. For air, take $c_p = 1.005$ kJ/kg K and $h = c_pT$, where c_p is constant.

9.18 Air is compressed by a centrifugal compressor at the rate of 15 kg/min from 1.5 to 3 bar. During the compression, the temperature of air raises from 15°C to 115°C. If the surrounding air temperature is 25°C, then determine the minimum and actual power required to run the compressor. Neglect the potential and kinetic energy changes and heat interaction between the compressor and surroundings.

9.19 A compression process of 1 kg of air is taking place polytropically from 1.5 bar and 320 K to 8 bar and 420 K. Determine the irreversibility of the process. Take for air $c_p = 1.005$ kJ/kg K, $c_v = 0.718$ kJ/kg K, and $R = 0.287$ kJ/K.

9.20 In an adiabatic steady-flow process, the air enters the system at a pressure of 8 bar and 175°C with a velocity of 80 m/s and leaves at 2 bar and 35°C with a velocity of 60 m/s. The temperature of the surroundings is 27°C and the pressure is 1 bar. Determine reversible work and also the irreversibility on the basis of 1 kg of airflow. Take for air $c_p = 1$ kJ/kg K and $R = 287$ J/kg K.

9.21 A piston-cylinder device, the refrigerant, R-134a at 0.8 MPa and 55 of 4 kg mass, is cooled at constant pressure until it exists as a liquid at 23°C. If the surroundings are at 80 kPa and 23°C, determine (i) the available energy of the refrigerant at the initial and the final states and (ii) the change in available energy during this process.

DESIGN AND EXPERIMENT PROBLEMS

9.22 Design an experiment to measure the irreversibilities associated with the air-conditioning system of your college and to determine its best possible efficiency.

9.23 Evaluate the first law and second-law efficiencies of the steam power plant nearby your location and determine the irreversibility associated with it and the work potential of the plant. Draw your conclusions based on the findings and suggest for improvement of the plant efficiency.

9.24 A steam boiler has to convert saturated liquid at 40 bar to a saturated vapor at this constant pressure. Assuming the combustion gases, entering into a steam boiler which is considered as a heat exchanger, as a stream of air, determine the temperature at which combustion gases must enter the unit in order to reduce the exergy loss minimum from combustion gases to boiling water.

9.25 Perform the second-law analysis for an application involving space heating with different methods used for it such as a convective heater, an electric-resistance heater, and a solar air heater. Determine entropy generation and exergy destruction in all of the above methods and compare them.

10 Vapor and Advanced Power Cycles

LEARNING OUTCOMES

After learning this chapter, students should be able to

- Analyze the Carnot and Rankine cycles to find their suitability for vapor power cycles.
- Evaluate the alternative approaches of improving the thermal efficiency of the Rankine cycle.
- Apply the first and second laws of thermodynamics in the analysis of best possible efficiency of the Rankine cycle and combined cycle systems.
- Demonstrate the knowledge of modifying the basic Rankine vapor power cycle such as reheat and regenerative cycles to increase the cycle thermal efficiency.
- Analyze power cycles that consist of two separate cycles known as combined cycles and binary cycles.
- Analyze power generation coupled with process heating called cogeneration.
- Analyze the reheat and regenerative vapor power cycles.

10.1 CARNOT VAPOR CYCLE

Thermodynamic cycles can be classified as power cycles and refrigeration cycles. Power cycles are used to analyze the devices such as engines that produce the net power output. Refrigeration cycles are used to analyze the devices such as refrigerators and heat pumps that produce the refrigeration effect. Power cycles can be divided into two categories: vapor and gas. Power is generated for propelling the automotives, locomotives, small aircrafts, and ships and for the electric power supply. The fundamental difference between vapor power and gas power cycles is that in the former case, the working fluid undergoes a phase change allowing more energy to be stored than can be stored by sensible heating, and in the latter case, the working fluid remains a gas and is used throughout the cycle. A heat engine is a device that converts the thermal energy into work. The performance of a heat engine is expressed by thermal efficiency defined as the ratio of net work output to the total heat input. The maximum thermal efficiency of a heat engine is obtained when heat is supplied at the highest possible temperature and rejected at the lowest possible temperature (temperature of surroundings).

Vapor power cycles use water as the working fluid, which is alternately vaporized and condensed. If the fluid is expanded as a vapor and compressed as a liquid with less specific volume, the back work ratio will be much less than that of gas cycle,

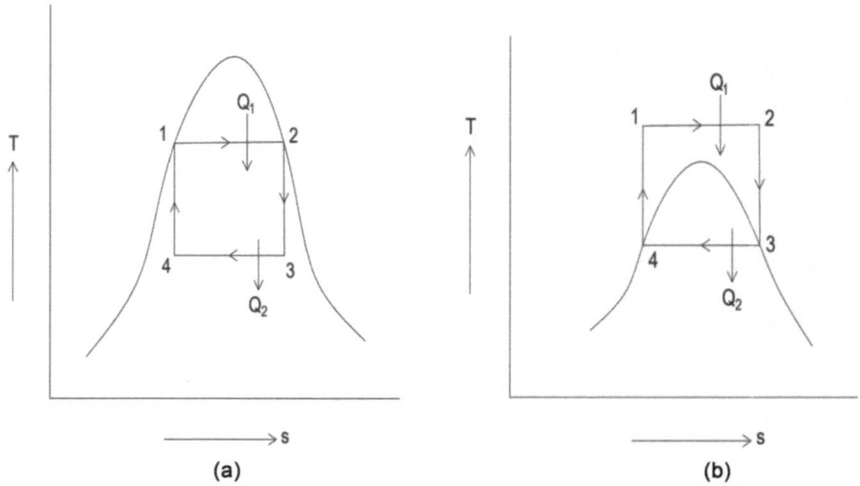

FIGURE 10.1 Carnot vapour cycles on T-s diagrams. (a) Within the saturation dome of a pure substance, (b) above the saturation dome of a pure substance.

since the work of compression is proportional to the specific volume. For this reason, vapor cycles have a long history compared to that of gas cycles.

The Carnot cycle has the maximum efficiency because heat is added isothermally at the source temperature (highest temperature) and rejected isothermally at the sink temperature (lowest temperature). Figure 10.1a shows the steady-flow Carnot cycle operating on wet steam (a liquid-vapor mixture). The working fluid is heated reversibly and isothermally in the boiler (1-2); it is then expanded reversibly and adiabatically in the turbine (2-3), condensed reversibly and isothermally in the condenser (3-4), and compressed reversibly and adiabatically in the compressor (4-1) and the cycle repeats. The temperature of steam during processes 1-2 and 3-4 can be maintained constant by maintaining the pressure constant since temperature and pressure are two dependent properties in the saturation region.

The limitations of the Carnot cycle are as follows:

1. The critical temperature of steam (374°C) is well below the metallurgical limits, which restricts the maximum temperature at which heat is added to the cycle and hence the thermal efficiency. This is due to that heat transfer processes are limited to two-phase systems since isothermal heat transfer to a single-phase system is not easy to achieve.
2. Since the products of combustion are cooled at the most to T_1, only a smaller portion of energy released on combustion is available for use.
3. Figure 10.1a shows that isentropic expansion in a turbine results in the steam of low quality, that is, steam with the higher moisture content is not safe for the turbine.
4. It is difficult to develop a pump that handles two-phase mixtures.

To overcome some of the above impracticalities, the cycle shown in Figure 10.1b may be used. In this, the heat addition process is performed at higher temperatures, which requires much higher pressures by compressing the liquid from state 4. Also isothermal heat addition (1-2) involves continual pressure variation causing additional control problems that is not common to the case shown in Figure 10.1a. Therefore, the Carnot cycle remains as a hypothetical cycle based on certain assumptions that can be used as a model to compare the other practical heat engines.

10.2 RANKINE CYCLE

The thermodynamic cycle for steam power plant is the Rankine cycle. Figure 10.2 shows the layout of a typical steam power plant used for the power generation. The heat liberated by burning the fuel in the furnace is transferred to water (working fluid) in the boiler in which steam is generated. The steam then expands in the turbine doing work on the rotor. The heat energy is continuously converted into shaft work (mechanical power). The turbine drives the generator that produces electric power. The steam after expansion in turbine condenses into liquid in the condenser, and then the condensate is circulated with the help of a pump into boiler and the cycle repeats.

The limitations of the Carnot cycle can be eliminated by superheating the steam in the boiler and condensing it completely in the condenser in the cycle known as the Rankine cycle. Each of the processes of the ideal cycle can be nearly achieved in an actual cycle.

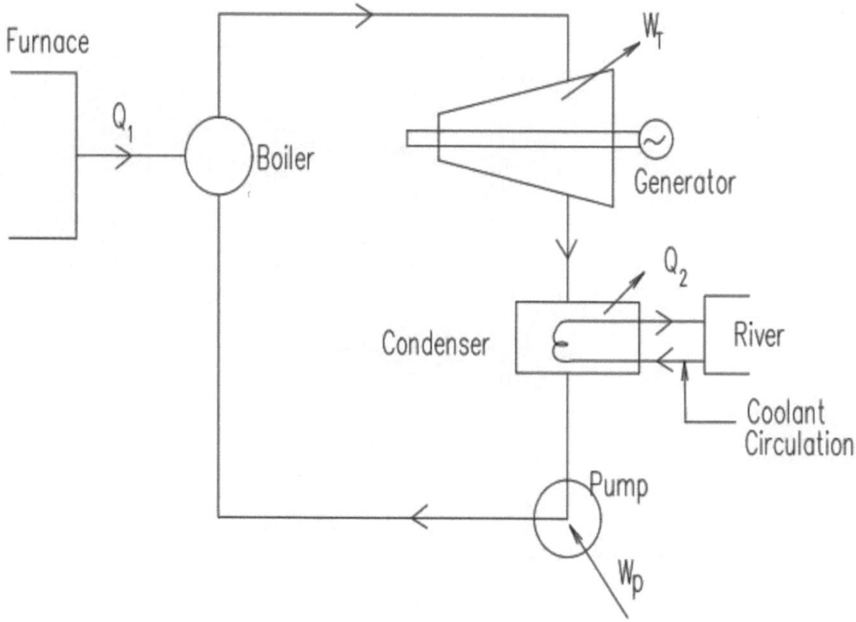

FIGURE 10.2 Schematic of simple steam power plant.

In an ideal Rankine cycle, there is no pressure drop during evaporation and condensation. Also in the absence of irreversibilities and heat interaction with the surroundings, the expansion and compression in the turbine approach the isentropic process.

An ideal Rankine cycle is illustrated in Figure 10.3. Figure 10.4a–c shows the p-υ, T-s, and h-s diagrams of the ideal Rankine cycle. The four processes involved in this complete cycle are different from each other, and each process requires a separate component. Moreover, it is to be noted that the working fluid is water that exists in liquid phase during the part of cycle and in vapor phase in the remaining part of cycle. The sequence of operations in the cycle is given below.

Process 1-2: Reversible adiabatic expansion (process 1-2, if the steam entering the turbine is dry-saturated steam) in the turbine from boiler pressure P_1 to condense pressure P_2 to produce the net work output. If the steam entering the turbine is wet steam, the expansion will be 1^1-2^1 and it will be 1^{11}-2^{11} if the steam is superheated one.

Process 2-3: Constant pressure heat rejection (process 2-3) from the steam in the condenser. Steam is condensed to saturated water at state 3.

Process 3-4: Reversible adiabatic compression (process 3-4) of saturated water in the pump. The pressure of water is raised from P_2 to P_1. The temperature of water at state 4 is less than the saturation temperature corresponding to the boiler pressure.

Process 4-1: Constant pressure heat addition (process 4-1) from a furnace to water in the boiler to produce steam.

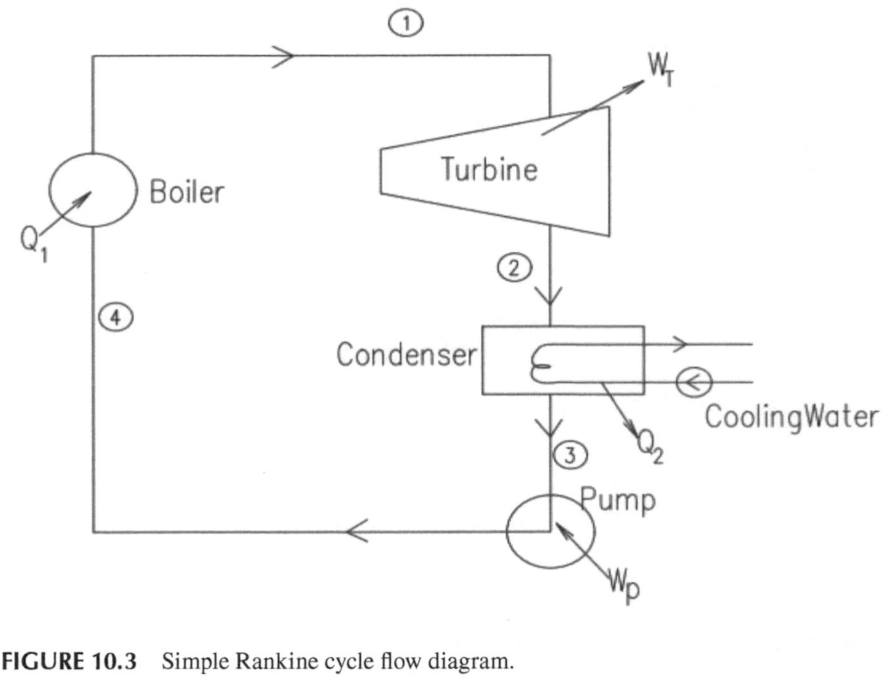

FIGURE 10.3 Simple Rankine cycle flow diagram.

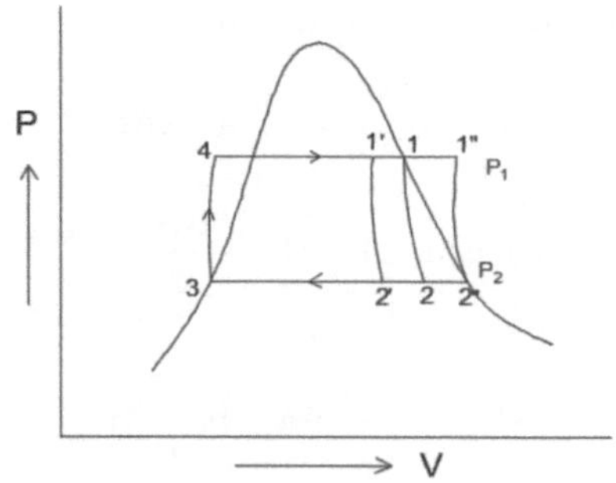

(a) p-v diagram

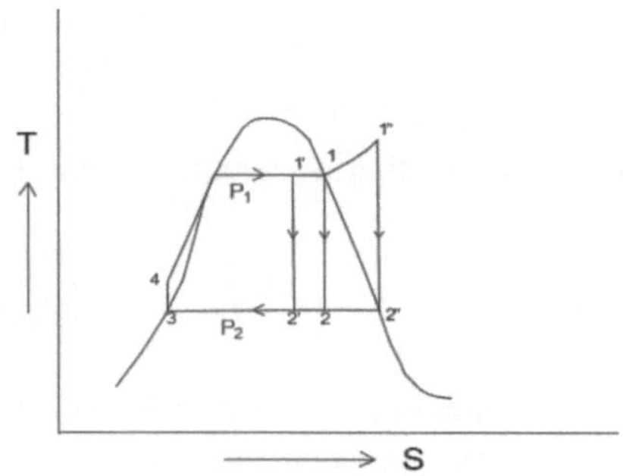

(b) T-s diagram

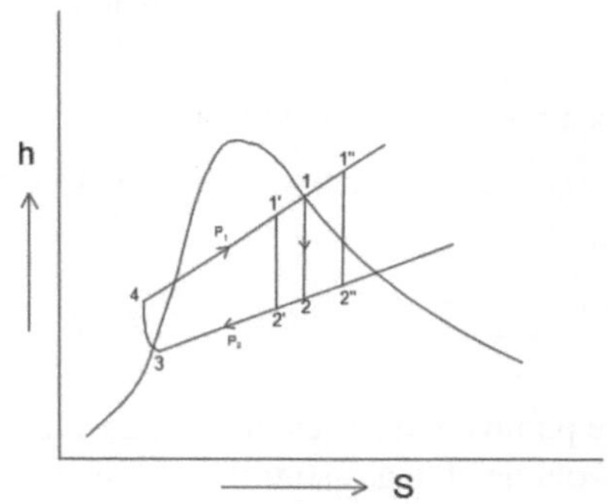

(c) h-s diagram

FIGURE 10.4 Ideal Rankine cycle. (a) p-v diagram, (b) T-s diagram, and (c) h-s diagram.

Considering a unit mass of working fluid, steady flow energy equation (SFEE) is applied to each of the devices of the Rankine cycle to evaluate the heat and work quantities neglecting the changes in kinetic and potential energy.

The general SFEE (Eq. 5.13 derived in Chapter 5) given below is applied to various control volumes such as boiler, turbine, condenser, and pump.

$$h_1 + \frac{V_1^2}{2} + z_1 g + \frac{\delta Q}{dm} = h_2 + \frac{V_2^2}{2} + z_2 g + \frac{\delta W}{dm}$$

Now we can apply SFEE to the control volumes with an assumption that the changes in kinetic and potential energies are negligible.

Heat added in the boiler, q_1 when work $w=0$ is

$$h_4 + q_1 = h_1 \Rightarrow q_1 = h_1 - h_4$$

For a well-insulated turbine, SFEE is reduced to $h_1 = h_2 + \dfrac{\delta W}{dt}$

Then the work output of turbine, $w_T = h_1 - h_2$

Heat rejected in the condenser, q_2 (work, w = 0)

$$h_2 = h_3 + q_2 \Rightarrow q_2 = h_2 - h_3$$

For a pump, the SFEE can be reduced to $h_1 = h_2 - \frac{\delta W}{dt}$

Work input for the pump, w_p

$$h_3 + w_P = h_4 \Rightarrow w_P = h_4 - h_3$$

The efficiency of the Rankine cycle is

$$\eta_{Rankine} = \frac{\text{Net work output}}{\text{Total heat input}} = \frac{w_T - w_P}{q_1} = \frac{(h_1 - h_2) - (h_4 - h_3)}{h_1 - h_4} \qquad (10.1)$$

It can also be expressed as $\eta_{Rankine} = \frac{q_1 - q_2}{q_1}$ (Since $w_T - w_P = q_1 - q_2$)

The pump work, w_P, can be derived from the basic thermodynamic property relation

$$Tds = dh - \upsilon dp \qquad (10.2)$$

The pump compresses the liquid water, which is incompressible; therefore its density or volume changes slightly with increase in pressure. For an isentropic compression (ds = 0), the Eq. 10.2 becomes

$$dh = \upsilon dp$$

$$\Delta h = \upsilon \Delta p \text{ (Since volume change is negligible)}$$

$$h_4 - h_3 = \upsilon_3 (p_1 - p_2), \qquad (10.3)$$

where enthalpy is in kJ/kg, p_1 and p_2 are boiler and condenser pressures, respectively, in kN/m^2, and specific volume is in m^3/kg.

Steam Rate

It is an important parameter used to express the capacity of the steam plant defined as the rate of steam flow required in kg/h to produce the unit shaft output given as

$$\text{Steam rate} = \frac{1}{w_T - w_P} \frac{kg}{kJ} \cdot \frac{1 kJ/s}{1 kW}$$

$$= \frac{1}{w_T - w_P} \frac{kg}{kWs} = \frac{3600}{w_T - w_P} \frac{kJ}{kWh} \qquad (10.4)$$

Heat Rate

The efficiency of the cycle is alternatively expressed as heat rate defined as the rate of heat input required to produce the unit work output given as

$$\text{Heat rate} = \frac{3600\ Q_1}{w_T - w_P} = \frac{3600}{\eta_{cycle}} \frac{kJ}{kWh} \tag{10.5}$$

10.3 COMPARISON OF RANKINE AND CARNOT CYCLES

The comparison of the Rankine cycle (12344^11) and the Carnot cycle (123^14^11) is depicted in Figure 10.5. The fundamental differences between Rankine and Carnot cycles are summarized below.

In the Carnot cycle, the heat addition is a reversible isothermal process, whereas in the Rankine cycle, it is a reversible constant pressure process; therefore the heat addition is at a lower temperature compared to that of the Carnot cycle, and consequently the mean temperature of the heat addition is lower. As a result, the efficiency of the Rankine cycle is lower than the Carnot cycle although the heat rejection is at the same temperature in both the cycles. The common features of both the cycles are reversible adiabatic compression in the pump and reversible adiabatic expansion in the turbine.

Referring to Figure 10.5, the thermal efficiency of ideal Rankine cycle 1-2-3-4-4^1-1 is less than that of the Carnot cycle 1-2-3^1-4^1-1 having the same maximum temperature T_1 and the minimum temperature T_2 in both the cycles since the mean temperature of the heat addition in the Rankine cycle between 4 and 4^1 is lower than that in the Carnot cycle, that is, T_1.

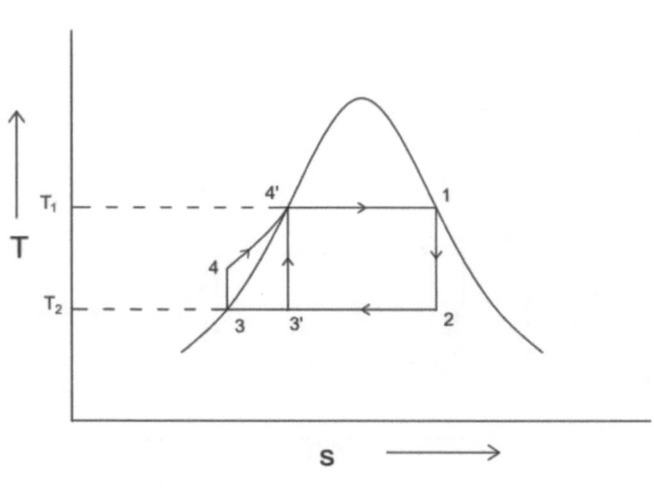

FIGURE 10.5 Comparison of Rankine and Carnot cycles.

10.4 MEAN TEMPERATURE OF THE HEAT ADDITION

In heat engine cycles, if the heat is added at higher temperatures or rejected at lower temperatures, the efficiency of the cycle will be high. In the Rankine cycle, the heat is added at constant pressure and at infinite temperatures. If T_m is the mean temperature of the heat addition, so that the area under 4-1 is equal to the area under 5-6 as shown in Figure 10.6, and then the heat added Q_1 is given by

$$q_1 = h_1 - h_4 = T_m(s_1 - s_4)$$

$$T_m = \frac{h_1 - h_4}{s_1 - s_4}$$

The heat rejected is given by $q_2 = h_2 - h_3 = T_2\,(s_1 - s_4)$

The efficiency of the Rankine cycle when temperature of the heat rejection is T_2 is given by $\eta_{Rankine} = 1 - \frac{q_2}{q_1} = 1 - \frac{T_2(s_1 - s_4)}{T_m(s_1 - s_4)}$

$$\therefore \eta_{Rankine} = 1 - \frac{T_2}{T_m} \tag{10.6}$$

For a given mean temperature of the heat addition T_m, the lower the temperature of the heat rejection T_2, the higher will be the efficiency of the cycle. Since the lowest possible temperature of the heat rejection is the temperature of surroundings, which is fixed, the efficiency of the cycle is a function of the mean temperature of the heat addition as given below:

$$\eta_{Rankine} = f(T_m) \tag{10.7}$$

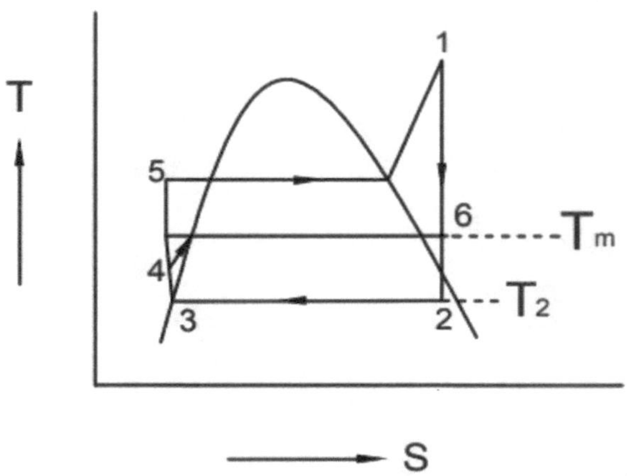

FIGURE 10.6 The mean temperature of the heat addition in Rankine cycle.

10.5 EFFICIENCY IMPROVEMENT OF THE RANKINE CYCLE

Electric power generation in the world largely depends on thermal power; thus it is essential to improve the thermal efficiency of steam power plants. A small increase in thermal efficiency can cause a large fuel savings and reduction in greenhouse gas emissions. Generally thermal efficiency of a heat engine cycle can be increased by either raising the temperature at which heat is added (considering the metallurgical limits) or lowering the temperature at which heat is rejected. The least possible temperature of the heat rejection is the temperature of surroundings.

There are three ways to increase the thermal efficiency of the Rankine cycle: they are (i) decreasing the condenser pressure, (ii) superheating the steam to a high temperature, and (iii) increasing the boiler pressure.

Decreasing the Condenser Pressure

It is mentioned earlier that lowering the temperature of the heat rejection improves thermal efficiency. The steam, after expansion in the turbine, enters the condenser as a saturated mixture at a saturation temperature corresponding to the condenser pressure. Thus lowering the condenser pressure automatically lowers the temperature of the heat rejection. The effect of lowering the condenser pressure on the Rankine cycle efficiency is shown on T-s diagram in Figure 10.7. The net work output increases and the amount of heat to be added also increases due to this reduction in the condenser pressure. However, this increase in heat input is so small when compared to the increase in thermal efficiency of the cycle.

Steam condensers can be effectively operated by creating partial vacuum that is lowering the pressure below the atmosphere, which results in lower temperature of the heat rejection so that the thermal efficiency can be improved. However, there is a limitation that the pressure in the condenser cannot be lowered below the saturation

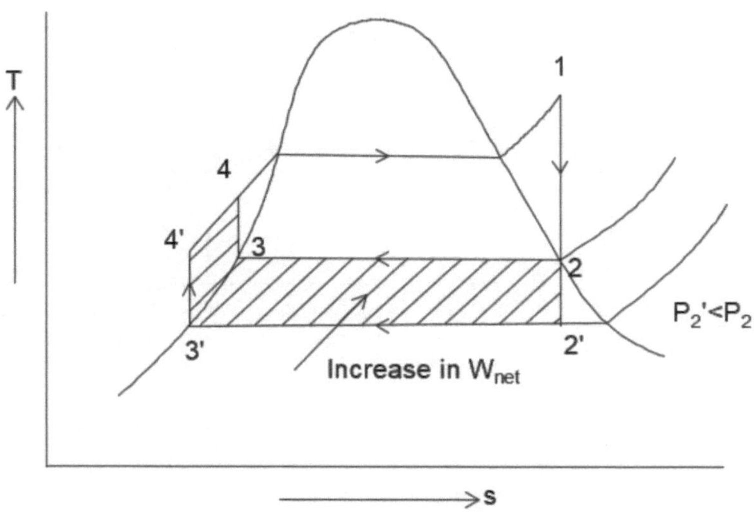

FIGURE 10.7 The effect of lowering the condenser pressure on the Rankine cycle.

pressure corresponding to the coolant temperature. When water at a temperature of 20°C is available as a cooling medium, a condenser temperature of 30°C is desirable to have better heat transfer between the condenser and the coolant. Then the pressure in the condenser must be above 4.3 kPa, which is the saturation pressure of water corresponding to a temperature of 30°C.

Lowering the condenser pressure, however, has some undesirable effects. Usually vacuum is created by the removal of air, which may result in the leakage of air into the condenser, thereby reducing its efficiency and also there is a possibility of more moisture to be present in the steam at the exhaust.

Superheating the Steam to a High Temperature

The mean temperature of the heat addition can be increased by superheating the steam to higher temperatures. Figure 10.8 shows on T-s diagram, the effect of superheating the steam to a high temperature on the Rankine cycle efficiency. It can be observed from the figure that the superheating can result in increased net work and increased heat input as well. The net effect with superheating is an increase in the thermal efficiency of the cycle. Metallurgical conditions put a limitation to the highest temperature (which is around 600°C at present) to which the steam can be superheated. However, with the advancements in material technology, it can be possible to increase the limit beyond this value.

The main advantage of superheating the steam is it can improve the quality of steam at turbine exhaust as the cycle shifts to the right (the expansion line 1-2 shifts to 1^1-2^1) on T-s diagram.

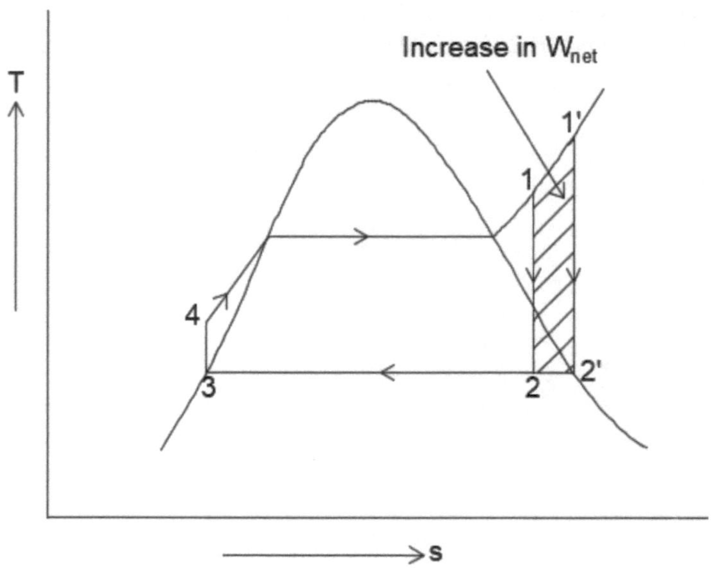

FIGURE 10.8 The effect of superheating the steam to a higher temperature on the Rankine cycle.

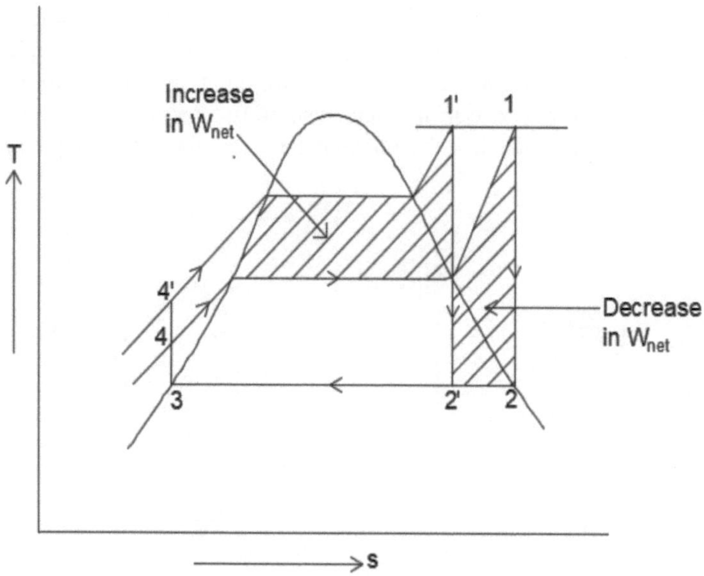

FIGURE 10.9 The effect of increasing the boiler pressure on the Rankine cycle.

Increasing the Boiler Pressure

The effect of increasing the boiler pressure on the Rankine cycle efficiency is presented on T-s diagram shown in Figure 10.9. It can be observed from the figure that with the increase in boiler working pressure, the mean temperature of the heat addition can be increased resulting from the increased boiling temperature of steam. However, increasing the boiler pressure results in the increased moisture content at turbine exhaust as seen on T-s diagram that the cycle shifts to the left. In order to maintain the desired quality at turbine exhaust when boiler pressure is raised, reheating the steam is essential.

There is a considerable increase in boiler operating pressures from about 30 to 300 bar during a period of almost one century. Nowadays so many boilers in steam power plants operate at a pressure higher than the critical pressure ($p_c = 220.6$ bar), and power plants are capable of generating 1000 MW electric power. Modern power plants have a thermal efficiency of around 40%, and it is even more with the combined cycle power plants that work on combined Rankine cycle and Brayton cycle.

10.6 REHEAT RANKINE CYCLE

When the boiler operating pressure is raised to take advantage of increased mean temperature of the heat addition, it results in excess moisture at the turbine exhaust, which is not a desirable effect as it reduces the overall efficiency of the plant. This side effect can be remedied by two different methods: one is by superheating the steam to high temperatures before it is fed to the turbine, as discussed in Section 10.4; however, metallurgical conditions put a limitation to the highest temperature used. Another alternative is to expand the steam in stages with reheating in between.

This method is comparatively a viable solution and practical when compared to that of superheating to reduce excess moisture at turbine exhaust. In a simple Rankine cycle, steam, after the isentropic expansion in turbine is directly fed into condenser for the condensation process. But in reheat system, two turbines (high pressure and low pressure turbines) are used, with a reheat in between, for improving the efficiency.

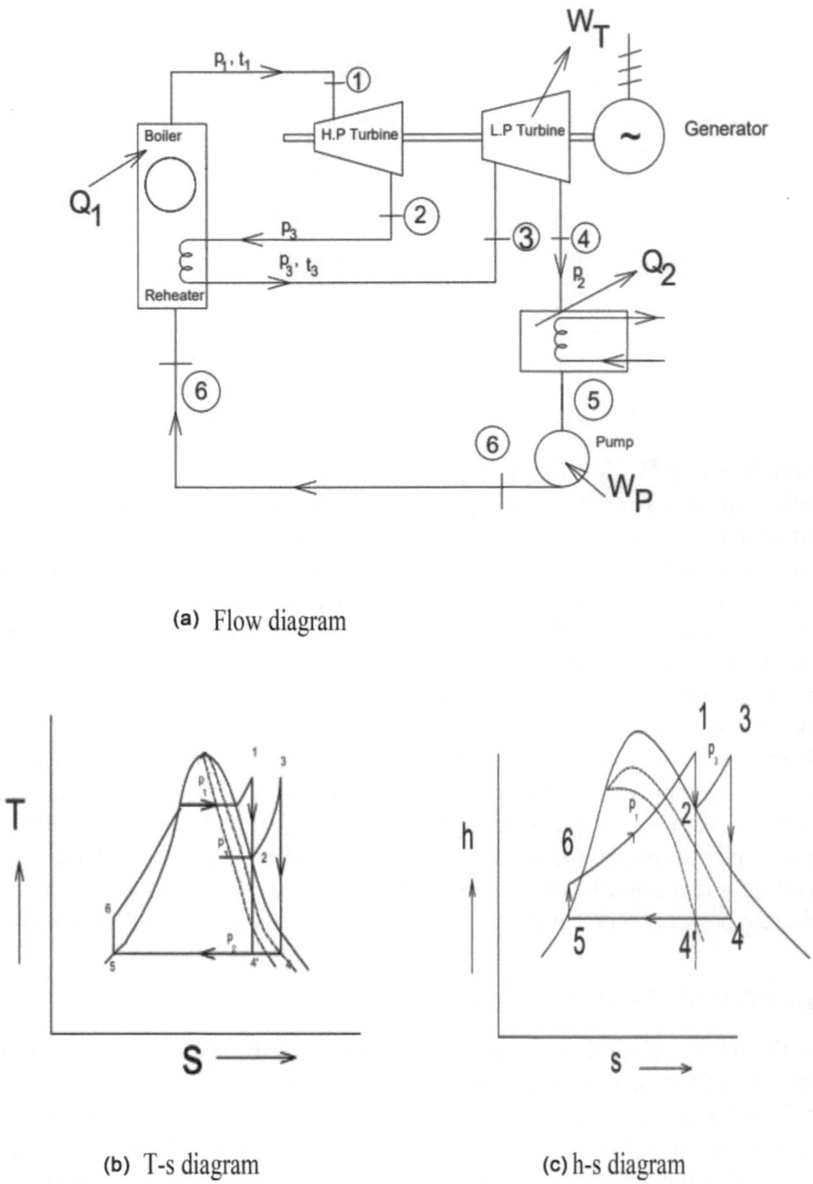

FIGURE 10.10 The reheat Rankine cycle. (a) Flow diagram, (b) T-s diagram, and (c) h-s diagram.

Figure 10.10 shows the schematic, T-s and h-s diagrams, of the reheat Rankine cycle. In reheat cycle, as shown in Figure 10.10a, steam after partial expansion in a high pressure turbine isentropically, is sent back to boiler where it is heated till it reaches superheated condition. It is then allowed to expand isentropically in a low pressure turbine to attain condenser pressure. Although the number of expansion and reheat stages can increase the mean temperature of the heat addition, more than two stages is not economical. Theoretically, the efficiency improvement in the second stage is half that attained in a single reheat, and improvement in the third stage is half that attained in the second stage; this gain can be offset by increased cost and complexity of the plant.

Applying the steady-flow energy equation for the reheat Rankine cycle

Total heat added, $q_1 = h_1 - h_6 + h_3 - h_2$

Heat rejected in the condenser, $q_2 = h_4 - h_5$

Work of turbine, $w_T = h_1 - h_2 + h_3 - h_4$

Pump work is $w_P = h_6 - h_5$

The efficiency of the reheat Rankine cycle is

$$\eta = \frac{w_T - w_P}{q_1} = \frac{(h_1 - h_2 + h_3 - h_4) - (h_6 - h_5)}{h_1 - h_6 + h_3 - h_2} \tag{10.8}$$

$$\text{Steam rate} = \frac{3600}{(h_1 - h_2 + h_3 - h_4) - (h_6 - h_5)} \text{kg/kWh} \tag{10.9}$$

10.7 REGENERATIVE RANKINE CYCLE

Regeneration is another method of improving the thermal efficiency of the Rankine cycle by raising the mean temperature of the heat addition. It is the process of heating the feedwater to raise its temperature before it enters the boiler so that the mean temperature of the heat addition can be increased. One practice to raise the temperature of feedwater is to install a heat exchanger built into a turbine and extracting the heat from expanding steam. This method, however, is not practical since designing such a heat exchanger is not that easy, and moreover, it creates additional problems of increased moisture in the exhaust. The other method of raising the temperature of feedwater is by extracting steam at different locations of the turbine and using this heat to heat the feedwater. The device used for heating the feedwater is termed feedwater heater or regenerator. There are two types of feedwater heaters: open feedwater heaters and closed feedwater heaters. In open feedwaters also known as direct contact heat exchangers, both the steam and feedwater mix with each other so that there is heat transfer from steam to water while in closed feedwater heaters, and both the fluid steams don't mix with each other.

Open Feedwater Heaters

In this, two fluid streams at different temperatures mix with each other to form a stream at an intermediate temperature. Figure 10.11 shows the schematic and T-s diagrams of the regenerative Rankine cycle with two open feedwater heaters.

Steam enters the turbine at state 1 and expands isentropically to state 2, where m_1 fraction of steam is extracted at a location where pressure is p_2 and is fed into the feedwater heater (heater-1) and m_2 fraction of steam is extracted at p_3 to heat the feedwater (heater-2) and the remaining steam $(1-m_1-m_2)$ kg continues to expand isentropically to the condenser pressure. The condensate $(1-m_1-m_2$ kg) then enters the pump-1 where its pressure is raised to the feedwater heater pressure and sent to the heater-2 where it mixes with m_2 kg of steam. The feedwater $(1-m_1$ kg) then enters the pump-2 and is pumped to the heater-2 where it mixes with m_1 kg of steam, and finally 1 kg of saturated steam is pumped to the boiler pressure with the help of pump-3 and routed to the boiler. The cycle is completed by adding heat from external source to heat the water in the boiler. The feedwater leaves the heaters as saturated water at the respective pressures from both the heaters.

For a unit mass of substance, applying SFEE for various control volumes such as boiler, turbine, condenser and pump, heat and work quantities can be estimated as given below:

The heat added in the boiler, $q_1 = (h_1 - h_{10})$

The heat rejected in the condenser, $q_2 = (1-m_1-m_2)(h_4 - h_5)$

Work of turbine, $w_T = 1(h_1 - h_2) + (1-m_1)(h_2 - h_3) + (1-m_1-m_2)(h_3 - h_4)$

Work of pump, $w_P = w_{P1} + w_{P2} + w_{P3}$

$$w_P = (1-m_1-m_2)(h_6 - h_5) + (1-m_1)(h_8 - h_7) + 1(h_{10} - h_9)$$

$$\text{Heater 1} \Rightarrow m_1 h_2 + (1-m_1)(h_8) = h_9$$

$$\Rightarrow m_1 = \frac{h_9 - h_8}{h_2 - h_8} \tag{10.10}$$

$$\text{Heater 2} \Rightarrow m_2 h_3 + (1-m_1-m_2)h_6 = (1-m_1)h_7$$

$$\Rightarrow m_2 = \frac{(1-m_1)(h_7 - h_6)}{(h_3 - h_6)} \tag{10.11}$$

$$T_m \text{ with regeneration} = \frac{h_1 - h_{10}}{s_1 - s_{10}} \tag{10.12}$$

$$T_m \text{ without regeneration} = \frac{h_1 - h_6}{s_1 - s_6} \tag{10.13}$$

The efficiency of regenerative Rankine cycle is higher than that of simple Rankine cycle since the mean temperature of the heat addition (T_m) with regeneration is greater than T_m without regeneration. By adding more number of feedwater heaters, the efficiency of Rankine cycle could be further improved; however, the optimum number of feedwater heaters is decided based on economical considerations.

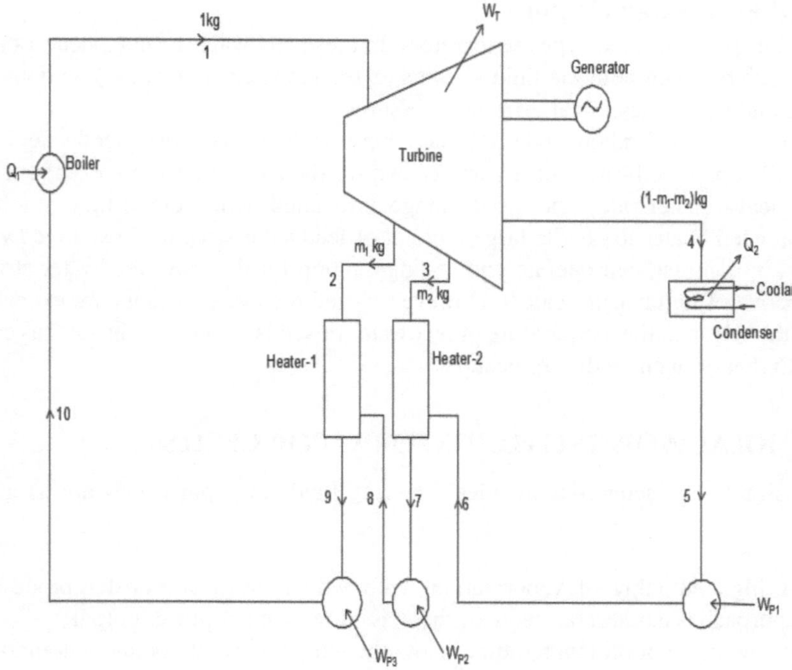

(a) The flow diagram

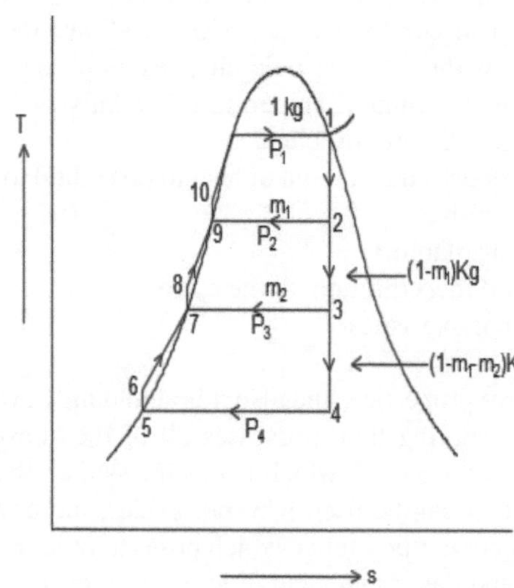

(b) T-s diagram

FIGURE 10.11 The regenerative Rankine cycle. (a) The flow diagram and (b) T-s diagram.

Closed Feedwater Heaters

These are shell-and-tube-type recuperators. In these feedwater heaters, heat transfer takes place between both the fluid streams (extracted steam and water) without any mixing and hence they are at different pressures.

There are some fundamental differences between open and closed feedwater heaters. While open feedwater heaters are known for their lower cost, compactness and higher heat transfer rates, the disadvantage associated with them is they require a pump at each heater to handle large amount of feedwater stream. Closed feedwater heaters, in contrast, can operate with a single pump for the main feedwater stream irrespective of number of heaters. However, closed feedwater heaters are expensive comparatively, and the temperature of feedwater raised is also somewhat less as compared to that of open feedwater heaters.

10.8 IDEAL WORKING FLUIDS FOR VAPOR CYCLES

The desirable characteristics for ideal working fluids in vapor cycles are as given below:

1. A high enthalpy of vaporization, with which the heat transfer process approaches isothermal, reduces the mass flow rate per power output.
2. A low triple point temperature below the temperature of coolant automatically prevents the possibility of any solidification of working fluid.
3. A saturated vapor line, close to the turbine expansion path to prevent any moisture present in the turbine exhaust, eliminates the use of reheat.
4. A higher critical temperature, above the maximum allowed temperature (based on metallurgical conditions, i.e., 620°C), allows the heat transfer almost isothermally as the working fluid changes its phase. Also the saturation pressure at this maximum temperature is relatively low, which can eliminate any material strength problems.
5. A low specific heat reduces the amount of heat to raise fluid to boiling point.
6. Nontoxic and noncorrosive.
7. Low cost and ready availability
8. Chemically stable and inert throughout the cycle
9. High heat transfer characteristics

Water has some of the above properties, and also it is abundantly available. However, including water, no other working fluid possesses all of the above characteristics. Critical temperature of water is 374°C, which is much less than the maximum temperature (around 620°C) that can be used in vapor cycles; moreover, its saturation pressure is also high at higher temperatures, which poses another challenge. Thus no liquid alone can make a candidate as an ideal fluid in vapor cycles.

10.9 BINARY VAPOR CYCLES

Water is the better working fluid in the low temperature range and may not be so in the high temperature range. There are other working fluids such as mercury, sodium,

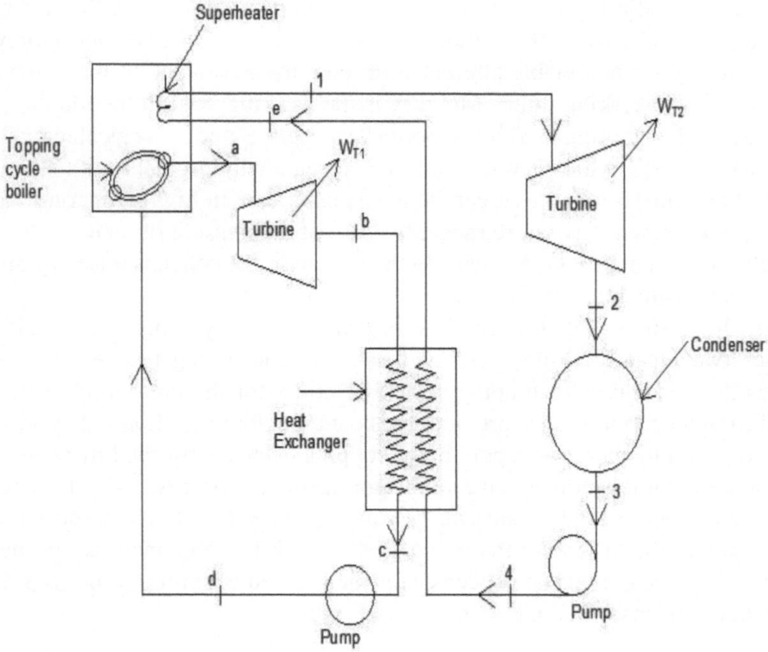

(a) The flow diagram

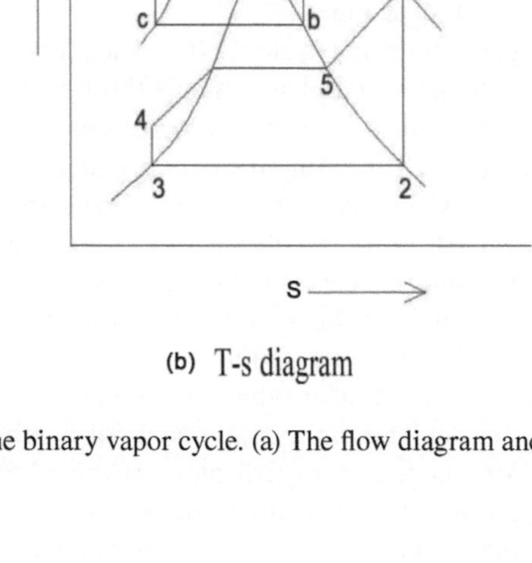

(b) T-s diagram

FIGURE 10.12 The binary vapor cycle. (a) The flow diagram and (b) T-s diagram.

potassium, and sodium-potassium mixtures, which perform better at high temperature ranges. Use of two working fluids in the combination of two cycles known as binary vapor cycles is a viable alternative to take the advantage of the better characteristics of two working fluids. Mercury is the working fluid in the topping cycle, while water is the working fluid in the bottoming cycle. Since the critical temperature of mercury is 898°C, which is well above the current metallurgical limit (620°C) and its critical pressure is 180 bar, it can be a best candidate in high-temperature cycle. However, in the low-temperature range, the saturation pressure of mercury becomes extremely low and maintaining such a high vacuum in the condenser is very difficult making it an unsuitable working fluid.

Figure 10.12 shows the flow and T-s diagrams of binary vapor cycle. This cycle combines two ideal Rankine cycles with an interconnecting heat exchanger that serves as the condenser for topping cycle and boiler for the bottoming cycle. Heat rejected from the topping cycle provides the heat input for the bottoming cycle so that the heat rejected by mercury vapors in the topping cycle is absorbed by water in the lower cycle so that it becomes saturated vapor during the process 4-5. The saturated vapor is then superheated by a superheater in the process 5-1. The superheated vapor then expands in the turbine in the process 1-2, which is then condensed in the condenser in the process 2-3. The condensate is then routed via pump (process 3-4) into the mercury condenser steam boiler.

10.10 ORGANIC RANKINE CYCLE

The continuously tightening emission norms and environmental concerns are the motivating factors for the development of engine waste heat recovery (WHR) technologies. Organic Rankine cycle (ORC) has become one of the most promising heat-driven technologies, that is capable of converting low- and medium-grade heat into electrical or mechanical power. An internal combustion engine (ICE) can only convert about 30% of the overall fuel energy into effective mechanical power and the rest of fuel energy is wasted as a loss in the form of engine exhaust system and cooling system. The integration of a well-designed ORC system to ICE can substantially improve the overall energy efficiency and also can reduce emissions with around 2–5 years payback period through fuel saving.

This cycle is named organic Rankine cycle (ORC), since it uses an organic, high molecular mass fluid with a liquid-vapor phase change (boiling point), which occurs at a comparatively lower temperature than that of water-steam phase change. ORC utilizes the organic substances such as ammonia, pentane, and some common refrigerants such as R12, R123, and R134-a as working fluid. The selection of these fluids depends on the specific application for which they are used. For the production of power from low-temperature sources such as industrial waste heat, internal combustion engine waste heat, geothermal hot water, and concentrating solar collectors typically use the working fluid with low boiling point.

Figure 10.13 shows the schematic of a typical combined IC engine waste heat recovery organic Rankine cycle. The main components of combined ORC system include (1) pump, (2) heat exchanger, (3) superheater, (4) valve, (5) turbine, (6) condenser, and (7) engine. There are two heat sources and has two kinds of cycle modes. In low temperature cycle, the working substance is evaporated by the engine coolant heat and then superheated by engine exhaust heat. In high temperature cycle, the working substance is preheated by engine coolant heat and then directly heated to the state of superheated gas by engine exhaust heat. The working fluid first undergoes a pressurization process in the pump so that it attains a certain working pressure. It is then preheated in the heat exchanger (process 2-3) into saturated vapor by IC engine coolant heat. The working fluid is further heated into superheated vapor in the superheater with the help of exhaust heat from IC engine. The liquid working medium, through these processes, is transformed into superheated vapor with a high pressure, which is capable of doing work. Next, the pressurized vapor is allowed to expand in the turbine, which can output the effective work. The work of turbine in the ORC is used to drive generator or directly output through coupling engine crankshaft. The combined ORC system thus recovers the waste heat of the IC engine and converts it into the effective power of turbine. The expanded vapor is finally allowed to condense into liquid state in the condenser so that it is ready for the next cycle.

Figure 10.14a and b shows the T-s diagrams of both low temperature and high temperature cycles, respectively. The process 1-2 is the actual compression and $1\text{-}2^1$ is the isentropic compression. The process 2-3 of Figure 10.14a represents the preheating and evaporation processes and is considered to be a constant pressure process, while process 2-3 of Figure 10.14b represents the preheating process only. During heat transfer process 6-7, IC engine coolant is cooled. Process 3-4 of Figure 10.14a and b represents overheating and evaporating and overheating processes, respectively. During process 8-9, IC engine exhaust gas is cooled due to the heat transfer process.

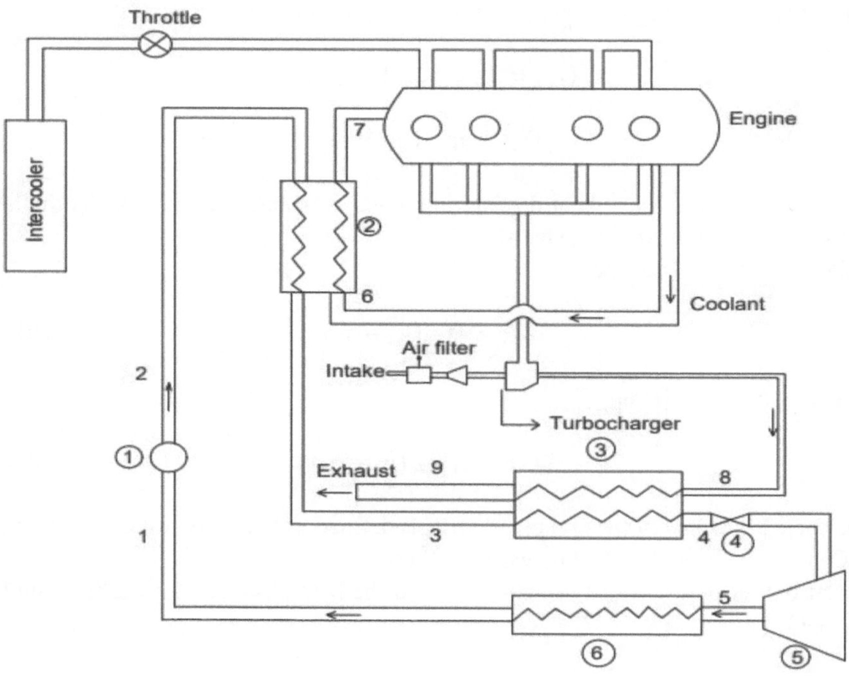

FIGURE 10.13 Schematic of combined ORC for IC engine waste heat recovery.

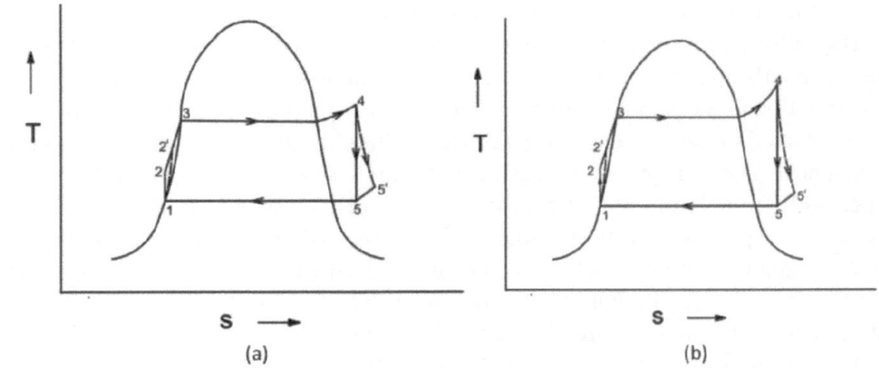

FIGURE 10.14 T-s diagrams of combined ORC for (a) low temperature cycle and (b) high temperature cycle.

10.10.1 Efficiency of the Cycle

Let $\dot{m}_{exh}, \dot{m}_{wf}, \dot{m}_{cool}$ be the mass flow rates of exhaust gas from IC engine, working fluid, and coolant, respectively, then

We can apply the SFEE to different control volumes of the cycle to evaluate the energy interactions.

The amount of coolant heat flowing in to the working fluid,

$$\dot{Q}_{2\text{-}3} = \dot{m}_{wf}(h_3 - h_2),$$

where $Q_{2\text{-}3}$ is the heat flux flowing out of IC engine coolant

$$\dot{Q}_{8\text{-}9} = \dot{m}_{exh}(h_8 - h_9) = Q_{3\text{-}4}$$

$$\dot{Q}_{3\text{-}4} = \dot{m}_{wf}(h_4 - h_3),$$

where $\dot{Q}_{3\text{-}4}$ is the exhaust heat flowing into ORC working fluid and $\dot{Q}_{8\text{-}9}$ is the heat transfer from the IC engine exhaust gas.

Power output of turbine,

$$\dot{W}_T = \dot{m}_{wf}(h_4 - h_5) \tag{10.14}$$

Power required for pump,

$$\dot{W}_P = \dot{m}_{wf}\left[\upsilon_1(p_2 - p_1)\right] \tag{10.15}$$

The cycle efficiency of the combined ORC is defined as the ratio of the output power of ORC to the heat absorbed by working fluid,

Net output power, $\dot{W}_{net,out} = \dot{W}_T - \dot{W}_P$ and the heat absorbed by working fluid is $\dot{Q}_{2\text{-}3} + \dot{Q}_{3\text{-}4}$

Efficiency of the cycle,

$$\eta_{cycle} = \frac{\dot{W}_{net,out}}{\dot{Q}_{2\text{-}3} + \dot{Q}_{3\text{-}4}} \tag{10.16}$$

Now, for waste heat recovery through this combined ORC, the recovery efficiency is given as

$$\eta_{recovery} = \frac{\dot{W}_{net,out}}{\dot{Q}_{cool} + \dot{Q}_{exh}}, \tag{10.17}$$

where $\dot{Q}_{cool}$ is the energy flow of IC engine coolant and $\dot{Q}_{exh}$ is the energy flow of IC engine exhaust gas.

The total fuel efficiency (η_{total}) of IC engine system after the combined ORC is coupled to IC engine can be calculated as

$$\eta_{total} = \frac{\dot{W}_{ICE} + \dot{W}_{net,out}}{\dot{m}_f \times C \cdot V}, \tag{10.18}$$

where $\dot{W}_{ICE}$ is the effective power of IC engine, $\dot{m}_f$ is the fuel consumption rate, and CV is the lower heating value of fuel.

The IC engine fuel efficiency improvement is based on the energy saving potential of this combined ORC given as

$$\eta_{imp} = \eta_{total} - \eta_{ICE}, \tag{10.19}$$

where η_{imp} and η_{ICE} are, respectively, the IC engine efficiency improvement and actual IC engine efficiency.

10.10.2 The Ideal Working Fluids for the Combined ORC

The ideal working fluid used in the combined ORC for waste heat recovery systems (WHRS) should have the following features:

1. The working fluid should be cheaply available so that it can lower the cost of WHRS.
2. The working fluid should have low boiling point and latent heat of evaporation so as to recover waste heat from low-temperature source, in particular coolant heat.
3. The working fluid should have low critical temperature, critical pressure, and high thermal conductivity.
4. The working fluid should have stable physical and chemical properties and thermal stability.
5. The working fluid should be nontoxic, nonflammable, nonexplosive, and environmental friendly.

TABLE 10.1
The Properties of the ORC Working Fluids [32]

Working Fluid	T_{cr} (K)	P_{cr} (Mpa)	Molecular Weight (g/mol)	GWP	ODP
Benzene	562.02	4.906	78.112	Very low	0
Cyclohexane	553.64	4.075	84.161	Very low	0
R245fa	427.16	3.651	134.05	950	0
Toluene	591.75	4.126	92.138	Very low	0

GWP, Global warming potential; ODP, Ozone depression potential.

Table 10.1 shows the thermal properties of the high and low temperature circulating working fluids. R245fa can be considered as the working fluid for the low temperature cycle according to the range of low temperature circulating heat source temperature.

10.11 COGENERATION

Cogeneration systems capture wasted heat energy from electric power-producing devices such as steam turbine, gas turbine, and diesel engine and use the same for applications such as industrial process heating, space heating, and water heating. The cogeneration systems differ from the conventional separate electrical and thermal energy systems (power plant and a boiler) in that it uses the cascading of energy use from high- to low-temperature uses.

The primary advantage with the cogeneration system is it can potentially improve the efficiency of fuel use in the production of electrical and thermal energy. For the production of a given amount of electrical and thermal energy, a single cogeneration unit requires less fuel than is needed to produce the same quantities of both types of energy with separate conventional methods. The reason for this can be regarded as the heat from the turbine-generator set, which uses a considerable quantity of fuel to fire the turbine, can be a useful heat in the form of process heat in the cogeneration system, which otherwise goes as a waste heat. Moreover, cogeneration offers environmental advantages. The reduction in fuel usage by cogeneration system, as compared to that of conventional processes for electrical and thermal energy production, contributes to the reduced greenhouse gas emissions (GHGs) by nearly 50% in some situations.

Cogeneration systems utilizing internal combustion engines and gas turbines in open cycle are the most utilized technologies worldwide. Heavy-fuel-fired diesel power plants run on relatively inexpensive diesel fuel, a low-grade product of oil refineries. Such power plants can be set up quickly, normally in less than 12 months, to generate hundreds of megawatts of energy.

It was mentioned earlier in Chapter 5 that even a best efficient heat engine requires about 250 kJ of energy input for producing 100 kJ of work. The remaining 150 kJ of energy is rejected to the surroundings in another form as waste heat but not as work. This waste heat, being of low quality, cannot be used for any purpose. However, there are many systems that require energy in the form of heat known as process heat as in the case of sugar industries, textiles, paper mills, refineries, steel manufacturing units, and some chemical industries. These systems require process heat supplied by the steam at a pressure of 5–7 bar and temperature around 200°C for heating and drying purposes. Saturated steam at the desired temperature is suitable for constant temperature heating as steam is good medium. Usually saturated steam is condensed at that temperature so that isothermal conditions are achieved. The power plant that is simultaneously producing the electric power and process heat is called a cogeneration plant. Cogeneration is producing more than one form of useful energy from the same source of energy.

Figure 10.15 shows the schematic of an ideal steam turbine cogeneration plant. The condenser of the Rankine cycle is replaced by a process heater in cogeneration plant so that no waste heat is rejected. Thus all the heat energy transferred to the steam in the steam generator is utilized as either process heat or electric power. For a cogeneration plant that produces electricity and heat, the first law (energy) efficiency is defined as the ratio of useful energy output (power output, $\dot{W}_T$ plus process heat, $\dot{Q}_H$) to the energy input $\left(\dot{Q}_1\right)$.

$$\eta_{cogen} = \frac{\text{Net work output} + \text{Process heat delivered}}{\text{Total heat input}} = \frac{\dot{W}_T + \dot{Q}_H}{\dot{Q}_1} \tag{10.20}$$

The above relation (Eq. 10.20) is referred to as the utilization efficiency to distinguish it from the thermal efficiency, commonly used in the case of a power plant with single output power. The utilization efficiency of actual steam turbine cogeneration plant can be upto 80%, while the ideal plant efficiency can be 100%.

We can also define the efficiency of the cogeneration plant based on the second-law aspect (exergy analysis) usually referred to as the second-law efficiency that is the ratio of exergy output to exergy input.

$$\eta_{II,cogen} = \frac{\dot{X}_{out}}{\dot{X}_{in}} = \frac{\dot{W}_T + \dot{X}_{heat}}{\dot{X}_{in}} \text{ or } \eta_{II,cogen} = 1 - \frac{\dot{X}_{destroyed}}{\dot{X}_{in}}, \tag{10.21}$$

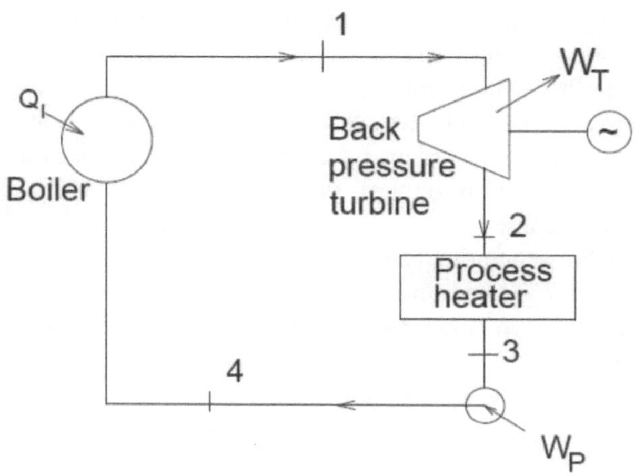

FIGURE 10.15 An ideal cogeneration plant.

where $\dot{X}_{destroyed}$ is the exergy destruction rate and $\dot{X}_{heat}$ is the exergy transfer associated with heat transfer given as

$$\dot{X}_{heat} = \int \delta\dot{Q}_H \left(1 - \frac{T_0}{T_1}\right), \tag{10.22}$$

where T_1 is the temperature at which heat is added. However, the Eq. 10.22 does not make any sense unless the functional relationship between the rate of heat transfer $\dot{Q}_H$ and T_1 is known. In many of the practical cases, heat is utilized by transferring it from the working fluid leaving the heat engine such as turbine to a secondary fluid in a heat exchanger. Then exergy rate of heating can be expressed as the exergy increase of the cold fluid in the heater, given as

$$\dot{X}_{heat} = \Delta\dot{X}_{cold} = \dot{m}_{cold}\left(\Delta h - T_0\Delta s\right)_{cold}, \tag{10.23}$$

where Δh and Δs are enthalpy and entropy changes of the cold fluid, respectively.

Then the second-law efficiency for the cogeneration becomes

$$\eta_{II,cogen} = \frac{\dot{W}_T + \dot{m}_{cold}\left(\Delta h - T_0\Delta s\right)_{cold}}{\dot{X}_{in}} \tag{10.24}$$

However, the ideal cogeneration plant cannot be implemented in practice since it cannot meet the requirements of process heat and electric power loads. A cogeneration plant uses either Rankine cycle or Brayton cycle as the power cycle. Figure 10.16 shows the schematic of an actual steam turbine cogeneration plant. Based on the process heat requirements, for example, the process heater requires a heat rate of 50 kW at a pressure of 5 bar, the steam is expanded in the turbine to a pressure of 5 bar producing the power at a rate of 25 kW. The steam leaves the process heater as a saturated liquid at 5 bar. With the help of a pump, the liquid is then pumped to the boiler pressure and heated in the boiler. The rate of heat input to the heater can be calculated from energy balance.

In cogeneration plant, since the same steam flows through the cycle (turbine and heater), the power output and heat output are coupled together. This type of plant gives better results when only the steam flow rate can be controlled based on heating load requirements, and power produced can be utilized. In case a single plant has to handle fluctuating heating and power loads, then a pressure regulating valve (PRV) is operated. Whenever the power load is zero, all the steam passes through the PRV and when the heating load is zero, all the steam expands through the turbine and condenses in the condenser.

If $\dot{m}$ is the steam generating capacity of boiler and $\dot{m}_1$ is the rate of steam flow required at the desired temperature in the process heater, then the rates of heat input, heat rejection, input to the heater, pump work and power produced by turbine can be given as follows:

The rate of heat input to the boiler,

$$\dot{Q}_1 = \dot{m}\left(h_1 - h_8\right) \tag{10.25}$$

The rate of the heat rejection in the condenser,

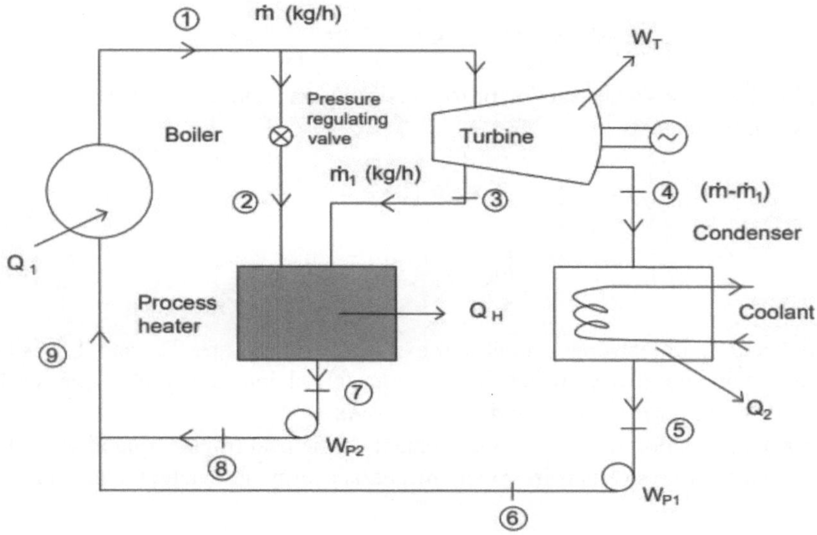

FIGURE 10.16 A cogeneration plant with adjustable loads.

$$\dot{Q}_2 = (\dot{m} - \dot{m}_1)(h_4 - h_5) \quad (10.26)$$

Rate of heater input,

$$\dot{Q}_H = \dot{m}_1 (h_3 - h_7) \quad (10.27)$$

Rate of pump work

$$\dot{W}_p = (\dot{m} - \dot{m}_1)(h_6 - h_5) + \dot{m}_1 (h_8 - h_7) \quad (10.28)$$

Rate of work produced by turbine,

$$\dot{W}_T = \dot{m}(h_1 - h_3) + (\dot{m} - \dot{m}_1)(h_3 - h_4) \quad (10.29)$$

10.12 EXERGY ANALYSIS OF VAPOR POWER CYCLES

The vapor power cycles operating on ideal Rankine cycle, unlike the Carnot cycle, are not totally reversible cycles since they involve irreversibilities. These irreversibilities present in each component result in available energy (exergy) loss. Boiler is the main source of the irreversibilities in steam power cycle as a substantial portion of exergy entering the plant with the fuel is destroyed in it. There are two primary reasons for this: one is the irreversible heat transfer that takes place between hot combustion gases and working fluid through the tubes carrying working fluid and the other is the combustion process itself. The second law analysis of the cycle such as available energy (exergy) and exergy destruction (available energy loss) provides a useful insight to quantify these irreversibilities. The available energy (exergy), ψ of a fluid stream at a specified state, is given as

$$\psi = (h - h_0) - T_0(s - s_0) + \frac{V^2}{2} + gz \tag{10.30}$$

From Eq. 9.24, exergy destruction or irreversibility on a unit mass basis for a steady-flow system is given as

$$x_{des} = T_0 s_{gen}$$

$$= T_0\left(s_e - s_i + \frac{q_2}{T_{b,2}} - \frac{q_1}{T_{b,1}}\right), \tag{10.31}$$

where s_i and s_e are entropy at inlet and exit, respectively, and $T_{b,1}$ and $T_{b,2}$ are the boundary temperatures at which heat is transferred into and out of the system, respectively, and T_0 is the surrounding temperature.

Exergy destruction for a cycle is dependent on the heat transfer quantities of high and low temperatures reservoirs. On a unit mass, exergy destruction is given as

$$x_{des} = T_0\left(\sum \frac{q_2}{T_{b,2}} - \sum \frac{q_1}{T_{b,1}}\right) \tag{10.32}$$

For a cycle with heat transfer with source at T_1 and sink at T_2, exergy destruction on a unit mass basis is

$$x_{des} = T_0\left(\sum \frac{q_2}{T_2} - \sum \frac{q_1}{T_1}\right) \tag{10.33}$$

Exergy destruction for a cycle that involves heat transfer with source at T_1 and sink at T_2 only is given as

$$x_{des} = T_0\left(\frac{q_2}{T_2} - \frac{q_1}{T_1}\right) \tag{10.34}$$

10.13 COMBINED CYCLE POWER PLANTS

To meet the increasing global demand for electric power, it is essentially important to sustain efforts at improving power generation technologies with high energy conversion efficiencies. Furthermore, global warming and depletion of fossil fuels have escalated the need and importance of integrating several energy production units to utilize a common primary energy input. The combined cycle power plant (CCPP) is one such attempt to this effect to improve the efficiency of power plant operating on single cycle. In this, the fossil fuels are used as the primary energy input to generate electricity in thermal power plants operating on gas turbine plant; the exhaust flue gases can be utilized to drive other thermal cycle operating on the Rankine cycle.

CCPP is a combination of the Rankine cycle and the Brayton cycle in which Brayton cycle (gas cycle) is topping the Rankine cycle (steam turbine cycle). The resulting cycle has higher thermal efficiency than either of the cycles executed separately. The gas turbine cycles usually operate at higher temperatures comparatively. The maximum temperature at the inlet to a steam turbine is around 600°C as opposed to that of around 1400°C in gas turbine cycles. Due to this higher average temperature at which heat is added in gas turbine cycles, they offer higher thermal efficiencies. However, the exhaust gases also leave the gas turbine cycle at higher temperatures that are of the order 500°C. This high temperature energy can be utilized in the bottoming cycle (steam turbine cycle), for heating the water to generate steam.

Figure 10.17a and b shows the schematic and T-s diagrams of combined gas-steam cycle plant. Fuel is burned in process 6-7 of gas turbine cycle (GTC) to heat the gas, the working fluid. The heat rejected in the process 8-9 serves as heat input (boiler) to the steam turbine cycle (STC). Some more modifications such as heating some fuel in the steam generator and various other combinations of enhancements to both steam and gas cycles may be possible. The advantage with the gas turbines is they can be started and brought to full-load conditions comparatively quickly, thus combining these with steam cycles can make it possible to meet fluctuating load requirements. This can minimize the combined fuel and fixed costs.

10.13.1 The Effect of Operating Parameters on Combined Cycle Performance

The major operating parameters that influence the combined cycle performance are as follows:

1. Turbine inlet temperature (TIT)
2. Compressor pressure ratio
3. Pinch point
4. Ambient temperature
5. Pressure levels

The increase of pressure levels of steam generation in heat recovery steam generators (HRSG) increases the heat recovery from the flue gas and consequently, the energetic efficiency of the cycle increases. Furthermore, the exergy destruction rate of the cycle decreases with the increase in number of pressure levels of steam generation in HRSG. The increase of the number of pressure levels of steam generation increases the total and specific investment cost of the plant as 6% and 4%, respectively. With an increase in ambient temperature from 273 to 333 K, the total power output increases about 7% for all configurations except the regenerative gas turbine. The overall thermal efficiency of the combined cycle obtained the maximum value with regenerative gas turbine configuration about 62.8% at ambient temperature of 273 K, and the minimum value of the overall thermal efficiency was about 53% for intercooler gas turbine configuration at ambient temperature of 333 K.

In combined cycle operation, the performance analysis of the new generation of gas turbines is indeed complex one and presents new problems, which need to be addressed. First, the new units operate at extremely high turbine firing temperatures;

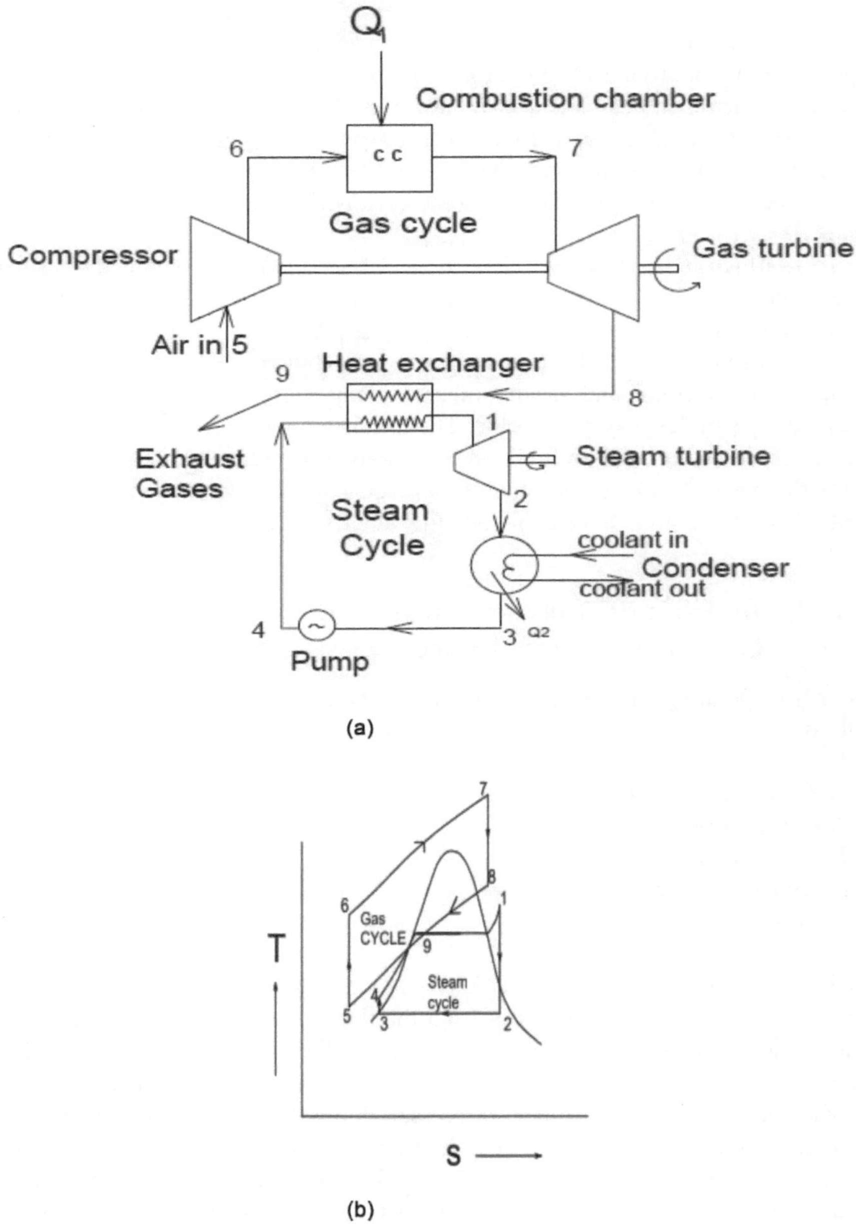

FIGURE 10.17 Combined gas-steam cycle. (a) Schematic diagram and (b) T-s diagram.

this variation in firing temperature considerably affects the performance and life of the components in the hot section of the turbine. Second, the pressure ratio in compressor is high, which leads to a very narrow operation margin, thereby keeping the turbine at a risk of compressor fouling. The performance of the combined cycle is also dependent on the steam-turbine performance. The steam turbine performance is dependent on the pressure, temperature, and flow generated in the heat-recovery steam generator, which, in turn, is dependent on the turbine firing temperature and the air mass flow through the gas turbine. If a life cycle analysis were conducted, the new costs of a plant are about 7%–10% of the life cycle costs. Maintenance costs are approximately 15%–20% of the life cycle costs. Operating costs, which essentially consist of energy costs, make up the remainder, between 70% and 80% of the life cycle costs, of any major utility plant. Thus, the performance evaluation of the turbine is one of the most important parameters in the operation of a plant.

10.13.2 Combined Cycle Power Plant Integrated with ORC

Nowadays, the technology of integrated multienergy generation has been gaining momentum and emerges as one of the effective means of improving the existing cycle performance and utilizing energy from waste heat.

A drawback of air-cooled condensers (ACCs) is that their performance can decline as ambient temperatures increase and result in a loss of steam turbine power output. Increased ambient temperature reduces the heat transfer (heat rejection) rate during steam condensation leading to rise in turbine back pressure. As the turbine back pressure increases, the output of the steam turbine decreases.

Although the combined cycle power plant is an improvement over the single cycle power plant, the primary drawback associated with it is it discharges usable low-grade waste heat. This exhaust heat from existing power plants can be used to drive energy production units such as organic Rankine cycle (ORC) for electricity and absorption refrigeration cycle (ARC) for cooling, desalination cycle and heating units, etc. The integration of the CCPP with ORC and absorption refrigeration cycle (ARC) resulted in the net power output increment by 9.1% of the integrated power plant. The thermal and exergy efficiencies increased by 8.7% and 8.8%, respectively, while the total exergy destruction rate and specific fuel consumption reduced by 13.3% and 8.4%, respectively. The first law and second law analyses of these CCPP show that average energy and exergy efficiencies of the plant are found to be 56% and 50.04%, respectively.

10.13.3 Combined Cycle Power Plant Integrated with Absorption Refrigeration System

The performance of gas turbine unit and hence the overall performance of the combined cycle power plant can be enhanced by reducing the inlet air temperature to the compressor. For this, injecting water into a gas turbine compressor proved to be an effective method in increasing the power production. Precooling of air before it enters the compressor can contribute to this supply of additional working generation by the combined-cycle gas turbine plant. An absorption refrigeration machine can be used as a source of cold. From energy and exergy analyses of this plant, it is revealed that with this arrangement, in summer season, it can be possible to develop an additional net power of 9440 kW with the corresponding increase in the thermal and exergy efficiencies of the plant by 1.193% and 1.133%, respectively. In winter, it can increase the power output by 400 kW.

10.14 INTEGRATED COAL GASIFICATION COMBINED CYCLE (IGCC) POWER PLANTS

The effective and efficient use of fossil fuels in energy industries is vital from the standpoint of energy security and environmental sustainability. The development of power plants such as integrated gasification combined cycle (IGCC) with advanced configurations is a move in this direction globally to use coal and biomass more efficiently and cleanly. Gasification is the main process within an IGCC system, which features the fuel flexibility of most advanced technologies for power production. Current IGCC technology can potentially use biomass and other low-value feedstocks that have high-ash residues. Gasification, besides regulating gaseous pollutant emissions, can be a cost-effective approach to concentrate carbon dioxide at high pressure to facilitate sequestration. The most important parameters for typical gasifier operating conditions are temperature, pressure, and oxygen-to-coal ratio.

The primary advantage with IGCC technology is its ability to use solid and liquid fuels in a power plant that can combine the environmental benefits of a natural gas fueled plant and thermal performance of a combined cycle. In this plant, the solid or liquid fuel is typically gasified with either oxygen or air, and the resulting raw gas (syngas) is cooled, cleaned of particulate matter and sulfur, and then fired in a gas turbine. Since the emission-forming constituents such as particulate matter and sulfur are removed from the gas under pressure prior to combustion in the power block, it can be possible for IGCC plants to meet stringent air emission norms. The hot exhaust gases from gas turbine are sent to a heat recovery steam generator (HRSG) where the steam is produced. The steam produced by HRSG drives a steam turbine. This power is produced from both the gas and steam turbines.

Integrated coal gasification combined cycle (IGCC) power plants can be the next-generation thermal power system that can meet the twin goals of enhanced power generation efficiency and environmental performance. To achieve this objective, it combines both coal gasification and the gas turbine combined cycle (GTCC) system. The flexibility with IGCC plant is its potential to use biomass and other low-value feedstocks with high-ash residues and also it is a low-cost approach. Large-scale IGCC systems are capable of improving the power generation efficiency by nearly 15% and reducing CO_2 emissions when compared to that of conventional coal-fired thermal power systems. IGCC plants, with the improved system efficiency, result in lower NO_x and SO_x and soot emissions per kilowatt hour of electric power generated. Moreover, they discharge 30% less waste water than that of conventional coal-fired thermal power systems. This is due to the reason that IGCC consumes much less amount of water because it treats fuel gas that is higher in pressure and smaller in volume as opposed to that in conventional coal-fired thermal power in which large amount of water is used for treating flue gas after fuel operation.

10.14.1 Working of IGCC Power Plant

The fuel for IGCC power plants is usually supplied through the gasification of coal. The option of having a steam-methane reforming process followed by precombustion carbon captures after coal gasification. Coal is obtained from an earlier process

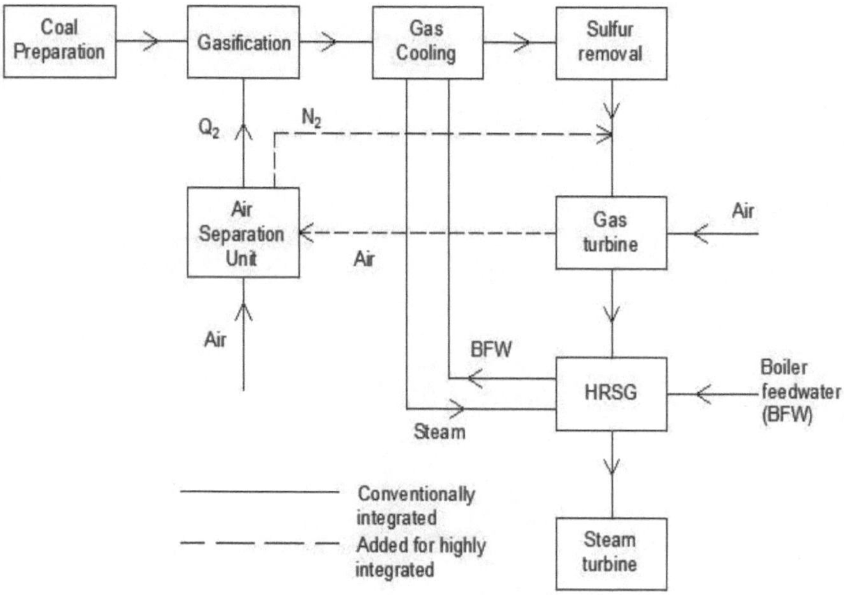

FIGURE 10.18 Schematic of integrated gasification combined cycle power plant.

of coal processing, which includes cleaning, pulverizing, and drying after being received from the export country. Usually, coal is mined from the ground either through an underground or open-pit mining method and transported to power plants by land and/or sea. Figure 10.18 shows the block diagram of integrated gasification combined cycle power plant. IGCC power plants make use of synthetic gas or syngas, which is converted from the gasification of coal. A typical IGCC power plant consists of a few key components, namely gasifier, gas turbine, steam turbine, and a heat recovery steam generator (HRSG). Syngas, which constitutes carbon monoxide, hydrogen, and methane, is produced by heating the coal with a mixture of steam and oxygen. The heat generated by the partial combustion of coal can be used to drive the steam turbine, which in turn drives the generator that generates electricity. In the first step, the coal is processed, with the air as the gasification agent, into coal gas in the gasifier. In the gas clean-up unit, coal gas undergoes desulfurization and dedusting treatment to meet the norms for gas turbine fuel and exhaust gas. The cleaned gas is supplied to the gas turbine combined cycle facility where the combustion of the cleaned gas takes place and power is generated. For the recovery of the heat of the exhaust gases, a heat recovery steam generator (HRSG) is used, which generates steam by exchanging heat with water and generated steam drives steam turbine for additional power generation.

10.14.2 Carbon Dioxide Capture from IGCC Power Plant

Fossil fuel burning, coal in particular, for power generation leads to carbon dioxide emissions, which is considered to be a major source of greenhouse gas (GHG)

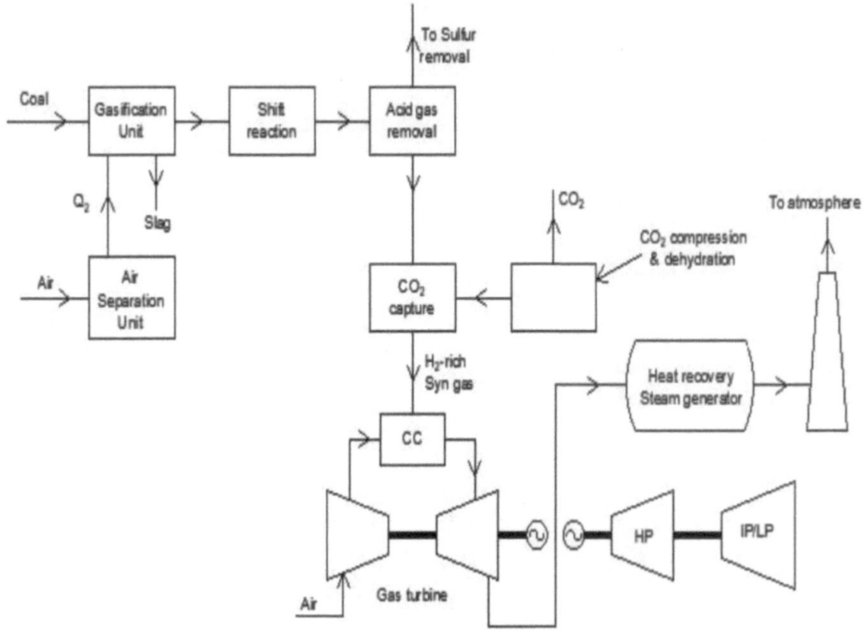

FIGURE 10.19 Schematic of an IGCC power plant with CO_2 capture.

emissions. There is a renewed interest to reduce these emissions across the world. This can be done by using fossil fuels more efficiently and avoiding wastefuel practices. Usage of low-carbon fuels, removal of CO_2 from the exhaust of the power plant and storing the captured CO_2 are some of the techiniques that can be used to reduce the emissions. The IGCC plant for CO_2 capture essentially consists of the following parts as shown in Figure 10.19.

Air Separation Unit (ASU)

The most widely used technology in the market for providing oxygen to the gasification block is cryogenic ASU, which produces large-scale O_2. This unit as a standalone unit can generate gaseous oxygen with purity as high as 95% from air supplied by an intercooled main air compressor.

Gasification Block Including Coal Milling and Drying (CMD), Gasification, Syngas Cooling and Scrubbing

Coal gasification occurs in a slagging entrained-flow, oxy-blown, and dry-fed gasifier on shell coal gasification process (SCGP). The reason for the selection of this technology is it offers several advantages such as high cold gas efficiency, desired working pressure, and higher calorific value of syngas produced. Moreover, SCGP maintains its consistency with a wide range of coal quality, that is, the syngas produced is of higher quality irrespective of the coal quality compared to that produced by other gasifiers such as slurry-fed slagging entrained-flow.

Sour Water-Gas Shift (SWGS) Reaction Unit

In SWGS process, carbon monoxide (CO) in the raw syngas is converted to CO_2 by shifting CO with water in the presence of H_2S over a catalytic bed. The following reaction shows the conversion

$$CO_{(g)} + H_2O_{(g)} \overset{44\frac{MJ}{kmole}}{\leftrightarrow} CO_{2(g)} + H_{2(g)} \tag{10.35}$$

Acid Gas Removal (AGR) Unit and CO_2 Capture Unit

For the removal of H_2S and capture of CO_2, a double-stage SELEXOL system that makes use of dimethyl ether of polyethylene glycol solvent is considered. Physical absorption of H_2S and CO_2 is preferred over chemical, amine-based absorption processes since the former offers better H_2S/CO_2 absorption selectivity and also partial pressures of acid gases are high in this case. For the removal of hydrogen sulfide present in the syngas, a counter-current flow of solvent in the first absorption column is used. The syngas leaving the H_2S absorber is then routed to the second stage where CO_2 is captured in a similar way as that of first stage. The CO_2 design capture rate that can be obtained is nearly 90% overall.

CO_2 Compression and Dehydration Unit

The captured CO_2 is compressed in a compressor and intercooler arrangement, dried from water, after-cooled and liquefied so that it can be delivered at a pressure of around 110 bar, which is essentially required for geological storage permanently. The drying process is carried in the transport pipeline to substantially reduce the risk of corrosion, and the water content in the captured CO_2 is reduced by the dehydration unit consisting of tri-ethylene glycol (TEG), which will maintain the water content less than 20 mg/kg.

Gas Turbine (GT)

The gas turbine block generating the electric power consists of compressor, combustor, expander, and generator. The fuel gas, typically undiluted hydrogen-rich syngas, has lower calorific value, compared to natural gas, and hence necessitates some engine modifications to suit the compressor stability requirements. A re-staggered expander is well suited to meet the fuel gas properties and to keep modifications at a reasonable level that can be performed relatively easy on the engine.

Heat Recovery Steam Generator (HRSG) and Steam Cycle

A triple pressure level heat recovery steam generator including a reheater and a steam turbine placed downstream of the gas turbine will generate steam and electric power.

10.15 POWER CYCLES FOR NUCLEAR PLANTS

10.15.1 Nuclear Power Plant

Nuclear power plant is another electric power generating device, which uses the process of nuclear fission. A nuclear reactor in combination with the Rankine cycle is used for the power generation. In this plant, the heat liberated by nuclear fission reaction in the reactor is transferred to water to convert it into steam, which drives the turbine and the generator. United States and France are the leading producers of the electric power from nuclear power plants. The electric power generated by nuclear power plants constitutes around 10% of the world's total electric power generation. Although nuclear power plants and coal-fired power plants are similar, the fundamental difference between the two lies in the heat source and safety measures for handling of the fuel and waste disposal. In nuclear power plants, thermal energy is released by splitting of nuclei of atoms in the reactor core. Uranium is the most widely used fuel, while thorium is also a potential candidate.

The reactor, like the boiler in coal-fired power plant, is the heart of the nuclear power plant. The fission reaction of uranium produces enormous amount of heat in the reactor, which is then transferred to the coolant of the reactor. The heat gained by the coolant supplies heat to the other parts of the power plant. The other important parts of the plant include turbine, generator, cooling tower, and safety systems. Figure 10.20 shows the schematic diagram of nuclear power plant cycle using sodium-cooled fast reactor.

Steam Generation

The pressurized water reactors and boiling water reactors are the commonest reactors used worldwide. In pressurized water reactors, two loops of circling water are used to generate steam. The first loop supplies hot liquid water to a heat exchanger in which water at lower pressure flows so that the water is converted to steam. The generated steam is supplied to the turbine. Boiling water reactors, in contrast, heat the water in the core directly to steam.

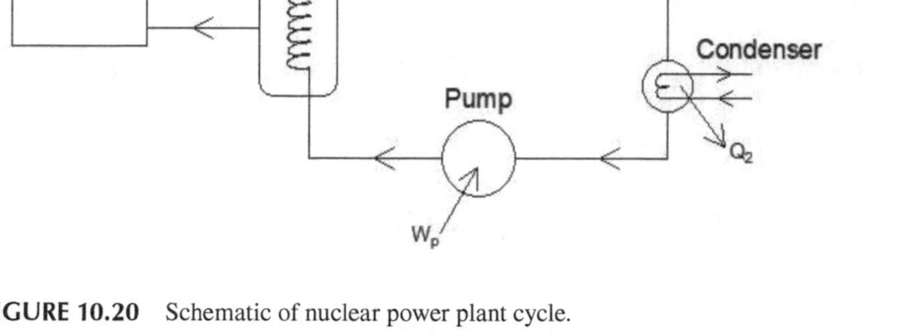

FIGURE 10.20 Schematic of nuclear power plant cycle.

Turbine and Generator

The steam generated in the reactor travels at high pressure and speeds through one or more turbines. This high pressure steam after expansion in the turbine condenses to liquid water. The turbine drives the generator that generates electric power.

Cooling Towers

The purpose of cooling toswer is to reject waste heat from hot water from the turbine section and transfer it to the cold outside air. Hot water by being in contact with the air eventually cools and only a marginal amount of it evaporates and raises upto through the top of cooling tower. There are certain similarities between nuclear power plants and coal-fired power plants since the process of turning heat into electricity is almost identical. Some nuclear power plants, instead of using cooling towers, reject the waste heat into a river, lake, or ocean. Many of the coal-fired power plants also have cooling towers.

The efficiency of nuclear power plants is in the range of 33%–37%, which is comparable to conventional fossil fuel-based plants. However, there is a possibility to improve further the efficiency beyond 45% with the innovative designs such as fourth-generation nuclear reactors. The capacity of global nuclear power generation grew at a faster pace from 1960 to the late 1980s reaching a share of 16% in 1986 and showed a slowdown thereafter. However, this share of 16% held steady for the next 20 years, which implies that the growth of nuclear power generation was steady at the same pace as that of overall worldwide electric power generation. As far as nuclear power generation is concerned, the issue of spent fuel is the topic of discussion. The countries such as China, France, India, Japan, and Russia reprocess the most of the spent fuel, while USA, Canada, Sweden, and Finland dispose it directly. The other countries generating nuclear power store the spent fuel looking forward to assess the better alternative of both the processes. Table 10.2 shows the data provided by U.S Energy Information Administration of top five nuclear electricity generation countries.

TABLE 10.2
Top Five Nuclear Electricity Generation Countries

Country	Nuclear Electricity Generation Capacity (million kW)	Nuclear Electricity Generation (billion kW h)	Nuclear Share of Country's Total Electricity Generation (%)
United States	99.6	805.0	19.8
France	63.1	381.8	71.5
China	34.5	232.8	3.7
Russia	26.1	190.1	18.4
South Korea	22.5	141.3	26.6

Source: U.S. Energy Information Administration, International Energy Statistics, as of April 16, 2020.

10.15.2 Nuclear Fuels

Nuclear fuel is a fuel used in a nuclear reactor to sustain a nuclear chain reaction. The general definition of the fuel that is a material used to produce heat is not applicable to nuclear fuel, since the heat in this case is generated by fission or disintegration of the fissile isotopes but not by the burning of fuel. Nuclear fuel is typically made up of one or more fissile isotopes such as ^{235}U, ^{239}Pu, and ^{233}U, often in combination with a fertile isotope, ^{238}U or ^{232}Th. These fuels are in the solid form of metals, alloys, oxides, carbides, nitrides of uranium, plutonium, and thorium. The radioactive nature of nuclear fuels and their fission products are extremely hazardous to health. To avoid harmful effects of radiation, nuclear fuels are hermetically sealed inside a nonradioactive structural material known as cladding. Cladding is essentially an integral part of the nuclear fuel element, which serves different purposes. First it acts as a primary containment for radioactive fission products, and then it acts as a barrier avoiding direct contact of the fuel with coolant, and finally it transfers fission heat energy from fuel to coolant. Usually, nuclear fuel elements are manufactured in different shapes such as plates, pins, or rods and assembled in specific geometric configurations using spacers, end fittings, and other supporting hardware. The package of fuel elements is called a fuel assembly.

EXAMPLE PROBLEMS

Example 10.1 Steam enters the turbine of a Rankine cycle at 2.5 MPa, 300°C and, after expansion, is condensed at a pressure of 15 kPa. Determine the (i) cycle efficiency, (ii) thermal efficiency if steam is superheated to 500°C instead of 300°C, and (iii) thermal efficiency if boiler pressure is raised to 5 MPa while turbine inlet temperature is maintained at 500°C.

Solution

The Figure Ex. 10.1 shows the T-s diagrams of the three different inlet conditions at turbine, 2.5 MPa, 300°C, 2.5 MPa, 500°C, and 5 MPa, 500°C.

From superheated steam tables,

$$\text{At} \left.\begin{aligned} p_1 &= 2.5\,\text{MPa} \\ t_1 &= 300°\text{C} \end{aligned}\right\} h_1 = 3009.6\,\text{kJ/kg} \qquad s_1 = 6.645\,\text{kJ/kg K}$$

From saturated steam tables,

$$\text{At}\, p_2 = 15\,\text{kPa} \Rightarrow s_{f2} = 0.7549\,\text{kJ/kg K} \qquad s_{fg2} = 7.252\,\text{kJ/kg K}$$

$$v_f = 0.001014\,\text{m}^3/\text{kg} (= v_3)$$

$$h_{f2} = 225.94\,\text{kJ/kg} \quad h_{fg2} = 2373.2\,\text{kJ/kg}$$

$$s_1 = s_2 = s_{f2} + x_2 \times s_{fg2}$$

$$s_2 = 6.645 = 0.7549 + x_2 \times 7.252 \rightarrow x_2 = 81.22\%$$

$$h_2 = h_{f2} + x_2 \times h_{fg2} = 225.94 + 0.8122 \times 2373.2 = 2153.46\ \text{kJ/kg}$$

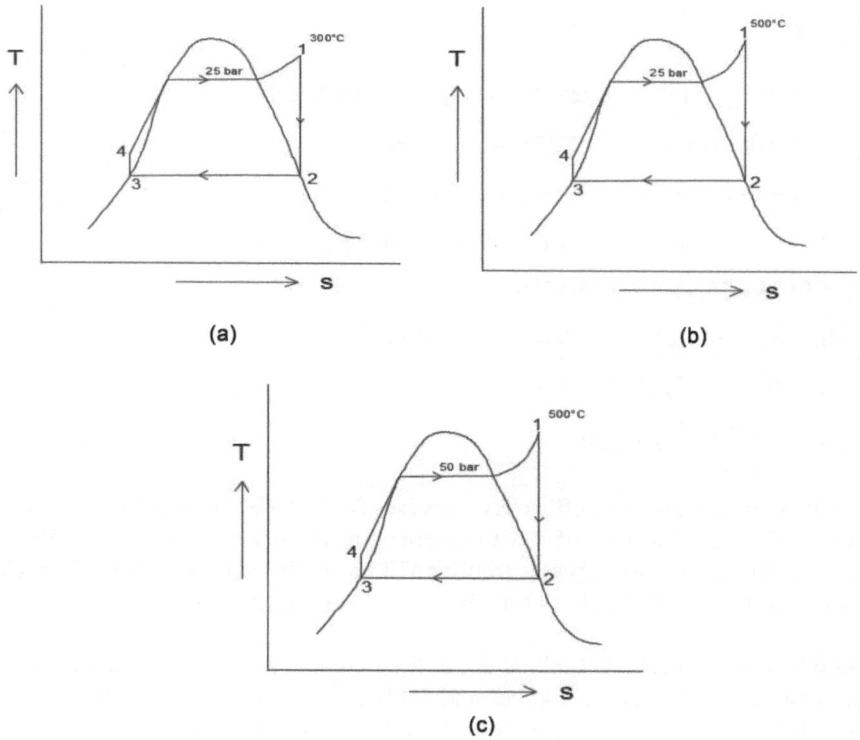

FIGURE EX. 10.1

$$h_{f3}\,@15\,\text{kPa} = 225.94\,\text{kJ/kg}$$

$$w_p = h_4 - h_3 \Rightarrow v_3(p_1 - p_2) = 0.001014\left(2.5\times10^3 - 15\right) = 2.519\,\text{kJ/kg}$$

$$\therefore h_4 = w_p + h_3 = 2.519 + 225.94 = 228.459\,\text{kJ/kg}$$

$$q_1 = h_1 - h_4 = 3009.6 - 228.459 = 2781.14\,\text{kJ/kg}$$

$$q_2 = h_2 - h_3 = 2153.46 - 225.94 = 1927.52\,\text{kJ/kg}$$

$$\eta_{cycle} = 1 - \frac{q_2}{q_1} = 1 - \frac{1927.52}{2781.14} = 30.69\%$$

$$\text{At}\left.\begin{matrix} p_1 = 2.5\,\text{MPa} \\ t_1 = 500°\text{C}\end{matrix}\right\} h_1 = 3462.8\,\text{kJ/kg} \quad s_1 = 7.33\,\text{kJ/kgK}$$

$$s_1 = s_2 = 7.33 = 0.7459 + \chi_2 \times 7.252 = 90.66\%$$

$$h_2 = 225.94 + 0.9066 \times 2373.2 = 2377.62\,\text{kJ/kg}$$

$$q_1 = h_1 - h_4 = 3462.8 - 228.459 = 3234.34\,\text{kJ/kg}$$

$$q_2 = h_2 - h_3 = 2377.62 - 225.94 = 2151.68\,\text{kJ/kg}$$

$$\eta_{cycle} = 1 - \frac{2151.68}{3234.34} = 33.47\%$$

At $p_1 = 5$ MPa, 500°C, $h_1 = 3434.7$ kJ/kg and $s_1 = 6.978$ kJ/kg K

$$s_1 = s_2 = 6.978 = 0.7549 + x_2 \times 7.252 \rightarrow x_2 = 85.81\%$$

$$h_2 = 225.94 + 0.8581 \times 2373.2 = 2262.43\,\text{kJ/kg}$$

$$w_p = h_4 - h_3 = 0.001014\left(5 \times 10^3 - 15\right) = 5.055\,\text{kJ/kg}$$

$$h_4 = 5.055 + 225.94 = 230.99\,\text{kJ/kg}$$

$$q_1 = h_1 - h_4 = 3434.7 - 230.99 = 3203.71\,\text{kJ/kg}$$

$$q_2 = h_2 - h_3 = 2262.43 - 225.94 = 2036.49\,\text{kJ/kg}$$

$$\eta_{cycle} = 1 - \frac{2036.49}{3203.71} = 36.43\%$$

It can be seen that the cycle efficiency increases from 30.69% to 33.47% and quality increases from 81.22% to 90.66% with superheating the steam from 300°C to 500°C. The cycle efficiency also increases from 33.47% to 36.43% with increasing the boiler pressure from 25 to 50 bar; however, quality decreases from 90.66% to 85.81%.

Example 10.2 In a reheat Rankine cycle, the initial steam pressure and the maximum temperature (T_{max}) are 150 bar and 550°C, respectively. The condenser pressure is 0.1 bar and the quality of steam at turbine exhaust is 95%. Assuming ideal processes, estimate (i) reheat pressure, (ii) η_{cycle}, and (iii) steam rate.

Solution

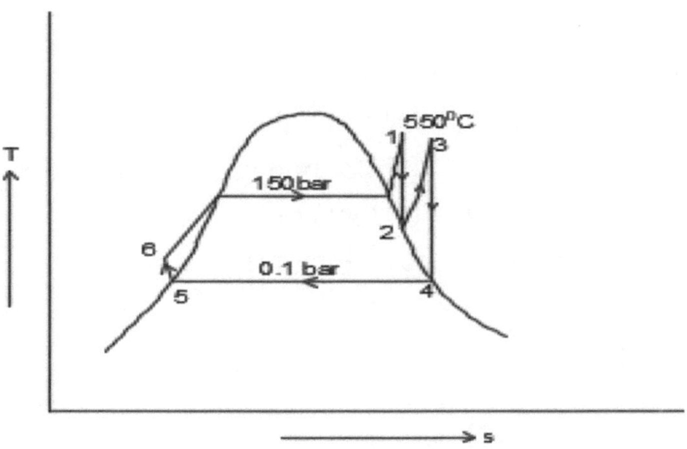

FIGURE EX. 10.2

The Figure Ex. 10.2 shows the T-s diagram of the reheat Rankine cycle turbine. From superheated steam tables,

At $p_1 = 150$ bar and $T_1 = 550°C$, $h_1 = 3448$ kJ/kg

From Mollier chart, $h_2 = 2740$ kJ/kg

At state 4: $P_4 = 0.1$ bar and quality is 0.95

(5% moisture) $\therefore x_4 = 0.95$

$s_{f4@0.1bar} = 0.6493$, $s_{fg4} = 7.602$ kJ/kg K

$h_{f4@0.1bar} = 191.83$ kJ/kg, $h_{fg4} = 2392.8$ kJ/kg

$s_4 = s_{f4} + x_4 \times s_{fg4}$

$s_4 = 0.6493 + 0.95 \times 7.602 = 7.87$ kJ/kg K

but $s_4 = s_3 \Rightarrow$ At state 3, $T_3 = 550°C$ and $s_3 = 7.87$ kJ/kg K

By interpolation, reheat pressure, $p_3 = 13$ bar (appox.) Ans.

$h_3 = 3500$ kJ/kg

And at 550°C $\Rightarrow s_g = 5.3$, since $s_g < s_4$, steam is in superheated state at 3

From superheated steam tables, at 13 bar and 550°C $\Rightarrow h_3 = 3500$ kJ/kg

Then $h_4 = h_{f4} + x_4 \times h_{fg4} = 191.83 + 0.95 \times 2392.8 = 2465$ kJ/kg

$h_5 = h_{f@0.1bar} \Rightarrow 191.83$ kJ/kg and $\nu_5 = \nu_{f@0.1bar} = 0.001$ m^3/kg

$w_P = \nu_5 \times (p_1 - p_4) = 0.001 \times 100(150 - 0.1) = 14.9$ kJ/kg

But $w_P = h_6 - h_5 \Rightarrow h_6 = w_P + h_5 = 14.9 + 191.83 = 205.4$

$\therefore h_6 = 205.4$ kJ/kg

$w_T = (h_1 - h_2) + (h_3 - h_4) = (3448 - 2740) + (3500 - 2465) = 1740$ kJ/kg

$w_{net,out} = w_T - w_P = 1740 - 14.9 = 1725.1$ kJ/kg

$q_1 = (h_1 - h_6) + (h_3 - h_2) = (3448 - 205.4) + (3500 - 2740) = 4002.6$ kJ/kg

$$\eta_{cycle} = \frac{w_T - w_P}{q_1} = \frac{1740 - 14.9}{3998} = 43.14\%$$ Ans.

$$\text{Steam rate} = \frac{3600}{w_{net,out}} = \frac{3600}{1725.1} = 2.086\,\text{kg/kWh}$$ Ans.

Example 10.3 In a regenerative Rankine cycle, steam enters the turbine at 30 bar and 300°C and leaves at a condenser pressure of 0.08 bar. The extraction points for two feedwater heaters are at 3.5 and 0.7 bar, respectively. Compute thermal efficiency of plant neglecting pump work.

Solution

The Figure Ex. 10.3 shows the T-s diagram of the regenerative Rankine cycle turbine. At 30 bar and 300°C $\Rightarrow h_1 = 2993$ kJ/kg

$s_1 = s_2 = s_3 = s_4 = 6.569$ kJ/kg K

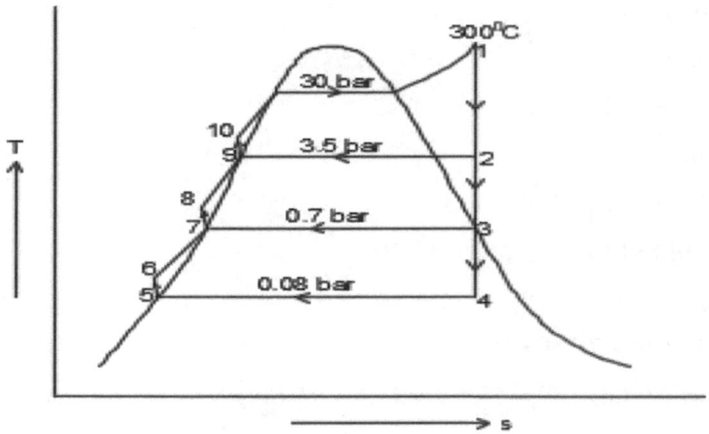

FIGURE EX. 10.3

At 3.5 bar, $s_{g(2.5bar)} = 6.9$ kJ/kg K, since $s_2 < s_g$, steam is a mixture at state 2

$s_f = 1.7275$ and $s_{fg} = 5.213$ kJ/kg K

$h_{f2} = 584$, $h_{fg2} = 2148$ kJ/kg

$s_2 = s_{f2} + x\ s_{fg2} \Rightarrow 6.569 = 1.7275 + x_2 \times 5.213$

$x_2 = 0.92$

$h_2 = h_{f2} + x_2\ h_{fg2} \Rightarrow 584 + 0.92 \times 2148 = 2560$ kJ/kg

At 0.7 bar, $s_f = 1.2$ and $s_{fg} = 6.2$ kJ/kg K

$h_{f3} = 384$, $h_{fg3} = 2278$ kJ/kg

$s_3 = s_{f3} + x\ s_{fg3}$

$s_3 = 6.569 = 1.2 + x_3 \times 6.2 \Rightarrow x_3 = 0.83$

$h_3 = 384 + 0.83 \times 2278 = 2270$ kJ/kg

At 0.08 bar, $s_f = 0.576$ and $s_{fg} = 7.67$ kJ/kg K

$h_{f4} = 168$, $h_{fg4} = 2406$ kJ/kg

$s_4 = 0.576 + x_4 \times 7.67 = 6.539 \Rightarrow x_4 = 0.77$

$h_4 = 168 + 0.77 \times 2406 = 2038$ kJ/kg

$h_5 = h_f$ at 0.08 bar $= 168$ kJ/kg

$h_5 = h_6$ (W_p is negligible)

$h_7 = h_8 = h_f$ at 0.7 bar $= 375$ kJ/kg

$h_9 = h_{10} = h_f$ at 3.5 bar $= 584.34$ kJ/kg

$$\text{Heater}\,1 \Rightarrow m_1h_2 + (1 - m_1)(h_8) = h_9 \Rightarrow m_1 = \frac{h_9 - h_8}{h_2 - h_8}$$

$$m_1 = \frac{584.33 - 375}{2560 - 375} = 0.095\,\text{kg}$$

$$\text{Heater}\,2 \Rightarrow m_2h_3 + (1 - m_1 - m_2)h_6 = (1 - m_1)h_7 \Rightarrow m_2 = \frac{(1 - m_1)(h_7 - h_6)}{(h_3 - h_6)}$$

$$m_2 = \frac{(1 - 0.095)(375 - 168)}{2270 - 168} = 0.089\,\text{kg}$$

Work of turbine $w_T = 1(h_1 - h_2) + (1 - m_1)(h_2 - h_3) + (1 - m_1 - m_2)(h_3 - h_4)$

(pump work is negligible)

$w_T = (2993 - 2560) + (1 - 0.095)(2560 - 2270) + (1 - 0.095 - 0.089)(2270 - 2038) = 884.76$ kJ/kg

Heat input, $q_1 = (h_1 - h_9) = 2993 - 584.34 = 2408.66$ kJ/kg

$$\eta = \frac{w_T - w_P}{q_1} = \frac{884.76}{2408.66} = 36.73\% \qquad \text{Ans.}$$

Example 10.4 A regenerative cycle operates with a feedwater heater operating at 7 bar. Steam enters the turbine at 50 bar and 350°C and leaves at 0.5 bar. Find (i) the efficiency of cycle, (ii) the steam rate, (iii) the mean temperature of the heat addition (T_m), and (iv) the increase in T_m and steam rate with regeneration (neglect pump work).

Solution

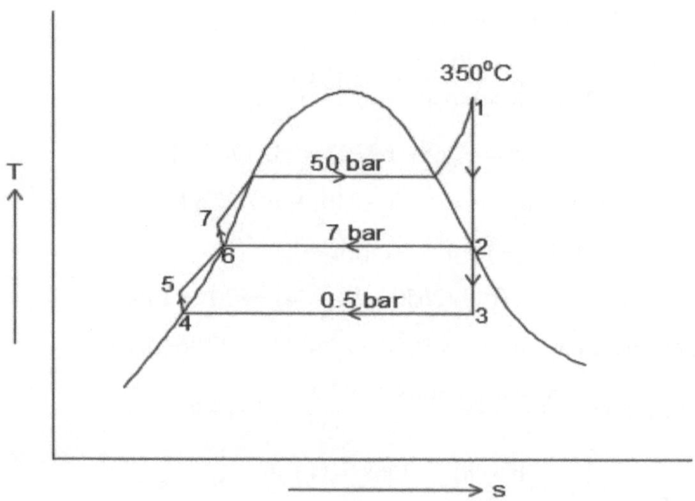

FIGURE EX. 10.4

The Figure Ex. 10.4 shows the T-s diagram of the regenerative cycle.

From steam tables, at 50 bar and 350°C,

$h_1 = 3068$ kJ/kg, $s_1 = s_2 = s_3 = 6.45$ kJ/kg K

At 7 bar, $s_g = 6.708$ kJ/kg K

since, $s_g > s_2$ steam at state 2 is a mixture

At 7 bar $h_f = 697.22$, $h_{fg} = 2066.3$ kJ/kg

$s_f = 1.992$; $s_{fg} = 4.716$ kJ/kg K (since $s_{fg} = s_g - s_f$)

$h_2 = h_{f2} + x_2 \times h_{fg2}$

$s_2 = s_{f2} + x_2 \times s_{fg2} \Rightarrow 6.45 = 1.992 + x_2 \times 4.716$

$x_2 = 0.945$

$h_2 = 697.22 + 0.945 \times 2066.3 = 2650.48$ kJ/kg

At 0.5 bar, $s_f = 1.091$, $s_g = 7.593 \Rightarrow s_{fg} = 6.502$

$s_3 = (6.45) = s_{f3} + x_3 \times s_{fg3} = 1.091 = x_3 \times 6.502 \Rightarrow x_3 = 0.824$

$h_{f3} = 340.49$, $h_{fg3} = 2305.4$ kJ/kg

$h_3 = h_{f3} + x_3 \times h_{fg3}$,

$h_3 = 340.49 + 0.824 \times 2305.4 = 2240.14$ kJ/kg

$h_4 = h_{f@0.5bar} = 340.49$ kJ/kg $= h_5$ $\quad s_4 = s_f$ at $0.5 = 1.091$ kJ/kg K

$h_6 = h_7$ at 7 bar $= h_{f6} = 697.22$ kJ/kg

Energy balance for heater $m_1 h_2 + (1 - m_1) h_5 = h_6$

$$\Rightarrow m_1 = \frac{h_6 - h_5}{h_2 - h_5} = \frac{384.39 - 340.49}{2650.48 - 340.49} = 0.019\,\text{kg}$$

q_1 with regeneration $= h_1 - h_6 = 3068 - 697.22 = 2370.78$ kJ/kg

q_1 without regeneration $= h_1 - h_4 = 3068 - 340.49 = 2727.5$ kJ/kg

w_T with regeneration $= 1(h_1 - h_2) + (1 - m_1)(h_2 - h_3)$

$= 1(3068 - 2650.48) + (1 - 0.019)(2650 - 2240.14) = 820.06$ kJ/kg

w_T without regeneration $= h_1 - h_3 = 3068 - 2240.14 = 827.86$ kJ/kg

$$T_m \text{ with regeneration} = \frac{h_1 - h_7}{s_1 - s_7} = \frac{3068 - 697.22}{6.45 - 1.992} = 531.8\,\text{K} \qquad \text{Ans.}$$

$$T_m \text{ without regeneration} = \frac{h_1 - h_4}{s_1 - s_4} = \frac{3068 - 340.49}{6.45 - 1.091} = 508.95\,\text{K} \qquad \text{Ans.}$$

$$\text{Efficiency of cycle with regeneration} = \frac{w_{net,out}}{q_1} = \frac{820.06}{2370.78} = 34.6\% \qquad \text{Ans.}$$

Efficiency of cycle without regeneration $= \dfrac{827.86}{2727.51} = 30.34\%$ Ans.

Increase in T_m with regeneration = 531.8 – 508.95 = 22.85°C Ans.

Steam rate with regeneration $= \dfrac{3600}{820.06} = 4.389$ kg/kW h Ans.

Steam rate without regeneration $= \dfrac{3600}{827.6} = 4.349$ kg/kW h Ans.

Increase in steam rate with regeneration = 4.389 – 4.349 = 0.039 kg/kW h Ans.

Example 10.5 A mercury cycle is superposed on the steam cycle operating between boiler outlet conditions of 40 bar and 400°C and condenser temperature of 40°C. The heat released by mercury (Hg) condensing at 0.2 bar is used to impart the latent heat of vaporization to water in the steam cycle. Mercury enters mercury turbine as saturated vapor at 10 bar. Compute (i) kg of Hg per kg of water and (ii) η of combined cycle.

P (bar)	t°C	h_f (kJ/kg)	h_g (kJ/kg)	s_f (kJ/kg K)	s_g (kJ/kg K)	ν_f (m³/kg)	ν_g (m³/kg)
10	515	72.23	363	0.1498	0.5167	80.9×10^{-6}	0.0333
0.2	277	38.35	336.55	0.0967	0.6385	77.5×10^{-6}	1.163

Solution

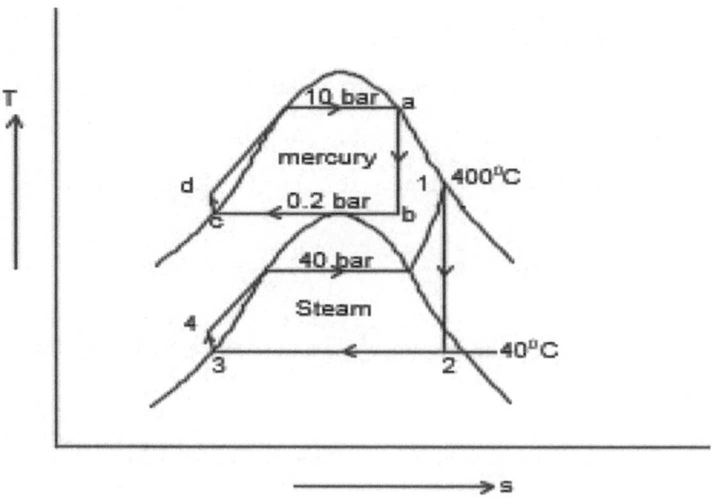

FIGURE EX. 10.5

The Figure Ex. 10.5 shows the T-s diagram of the binary cycle.

For mercury cycle, at 10 bar, h_a (= h_g) = 363 kJ/kg

$$s_a\ (= s_b) = 0.5167 \text{ kJ/kg K}$$

at 0.2 bar, $s_b = s_f + x_m \times s_{fg} \Rightarrow 0.5167 = 0.0967 + x_m(0.6385 - 0.0967)$

$$x_m = 0.775 \text{ or } 77.5\%$$

$$h_b = h_f + x \times h_{fg} = 38.35 + 0.775 \times 336.55 = 269 \text{ kJ/kg}$$

$$h_c = h_d = h_{f@0.2bar} = 38.35 \text{ kJ/kg}$$

$$q_1 = h_a - h_d = 363 - 38.36 = 324.64 \text{ kJ/kg}$$

$(w_{net,out})_m = h_a - h_b = 363 - 299.17 = 63.83$ kJ/kg (since pump work is negligible)

$$\text{Efficiency of mercury cycle, } \eta_m = \frac{(w_{net,out})_m}{q_1} = \frac{63.83}{324.64} = 19.66\%$$

For steam cycle, at 40 bar and 400°C, $h_1 = 3273.4$ kJ/kg, $s_1 = 6.77$ kJ/kg K

At 40°C $h_f = 167.57$ kJ/kg $h_{fg} = 2406.7$ kJ/kg

$s_f = 0.5725$ kJ/kg K $s_{fg} = 8.257$ kJ/kg

$$(s_1) = s_2 = s_f + x_w \times s_{fg} \Rightarrow 6.77 = 0.5725 + x_w \times 8.257$$

$$x_w = 0.7505$$

Then $h_2 = h_f + x_w \times h_{fg}$

$$= 167.57 + 0.75 \times 2406.7 = 1972.59 \text{ kJ/kg}$$

$h_3 = h_4 = 167.57$ kJ/kg (since pump work is negligible)

$$q_2 = h_1 - h_4 = 3273.4 - 167.57 = 3105.83 \text{ kJ/kg}$$

$$(w_{net,out})_{st} = h_1 - h_2 = 3273.4 - 1972.59 = 1302 \text{ kJ/kg}$$

Mass of mercury per kg of water, $m(h_b - h_c) = (h_1 - h_4)$

$\Rightarrow m(299.17 - 38.35) = (3273.4 - 167.57) = 11.9$ kg/kg of water **Ans.**

$$\text{Efficiency of steam cycle, } \eta_{st} = \frac{(w_{net,out})_{st}}{q_2} = \frac{1302}{3105.83} = 41.92\%$$

where q_2 is heat rejected by the topping cycle and received by bottoming cycle

Overall efficiency of cycle,

$$\eta_o = \eta_m + \eta_{st} - \eta_m\eta_{st} = 19.66 + 41.92 - 19.66 \times 41.992 = 53.34\%$$ **Ans.**

Example 10.6 A steam power cycle operates with solar energy as heat input. Water enters the pump as a saturated liquid at 30°C and is pumped to boiler at 2.5 bar, it then enters the turbine as a saturated vapor. The conditions at turbine exhaust are 30°C and 85% dry. Estimate (i) $w_{net,out}$, (ii) η_{cycle}, (iii) steam rate, and (iv) isentropic efficiency of turbine.

Solution

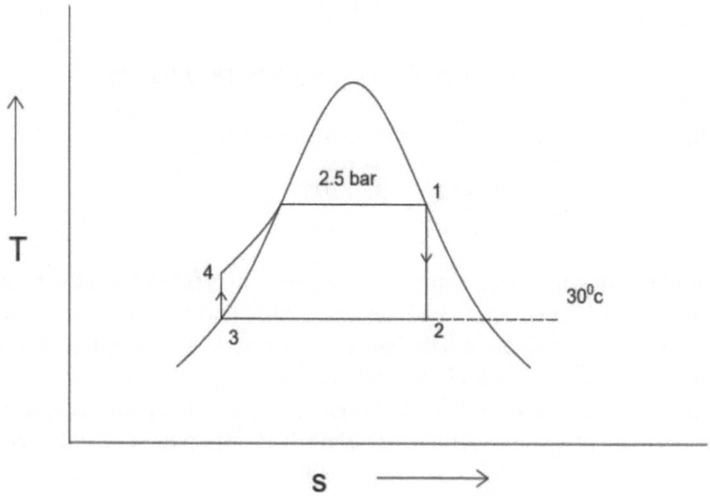

FIGURE EX. 10.6

The Figure Ex. 10.6 shows the T-s diagram of the Rankine cycle.

From steam tables, at 2.5 bar, $h_1 = h_2 = 2716.9\,kJ/kg$

$$s_1 = s_2 = s_g = 7.052\,kJ/kgK$$

At 30°C, $h_f = 125.77$ kJ/kg and $h_{fg} = 2430.5$ kJ/kg

$$\therefore h_2 = h_f + \chi h_{fg} = 125.77 + 0.85 \times 2430.5 = 2191.69\,kJ/kg$$

$$h_3 = h_f \text{ at } 30°C = 125.77\,kJ/kg$$

p_2 is the saturation pressure at 30°C = 0.0562 bar

$$w_p = h_4 - h_3 = v_{f@p2}(p_1 - p_2)$$

$$= 1.006 \times 10^{-3}(2.5 - 0.0562) = 0.0245\,kJ/kg$$

$$v_f \text{ at } 30°C = 1.006 \times 10^{-3}\,m^3/kg$$

$$s_1 = s_2^1 = s_f + xs_{fg} \text{ (at } 30°C)$$

$$\text{At } 30°C \Rightarrow s_f = 0.4369, \quad s_{fg} = 8.0164\,kJ/kgK$$

$$s_2' = 7.052 = 0.4369 + \chi_2' \times 8.0164$$

$$\Rightarrow \chi_2' = 0.825$$

$$\therefore h_2' = h_f + \chi_2' \times h_{fg} = 125.77 + 0.825 \times 2430.5 = 2130.93\,kJ/kg$$

$$h_4 = w_p + h_3 \Rightarrow 0.0245 + 125.77 = 125.795\,kJ/kg$$

$$q_1 = h_1 - h_4 = 2716.9 - 125.795 = 2591.105\,kJ/kg$$

$$w_T = h_1 - h_2 = 2716.9 - 2191.69 = 525.205\,\text{kJ/kg}$$

$$\eta_{cycle} = \frac{w_{net,out}}{q_1} = \frac{525.18}{2591.105} = 20.26\% \qquad \text{Ans.}$$

$$w_{net,out} = w_T - w_P = 525.205 - 0.0245 = 525.18\,\text{kJ/kg}$$

$$\text{Steam rate} = \frac{3600}{w_{net,out}} = \frac{3600}{525.18} = 6.854\,\text{kg/kW h} \qquad \text{Ans.}$$

$$\eta_{isentropic} \text{ of turbine} = \frac{h_1 - h_2}{h_1 - h_2'} = \frac{525.205}{585.97} = 89.6\% \qquad \text{Ans.}$$

Example 10.7 In a cogeneration plant, the power load is 5.6 MW and the heating load is 1.163 MW. Steam is generated at 40 bar and 500°C and is expanded isentropically through a turbine to a pressure of 0.06 bar. The heating load is supplied by extracting steam from turbine at 2 bar, which is condensed in the process heater to a saturated liquid at 2 bar and then pumped back to boiler. Compute (i) steam generation capacity of boiler, (ii) rate of heat input to boiler, and (iii) rate of heat rejected to condenser.

Solution

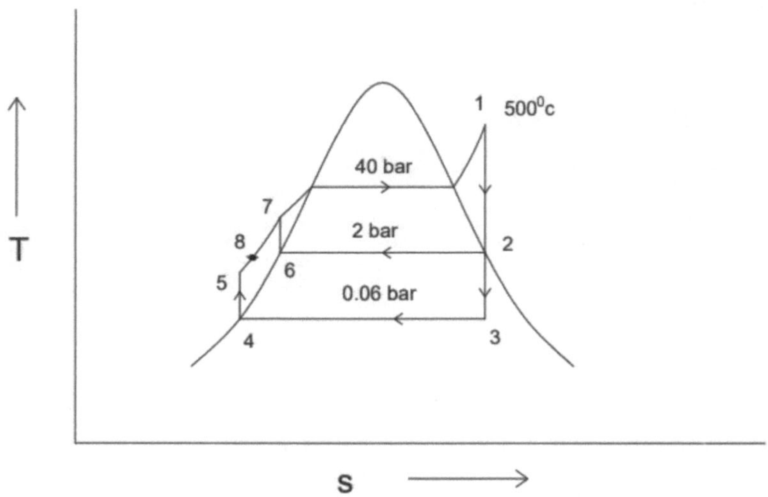

FIGURE EX. 10.7

The Figure Ex. 10.7 shows the T-s diagram of the cogeneration plant.

Power of the cogeneration plant, $\dot{W}_{net,out} = 5600\,\text{kW}$

At 40 bar and 500°C $\Rightarrow h_1 = 3450$ kJ/kg

From Mollier diagram, by drawing a vertical line from point 1, that is, at 40 bar and 500°C to 2 bar, the enthalpy at state 2 can be found as

$$h_2 = 2700\ \text{kJ/kg}$$

By extending the same vertical line from 2 to 0.06 bar,

$$h_3 = 2230 \text{ kJ/kg}$$

At 0.06 bar, from saturated steam tables

$$h_4 = h_f = 145 \text{ kJ/kg}, \qquad v_4 = 0.0010055 \text{ m}^3\text{/kg}$$

At 2 bar, $h_6 = h_f = 504$ kJ/kg, $\quad v_6 = v_f = 0.001061 \text{ m}^3\text{/kg}$

Pump work:

$$w_{p_1} = v\Delta p = v_4(p_5 - p_4) = 0.0010055(40 - 0.06) = 4.016 \text{ kJ/kg}$$

$$w_{p_2} = v\Delta p = v_6(p_7 - p_6) = 0.001061(40 - 2) = 4.032 \text{ kJ/kg}$$

Rate of heat input to boiler, $\dot{Q}_1 = \dot{m}(h_1 - h_8)$

$$h_5 = h_4 + w_{p_1} = 145 + 4.016 = 149.016 \text{ kJ/kg}$$

$$h_7 = h_6 + w_{p_2} = 504 + 4.032 = 508.032 \text{ kJ/kg}$$

$$\text{Heater input, } \dot{Q}_H = \dot{m}_1(h_2 - h_6) \Rightarrow \dot{m}_1 = \frac{1163}{2700 - 504} = 0.53 \text{ kg/s},$$

where $\dot{m}_1$ is the rate of steam flow required at the desired temperature in the process heater

i. Steam generation capacity of boiler, $\dot{m}$ can be found from

$$\dot{W}_T = \dot{m}(h_1 - h_2) + (\dot{m} - \dot{m}_1)(h_2 - h_3)$$

$$= \dot{m}(3450 - 2700) + (\dot{m} - 0.53)(2700 - 2230)$$

$$= 5600 = (1220\dot{m} - 249.1) - (4.016 + 4.032) \Rightarrow \dot{m} = 4.8 \text{ kg/s}$$

ii. Rate of heat input to boiler, $\dot{Q}_1 = 4.8(3450 - 149.016) = 15.844 \text{ MW}$ Ans.

iii. Rate of heat rejected in the condenser, $\dot{Q}_2 = (\dot{m} - \dot{m}_1)(h_3 - h_4)$

$$= (4.8 - 0.53)(2230 - 145)$$

$$= 8.902 \text{ MW}$$ Ans.

Example 10.8 In a steam power plant operating on ideal Rankine cycle, steam flows through a turbine entering at 2.5 MPa, 300°C and condenses at 15 kPa in the condenser. Heat input to the steam is supplied from a source maintained at 500°C and waste heat is rejected to the surroundings maintained at 25°C. Calculate (i) exergy destruction and (ii) the second-law efficiency.

Solution

From superheated steam tables,

$$\left.\begin{aligned} p_1 &= 2.5 \text{ MPa} \\ t_1 &= 300°\text{C} \end{aligned}\right\} h_1 = 3009.6 \text{ kJ/kg} \qquad s_1 = 6.645 \text{ kJ/kgK}$$

From saturated steam tables,

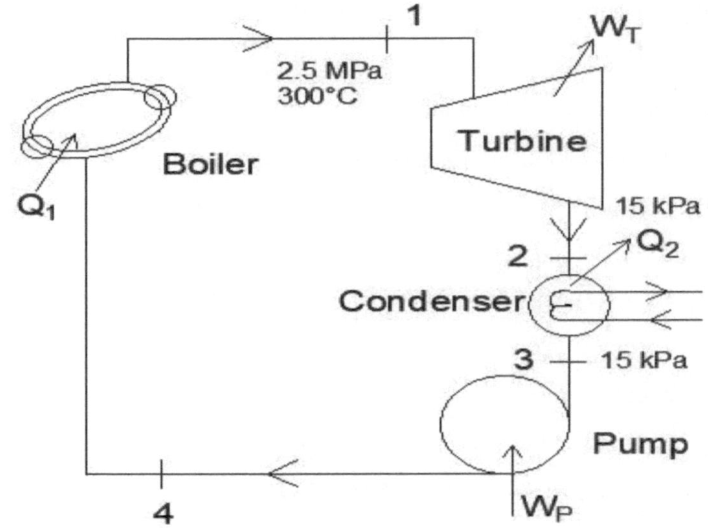

FIGURE EX. 10.8

At $p_2 = 15\,\text{kPa} \Rightarrow s_{f2} = 0.7549\,\text{kJ/kgK} \quad s_{fg2} = 7.252\,\text{kJ/kgK}$

$$h_{f2} = 225.94\,\text{kJ/kg} \quad h_{fg2} = 2373.2\,\text{kJ/kg}$$

$$s_1 = s_2 = s_{f2} + x_2 \times s_{fg2}$$

$s_2 = 6.645\ \text{kJ/kg K} = 0.7549 + x_2 \times 7.252 \rightarrow x_2 = 81.22\%$

$h_2 = h_{f2} + x_2 \times h_{fg2} = 225.94 + 0.8122 \times 2373.2 = 2153.46\ \text{kJ/kg}$

$h_{f3}\,@15\,\text{kPa} = 225.94\,\text{kJ/kg}$

$s_3 = s_4 = 0.7549\ \text{kJ/kg K}$

Pump work, $w_p = h_4 - h_3 \Rightarrow v_3(p_1 - p_2) = 0.001014(2.5 \times 10^3 - 15) = 2.519\,\text{kJ/kg}$

$\therefore h_4 = w_p + h_3 = 2.519 + 225.94 = 228.459\,\text{kJ/kg}$

Turbine work, $w_T = h_1 - h_2 = 3009.6 - 2153.46 = 856.14\ \text{kJ/kg}$

$q_1 = h_1 - h_4 = 3009.6 - 228.459 = 2781.14\,\text{kJ/kg}$

$q_2 = h_2 - h_3 = 2153.46 - 225.94 = 1927.52\,\text{kJ/kg}$

i. Processes 1-2 and 3-4 are isentropic expansion and isentropic compression, respectively, and hence involve no irreversibilities; therefore exergy destruction is zero in either case.

   Processes 4-1 and 2-3 are constant pressure heat addition in boiler and constant pressure heat rejection in condenser, respectively. These processes are irreversible since they involve heat transfer through a finite temperature difference.

   Exergy destruction for process 2-3 is found as

$$x_{destroyed,23} = T_0\left(s_3 - s_2 + \frac{q_2}{T_2}\right) = 298\left(0.7549 - 6.645 + \frac{1927.52}{298}\right) = 172.24\,\text{kJ/kg}$$

Exergy destruction for process 4-1 is found as

$$x_{destroyed,41} = T_0\left(s_1 - s_4 - \frac{q_1}{T_1}\right) = 298\left(6.645 - 0.7549 - \frac{2781.14}{773}\right) = 683.34\,\text{kJ/kg}$$

Therefore the total exergy destruction (irreversibility) of the cycle

$$x_{destroyed} = x_{destroyed,12} + x_{destroyed,23} + x_{destroyed,34} + x_{destroyed,41}$$

$$= 0 + 172.24 + 0 + 683.34 = 855.58\text{ kJ/kg}$$ Ans.

From the above result, it is found that exergy destruction is highest in boiler.

ii. The second-law efficiency is defined as, $\eta_{II} = 1 - \frac{\text{Exergy destroyed}}{\text{Exergy supplied}}$

$$x_{heat,in} = \left(1 - \frac{T_0}{T_1}\right) = \left(1 - \frac{2781.14}{773}\right) = 1708.98\,\text{kJ/kg}$$

$$x_{supplied} = x_{heat,in} + x_{pump,in} = 1708.98 + 2.519 = 1711.49\,\text{kJ/kg}$$

$$\therefore \eta_{II} = 1 - \frac{855.58}{1711.49} = 50\%$$

REVIEW QUESTIONS

10.1 Why is excessive moisture in steam undesirable in steam turbines? What is the highest moisture content allowed?

10.2 Compare the pressures at the inlet and the exit of the boiler for (i) actual and (ii) ideal cycles.

10.3 What is the difference between cogeneration and regeneration?

10.4 Why is the Carnot cycle not a realistic model for steam power plants?

10.5 In combined gas–steam cycles, what is the energy source for the steam?

10.6 Why is the combined gas–steam cycle more efficient than either of the cycles operated alone?

10.7 Why is steam not an ideal working fluid for vapor power cycles?

10.8 Why is mercury a suitable working fluid for the topping portion of a binary vapor cycle but not for the bottoming cycle?

10.9 What is the difference between the binary vapor power cycle and the combined gas–steam power cycle?

10.10 Explain various processes in the Rankine cycle.

10.11 What is organic Rankine cycle and what are the working fluids used in it?

10.12 What are the uses of organic Rankine cycle?

10.13 Why is reheating used in the Rankine cycle?

10.14 Define steam rate and heat rate

10.15 Why is regeneration used in the Rankine cycle?

10.16 With the help of p-v and T-s diagram, explain the process of reheating.

10.17 With the help of p-v and T-s diagram, explain the process of reheating.

10.18 What are the operating parameters that affect the combined cycle performance?
10.19 Derive the expression for efficiency of the Rankine cycle.
10.20 What is the mean temperature of the heat addition?
10.21 How do you compare the Rankine and Carnot cycles?
10.22 What are the ways to increase the thermal efficiency of the Rankine cycle?
10.23 What is the difference between open feedwater heaters and closed feedwater heaters?
10.24 What are the desirable characteristics for ideal working fluids in vapor cycles?
10.25 What are integrated coal gasification combined cycle (IGCC) power plants?
10.26 What are the desirable characteristics for an ideal working fluid for organic Rankine cycle?
10.27 How can you improve the performance of organic Rankine cycle?
10.28 What are the steps involved in the preparation of nuclear fuel?
10.29 What is the heat source for nuclear power generation?
10.30 What are the various nuclear fuels used in power generation?

EXERCISE PROBLEMS

10.1 A steam power plant whose net power output is 200 MW works on ideal Rankine cycle. The condition of steam at inlet to the turbine is at 15 MPa and 505°C and is cooled in the condenser at a pressure of 15 kPa. Show the cycle on a T-s diagram with respect to the saturation lines and calculate (i) quality of steam at the turbine exit, (ii) thermal efficiency of the cycle, and (iii) mass flow rate of the steam.

10.2 The net power output of a steam power plant operating on ideal Rankine cycle is 48 MW. Steam enters the turbine at 8 MPa and 510°C and is cooled in the condenser to a pressure of 15 kPa by running cooling water from a lake through the condenser at a rate of 2000 kg/s. Show the cycle on a T-s diagram with respect to the saturation lines and determine (i) thermal efficiency of the cycle, (ii) mass flow rate of the steam, and (iii) temperature rise of the cooling water.

10.3 A geothermal hot water is to be used as the energy source in an ideal Rankine cycle, with R-134a as working fluid in the cycle. Saturated vapor R-134a leaves the boiler at a temperature of 80°C, and the condenser temperature is 35°C. Calculate the thermal efficiency of this cycle.

10.4 In an ideal Rankine cycle working with superheat and reheat, steam enters the first-stage turbine at 8.3 MPa, 485°C, and expands to 0.8 MPa. It is then reheated to 440°C before entering the second-stage turbine, where it expands to the condenser pressure of 0.008 MPa. The 100 MW is net output of the cycle. Find (i) the thermal efficiency of the cycle, (ii) the mass flow rate of steam, in kg/h, and (iii) the rate of heat transfer from the condensing steam as it passes through the condenser, in MW.

10.5 A steam power plant operates with a high pressure of 4 MPa and has a boiler exit of 500°C receiving heat from a 750°C source. The ambient at

25°C provides cooling to maintain the condenser at 50°C. All components are ideal except for the turbine, which has an isentropic efficiency of 90%. Find (i) ideal and actual turbine exit qualities and (ii) actual specific work and specific heat transfer in all four components.

10.6 Find the efficiency of the cycle and specific steam consumption of a reheat cycle working between pressures of 3.5 and 0.006 MPa, with superheat temperature of 455°C. Assume that the steam first expands till dry saturation point and then the steam is reheated to the original superheat temperature. Neglect feed pump work.

10.7 A power plant produces 30 kg/s of steam at 4 MPa, 610°C, in the boiler. The condenser gets cooled with ocean water, and the condenser exits temperature is 48°C. Reheat is done at 505 kPa up to 405°C, and then expansion takes place in the low pressure turbine. Find heat transfer in the boiler.

10.8 Steam must be generated at 140°C for process heat in a food production facility. This is done in a combined heat and power system as extraction steam from the turbine. Assume the standard cycle has a turbine inlet of 2.5 MPa, 400°C and 50°C in the condenser. What pressure should be used for the extraction so that a maximum of process heat at 140°C is available and the least amount of turbine work is lost?

10.9 A steam power plant operates on an ideal regenerative Rankine cycle. Steam enters the turbine at 6 MPa and 450°C and is condensed in the condenser at 20 kPa. Steam is extracted from the turbine at 0.4 MPa to heat the feedwater in an open feedwater heater. Water leaves the feedwater heater as a saturated liquid. Show the cycle on a T-s diagram. Find the following: (i) net work output per kilogram of steam flowing through the boiler and (ii) thermal efficiency.

10.10 A regenerative Rankine cycle has steam entering turbine at 205 bar, 655°C and leaving at 0.08 bar. Considering feedwater heaters to be of open type, estimate the thermal efficiency for the following conditions: (i) without feedwater heater, (ii) one feedwater heater working at 9 bar, and (iii) two feedwater heaters working at 45 and 6 bar, respectively. Also give layout and T-s representation for each of the cases described above.

10.11 A regenerative vapor power cycle operates with one open feedwater heater. The condition of steam at inlet to the turbine is 8.5 MPa, 490°C and expands to 0.8 MPa, and some of the steam is extracted and diverted to the open feedwater heater working at 0.8 MPa. Steam gets expanded through the second-stage turbine to the condenser pressure of 0.008 MPa. Saturated liquid exits the open feedwater heater at 0.8 MPa. The isentropic efficiency of each turbine stage is 88%, and each pump operates isentropically. If the net power output of the cycle is 100 MW, determine (i) thermal efficiency and (ii) mass flow rate of steam entering the first turbine stage, in kg/h.

10.12 A power plant operates on a regenerative vapor power cycle with one closed feedwater heater. Steam enters the first turbine stage at 100 bar, 520°C and expands to 10 bar, where some of the steam is extracted and diverted to a closed feedwater heater. Condensate exiting the feedwater heater as saturated liquid at 10 bar passes through a trap into the condenser. The feedwater exits the heater at 100 bar with a temperature of 130°C. The condenser pressure is 0.5 bar. For isentropic processes in each turbine stage and the pump, determine for the cycle: (i) the thermal efficiency and (ii) the mass flow rate into the first-stage turbine, if the net power developed is 250 MW.

10.13 A steam power plant operates on the simple ideal Rankine cycle. Steam enters the turbine at 4.5 MPa, 505°C and is condensed in the condenser at a temperature of 40°C. Assuming the mass flow rate as 10 kg/s, determine (i) the thermal efficiency of the cycle and (ii) net power output in MW. Show the cycle on a T-s diagram.

10.14 In an ideal Rankine cycle, water is the working fluid. Inlet to the turbine is saturated vapor that enters at 7 MPa. The condenser pressure is 7 kPa. Estimate (i) net work per unit mass of steam flow in kJ/kg, (ii) heat transfer to the steam passing through the boiler in kJ/kg, (iii) thermal efficiency, and (iv) back work ratio.

10.15 In an ideal Rankine cycle with working fluid as steam, saturated vapor enters the turbine inlet at 10 MPa and saturated liquid exits the condenser at 0.01 MPa. The net power output of the cycle is 100 Mw. Find (i) thermal efficiency, (ii) back work ratio, (iii) mass flow rate of steam, (iv) heat transfer into the working fluid as it passes through the boiler, and (v) the heat transfer from the condenser to the steam as it passes through the condenser.

10.16 A power plant operates on a regenerative vapor power cycle with two feedwater heaters. Steam enters the first turbine stage as 10 MPa, 510°C and expands in three stages to the condenser pressure of 5 kPa. Between the first and second stages, some steam is diverted to a closed feedwater heater at 1.4 MPa, with saturated liquid condensate being pumped ahead into the boiler feedwater line. The feedwater leaves the closed heater at 10 MPa, 170°C. Steam is extracted between the second and third turbine stages at 0.15 MPa and fed into an open feedwater heater operating at that pressure. Saturated liquid at 0.14 MPa leaves the open feedwater heater. For isentropic processes in the pumps and turbines, determine for the cycle: (i) the thermal efficiency and (ii) the mass flow rate into the first-stage turbine, in kg/h, if the net power developed is 300 MW.

10.17 Steam at 30 MPa, 510°C enters the first stage of a supercritical reheat cycle including three turbine stages. Steam exiting the first-stage turbine at pressure p is reheated at constant pressure to 420°C, and steam exiting the second-stage turbine at 0.5 MPa is reheated at constant pressure to 350°C. Each turbine stage and the pump have an isentropic efficiency of 80%. The condenser pressure is 7 kPa. (i) For $p = 5$ MPa, determine the net work per unit mass of steam flowing, in kJ/kg, and the thermal efficiency.

10.18 A steam power plant works on an ideal regenerative Rankine cycle. Steam at inlet to the turbine is at 8 MPa and 400°C and is condensed in the condenser at 25 kPa. Steam is extracted from the turbine at 0.4 MPa to heat the feedwater in an open feedwater heater. Water leaves the feedwater heater as a saturated liquid. Show the cycle on T-s diagram, and find (i) mass flow rate of steam through the boiler and (ii) thermal efficiency.

10.19 In a Rankine cycle, the working fluid used is water, boiler working pressure is 5 MPa and the condenser pressure is 50 kPa. At the inlet to the turbine, steam temperature is 500°C. The isentropic efficiency of the turbine is 95%, and water leaving the condenser is subcooled by 6.5°C. The boiler is sized for a mass flow rate 25 kg/s. Estimate (i) the rate at which heat is added in the boiler, (ii) the power required to operate the pumps, (iii) net power produced, and (iv) thermal efficiency.

10.20 An ideal Rankine cycle with reheat uses water as the working fluid. The conditions at the inlet to the first-stage turbine are 15 MPa, 580°C and the steam is reheated between the turbine stages to 580°C. For a condenser pressure of 5 kPa, plot the cycle thermal efficiency versus reheat pressure for pressures ranging from 2 to 10 MPa.

10.21 A coal-fired steam power plant produces 170 MW of electric power. It works on the principle of ideal Rankine cycle. Steam at inlet to the turbine is at 8 MPa and 600°C and condensed to a pressure of 16 kPa. The coal has a heating value of 30 MJ/kg. Assuming that 80% of this energy is transferred to the steam in the boiler and the electric generator efficiency is 90%. Find (i) plant efficiency and (ii) required rate of coal supply.

10.22 Water is the working fluid in a Rankine cycle modified to include one closed feedwater heater and one open feedwater heater. Superheated vapor enters the turbine at 15 MPa, 540°C, and the condenser pressure is 7 kPa. The mass flow rate of steam entering the first-stage turbine is 110 kg/s. The closed feedwater heater uses extracted steam at 4 MPa, and the open feedwater heater uses extracted steam at 0.3 MPa. Saturated liquid condensate drains from the closed feedwater heater at 4 MPa and is trapped into the open feedwater heater. The feedwater leaves the closed heater at 16 MPa and a temperature equals to the saturation temperature at 4 MPa. Saturated liquid leaves the open heater at 0.3 MPa. Assume all turbine stages and pumps operate isentropically. Determine (i) net power developed, in kW, (ii) the rate of heat transfer to the steam passing through the steam generator, in kW, and (iii) thermal efficiency.

10.23 A binary vapor power cycle, consisting of two ideal Rankine cycles, operates with steam and ammonia as the working fluids. In the steam cycle, superheated vapor enters the turbine at 5 MPa, 580°C, and saturated liquid exits the condenser at 45°C. The heat rejected from the steam cycle is provided to the ammonia cycle, producing saturated vapor at 40°C, which enters the ammonia turbine. Saturated liquid leaves the ammonia condenser at 1 MPa. The net power output of the binary cycle is 20 MW. Evaluate (i) the power output of the steam and ammonia turbines, respectively, in MW, (ii) the rate of heat addition to the binary cycle, in MW and, (iii) the thermal efficiency.

10.24 A steam power plant, whose net power output is 200 MW, works on the ideal Rankine cycle. The condition of steam at inlet to the turbine is at 15 MPa and 550°C and is condensed in the condenser at a pressure of 15 kPa. Assume a source temperature of 1500 K and a sink temperature of 300 K. Determine (i) exergy destruction associated with the cycle and (ii) the second-law efficiency.

10.25 An ideal Rankine cycle with reheat uses water as the working fluid. The conditions at the inlet to the first-stage turbine are 15 MPa and 580°C, and the steam is reheated between the turbine stages to 580°C. For a condenser pressure of 25 kPa, a source temperature of 1500 K, and a sink temperature of 300 K, determine (i) exergy destruction associated with the cycle and (ii) the second-law efficiency.

10.26 A steam power plant operates on an ideal regenerative Rankine cycle. Steam enters the turbine at 5 MPa and 450°C and is condensed in the condenser at 20 kPa. Steam is extracted from the turbine at 0.4 MPa to heat the feedwater in an open feedwater heater. Water leaves the feedwater heater as a saturated liquid. Assume a source temperature of 1450 K and a sink temperature of 295 K. Determine (i) exergy destruction associated with the cycle and (ii) the second-law efficiency.

DESIGN AND EXPERIMENT PROBLEMS

10.27 Compare the alternative energy sources such coal, natural gas and biomass for a 500 MW steam power plant in terms of fuel availability, cost of fuel, material handling, cost of energy, environmental considerations, and global warming potential (GWP). How do you compare these for the same capacity power plant, operating on combined cycle using both Rankine and Bryton cycles?

10.28 A power plant is built to provide district heating of buildings that requires 90°C liquid water at 150 kPa. The district heating water is returned at 50°C, 100 kPa in a closed loop in an amount such that 20 MW of power is delivered. This hot water is produced from a steam power cycle with a boiler making steam at 5 MPa, 600°C, delivered to the steam turbine. The steam cycle could have its condenser operate at 90°C, providing the power to the district heating. It could also be done with extraction of steam from the turbine. Suggest a system and evaluate its performance in terms of the cogenerated amount of turbine work.

10.29 A supercritical steam power plant cycle has to be designed for the maximum pressure of 25 MPa and the maximum temperature of 640°C. The temperature of the cooling water circulated in the condenser should be at pressure of 15 kPa and a turbine should have an isentropic efficiency of 90%. For this plant, (i) do you recommend any reheat for this cycle? If so, how many stages of reheat and at what pressures, (ii) do you recommend feedwater heaters? If so, how many and at what pressures would they operate? Should they be open or closed feedwater heaters? (iii) Estimate the thermal efficiency of the cycle that you recommend.

10.30 In steam power plants, large quantities of water are circulated through the condensers to condense the steam. Water in these plants exits at temperatures slightly above ambient temperature. Design a water cooled condenser for a 500-MW power plant to effectively reduce the quantity of water circulated. For this, estimate the annual economic benefit provided by the reduced water consumption in condenser.

11 Gas Power Cycles

LEARNING OUTCOMES

After learning this chapter, students should be able to

- Demonstrate the knowledge of idolized hot air cycles such as Carnot, Stirling, and Ericsson cycles and their limitations as practical cycles.
- Analyze air-standard internal combustion engines working on Otto and Diesel cycles.
- Evaluate net power output, thermal efficiency, and mean effective pressure of Otto and Diesel cycles.
- Analyze air-standard analyses of gas turbine power plants based on the Brayton cycle and its modifications for performance improvement.
- Evaluate net power output, thermal efficiency, back work ratio, and the effects of compressor pressure ratio on performance of gas turbine power cycles.
- Evaluate second-law analysis of gas power cycles.

11.1 GENERAL ANALYSIS OF CYCLES

Heat engines operating on gas cycles are of two types: internal combustion and external combustion engines. They may be cyclic or noncyclic. The reciprocating internal combustion engines in which combustion takes place inside the cylinder are noncyclic heat engines. Spark ignition and compression ignition engines belong to this category. Gas turbines are rotary engines in which combustion is either inside (open cycle) or outside (closed cycle) and are termed either internal combustion or external combustion engines. Spark-ignition, compression-ignition, and gas turbine engines operate on gas cycles. In external combustion engines, combustion takes place outside the cylinder as opposed to that of spark ignition and compression ignition engines. External combustion engines present several benefits over the internal combustion engines. Various fuels (even cheaper) can be burned in external combustion engines; complete combustion takes place due to the time available for combustion is more; and in addition to these advantages, since they operate on a closed cycle, the working fluid with best features can be utilized. For example, hydrogen and helium are the two prominent gases that are used in closed-cycle gas turbine plants used in nuclear power plants for the electrical power generation.

It was mentioned in Chapter 5 that of all the heat engines operating between same temperature source and same temperature sink, none has a higher thermal efficiency than a reversible heat engine (Carnot heat engine). Carnot cycle is a totally reversible cycle unlike other ideal cycles that are internally reversible but not externally reversible as there are certain external irreversibilities associated with these cycles.

For this reason, the thermal efficiencies of all other ideal cycles are less than those of the Carnot cycle. However, the efficiencies of ideal cycles are higher than those of actual cycles.

11.2 CARNOT CYCLE

The Carnot cycle, as discussed in Chapter 5, is a reversible cycle consisting of two reversible adiabatic and two reversible isothermal processes. It is the highest efficient cycle operating between the same temperature source and the same temperature sink. However, the Carnot cycle cannot be a reality due to certain impracticalities associated with it. Figure 11.1 shows the Carnot cycle on p-v and T-s diagrams.

11.3 AIR-STANDARD CYCLES—ASSUMPTIONS

The complete analysis of reciprocating internal combustion (IC) engines essentially requires the study of combustion process in the cylinder and heat transfer from the gases to the cylinder walls and work required to induct the charge into the cylinder and pump the exhaust gases out of the cylinder. The irreversibilities due to the friction and the heat transfer due to the temperature difference are also to be considered. The fuel and air after combustion change the phase as gas. Thus the analysis of gas cycles is a complex job, and a simplified model is the *air-standard analysis* with the following assumptions:

1. Air, assumed as ideal gas, is the working fluid
2. The combustion process is replaced by the reversible heat addition
3. Compression and expansion processes are reversible and adiabatic
4. Exhaust process of actual IC engine is replaced by constant volume heat rejection since the cycle is assumed to be a closed cycle

In addition to the air-standard analysis, if the specific heats of the fluid are assumed to be constant at their ambient temperature values, then the analysis is termed the

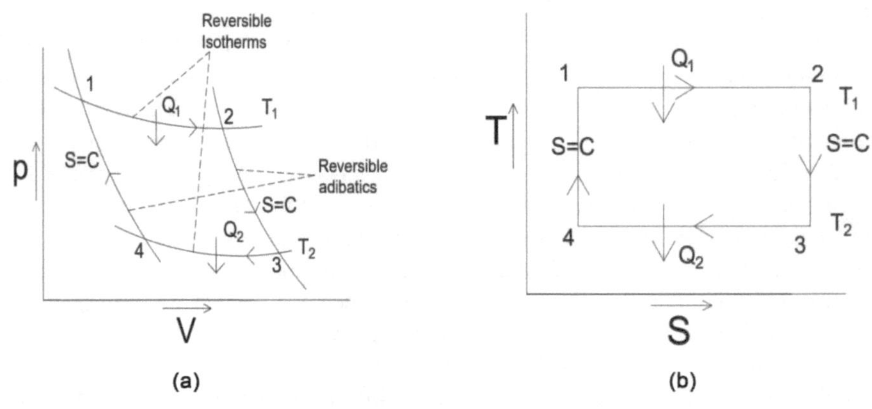

FIGURE 11.1 The Carnot cycle. (a) p-v diagram and (b) T-s diagram.

cold air-standard analysis. It is to be noted here that the air-standard analysis makes the study of internal combustion engines simpler and provides the means of analyzing the influence of various parameters on the performance of actual engines. However, the actual cycle analysis deviates slightly from this approach.

11.4 RECIPROCATING ENGINES—AN OVERVIEW

Reciprocating engines are being widely used as prime movers in transportation, power generation, and propulsion of locomotives, ships, and aircraft. Figure 11.2 shows the basic parts of a reciprocating engine. These are the piston engines in which the piston reciprocates inside the cylinder between two fixed positions called the top dead center and the bottom dead center. The position of the working piston and moving parts associated with it, at the moment when the direction of piston motion is reversed at either end of the stroke is called *dead center. Top dead center (TDC)* is the position in the cylinder when the piston is farthest from the crankshaft or the volume of the cylinder is smallest. *Bottom dead center (BDC)* is the position in the cylinder when the piston is nearer to the crank shaft or the volume of the cylinder is highest. The reciprocating motion of the piston is converted into rotary motion via a connecting rod and crankshaft. The distance traversed by the piston while moving from one dead center to another dead center is called *stroke*. The nominal inner diameter of the cylinder is called *bore*. The air–fuel mixture (in SI engines) or air (in CI engines) is admitted into the cylinder through the *intake valve* and products of combustion (exhaust gases) are expelled through the *exhaust valve*. The volume swept by the piston while moving between TDC and BDC is called *displacement volume or swept volume* (V_s), and the minimum volume formed when the piston is at the TDC is called *clearance volume* (V_c). The displacement volume of a cylinder multiplied by number of cylinders in a given engine is called *cubic capacity or engine capacity*. Then the *compression ratio* (r_k) of the engine is the ratio of total volume of the cylinder to the clearance volume, that is

$$r_k = \frac{\text{Total volume of the cylinder}}{\text{Clearance volume}} = \frac{V_s + V_c}{V_c} \tag{11.1}$$

$$\text{or} \quad r_k = \frac{\text{Volume at the beginning of compression}}{\text{Volume at the end of compression}} = \frac{V_1}{V_2}$$

Mean Effective Pressure (MEP)

The mean effective pressure is defined as the hypothetical pressure acting on the piston during its expansion stroke producing the same amount of net work as that would be produced during the actual cycle.

$$W_{net} = \text{MEP} \times \text{Displacement volume}$$

Then

$$\text{MEP} = W_{net}/V_s \tag{11.2}$$

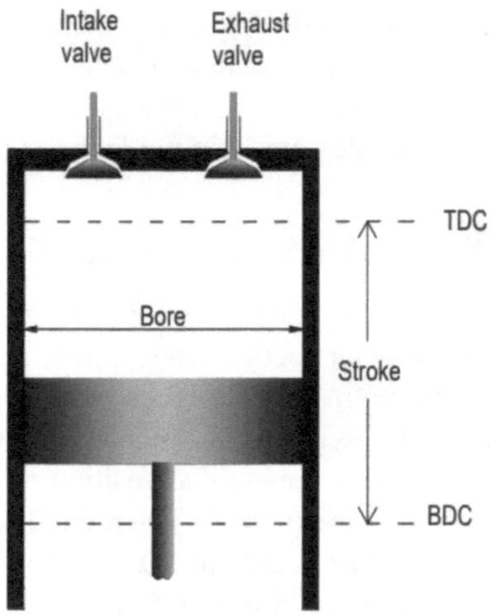

FIGURE 11.2 Reciprocating engine nomenclature.

Reciprocating engines are classified as spark ignition (SI) engines and compression ignition (CI) engines. Both SI and CI engines are further classified as two-stroke and four-stroke engines.

11.5 OTTO CYCLE

The Otto cycle, named after Nikolaus August Otto (1832–1891), is an idealized thermodynamic cycle that describes the functioning of a typical spark ignition (SI) piston engine known as petrol engine. The air-standard Otto cycle consists of four processes, two reversible adiabatic processes (compression and expansion) and two reversible constant volume heat transfer processes (heat addition and rejection). In a typical four-stroke spark ignition engine, the piston executes four complete strokes, namely suction, compression, expansion (power stroke), and exhaust strokes within the cylinder and the crankshaft completes two revolutions for each thermodynamic cycle. Figure 11.3a and b shows the actual Otto cycle and its indicator diagram, respectively.

In an actual SI engine, the air–fuel mixture is inducted as the inlet valve is opened during the intake stroke when the piston descends, compressed by the upward or inward motion of the piston to high pressure, then a spark generated by the spark plug ignites the compressed mixture. The hot gases then expand pushing the piston downward or outward, and the work output is generated during the power stroke. The exhaust gases are thrown out to the atmosphere so that the fresh charge is inducted. Figure 11.4 shows the p-v and T-s diagrams of the Otto cycle.

The thermal efficiency of the ideal Otto cycle is evaluated considering 'm' kg mass of working fluid.

The heat added to the working fluid $(Q_1) = mc_v\,(T_3 - T_2)$ or $q_1 = c_v\,(T_3 - T_2)$.

The heat rejected from the working fluid $(Q_2) = mc_v\,(T_4 - T_1)$ or $q_2 = c_v\,(T_4 - T_1)$.

Then the thermal efficiency of the Otto cycle based on air-standard assumptions

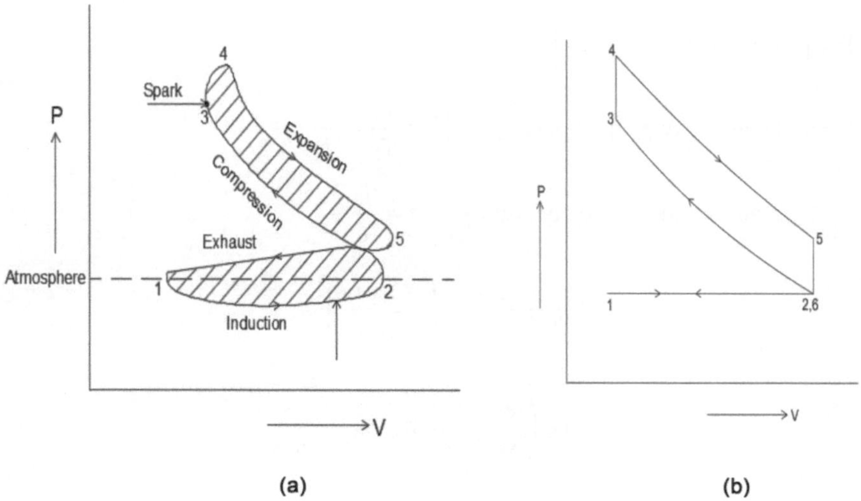

FIGURE 11.3 (a) Actual Otto cycle and (b) its indicator diagram.

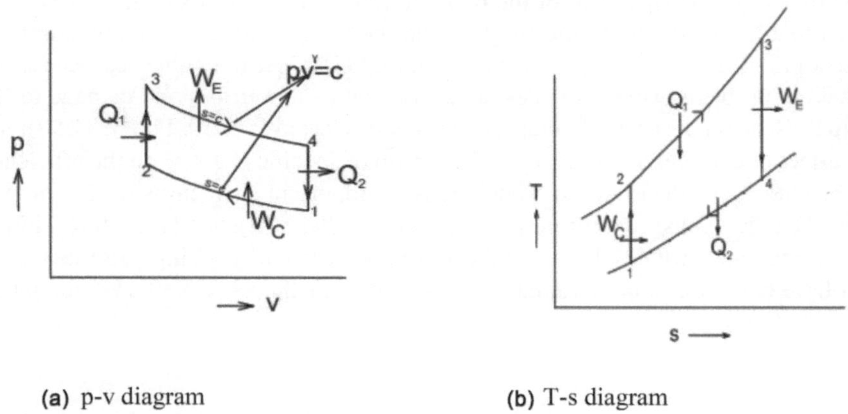

(a) p-v diagram **(b)** T-s diagram

FIGURE 11.4 The Otto cycle. (a) p-v diagram and (b) T-s diagram.

$$\eta_{otto} = \frac{W_{net}}{Q_1} = 1 - \frac{Q_2}{Q_1} = 1 - \frac{mc_v(T_4 - T_1)}{mc_v(T_3 - T_2)} = 1 - \frac{(T_4 - T_1)}{(T_3 - T_2)}$$

On a unit mass basis, the thermal efficiency of the Otto cycle can also be expressed as

$$\eta_{otto} = \frac{w_{net}}{q_1}$$

For process 1-2, $\frac{T_2}{T_1} = \left(\frac{v_1}{v_2}\right)^{\gamma-1}$

For process 3-4, $\frac{T_3}{T_4} = \left(\frac{v_4}{v_3}\right)^{\gamma-1} = \left(\frac{v_1}{v_2}\right)^{\gamma-1}$

From the above two equations, $\frac{T_2}{T_1} = \frac{T_3}{T_4}$ or $\frac{T_3}{T_2} = \frac{T_4}{T_1}$

On rearranging the above equation $\frac{T_3}{T_2} - 1 = \frac{T_4}{T_1} - 1$

$$\frac{(T_4 - T_1)}{(T_3 - T_2)} = \frac{T_1}{T_2} = \left(\frac{v_2}{v_1}\right)^{\gamma-1}$$

$$\eta = 1 - \left(\frac{v_2}{v_1}\right)^{\gamma-1} \text{ or } \eta_{otto} = 1 - \frac{1}{r_k^{\gamma-1}} \tag{11.3}$$

Thus the thermal efficiency of the Otto cycle is mainly a function of compression ratio, r_k, and ratio of specific heats, γ. As γ is assumed to be constant for any working fluid, the efficiency increases with the increase in the compression ratio. However, the efficiency is independent of the pressure ratio and the heat supplied. It is to be noted that the increase in efficiency beyond the certain value of compression ratio is so small. Further, the higher compression ratios in petrol engines can cause the knocking problem, which is undesirable as it causes power loss and damage to the engine. Hence, the range of compression ratio preferred is 6–10. Figure 11.5 shows the effect of compression ratio, r_k, and the ratio of specific heats, γ, on the efficiency of the Otto cycle. Resistance to knocking is an important characteristic of SI engine fuels. Octane number determines the anti-knock characteristics of the fuel. Higher the octane number (ON), higher is the anti-knock capability. Using gasoline blends that have better anti-knock characteristics such as methanol (ON of 111), tetraethyl

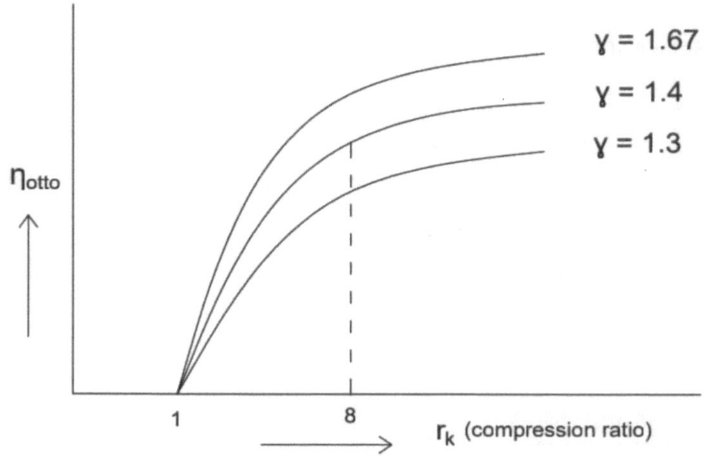

FIGURE 11.5 The effect of r_k and γ on efficiency of the Otto cycle.

lead and ethanol (ON of 108), the thermal efficiency of the Otto cycle can be improved by operating the engine at the higher compression ratios without the possibility of knocking. However, the use of tetraethyl lead was ruled out in 1970s due to its harmful effects on human beings and environment.

Recent Advances in SI Engines

The evolutionary changes that occurred in case of SI engines from the last two to three decades will be from the carburettor engines to gasoline direct injection (GDI) through multipoint fuel injection (MPFI) engines. Carburettor of conventional SI engines is being replaced by multipoint fuel injection (MPFI) system in which fuel is injected into the intake manifold (called port injection), instead of being injected into the combustion chamber as in CI engines. GDI, also known as petrol direct injection, is a mixture formation system for SI engines that run on gasoline, in which fuel is directly injected into the combustion chamber, in contrast to the older port fuel injection systems or MPFI system. The GDI is proving to be a promising technology to simultaneously increase engine efficiency and specific power output and to potentially reduce exhaust emissions.

GDI engine was first introduced for the production in 1925 for a low-compression truck engine. Later, in 1950s, a Bosch mechanical GDI system was used by so many German cars. Thereafter, the use of this technology was kept aside for some period until an electronic GDI system was introduced in 1996 by Mitsubishi for mass-produced cars. In recent years, there has been a tremendous growth in the usage of GDI by the automotive industry. In United States, the production of GDI increased from 2.3% for model year 2008 vehicles to nearly 50% for model year 2016.

It is predicted that, in the future, the refineries will produce relatively inferior petroleum products from heavier crude oil to power the IC engines. Also, there will be a shift in fuel demand toward diesel and jet fuels, leading to the excess amounts of low octane gasoline in the market, with little apparent use for operating the engines. Moreover, due to improvement in the fuel economy of advanced gasoline fueled vehicles, the demand for gasoline will drop further. This allows the low octane gasoline to become cheaper and thereby making it available in excess quantities in future. Gasoline compression ignition (GCI) engine technology is being developed to address the above issues and to take the advantage of being powered by low octane gasoline. GCI is a futuristic engine technology that has the potential to combine the benefits of SI engines such as higher volatility and higher auto-ignition temperature of gasoline and higher compression ratio (CR) of a diesel engine to address the issues of soot and NO_x emissions without compromising diesel engine like efficiency.

11.6 DIESEL CYCLE

The Diesel cycle, named after Rudolf Christian Karl Diesel (1858–1913), is an ideal air-standard cycle that describes the functioning of a compression ignition piston (CI) engine. The air-standard Diesel cycle consists of four processes, two reversible adiabatic processes (compression and expansion) and one reversible constant pressure heat addition and constant volume heat rejection process. Just like in a four-stroke SI engine, the piston executes four complete strokes within the cylinder, and

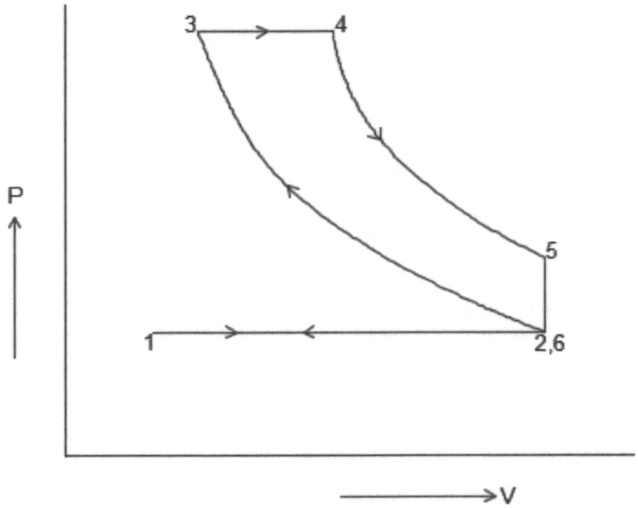

FIGURE 11.6 Indicator diagram of the Diesel cycle.

the crankshaft completes two revolutions for each thermodynamic cycle. Figure 11.6 shows the indicator diagram of the Diesel cycle.

In actual SI engines, the upper limit of compression ratio is limited by the self-ignition temperature of fuel to avoid any possibility of knocking. This limitation is overcome in diesel engines by compressing the air and fuel separately. Air alone is inducted during the intake stroke, and then the fuel is injected into the compressed hot air stream. The fuel is auto-ignited by the high temperature air, the temperature of which is higher than the self-ignition temperature of the fuel. Thus the fuel auto-ignites without any special device as in SI engines. Combustion takes place and the hot gases then expand doing work on the piston. The heat addition in diesel engines is at constant pressure unlike at constant volume as in the case of petrol engines. The exhaust gases are thrown out to the atmosphere so that the fresh charge is inducted. Figure 11.7 shows the p-v and T-s diagrams of the Diesel cycle.

The thermal efficiency of the ideal Diesel cycle is evaluated considering 'm' kg mass of working fluid.

The heat added to the working fluid $(Q_1) = mc_p\,(T_3 - T_2)$ or $q_1 = c_p\,(T_3 - T_2)$

The heat rejected from the working fluid $(Q_2) = mc_v\,(T_4 - T_1)$ or $q_2 = c_v\,(T_4 - T_1)$

Then the thermal efficiency of the Diesel cycle based on air-standard assumptions

$$\eta_{\text{diesel}} = \frac{W_{\text{net}}}{Q_1} = 1 - \frac{Q_2}{Q_1} = 1 - \frac{mc_v(T_4 - T_1)}{mc_p(T_3 - T_2)} = 1 - \frac{(T_4 - T_1)}{\gamma(T_3 - T_2)}$$

Compression ratio,

$$r_k = \frac{v_1}{v_2} \tag{11.4}$$

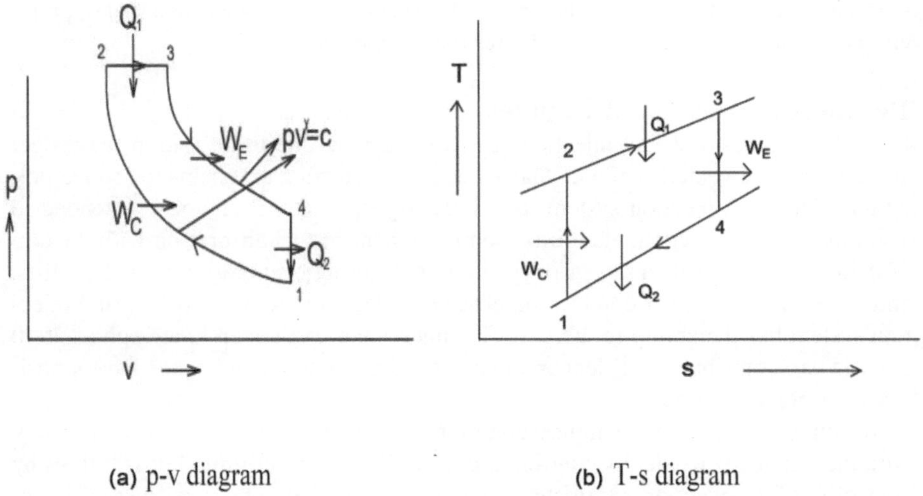

FIGURE 11.7 The Diesel cycle. (a) p-v diagram and (b) T-s diagram.

Cutoff ratio,

$$r_c = \frac{v_3}{v_2} \tag{11.5}$$

Expansion ratio,

$$r_e = \frac{v_4}{v_3} \tag{11.6}$$

For process 3-4, $\frac{T_4}{T_3} = \left(\frac{v_3}{v_4}\right)^{\gamma-1} = \frac{1}{r_e^{\gamma-1}} \Rightarrow T_4 = T_3 \frac{r_c^{\gamma-1}}{r_k^{\gamma-1}} \quad \left(\text{since } r_k = r_c \cdot r_e\right)$

For process 2-3, $\frac{T_2}{T_3} = \frac{p_2 v_2}{p_3 v_3} = \frac{v_2}{v_3} = \frac{1}{r_c} \Rightarrow T_2 = \frac{T_3}{r_c}$

$$\frac{T_2}{T_1} = \left(\frac{v_1}{v_2}\right)^{\gamma-1} = r_k^{\gamma-1} \Rightarrow T_1 = \frac{T_2}{r_k^{\gamma-1}} = \frac{T_3}{r_c r_k^{\gamma-1}}$$

$$\eta = 1 - \frac{1}{\gamma} \cdot \frac{1}{r_k^{\gamma-1}} \cdot \frac{r_c^{\gamma} - 1}{r_c - 1} \text{ or } \eta = 1 - \frac{1}{r_k^{\gamma-1}} \left[\frac{1}{\gamma} \cdot \frac{r_c^{\gamma} - 1}{r_c - 1}\right] \tag{11.7}$$

From Eqs. 11.3 and 11.7, it is seen that the thermal efficiency of the diesel cycle differs from the Otto cycle only in the bracketed term, which is always greater than unity. Thus for the same compression ratio, the Otto cycle is more efficient than the Diesel cycle counterpart. The fuel cutoff ratio, r_c, is directly proportional to the output in the Diesel cycle; therefore air-standard efficiency depends on output. The range of compression ratio in Diesel engines is 16–20, which is much higher than petrol engines, and hence the efficiency of Diesel engines is higher than that of petrol engine counterpart and also offers better fuel efficiency.

Recent Advances in CI Engines

In Diesel engines, the fuel injection pressure plays an important role in the proper atomization of injected fuel and enables complete burning and helps to reduce pollutants. The fuel injection system in a direct injection diesel engine is intended to accomplish a large degree of atomization for enhanced fuel absorption with the aim of using the highest quantity of oxygen and promoting evaporation in a short time and achieving the higher combustion efficiency. The common rail direct fuel injection system has the ability to increase fuel injection pressures up to as high as 2000 bar and can contribute to a cleaner and quieter diesel engine and hence have a positive environmental impact.

In compression ignition engines, common rail direct injection (CRDI) technology with the aid of electronically controlled unit (ECU) offers unlimited possibilities by controlling fuel injection parameters such as fuel injection pressure, start of injection (SOI) timing, rate of fuel injection, and injection duration for improving engine performance, combustion, and controlling emissions. Commercially available CRDI systems at present are quite complex as they use a large number of sensors, hardware and analytical circuits, making them very expensive and unrealistic for cheaper single cylinder engines, typically used in agricultural sector and decentralized power sector.

Although the conventional diesel engines maintain lower exhaust emissions such as carbon monoxide (CO) and hydrocarbons (HC) comparatively due to their better efficiency, still they suffer from an inherent drawback of higher amounts of emissions such as NO_x and particulates. This has led to the development of new combustion technologies such as lean burn engines, homogeneous charge compression ignition (HCCI) engines, and premixed charge compression ignition (PCCI) engines.

Lean Burn Engines

Lean-burn combustion is basically burning the fuel with excess air in an IC engine. In lean-burn engines, the air fuel ratio may be as lean as 65:1 (by mass) in contrast to the stoichiometric air/fuel ratio (14.6:1) required to combust gasoline. The excess air in a lean-burn engine leads to complete combustion far less hydrocarbons. High air/fuel ratios will also help in reducing the losses caused by other engine power management systems such as throttling losses. The knock resistance capability of lean mixtures is higher than that of stoichiometric mixtures, permitting the use of high compression ratios. The fuels with high research octane number such as natural gas in a lean burn engine with a high compression ratio can achieve a high thermal efficiency, due to the increased specific heat ratio, lower combustion temperature, and reduced throttling losses.

HCCI Engines

HCCI strategy proves to be a viable alternative for reducing both nitrogen oxides (NO_x) and soot emissions simultaneously without compromising the efficiency.

HCCI combines the working principles of both spark ignition (SI) and compression ignition (CI) engines. It takes the advantage of the homogeneous mixture preparation like an SI engine and the combustion like a CI engine and hence is a hybrid between both SI and CI engines. HCCI is described as the one in which the fuel and air are mixed before combustion starts and the mixture auto-ignites as a result of the temperature increase in the compression stroke. It is therefore similar to SI in the sense that both engines use premixed charge and similar to CI as both rely on auto-ignition to initiate combustion.

11.7 DUAL CYCLE

The dual combustion cycle or mixed cycle is the more appropriate cycle for modern high speed compression ignition engines. In this cycle, part of heat is added at constant volume and the rest is added at constant pressure, and hence the cycle is termed the mixed cycle. Whereas in the case of the Diesel cycle, the fuel injection takes place just before the end of the compression stroke, in the dual cycle the fuel is injected somewhat early into the combustion chamber so that it starts igniting at the end of the compression stroke, which leads to the combustion to take place partly at constant volume. The high pressure resulting from the combustion of the fuel continues through the expansion stroke as the injection is carried till the piston reached the top dead center and hence the cycle is a combination of constant volume and constant pressure Figure 11.8 shows the p-v and T-s diagrams of the dual cycle.

Thermal efficiency of the ideal dual cycle is evaluated considering 'm' kg mass of working fluid.

The heat added to the working fluid, $Q_1 = mc_v\,(T_3 - T_2) + mc_p\,(T_4 - T_3)$

The heat rejected from the working fluid, $Q_2 = mc_v\,(T_5 - T_1)$

Then the thermal efficiency of the dual cycle based on air-standard assumptions

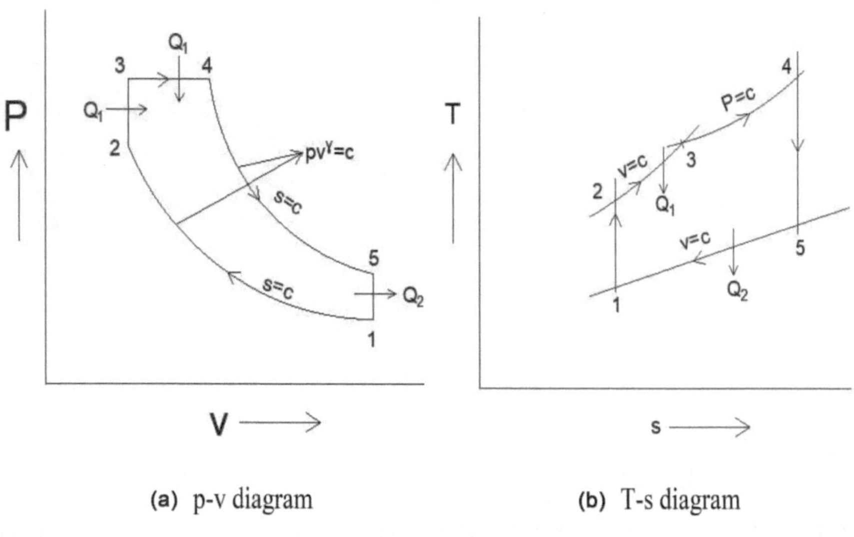

FIGURE 11.8 The dual cycle. (a) p-v diagram and (b) T-s diagram.

$$\eta_{dual} = \frac{W_{net}}{Q_1} = 1 - \frac{Q_2}{Q_1} = 1 - \frac{mc_v(T_5 - T_1)}{mc_v(T_3 - T_2) + mc_p(T_4 - T_3)}$$

$$= 1 - \frac{(T_5 - T_1)}{(T_3 - T_2) + \gamma(T_4 - T_3)} \tag{11.8}$$

11.8 COMPARISON OF OTTO, DIESEL, AND DUAL CYCLES

The three cycles such as Otto, Diesel, and dual cycles can be compared on the basis of same compression ratio (CR) and same maximum pressure and temperature.

11.8.1 Based on Same Compression Ratio and Heat Rejection

Figure 11.9 shows the comparison of three cycles for the same compression ratio and heat rejection with the use of p-v and T-s diagrams.

On p-v diagram, 1-2-6-5 is for Otto, 1-2-7-5 is for Diesel, and 1-2-3-4-5 is for dual cycles. In T-s diagram, for the same heat rejection (Q_2), the area under 2-6 represents Q_1 for Otto while under 2-7 for diesel and 2-3-4 for the dual cycle. From this, it can be concluded that the Otto cycle has the highest efficiency and the Diesel cycle the lowest since Q_1 is higher for the Otto cycle as shown below:

$$\eta_{Otto} > \eta_{Dual} > \eta_{Diesel} \tag{11.9}$$

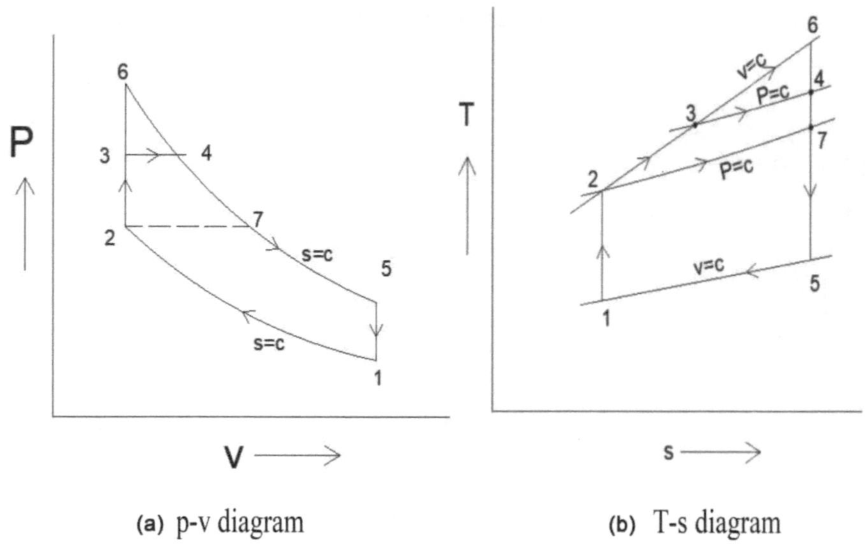

FIGURE 11.9 Comparison of Otto, Diesel, and dual cycles for the same compression ratio and heat rejection. (a) p-v diagram and (b) T-s diagram.

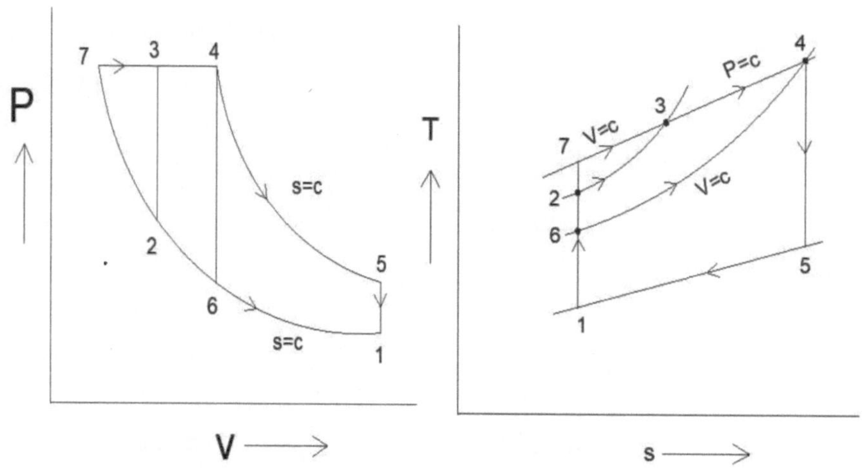

FIGURE 11.10 Comparison of Otto, Diesel, and dual cycles for the same maximum pressure and temperature. (a) p-v diagram b) T-s diagram.

11.8.2 Based on Same Maximum Pressure and Temperature

Figure 11.10 shows the comparison of three cycles for same maximum pressure and temperature. For the Otto cycle, Q_1 is given under area 6-4, for Diesel 7-4, and for dual it is 2-3-4. From this, it can be found that the Diesel cycle is more efficient than the dual cycle, which in turn is more efficient than the Otto cycle given as

$$\eta_{\text{Diesel}} > \eta_{\text{Dual}} > \eta_{\text{Otto}} \tag{11.10}$$

11.9 STIRLING AND ERICSSON CYCLES

Stirling and Ericsson cycles, like Carnot cycle, are totally reversible heat engine cycles. These cycles also involve isothermal heat addition and heat rejection; however, the difference lies in that isentropic compression and expansion processes of the Carnot cycle are replaced by constant volume regeneration processes in the Stirling cycle and constant pressure regeneration processes in the Ericsson cycle, respectively. In the reversible cycles, the heat transfer between the constant temperature source and the system can be possible during the isothermal process only, while during nonisothermal processes, the heat transfer takes place between the system and a regenerative storage device known as regenerator. The regeneration is basically a process during which heat is transferred to the regenerator during one part of the cycle and transferred back to the working fluid in another part of the cycle, which is widely used in many modern steam and gas turbine power plants.

Reverend R. Stirling (1790–1878) proposed the regenerative Stirling cycle. Figure 11.11 shows the Stirling engine, p-v, and T-s diagrams of the Stirling cycle in which an ideal gas is working fluid. The engine constitutes a cylinder with double-acting piston and a regenerator in between the two pistons. The regenerator may be a plug of wire gauge or a porous plug. The regenerator is assumed to be a poor

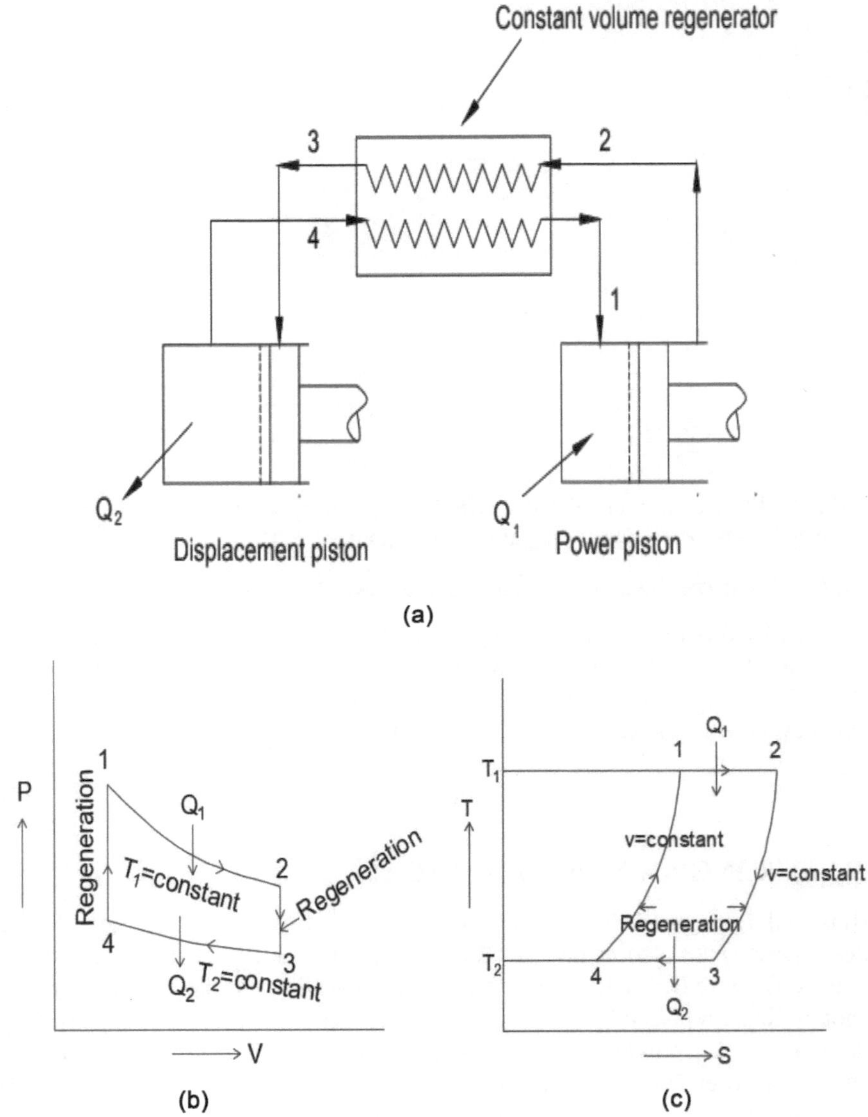

FIGURE 11.11 (a) The Stirling engine, (b) p-v, and (c) T-s diagrams of the Stirling cycle.

conductor of heat, and therefore the heat conducted in the direction of cylinder axis is negligible although there is a temperature gradient.

The cycle consists of four reversible processes as given below:

Process 1-2: Heat is added to the gas reversibly and isothermally from a reservoir at T_1, the left piston moves outward doing work. During this process, the system volume increases and the pressure decreases.

Process 2-3: Constant volume regeneration in which both the pistons are moved to the right. Heat transfer takes place from working fluid to the regenerator and gas temperature falls to T_2. There is no heat transfer with either reservoir, and no work is done in this process as the volume is constant.

Process 3-4: Heat is rejected reversibly and isothermally from the gas at T_2 to the reservoir at T_2. To maintain the gas temperature constant, the right piston moves to the left (inward), doing work on the gas and its pressure rises.

Process 4-1: Constant volume regeneration in which both the pistons are moved to the left. Heat transfer takes place from regenerator to the working fluid. There is no heat transfer with either reservoir, and no work is done in this process as the volume is constant. As the gas returns through the regenerator, the energy stored in the regenerator is returned to the gas and the gas comes out from the left end.

John Ericsson (1803–1889) proposed the Ericsson cycle. Figure 11.12 shows the steady-flow power plant working on the Ericsson cycle, p-v, and T-s diagrams of the Ericsson cycle in which an ideal gas is working fluid.

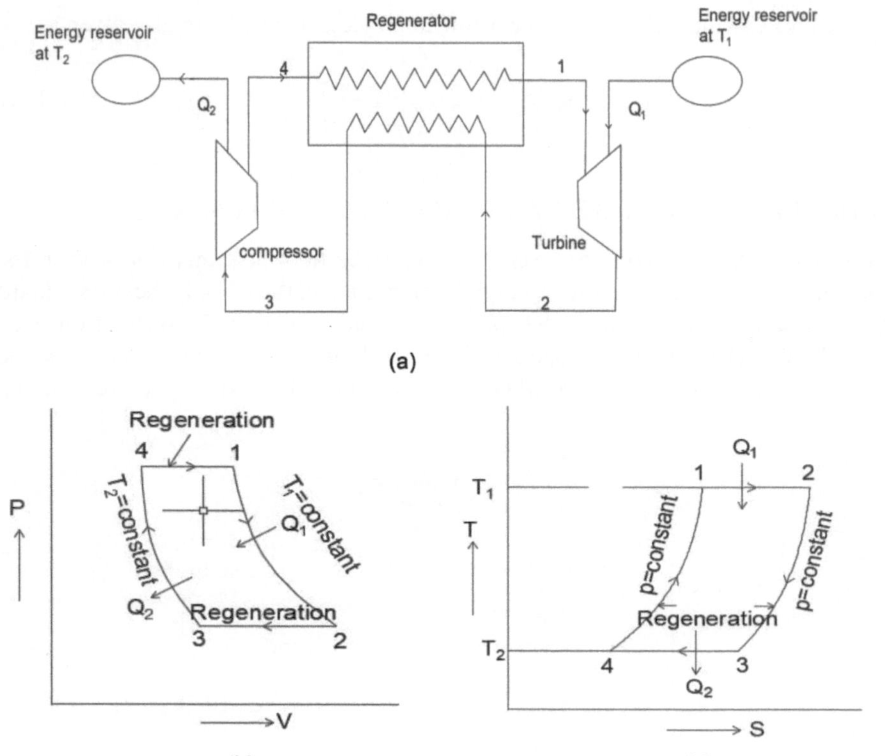

FIGURE 11.12 (a) The Ericsson engine, (b) p-v, and (c) T-s diagrams of the Ericsson cycle.

The power plant consists of three important parts as given below:

1. A turbine in which the gas expands isothermally (process 1-2) doing work. The gas absorbs the heat from a reservoir at T_1.
2. A compressor (process 3-4) that compresses the gas isothermally. Heat is removed from the gas at T_2 to the reservoir at T_2.
3. A counter-flow heat exchanger that acts as a regenerator in which constant pressure regeneration takes place. The gas from the compressor is heated (process 4-1) and the gas from the turbine is cooled (process 2-3).

Since there always exists an infinitesimal temperature difference, the cycle is a reversible one. Two energy reservoirs maintained at temperatures T_1 and T_2 are the external devices.

Since both the Stirling and Ericsson cycles are totally reversible cycles, the thermal efficiencies of these cycles are same as that of the Carnot cycle operating between the same temperature limits according to the Carnot theorem.

The thermal efficiency of the Carnot cycle is proved in Section 6.7 (Eq. 6.21) $\eta_{Carnot} = 1 - \frac{T_2}{T_1}$.

Thus the thermal efficiencies of Stirling and Ericsson cycles are also given as

$$\eta_{Stirling} = \eta_{Ericsson} = 1 - \frac{T_2}{T_1} \tag{11.11}$$

11.10 BRAYTON CYCLE-GAS TURBINE POWER PLANTS

Gas turbines are used as prime movers in power generation and propulsion of aircrafts. The Brayton cycle, named after George B. Brayton (1830–1892), is the air-standard cycle for simple gas turbine power plants. Gas turbines typically operate on open cycle. Figure 11.13 shows the open-cycle gas turbine engine in which atmospheric air is drawn into the compressor where its pressure is raised with the corresponding

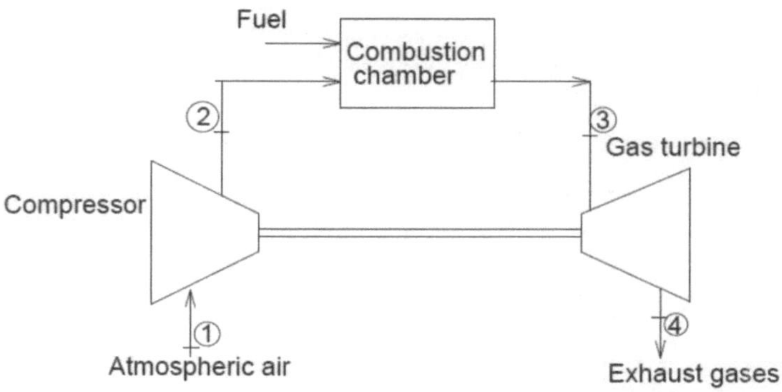

FIGURE 11.13 Schematic of an open-cycle gas turbine plant.

rise in temperature. Fuel is then injected into the combustion chamber consisting of the pressurized hot air stream. Products of combustion expand in a turbine, produce net work output, and are thrown out without being reused. Open-cycle gas turbines can be modeled as a closed cycle based on air-standard assumptions. Based on them,

1. The working substance is air, which is assumed as an ideal gas throughout the cycle of operation.
2. The combustion process is replaced by an isobaric (constant pressure) heat addition process.
3. The intake and exhaust processes are replaced by a constant pressure heat rejection.

The p-ν and T-s diagrams of the ideal Brayton cycle are shown in Figure 11.14. Gas turbine cycle consists of four processes.

Process 1-2: isentropic compression in a compressor
Process 2-3: constant pressure heat addition
Process 3-4: isentropic expansion in a turbine
Process 4-1: constant pressure heat rejection

Considering 'm' kg of working fluid,

The heat added to the working fluid, $Q_1 = m(h_3 - h_2) = mc_p(T_3 - T_2)$ (For ideal gas, $\Delta h = c_p\, dt$)

The heat rejected from the working fluid, $Q_2 = m(h_4 - h_1) = mc_p(T_4 - T_1)$

Then the thermal efficiency of the Brayton cycle,

$$\eta_{th,Brayton} = \frac{W_{net}}{Q_1} = 1 - \frac{Q_2}{Q_1} = 1 - \frac{T_1(T_4 / T_1 - 1)}{T_2(T_3 / T_2 - 1)}$$

For isentropic processes 1-2 and 3-4, $\frac{T_2}{T_1} = \left(\frac{P_2}{P_1}\right)^{\frac{\gamma-1}{\gamma}} = \left(\frac{P_3}{P_4}\right)^{\frac{\gamma-1}{\gamma}} = \frac{T_3}{T_4}$

$$\eta_{th,Brayton} = 1 - \frac{1}{r_p^{\left(\frac{\gamma-1}{\gamma}\right)}}, \tag{11.12}$$

where r_p is the pressure ratio $= \frac{P_2}{P_1}$ and γ is the ratio of specific heats.

Thus the thermal efficiency of the Brayton cycle is a function of pressure ratio and specific heat ratio of working fluid; it increases with increase in either of these parameters. However, the thermal efficiency of the cycle is independent of peak temperature, and for the special case of constant specific heats, it is a function of pressure ratio only. Atmospheric air is usually the working fluid in gas turbines. Other gases such as helium can be used as working fluid for some applications in a closed cycle so that the working fluid can be used repeatedly. A gas with different specific heat ratio from that of air can increase the thermal efficiency of the cycle.

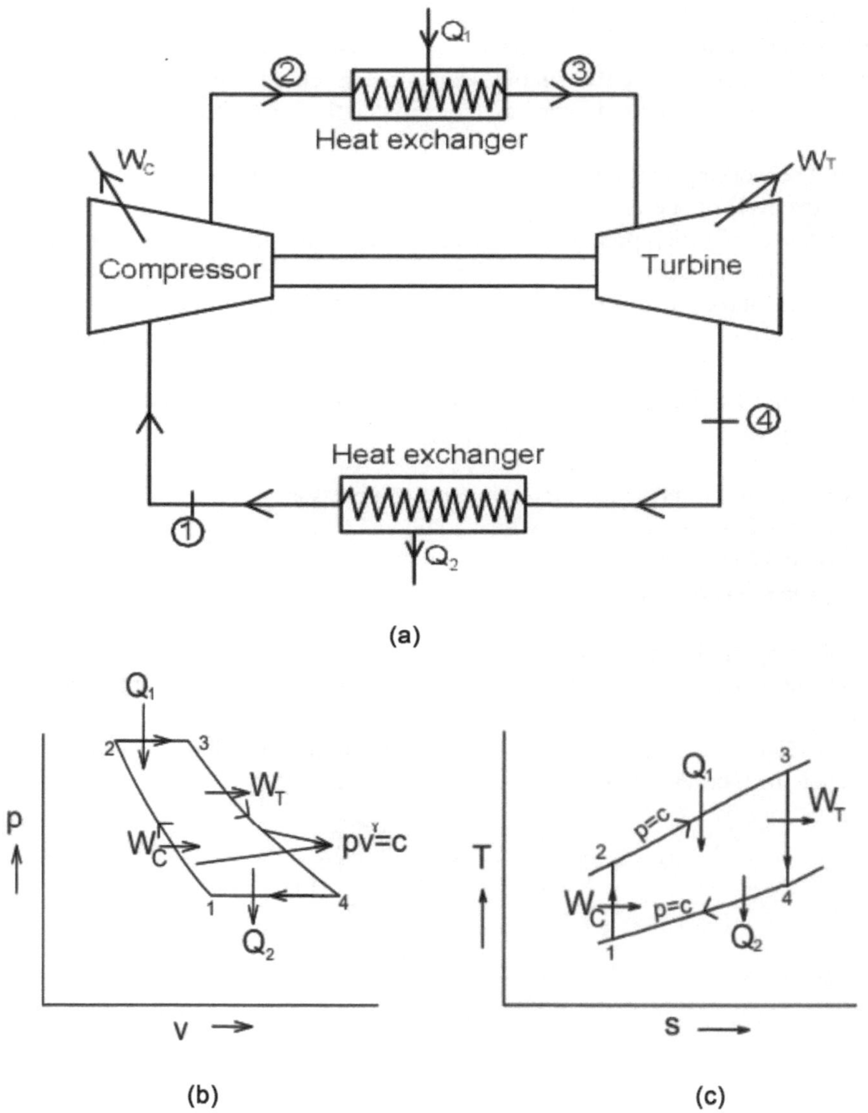

FIGURE 11.14 Schematic of (a) p-v and (b) T-s diagrams of closed-cycle gas turbine plant.

Comparison of Brayton and Carnot cycles

Figure 11.15a and b shows the comparison of both Brayton and Carnot cycles on T-s diagram. The thermal efficiency of the Brayton cycle is less than that of the Carnot cycle operating between the same two temperature limits. The Brayton cycle, as shown in Figure 11.15a, can be considered to be composed of several Carnot cycles with very small heat added at constant temperature to each cycle. It is seen that the efficiencies of all these Carnot cycles are less than the efficiency of the Carnot cycle operating between the maximum and the minimum temperature limits. Therefore

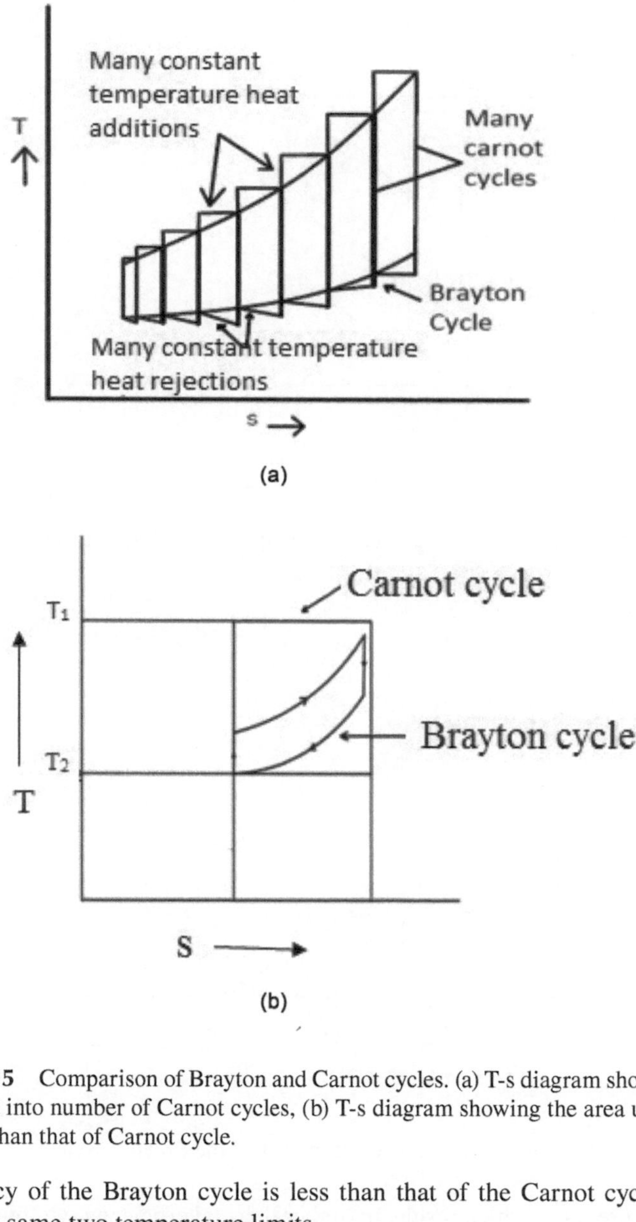

FIGURE 11.15 Comparison of Brayton and Carnot cycles. (a) T-s diagram showing Brayton cycle divided into number of Carnot cycles, (b) T-s diagram showing the area under Brayton cycle is less than that of Carnot cycle.

the efficiency of the Brayton cycle is less than that of the Carnot cycle operating between the same two temperature limits.

Figure 11.16 shows how the pressure ratio affects the thermal efficiency of the Brayton cycle for a constant value of $\gamma = 1.4$ (the specific heat ratio at room temperature). The maximum temperature in the cycle T_3 (T-s diagram) is limited by the withstanding capacity of turbine blades to the higher temperatures, which in turn limits the pressure ratio used in the cycle. Figure 11.17 shows how the pressure ratio affects net work output. For a given temperature in the cycle T_3, the net work output increases first, reaching a maximum, and then decreases. Thus there is an optimum

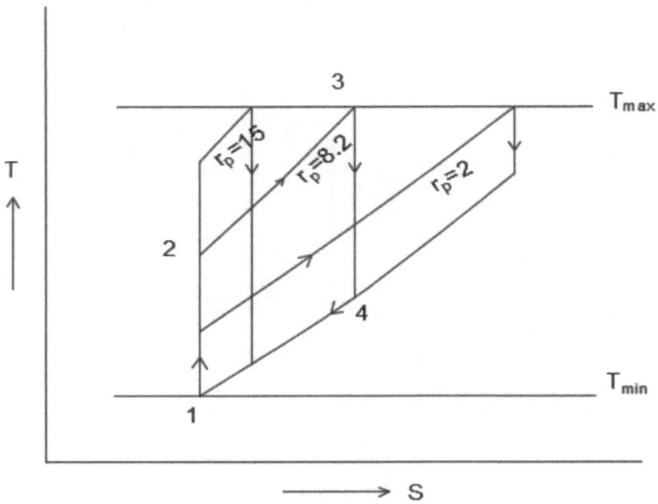

FIGURE 11.16 Effect of pressure ratio on work output of the Brayton cycle.

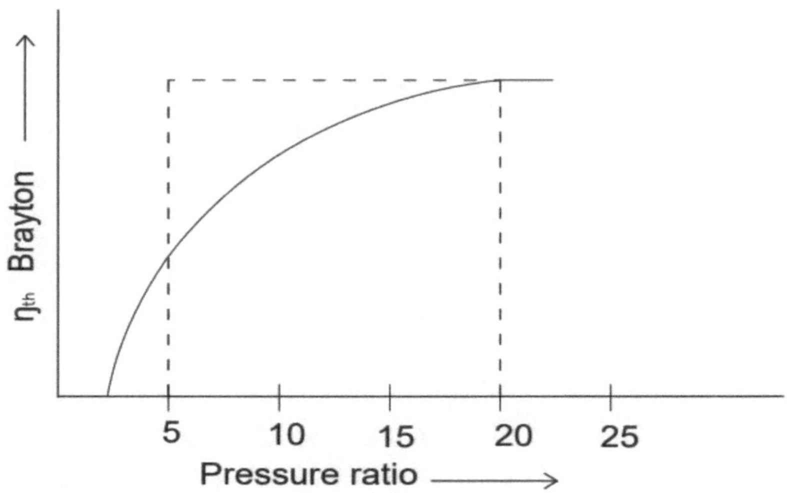

FIGURE 11.17 Effect of pressure ratio on efficiency of the Brayton cycle.

value of pressure ratio, which results in higher thermal efficiency without a compromise in work output. Usually the pressure ratio of gas turbines is 11–16 in most of the designs.

Back Work Ratio

It is defined as the ratio of compressor work input to the gross work output of the turbine. Usually the net work output of the gas turbine is very small as large amount of work produced by turbine is utilized in running the compressor and other auxiliary devices. For example, of the 1000 kW gross work output of the cycle, about 650 kW is required to run the compressor and remaining 350 kW is the net work output. Also if there is any reduction in compressor efficiency, it will reduce the net work output further.

$$\text{Back work ratio} = \frac{\text{Compressor work input}}{\text{Gross work output of the turbine}} \tag{11.13}$$

For gas turbines, back work ratio is very high compared to that of steam turbines; thus they are not preferred in steam power plants as large gas turbine is required for the same capacity plant than steam turbine.

11.11 BRAYTON CYCLE WITH REGENERATION

Gas turbine plants operate at higher temperatures, due to which the temperature of exhaust gases leaving the turbine is also high, even higher than the temperature of air leaving the compressor resulting in a large amount of heat loss to the surroundings. This heat energy can be utilized to heat the high pressure air leaving the compressor using a counter-flow heat exchanger called regenerator or recuperator. However, for the gas turbines operating at higher pressure ratios, the regeneration may not be suitable since the temperature of exhaust gases in this case will be lower than the temperature of air leaving the compressor.

Figure 11.18 shows the flow and T-s diagrams of the air-standard regenerative Brayton cycle. Steady flow energy equation can be applied to the regenerator, assuming that the kinetic energy changes and heat loss to the surrounding are negligible, which gives that

$$h_3 - h_2 = h_5 - h_6 \tag{11.14}$$

Effectiveness of a regenerator is defined as a ratio of actual temperature rise of air to the maximum possible temperature rise. In ideal case, when regenerator is used, the heat supplied and heat rejected are reduced by the same amount, that is

$$Q_{2\text{-}3} = -Q_{5\text{-}6}$$

$$\text{Regenerator effectiveness, } \varepsilon = \frac{T_3 - T_2}{T_5 - T_2} \tag{11.15}$$

The thermal efficiency of the cycle can be increased due to this regeneration since the air entering the combustion chamber can be preheated so that the heat input requirements and the amount of fuel added for the same net work output can be reduced. Regeneration in gas turbines is usually preferred in the applications such as driving the gas-line compressors in the areas where there is no demand for power. However, regeneration increases maintenance problems since gas turbines

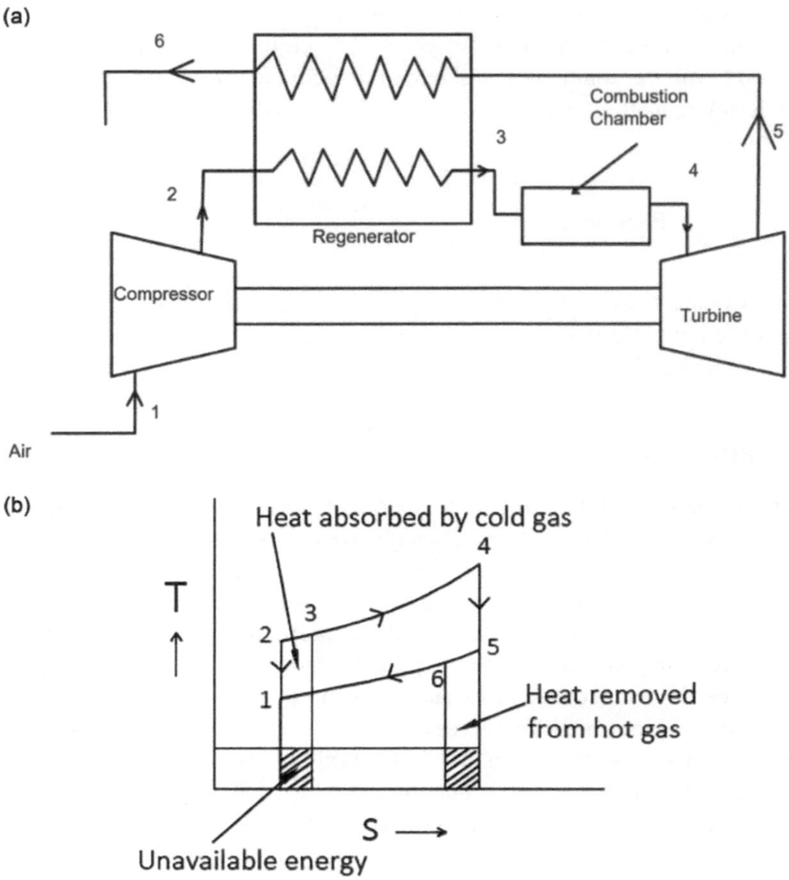

FIGURE 11.18 (a) The flow and (b) T-s diagrams of the regenerative Brayton cycle.

are often used for meeting the peak load demands on electric power generation systems that result in fluctuating temperature changes. Moreover, for higher effectiveness, larger regenerator is required, which increases cost significantly and causes large pressure drop. Most of the regenerators used have an effectiveness of around 0.85.

The thermal efficiency of the air-standard regenerative Brayton cycle is

$$\eta_{reg,Brayton} = 1 - \frac{T_1}{T_4} r_p^{\frac{\gamma-1}{\gamma}}, \tag{11.16}$$

where T_1 and T_4 are the minimum and maximum temperatures in the cycle. Thus the thermal efficiency of the air-standard regenerative Brayton cycle is a function of the minimum and maximum temperatures of the cycle and pressure ratio.

11.12 BRAYTON CYCLE WITH INTERCOOLING, REHEATING, AND REGENERATION

11.12.1 Brayton Cycle with Intercooling

The net work output of a gas turbine cycle can be increased by either reducing the compressor work or by increasing the turbine work or by both. The work of compression can be reduced by a multistage compression, that is, compressing the gas into two or more stages, with an intercooler in between. Similarly the work of turbine can be increased by expanding the gas in stages with a reheat in between. As the number of stages in both the compression and expansion increases, they approach isothermal processes. The work of compression or the work of expansion in a steady-flow process is very much related to the specific volume. Intercooling in multistage compression reduces the specific volume, thereby reducing the work required, while reheating in multistage expansion process increases the specific volume thereby increasing the work output.

Figure 11.19 shows the flow, p-v, and T-s diagrams of the Brayton cycle with intercooling. Gas is first compressed in process 1-2 in the first stage of compression and then enters the intercooler (2-3) where it is cooled at constant pressure to its initial temperature and again compressed in process 3-4 in the second stage of compression. Staging the compression with intercooler in between reduces the back work ratio and can increase the net work without changing the turbine work. However, the thermal efficiency of ideal gas turbine cycle is lowered when intercooling is added, which can be explained using T-s diagram in Figure 11.19c. On T-s diagram, the simple ideal gas turbine cycle is shown by $1\text{-}2^1\text{-}5\text{-}6\text{-}1$, while the cycle with intercooling is shown by1-2-3-4-5-6-1. If we divide the simple ideal gas turbine cycle $1\text{-}2^1\text{-}5\text{-}6\text{-}1$ into number of cycles like a-b-c-d-a and p-q-r-s-p, these cycles come close to Carnot cycles. For constant specific heats,

$$\frac{T_5}{T_6} = \frac{T_a}{T_d} = \frac{T_p}{T_s} = \frac{T_2^1}{T_1} = \left(\frac{P_2}{P_2}\right)^{\frac{\gamma-1}{\gamma}} \tag{11.17}$$

From Eq. 11.14, it is clear that all the Carnot cycles that make up the simple ideal gas turbine cycle will have the same efficiency. If we divide the cycle $2\text{-}3\text{-}4\text{-}2^1$ into similar Carnot cycles, the efficiencies of all these cycles are lower than the Carnot cycles that comprise the $1\text{-}2^1\text{-}5\text{-}6\text{-}1$. Therefore the addition of intercooler $(2\text{-}3\text{-}4\text{-}2^1)$ to ideal gas turbine cycle lowers the efficiency. The regenerative intercooling improves the thermal efficiency of the cycle.

11.12.2 Brayton Cycle with Reheating

Figure 11.20 shows the flow, p-v, and T-s diagrams of the Brayton cycle with reheating. The gas is first expanded in the first stage turbine (3-4), reheated in reheater (4-5), and then expanded further in the second stage turbine (5-6). However, the thermal efficiency of ideal gas turbine cycle is lowered when reheating is added similar

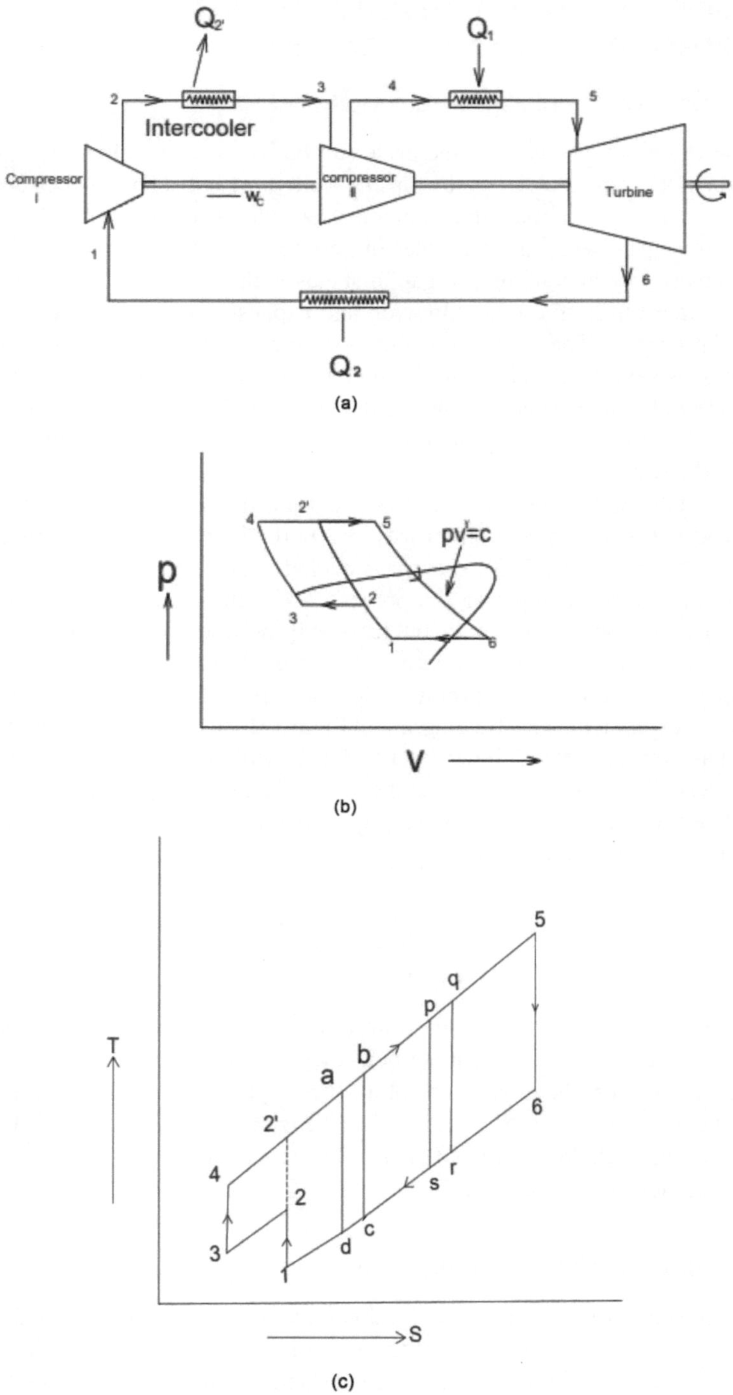

FIGURE 11.19 (a) The flow, (b) p-v, and (c) T-s diagrams of the Brayton cycle with intercooling.

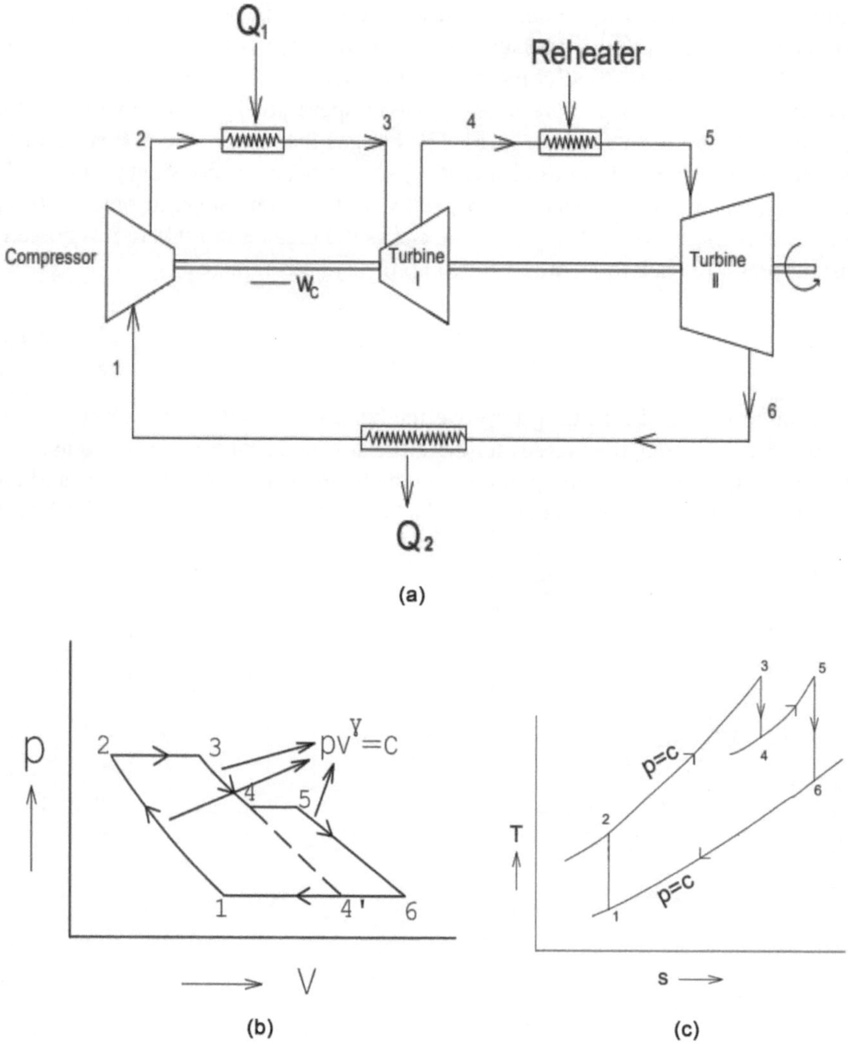

FIGURE 11.20 (a) The flow, (b) p-v, and (c) T-s diagrams of the Brayton cycle with reheating.

to that of gas turbine cycle with intercooling. The regenerative reheating improves the thermal efficiency of the cycle.

11.12.3 Brayton Cycle with Intercooling, Reheating, and Regeneration

Figure 11.21 shows the schematic and T-s diagrams of a gas turbine cycle with intercooling, reheating, and regeneration. Because of this intercooling and reheating, air leaves the compressor at lower temperatures and exhaust gases leave the turbine at higher temperatures, respectively. The regeneration effect can be best utilized with this arrangement as there is more scope for heat transfer between the high temperature exhaust gases leaving the turbine and low temperature air leaving the compressor. The gas is first compressed isentropically in compressor stage-I (1-2) to pressure p_2, cooled at this pressure to state 3, and then compressed isentropically in stage-II compressor (3-4) to final pressure p_4. The gas then enters the regenerator where it is heated at constant pressure (4-5) to temperature T_5. The combustion takes place in the combustion chamber (5-6). The hot gas then enters the first stage of turbine where it is expanded isentropically (6-7), then reheated at constant pressure (7-8) in reheater, and again expanded isentropically in the second stage turbine (8-9). The gas after leaving the second stage turbine enters the regenerator where it is cooled at constant pressure back to its initial state and the cycle completes.

$$\frac{p_2}{p_1} = \frac{p_4}{p_3} \text{ and } \frac{p_6}{p_7} = \frac{p_8}{p_9} \tag{11.18}$$

Both intercooling and reheating improve the back work ratio of gas turbines, however, with intercooling, the average temperature at which heat is added decreases, and with reheating, the average temperature at which heat is rejected increases, and consequently thermal efficiency decreases. For this reason, intercooling and reheating

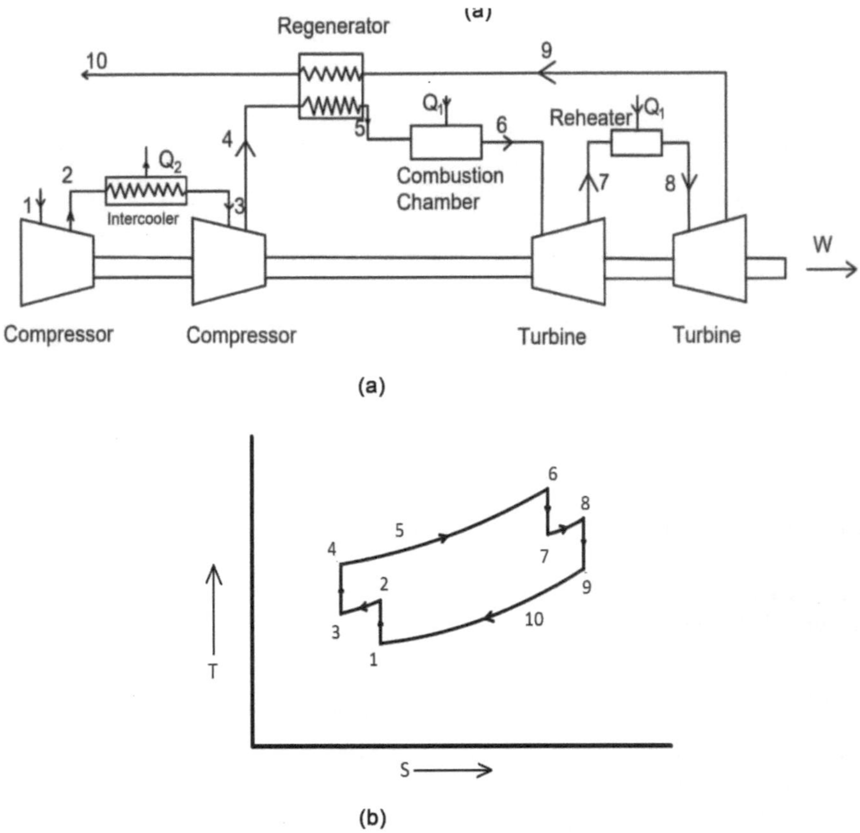

FIGURE 11.21 (a) The flow and (b) T-s diagrams of the Brayton cycle with intercooling, reheating and regeneration.

are coupled with regeneration in gas turbines to take the advantage of both intercooling and reheating effects. With more number of intercools and reheats and increased regenerator effectiveness, the cycle approaches the Ericsson cycle.

11.13 GAS TURBINES FOR JET PROPULSION

A propulsion system that generates the thrust required to move the airplane through the air is called jet propulsion system. It is based on Sir Isaac Newton's third law, which is stated as "for every force acting on a body, there is an opposite and equal reaction." For jet propulsion, the atmospheric air (considered as body) is accelerated as it passes through the engine. The force, which is needed to accelerate the atmospheric air, will have an equal effect acting in the opposite direction on the apparatus that produces the acceleration.

For propelling the aircraft, gas turbines are well suited due to their higher power-to-weight ratios and compactness. There are primarily four types of gas turbine engines: turbojet, afterburning turbojet, turbofan, and turboprop engines. For the analysis of jet propulsion, the commonly used one is turbojet engine. These cycles operate on an open cycle and are called jet propulsion cycles. The gas turbines used in the aircraft operate at higher pressure ratios comparatively (between 10 and 25).

The turbojet engine as shown in Figure 11.22a typically consists of three primary sections:diffuser, gas generator, and nozzle. The gas generator section consists of compressor, burner section, and turbine. The diffuser decelerates the incoming air and raises the pressure slightly before it enters the compressor. The air pressure is further raised by the compressor, which then mixes with the fuel in the burner section and the mixture is burned at constant pressure. The combustion gases then expand in the turbine, since in the jet, propulsion cycles are not expanded to the ambient pressure, the power developed by the turbine is just sufficient to drive the compressor and other auxiliaries and hence the net work output of the cycle is zero. Figure 11.22b shows the flow diagram of turbojet engine. The gases, after leaving the turbine, expand in a nozzle where they are accelerated to higher velocity before being discharged to the atmosphere. The difference in low velocity incoming air and high velocity exhaust gases leaving the engine causes an unbalanced force, which will generate the *thrust* to propel the aircraft.

The T-s diagram of the ideal turbojet cycle is shown in Figure 11.22c. Process 1-2 is the pressure rise in the diffuser followed by isentropic compression in process 2-3 in the compressor. Heat is added at constant pressure in the process 3-4. Isentropic expansion during process 4-5 takes place in the turbine in which work is developed, and in isentropic expansion during process 5-6 in the nozzle, air is accelerated to higher velocities at the expense of pressure drop. However, due to the irreversibilities in actual engine, the specific entropy of the diffuser, compressor, turbine, and nozzle increases. Also there is a pressure drop through the combustor due to combustion irreversibility.

Propulsive Power, $\dot{W}_P$

Propulsive power of the turbojet engine is defined as that power developed from the thrust of the engine. It is the thrust times the velocity of the aircraft. Thrust (F) of the engine can be calculated from Newton's second law, that is

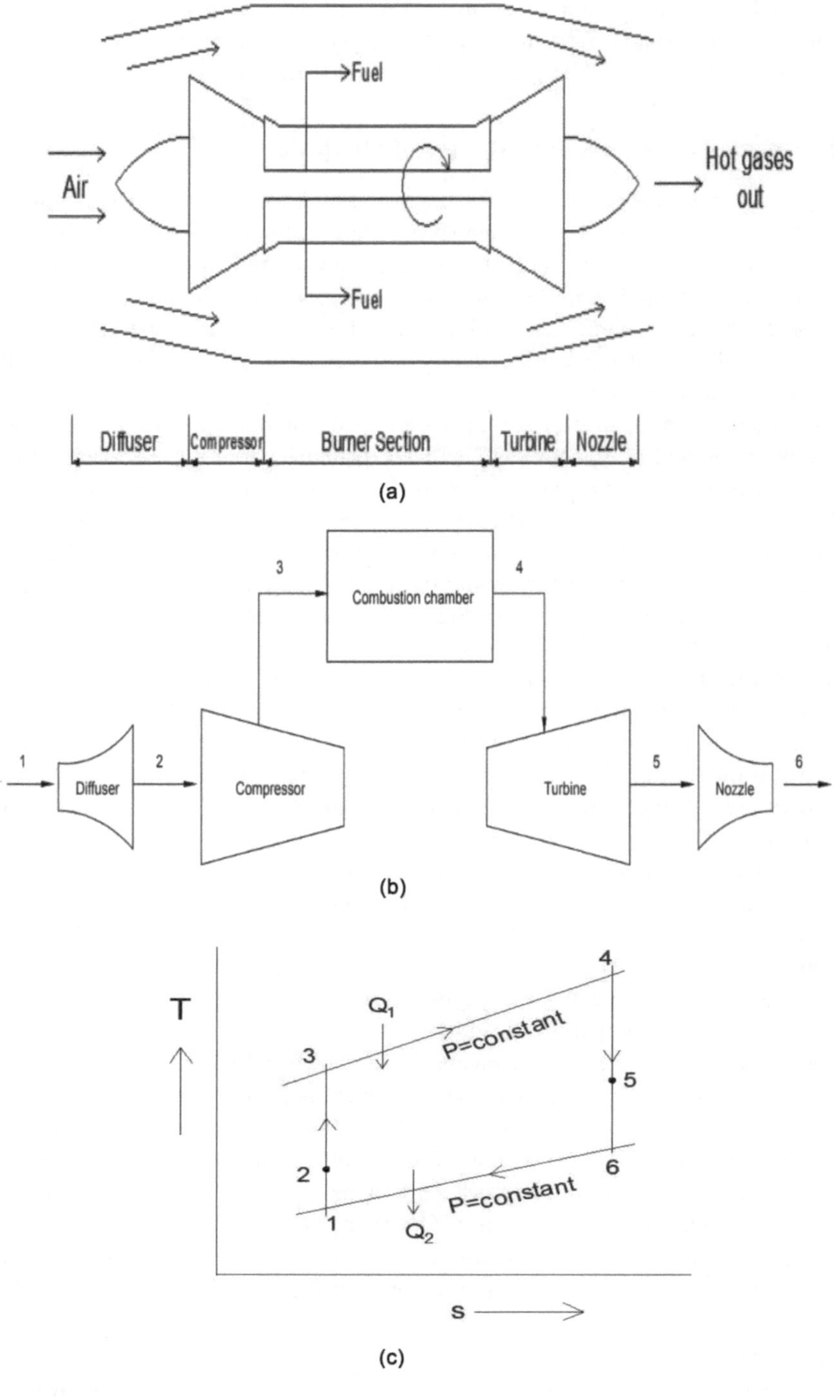

FIGURE 11.22 The turbojet engine components and its T-s diagram. (a) Basic components of turbojet engine, (b) Flow diagram of turbojet engine, (c) T-s diagram of turbojet cycle.

$$F = \dot{m}(V_e - V_i), \tag{11.19}$$

where V_i and V_e are the velocities of air at inlet and exhaust gases at exit of the aircraft, respectively, with respect to aircraft and $\dot{m}$ is the mass flow rate of air flowing through the aircraft.

Now the propulsive power of the turbojet engine is given as

$$\dot{W}_P = FV = \dot{m}(V_e - V_i)V, \tag{11.20}$$

where V is velocity of the aircraft.

Propulsive Efficiency, η_p

The *propulsive efficiency* of the turbojet engine is defined as the ratio of power produced by the engine to propel the aircraft to the heating value of the fuel, given as

$$\eta_P = \frac{\text{Propulsive power}}{\text{Rate of heat input}} = \frac{\dot{W}_P}{\dot{Q}_1} \tag{11.21}$$

Different jet engine designs have their own advantages and disadvantages. A turbojet engine is an ideal candidate for very high speed flights but not that efficient at low speeds and low altitudes. A turbofan engine, in contrast, is suitable for moderate to high speeds, and turboprop is good at low speeds but not at high speeds.

A *ramjet engine* is a type of air-breathing jet engine, which makes use of vehicle's forward motion to compress the incoming air and requires no rotating compressor. Fuel is injected into the combustion chamber in which mixing of fuel and hot compressed air takes place and ignition occurs. Ramjet operates in the supersonic speeds, typically in the Mach number range of 3–6. However, they become less efficient at hypersonic speeds.

Scramjet, a supersonic combustion ramjet, is an improvement over the ramjet engine, which can operate at hypersonic speeds.

A *dual mode ramjet (DMRJ)* is a transformation of ramjet into scramjet with a Mach number range of 4–8, meaning that it can operate efficiently in subsonic as well as in supersonic mode.

11.13.1 Rocket Engine

Turbine and propeller engines use air from atmosphere, while rocket engines contain oxygen (oxidizer) within itself. Fuel and oxidizer are mixed and exploded in the combustion chamber, and the hot exhaust gases produced during combustion pass through the nozzle to accelerate the flow and produce the thrust required to propel the rocket. Turbine and propeller engines cannot operate in outer space since there is no atmosphere, but a rocket works in space. There are basically two types of rocket engines: liquid propellant and solid propellant. In the former case, the propellants (fuel and oxidizer) are stored separately as liquid and are sent to the combustor of the nozzle in which combustion takes place. In the latter case, the propellants are mixed together and packed into a solid cylinder. Burning takes place when propellants are exposed to a source of heat supplied by an igniter. The burning continues until all the propellants are exhausted. Due to the pumps and storage tanks, liquid propellant rockets tend to be heavier and complex compared to that of solid propellant that can be handled with ease.

11.13.2 Compressors Used in Jet Engines

Although there are so many different types of gas turbine engines being used in passenger and military aircrafts, they have some common parts including compressor used to raise the pressure of incoming air. The performance of compressor plays a vital role in overall engine performance. The earlier jet engines used centrifugal compressors, which are now replaced by axial-flow type in all modern turbojet and turbofan engines. However, centrifugal compressors are still used in small turbojet engines and rocket engines as pumps. Whereas in axial-flow compressors, the flow through the compressor travels parallel to the axis of rotation, in centrifugal compressors, it is turned perpendicular to the axis of rotation.

A single-stage centrifugal compressor can increase the pressure of air by a factor of 4 compared to that of 1.2 by a similar axial-flow compressor. However, it is comparatively easy to produce a multistage axial-flow compressor in which pressure can be multiplied from row to row (e.g., 8 stages with 1.2 per stage gives a factor of 4.3). In contrast, it is quite difficult to produce a multistage centrifugal compressor since the flow must be deducted back to the axis at each stage. Since the flow is turned perpendicular to the axis with the consequent increase in cross-sectional area, the engine becomes bulky. Thus the use of multistage axial-flow compressors will be a viable option for high-performance and high-compression turbine engines.

11.14 EXERGY ANALYSIS OF GAS POWER CYCLES

The gas power cycles such as Otto, Diesel, and Brayton cycles are not totally reversible cycles since they involve irreversibilities such as heat transfer through a finite temperature difference. The second law analysis of these cycles such as available energy (exergy) and exergy destruction (available energy loss) provides a useful insight to quantify these irreversibilities. Gas power cycles operate on both open cycle (all internal combustion engines) and closed cycle (all external combustion engines). The available energy (exergy) of steady-flow system (ψ) and that of a closed system (ϕ) at a specified state are given as

$$\psi = (h - h_0) - T_0(s - s_0) + \frac{V^2}{2} + gz \tag{11.22}$$

$$\phi = (u - u_0) - T_0(s - s_0) + P_0(v - v_0) + \frac{V^2}{2} + gz \tag{11.23}$$

Exergy destruction or irreversibility on a unit mass basis for a steady-flow system is given as

$$x_{des} = T_0 s_{gen} = T_0\left(s_e - s_i + \frac{q_2}{T_{b,2}} - \frac{q_1}{T_{b,1}}\right), \tag{11.24}$$

where s_e and s_i are entropy at exit and inlet, respectively, and T_1 and T_2 are the boundary temperatures into and out of the system, respectively, and T_0 is the surrounding's temperature.

Exergy destruction on a unit mass basis for a cycle is given as

$$x_{des} = T_0\left(\sum \frac{q_2}{T_{b,2}} - \sum \frac{q_1}{T_{b,1}}\right)$$

For a cycle with heat transfer with source at T_1 and sink at T_2, exergy destruction on a unit mass basis is

$$x_{des} = T_0\left(\sum \frac{q_2}{T_2} - \sum \frac{q_1}{T_1}\right) \tag{11.25}$$

11.15 NEW COMBUSTION SYSTEMS FOR GAS TURBINES

The future gas turbines would operate comparatively at higher pressure ratios as well as higher turbine inlet temperatures and may result in higher nitrogen oxide (NO_x) emissions since the formation of NO_x is a temperature-dependent phenomenon as the nitrogen has a tendency to react with oxygen at higher combustion temperatures leading to so-called NO_x emissions. To comply with the future air quality requirements, it is essentially required that a combustion technology that will lower the emissions. The following are the new combustion technologies to meet the above requirements.

11.15.1 Trapped Vortex Combustion (TVC)

This combustion technology makes use of the cavity stabilization concept that is mixing hot combustion products and reactants at a high rate. The earlier studies of TVC (early 1990s) focused on liquid fuel applications for aircraft combustors. To achieve the flame stability, it uses the recirculation zones to provide a continuous ignition source by which it is possible to mix the hot combustion products with the incoming fuel and the air mixture. TVC induces the turbulence in the combustion chamber that is trapped within a cavity so that the reactants are injected in to this cavity and efficiently mixed. A flameless regime is typically achieved since a part of the combustion occurs within the recirculation zone while it can be possible to provide a significant pressure drop reduction with the trapped turbulent vortex. Moreover, TVC has the potential to operate as a staged combustor if the fuel is injected into both the cavities and the main airflow. The staged combustion systems are capable of achieving about 10%–40% reduction in NO_x emissions. The advantages of TVC are given below:

1. Variety of fuels with low and medium calorific value can be burnt.
2. Combustion can be carried with high excess air with premixed regime, which can eliminate flashback.

3. NO_x emissions can be reduced significantly without any postcombustion treatments.
4. Flammability limits can be extended and hence flame stability can be improved.

11.15.2 Rich Burn, Quick-Mix, Lean Burn (RQL)

RQL and lean burn are the most promising technologies to curtail the NO_x emissions in gas turbines. The RQL relies on the principle of rich burn emission controlling mechanism. In this concept, combustion is initiated by a fuel-rich mixture in the primary zone with equivalence ratio typically in the range of 1.2–1.8 as shown in Figure 11.23. The advantages with rich burn are twofold: one is the enhancement of combustion stability due to rich burn producing a high concentration of energetic hydrogen and hydrocarbon radical species and the other is the reduced NO_x concentration due to comparatively low flame temperatures and low concentration of oxygen containing intermediate species.

A quench section is employed downstream of the rich zone for further processing of the high amount of carbon monoxide (CO), unburned hydrocarbons (UHC), and smoke contained in the hot efflux gas from the primary zone. To oxidize CO, hydrogen, a large proportion of dilution airflow is admitted into the quench section. The addition of airflow may, however, leads to a zonal equivalence ratio, which is close to the stoichiometric value, causing rapid formation of thermal NO_x. Therefore, to overcome this shortcoming, the air has to mix rapidly with the primary zone effluent in order that it can be quickly switched from the rich burn to

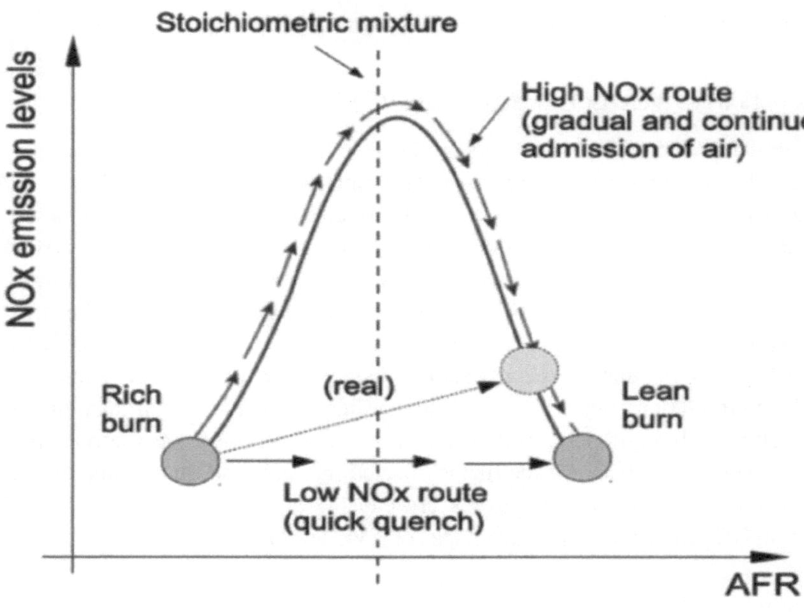

FIGURE 11.23 RQL working principle schematics and NO_x formation routes [31].

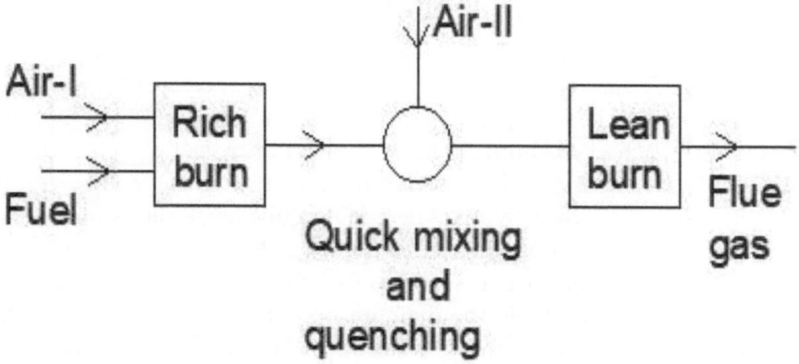

FIGURE 11.24 Schematic of RQL combustor.

the lean burn mode, thereby reducing the tendency of thermal NO_x formation. This process is then followed by the lean burn section, which will further consume CO and UHC such that an exhaust at the exit of the combustor contains a major combustion production of CO_2, N_2, O_2, and H_2O. Moreover, the lean burn section also controls the combustor outlet temperature distribution quality. The equivalence ratio employed in lean burn is in the range of 0.5–0.7. Figure 11.24 shows the schematic of RQL combustor.

There are several benefits of RQL as given below:

1. Flame stability is high.
2. Low overall temperature distributions and hence lower NO_x emissions.
3. It provides higher flexibility for fuel type and fuel compositions.

11.15.3 Double Annular Combustor (DAC)

The double annular combustor makes use of radially staging strategy, in which the simultaneous control of combustion stoichiometry and hence temperature can be possible through the use of fuel injection in multiple combustion locations. At low power settings, part of combustion zones usually referred to as the pilot zone (the outer annulus) operates to raise the equivalence ratio up to nearly 0.8 to improve the combustion efficiency and reduce CO and UHC. The local high combustion stoichiometry also lessens the risk of lean instability. At higher power settings, zones that are referred to as the main (the inner annulus) get fueled and ignited. The equivalence ratios for both zones are typically kept at 0.6 with an intent of achieving lean combustion for lower NO_x and smoke at high power. At midpower settings, part of the main zone is fueled and operated, the objective of which is to increase the transition efficiency and more than one staging point is adopted. The DAC combustor operating with the main stage could reduce the NO_x emissions by nearly 40% as opposed to that of conventional combustor on the CFM56-5B4 engine.

11.15.4 Axially Staged Combustors (ASC)

The concept of ASC was developed by Pratt and Whitney in the Experimental Clean Combustor Program (ECCP) of NASA and conceived in the 1970s. The working principle of axially staged combustors is similar to the DAC except that the fuel staging is achieved through the fuel injection zones placed in the axial direction. In this, the pilot zone is placed at the upstream of the combustor, while the main zone is placed downstream. The zonal equivalence ratios are similar to the DAC for different power conditions, and unlike the DAC, the pilot and main zones have an arrangement for two separate fuel delivering systems. The separation arrangement of the pilot and the main ensures the efficiency and stability at low power and stability at all operating conditions. The heat released is distributed axially in ASC, which will reduce the susceptibility to acoustics. The advantages with this technology are as follows:

1. High combustion efficiency from the main stage even at low equivalence ratios.
2. Since the main stage can burn efficiently, reduced NO_x emissions at high power can be achieved for reduced residence time.

11.15.5 Twin Annular Premixing Swirler Combustors (TAPS)

Basically, TAPS technology is categorized as partially premixed combustion based on the mechanisms although it is also referred to as lean direct injection (LDI). The configuration of TAPS looks much alike the conventional single annular combustor (SAC). However, the fundamental difference between TAPS and SAC lies in the fuel injector heads, as the former makes use of internally staged partially premixed technology. This system has both pilot and main stages, which are mounted concentrically. A simplex atomizer is used in the pilot stage, which sprays the fuel onto the prefilm lip so that it is atomized in an air blast mode between the two axial air streams. To stabilize the pilot flame, the fuel sprays interact with the surrounding counter rotating swirl to generate a pilot recirculation zone.

At lower power, the pilot-only mode is operated to maintain sufficiently high combustion efficiency and stability including ignition through to idle. At higher power, the main is also turned on, which allows partial premixing through the mixing of the discrete liquid jets issuing radially outward into the premixing channel (i.e., cavity) with high swirling air stream generated from cyclone swirlers. A mixing layer between the pilot and the main ensures the stability of the main flame. The main flame stability is also achieved through the small recirculation that stores radicals from the pilot combustion. The airflow for GEnx TAPS is split in such a way that 70% of the air flows through the mixer and the rest is for dome and liner cooling without any air for dilution. The development of TAPS combustor is based on lessons learnt from fuel staging of DAC and also benefited from the experience of lean premixing combustors in aero-derivative industrial gas turbines with dry low emissions.

11.15.6 Lean Direct Injection (LDI)

The working of principle of LDI is based on injecting the fuel directly into the combustor chamber and to allow quick mixing the fuel with a large fraction of air. This combustion technology can reduce peak flame temperatures at medium to high power if thorough mixing of fuel and air takes place before the reaction is completed. Dilution air is significantly reduced since a large proportion of air has to be introduced through the injector. The injector used in LDI occupies a large portion of the dome area in order to pass the large fraction of the total air. The air split for LDI combustor is designed in such a way that most of the air (60%–70%) passes through the injector system and the rest for cooling. The LDI combustor, like TAPS, has also built-in staged singular annular configuration internally. And also pilot and main are concentrically mounted. The two types of pilot nozzle used are pressure atomizing and air blast types. The main nozzle uses air blast. A splitter typically separates the pilot and main flow field by which there sets up a wake, called bifurcated flow field, which leads to the separated pilot and main flame.

At low power, the pilot operates and the flame is stabilized in the bifurcated flow field to sustain the combustion efficiency and stability. As power is increased, the main mixture is injected into the main bifurcated flow field, and the wide cone angle leads to the rapid fuel evaporation and low residence time regions for low NO_x at high power. A lean blowout at idle at equivalence ratio of 0.04 has been achieved. As power is increased to approach, the fuel flow at the pilot is reduced and the part of the main is fueled. At cruise to full power, approximately 10% of fuel enters into the pilot and 90% goes to the main. At full power, the local equivalence ratio at the main is 0.6–0.65 to minimize NO_x emissions.

EXAMPLE PROBLEMS

Example 11.1 An air-standard Otto cycle has cycle efficiency of 45%. The compression begins at 1.1 bar and 50°C. The cycle rejects 0.8 MJ of heat per kg of air. Determine (i) the work done per kg of air, (ii) the compression ratio, (iii) the pressure and temperature at the end of compression, and (iv) the maximum pressure in the cycle.

Solution

The working fluid is assumed as an ideal gas. The processes of air-standard Otto cycle are shown on p-v and T-s diagrams.

The pressure and temperature before the start of compression

$$p_1 = 1.1 \text{ bar} = 101 \text{ kPa}, \ T_1 = 323 \text{ K}$$

$$\eta_{cycle} = 45\%,$$

Heat rejected $(q_2) = 800$ kJ/kg

Heat added (q_1) can be found from the efficiency of cycle

$$\eta_{cycle} = 1 - \frac{q_2}{q_1} \Rightarrow 0.45 = 1 - \frac{800}{q_1} \rightarrow q_1 = 1454.54 \text{ kJ/kg}$$

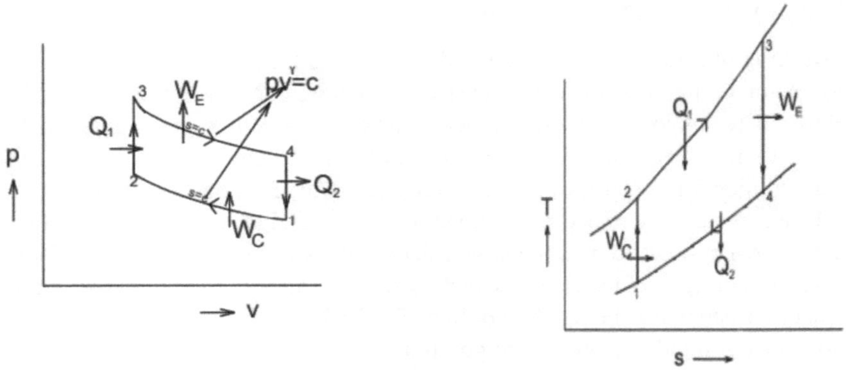

FIGURE EX. 11.1

Net work done per kg can be found from $\eta_{cycle} = \dfrac{w_{net}}{q_1}$

$$\Rightarrow w_{net} = 0.5 \times 1454.54 = 727.27 \text{ kJ/kg} \qquad \text{Ans.}$$

Compression ratio can be found from the efficiency of air-standard Otto cycle

$$\eta_{otto} = 1 - \frac{1}{(r_k)^{\gamma-1}} \Rightarrow 0.45 = 1 - \frac{1}{(r_k)^{1.4-1}}$$

$$\text{Compression ratio}(r_k) = 5.6 = \frac{V_1}{V_2} \qquad \text{Ans.}$$

For ideal gas for an isentropic process 1-2, $\dfrac{T_2}{T_1} = \left(\dfrac{v_1}{v_2}\right)^{\gamma-1} \Rightarrow T_2 = 323 \times (5.6)^{0.4} \Rightarrow 643.4\,\text{K}$

Again $\quad \dfrac{p_2}{p_1} = \left(\dfrac{v_1}{v_2}\right)^{\gamma} \Rightarrow p_2 = 101 \times (5.6)^{1.4} \Rightarrow 1126\,\text{kPa}$

$$q_1 = c_v(T_3 - T_2) \Rightarrow 1454.54 = 0.783(T_3 - 643.4)$$

Then $\quad T_3 = 2501.05$ K

For process 2-3, $\dfrac{p_2 v_2}{T_2} = \dfrac{p_3 v_3}{T_3}$

The maximum pressure in the cycle, $p_3 = \dfrac{1126 \times 2501.5}{643.4} = 4377.03\,\text{kPa}$ Ans.

Example 11.2 In an air-standard Diesel cycle, compression begins at 1.1 bar and 35°C. Heat added is 1.8 MJ/kg and compression ratio is 17:1. Estimate (i) the maximum temperature of the cycle, (ii) the work done per kilogram of air, (iii) the maximum pressure, (iv) η_{cycle}, (v) the temperature at the end of isentropic expansion, (vi) the cutoff ratio, and (vii) the MEP of cycle.

Solution

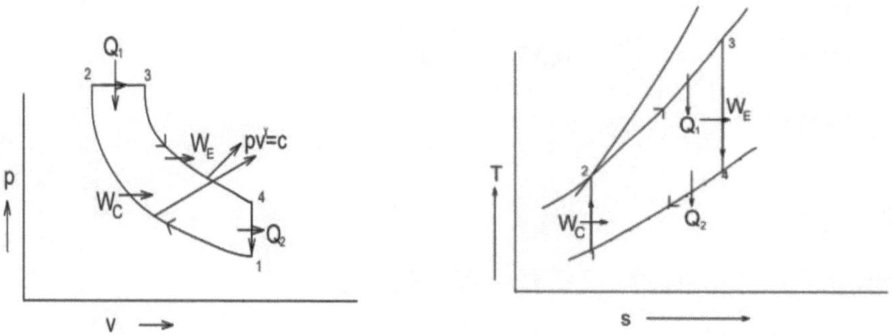

FIGURE EX. 11.2

The working fluid is assumed as an ideal gas. The processes of air-standard Diesel cycle are shown on p-v and T-s diagrams.

Compression ratio $(r_k) = 17:1$

The pressure and temperature before the start of compression

$$p_1 = 1.1 \text{ bar} = 101 \text{ kPa}, \; T_1 = 273 + 35 = 308 \text{ K}$$

The maximum temperature of the cycle $(T_{max}) = T_3$

For ideal gas for an isentropic process, 1-2, $\frac{T_2}{T_1} = \left(\frac{v_1}{v_2}\right)^{\gamma-1} \Rightarrow T_2 = 308(17)^{1.4-1} = 956.6\,\text{K}$

Heat supplied, $q_1 = c_p(T_3 - T_2) \Rightarrow 1800 = 1.005(T_3 - 956.6)$

$$\therefore T_3 = 2747.65 \text{ K}$$ Ans.

For process 2-3, $\frac{p_2 v_2}{T_2} = \frac{p_3 v_3}{T_3} \Rightarrow \frac{T_3}{T_2} = \frac{v_3}{v_2} \; (\because p_2 = p_3)$

Cutoff ratio, $\frac{v_3}{v_2} = \frac{T_3}{T_2} = \frac{2747.65}{956.6} = 2.87$ Ans.

For process 3-4, $\frac{T_3}{T_4} = \left(\frac{v_4}{v_3}\right)^{\gamma-1} = \left(\frac{v_1}{v_2} \times \frac{v_2}{v_3}\right)^{\gamma-1}$

$$\frac{T_3}{T_4} = \left(17 \times \frac{1}{2.87}\right)^{1.4-1} = 2.037 \Rightarrow T_4 = 1348.87\,\text{K}$$

Heat rejected $(q_2) = c_v(T_4 - T_1) = 0.718(1348.87 - 308) = 747.35\,\text{kJ/kg}$

Work done per kilogram of air $w_{net,out} = q_1 - q_2 = 1800 - 747.35 = 1052.65$ kJ/kg Ans.

$\therefore \eta_{cycle} = 1 - \frac{q_2}{q_1} = 1 - \frac{747.35}{1800} = 58.48\%$ Ans.

$$p_2 v_2 = RT_2 \Rightarrow p_2 = \frac{RT_2}{v_2}$$

$$v_1 = \frac{RT_1}{p_1} = \frac{0.283 \times 308}{1.1 \times 10^2} = 0.863\,m^3/kg$$

$$\frac{v_1}{v_2} = 17 \Rightarrow v_2 = \frac{v_1}{17} = \frac{0.863}{17} = 0.0507\,m^3/kg$$

$$p_2 = \frac{0.283 \times 956}{0.0507} = 5.33\,MPa$$ Ans.

$$MEP = \frac{w_{net}}{v_1 - v_2},$$

where $w_{net} = q_1 - q_2 = 1800 - 747.35 = 1052.65\,kJ/kg$

$$MEP = \frac{1052.65}{0.863 - 0.0507} = 1295.88\,kPa\ or = 1.295\,MPa$$ Ans.

Example 11.3 In a Brayton cycle gas turbine plant, air enters at 25°C and 1 bar. The maximum temperature in the cycle is 950°C and pressure ratio is 8. The turbine and compressor have isentropic efficiency of 85% each. Compute (i) the turbine work per kilogram of air, (ii) the compressor work per kilogram of air, (iii) the cycle efficiency, (iv) the heat supplied per kilogram of air, (v) the turbine exhaust temperature, and (vi) the back work ratio.

Solution

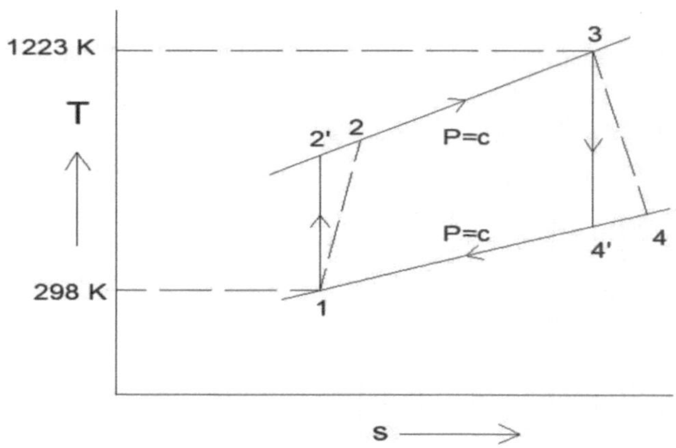

FIGURE EX. 11.3

The working fluid is assumed as an ideal gas. The processes of the ideal Brayton cycle are shown on T-s diagram.

Given $p_1 = 1$ bar, $T_1 = 298$ K

Pressure ratio $r_p = \frac{p_2}{p_1} = 8$

The isentropic efficiency of turbine, $\eta_T = 85\%$

The isentropic efficiency of compressor, $\eta_c = 85\%$

For process 1-2, $\frac{T_2^1}{T_1} = \left(\frac{P_2}{P_1}\right)^{\frac{\gamma-1}{\gamma}} \Rightarrow T_2^1 = 298 \times (8)^{\frac{0.4}{1.4}}$

$$T_2^1 = 539.79\,\text{K}$$

The maximum temperature is $T_3 = 950°C = 1223$ K Ans.

For process 3-4 $\Rightarrow \frac{T_4^1}{T_3} = \left(\frac{p_4}{p_3}\right)^{\frac{\gamma-1}{\gamma}} = \left(\frac{p_1}{p_2}\right)^{\frac{\gamma-1}{\gamma}} = \left(\frac{1}{8}\right)^{\frac{0.4}{1.4}}$

$$T_4^1 = 1223 \times \left(\frac{1}{8}\right)^{0.2857} = 675.17\,\text{K}$$

The isentropic efficiency of compressor, $\eta_C = \frac{\text{Isentropic work input}}{\text{Actual work input}} = \frac{T_2^1 - T_1}{T_2 - T_1}$

$$0.85 = \frac{539.79 - 298}{T_2 - 298} \Rightarrow T_2 = 582.46\,\text{K}$$

Work input to compressor, $w_c = h_2 - h_1 = c_p\,(T_2 - T_1)$

$$= 1.005(582.46 - 298) = 285.88 \text{ kJ/kg}$$ Ans.

The isentropic efficiency of turbine, $\eta_T = \frac{T_3 - T_4}{T_3 - T_4^1} \Rightarrow 0.85 = \frac{1223 - T_4}{1223 - 675.17}$

$$\Rightarrow T_4 = 757.345\,\text{K}$$

Work output of turbine, $w_T = h_3 - h_4 = c_p(T_3 - T_4)$

$$w_T = 1.005(1223 - 757.345) = 467.98\,\text{kJ/kg}$$ Ans.

$$q_1 = h_3 - h_2 = c_p(T_3 - T_2)$$

$$1.005(1223 - 582.46) = 643.74\,\text{kJ/kg}$$

Heat supplied $(q_1) = 643.74$ kJ/kg

Cycle efficiency $(\eta_{cycle}) = \frac{w_T - w_c}{q_1} = \frac{467.98 - 285.88}{643.74} = 28.28\%$ Ans.

Turbine exhaust temperature $(T_4) = 757.345$ K Ans.

Back work ratio $= \frac{\text{Compressor work}}{\text{Turbine work}} = \frac{285.88}{467.98} = 0.6108$ Ans.

Back work ratio indicates that 61.08% of turbine work is used for running the compressor.

Example 11.4 Air enters the compressor of a Brayton cycle gas turbine plant at 103 kPa and 27°C. The pressure ratio is 5, and the heat added is 700 kJ/kg. The regenerator effectiveness is 80%, and both turbine and compressor each have an efficiency of 75%. Compute (i) the maximum temperature in the cycle and (ii) the percentage increase in the efficiency of the cycle due to regeneration.

Solution

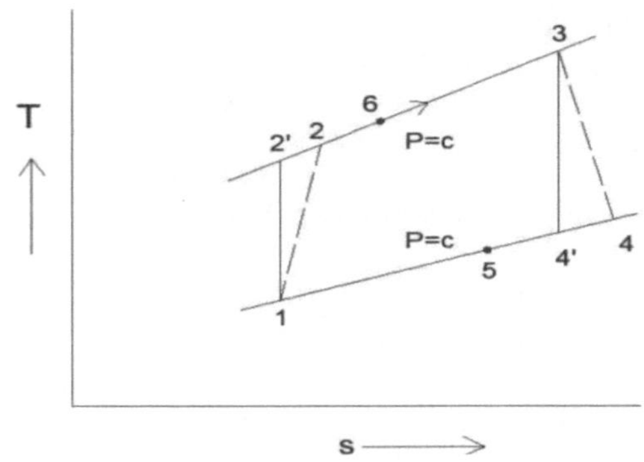

FIGURE EX. 11.4

The working fluid is assumed as ideal gas. The processes of ideal Brayton cycle are shown on T-s diagram.

$p_1 = 103$ kPa $\quad T_1 = 300$ K

$q_1 = 700$ kJ/kg $\quad r_p = 5$

The turbine and compressor have same efficiency, that is, $\eta_T = \eta_c = 75\%$.

Without Regenerator

$$\text{For process 1-2,} \frac{T_2^1}{T_1} = \left(\frac{p_2}{p_1}\right)^{\frac{\gamma-1}{\gamma}}$$

$$\Rightarrow T_2^1 = 300(5)^{\frac{0.4}{1.4}} = 474.59\,\text{K}$$

$$\text{For process 3-4,} \frac{T_3}{T_4^1} = \left(\frac{p_3}{p_4}\right)^{\frac{\gamma-1}{\gamma}} = \left(\frac{p_2}{p_1}\right)^{\frac{\gamma-1}{\gamma}} \Rightarrow T_4^1 = \frac{1229.29}{(5)^{\frac{0.4}{1.4}}} = 777.28\,\text{K}$$

$$(\text{Since } p_3 = p_2 \text{ and } p_4 = p_1)$$

$$q_1 = h_3 - h_2 = c_p(T_3 - T_2) \Rightarrow 700 = 1.005(T_3 - 532.78)$$

The maximum temperature in the cycle, $T_3 = 1229.29$ K Ans.

$$\text{The compressor efficiency,} \eta_c = \eta_c = \frac{T_2^1 - T_1}{T_2 - T_1} \Rightarrow 0.75 = \frac{474.59 - 300}{T_2 - 300}$$

$T_2 = 532.78$ K

$$\text{Turbine efficiency } (\eta_T) = \frac{T_3 - T_4}{T_3 - T_4^1} \Rightarrow 0.75 = \frac{1229.29 - T_4}{1229.29 - 777.28}$$

$T_4 = 890.28$ K

$$\text{Cycle efficiency } (\eta_{cycle}) = \frac{w_T - w_c}{q_1}$$

Where turbine work, $w_T = h_3 - h_4 = c_p(T_3 - T_4) = 1.005(1229.29 - 890.28) = 340.70\,\text{kJ/kg}$

Work input for compressor,

$$w_c = h_2 - h_1 = c_p(T_2 - T_1) = 1.005(532.78 - 300) = 233.94\,\text{kJ/kg}$$

$$\therefore \eta = \frac{340.70 - 233.94}{700} = 15.25\%$$ Ans.

With Regenerator

$$\text{Effectiveness of regenerator} = \frac{T_6 - T_2}{T_4 - T_2} = 0.80$$

$$\Rightarrow \frac{T_6 - 532.78}{890.28 - 532.78} = 0.8 \Rightarrow T_6 = 818.78\,\text{K}$$

$$q_1 = h_3 - h_6 = c_p(T_3 - T_6) = 1.005(1229.29 - 818.78) = 412.56\,\text{kJ/kg}$$

$$\text{Cycle efficiency}, \eta = \frac{w_{net}}{q_1} = \frac{340.70 - 233.94}{412.56} = 25.87\% \ (w_{net}\ \text{remains same})$$

$$\%\ \text{Increase in efficiency due to regeneration} = \frac{25.87 - 15.25}{15.25} = 69.63\%$$ Ans.

Regenerator recovers some of the thermal energy of the exhaust gases, and as a result, it helps increasing the thermal efficiency of the plant.

Example 11.5 Air enters the gas turbine plant at 101.32 kPa and 27°C. The air, after compression, is passed to the combustion chamber where heat is added to it so that its temperature becomes 750°C. The exhaust gases, after expansion in a high pressure (H.P) turbine, are reheated to 750°C and then expanded in a low pressure (L.P) turbine. The pressure ratio is 7. The compressor has an efficiency of 80% and both turbines each has an efficiency of 85%. Compute the efficiency of the cycle.

Solution

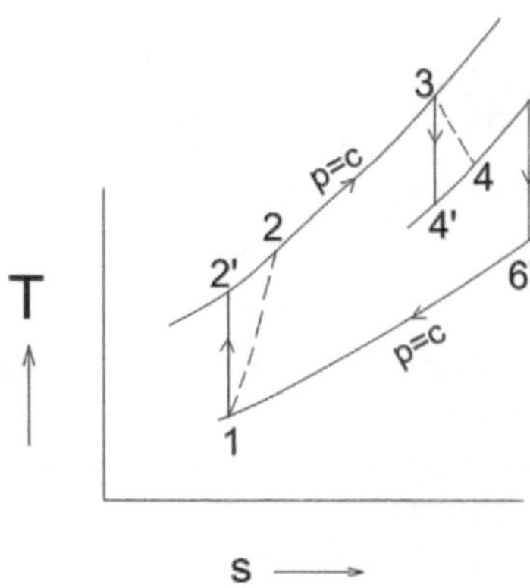

FIGURE EX. 11.5

The working fluid is assumed as an ideal gas. The processes of the ideal Brayton cycle are shown on T-s diagram (Figure Ex. 11.5).

Given $p_1 = 101.32$ kPa $T_1 = 300$ K

$r_p = 7$ $T_3 = T_5 = 1023$ K

$\eta_{T,H.P} = \eta_{T,L.P} = 85\%$ $\eta_c = 80\%$

For process 1-2, $\frac{T_2^1}{T_1} = \left(\frac{p_2}{p_1}\right)^{\frac{\gamma-1}{\gamma}} \Rightarrow T_2^1 = T_1\left(\frac{p_2}{p_1}\right)^{\frac{\gamma-1}{\gamma}} = 300(7)^{\frac{0.4}{1.4}} = 523\,\text{K}$

The isentropic efficiency of compressor, $(\eta_c) = \frac{T_2^1 - T_1}{T_2 - T_1} \Rightarrow 0.80 = \frac{523 - 300}{T_2 - 300} \Rightarrow T_2 = 578\,\text{K}$

Work of compression $(w_c) = c_p(T_2 - T_1) = 1.005(578 - 300) = 278\,\text{kJ/kg}$

Work of compression (w_c) = Work output of high pressure turbine, $w_{out,H.P}$

278 kJ/kg $= c_p\,(T_3 - T_4) = 1.005\,(1023 - T_4) \Rightarrow T_4 = 747$ K

The isentropic efficiency of high pressure turbine, $\eta_{T,H.P} = \frac{T_3 - T_4}{T_3 - T_4^1} \Rightarrow 0.85$

$= \frac{1023 - 747}{1023 - T_4^1} \Rightarrow T_4^1 = 699\,\text{K}$

For process 3-4, $\frac{T_3}{T_4^1} = \left(\frac{p_3}{p_4}\right)^{\frac{\gamma-1}{\gamma}} \Rightarrow p_4 = \frac{p_3}{\left(\frac{T_3}{T_4}\right)^{\frac{\gamma}{\gamma-1}}} \Rightarrow p_4 = 1.82\,\text{bar}$

Where $p_4 = p_5$, $p_2 = p_3$ and $p_1 = p_6$

For process 5-6, $\frac{T_5}{T_6^1} = \left(\frac{p_5}{p_6}\right)^{\frac{\gamma-1}{\gamma}} \Rightarrow T_6^1 = 865\,\text{K}$

The isentropic efficiency of low pressure turbine, $\eta_{T,L.P} = \frac{T_5 - T_6}{T_5 - T_6^1} \Rightarrow 0.85$

$= \frac{1023 - T_6}{1023 - 865} \Rightarrow T_6 = 889\,\text{K}$

Work output of low-pressure turbine

$w_{out,L.P} = c_p(T_5 - T_6) = 1.005(1023 - 889) = 134.67\,\text{kJ/kg}$

$q_1 = c_p\left[(T_3 - T_2) + (T_5 - T_4)\right] = 1.005\left[(1023 - 578) + (1023 - 747\right] = 724.605\,\text{kJ/kg}$

Cycle efficiency, $\eta = \frac{w_{net,out}}{q_1} = \frac{w_{out,L.P}}{q_1} = \frac{134.87}{724.605} = 18.61\%$ Ans.

Example 11.6 In a gas turbine plant, air is taken at 27°C and 1 bar and compressed in two stages with an intercooler in-between. The maximum temperature, T_{max}, in the cycle is limited to 723°C. Assuming isentropic efficiency of each stage of compressors as 85% and that of turbine as 90%, determine efficiency of the plant if air flow is 1.2 kg/s and pressure ratio in each stage is 6.

Solution

The working fluid is assumed as ideal gas. Figure Ex. 11.6 shows the processes of the ideal Brayton cycle on T-s diagram.

$$T_1 = T_3, T_2^1 = T_4^1$$

The maximum temperature $(T_{max}) = T_5 = 996$ K

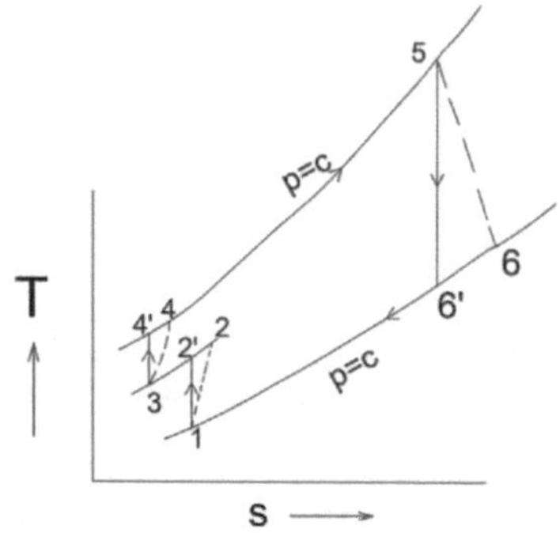

FIGURE EX. 11.6

The pressure ratio in each stage is same $\Rightarrow \frac{p_2}{p_1} = \frac{p_4}{p_3} = \sqrt{\frac{p_4}{p_1}} = \sqrt{6} = 2.5$

For process 1-2, $\frac{T_2^1}{T_1} = \left(\frac{p_2}{p_1}\right)^{\frac{\gamma-1}{\gamma}} \Rightarrow T_2^1 = 300 \times (2.5)^{\frac{0.4}{1.4}} = 388\,\text{K}$

$$\eta_{\text{comp,L.P}} = \frac{T_2^1 - T_1}{T_2 - T_1} \Rightarrow 0.85 = \frac{388 - 300}{T_2 - 300} \rightarrow T_2 = 410\,\text{K}$$

The rate of work input, $\dot{W}_C$ to the two-stage compression

$= 2 \times m \times c_p (T_2 - T_1) = 2 \times 1.2 \times 1.005 \times (410 - 300) = 217.08\,\text{kW}$

$$\frac{T_5}{T_6^1} = \left(\frac{p_5}{p_6}\right)^{\frac{\gamma-1}{\gamma}} \Rightarrow \frac{996}{T_6^1} = (6)^{\frac{0.4}{1.4}} \rightarrow T_6^1 = 596.4\,\text{K}$$

$$\eta_T = \frac{T_5 - T_6}{T_5 - T_6^1} \Rightarrow 0.90 = \frac{996 - T_6}{996 - 596.4} \text{ then } T_6 = 637\,\text{K}$$

The power output of turbine $\dot{W}_T = mc_p(T_5 - T_6) = 1.2 \times 1.005 \times (996 - 637) = 432.95\,\text{kW}$

$\dot{W}_{\text{net}} = \dot{W}_T - \dot{W}_C = 432.95 - 217.08 = 215.874\,\text{kW}$

The efficiency of low pressure compressor, $\eta_{\text{comp,L.P}} = \frac{T_4^1 - T_3}{T_4 - T_3} \Rightarrow 0.85 = \frac{410 - 300}{T_4 - 300}$

$\Rightarrow T_4 = 429.41\ \text{K}$

The rate of heat input, $\dot{Q}_1 = mc_p\,(T_5 - T_4) = 1.2 \times 1.005 \times (996 - 429.41) = 683.307\ \text{kJ/s}$

Cycle efficiency $(\eta) = \frac{\dot{W}_{\text{net}}}{\dot{Q}_1} = \frac{215.874}{683.307} = 31.59\%$ Ans.

Example 11.7 A turbojet aircraft flies with a velocity 275 m/s at an altitude where temperature and pressure are 240 K and 35 kPa. Air leaves the diffuser at 45 kPa with a velocity of 20 m/s and hot combustion gases enter the turbine at 400 kPa and 1200 K. Turbine produces 450 kW of power. Assuming that the power developed by turbine is used to drive the compressor only, find (i) the pressure of exhaust gases at exit of turbine, (ii) the mass flow rate of air through compressor, (iii) thrust, and (iv) the propulsive efficiency.

Solution

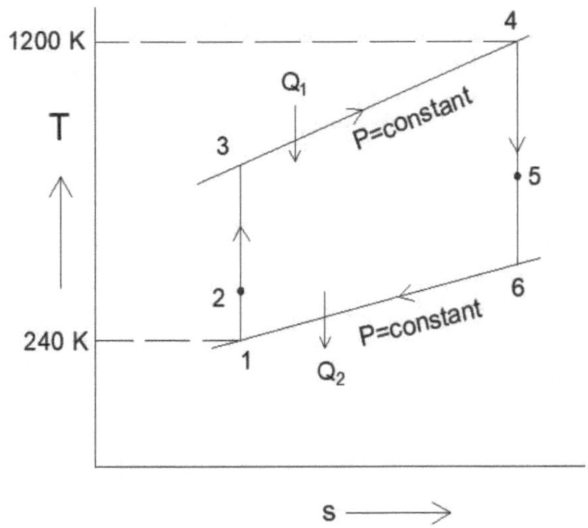

FIGURE EX. 11.7

Figure Ex. 11.7 shows the T-s diagram of ideal jet propulsion cycle.

$$P_1 = 35 \text{ kPa} \qquad T_1 = 240 \text{ K}$$

$$P_3 = P_4 = 400 \text{ kPa}, \ T_4 = 1200 \text{ K}$$

For process 1-2, that is, isentropic compression in diffuser

$$\frac{T_2}{T_1} = \left(\frac{P_2}{P_1}\right)^{\frac{\gamma-1}{\gamma}} \Rightarrow T_2 = 240\left(\frac{45}{35}\right)^{\frac{0.4}{1.4}} \to T_2 = 257.86\text{K}$$

For process 2-3, that is, isentropic compression in compressor

$$\frac{T_3}{T_2} = \left(\frac{P_3}{P_2}\right)^{\frac{\gamma-1}{\gamma}} \Rightarrow T_3 = 257.86\left(\frac{400}{45}\right)^{\frac{0.4}{1.4}} = 481.36\text{K}$$

Work of compression = work of turbine

$$\Rightarrow c_p(T_3 - T_2) = c_p(T_4 - T_5) \Rightarrow 481.36 - 257.86 = 1200 - T_5 \to T_5 = 976.5\text{K}$$

i. Pressure of exhaust gases at exit of turbine, P_5:

For process 4-5, that is, isentropic expansion in turbine

$$\frac{P_5}{P_4} = \left(\frac{T_5}{T_4}\right)^{\frac{\gamma}{\gamma-1}} \Rightarrow P_5 = 400\left(\frac{976.5}{1200}\right)^{\frac{1.4}{0.4}} \Rightarrow P_5 = 194.4\text{kPa}$$

ii. Mass flow rate of air through the compressor, $\dot{m}$:

Power input to compressor, $\dot{W}_c = c_p(h_3 - h_2) = c_p(T_3 - T_2)$

(Since Power output of turbine is equal to power input to compressor)

$450 = \dot{m}\ (481.36 - 257.86) \rightarrow \dot{m} = 2.003$ kg/s

For process 5-6, that is, isentropic expansion in nozzle

$$h_6 + \frac{V_6^2}{2} = h_5 + \frac{V_5^2}{2} \qquad (V_5 = 0)$$

$$\frac{V_6^2}{2} = h_5 - h_6 \Rightarrow V_6 = \sqrt{2c_p(T_5 - T_6)}$$

Velocity at nozzle exit, $V_6 = \sqrt{2 \times 1.005(976.5 - 598.32) \times 1000} = 871.86\,\text{m/s}$

Also for process 5-6, $\frac{T_6}{T_5} = \left(\frac{P_6}{P_5}\right)^{\frac{\gamma-1}{\gamma}} \Rightarrow T_6 = 976.5\left(\frac{35}{194.4}\right)^{\frac{0.4}{1.4}} = T_6 = 598.32\,\text{K}$

iii. Thrust $(F) = \dot{m}(V_e - V_i) = 2.003(871.86 - 275) = 1195.51\,\text{N}$

Propulsive power $(\dot{W}_P) = \dot{m}\,(V_e - V_i) \times V = 2.003\,(871.86 - 275) \times 275 \times \frac{1}{1000}$

$= 328.76\,\text{kW}$

Rate of heat input $(\dot{Q}_1) = \dot{m}\,(h_4 - h_3) = \dot{m}c_p\,(T_4 - T_3)$

$= 2.003 \times 1.005\,(1200 - 481.36) = 1446.63\,\text{kW}$

Propulsive efficiency $(\eta_P) = \frac{\text{Propulsive power}}{\text{Rate of heat input}} = \frac{\dot{W}_P}{\dot{Q}_1}$

$$\eta_p = \frac{\dot{W}_P}{\dot{Q}_1} = \frac{328.76}{1446.63} = 22.72\%$$ Ans.

It can be concluded from the above result that 22.72% of the energy input is used to propel the aircraft.

Example 11.8 A simple Brayton cycle using air as the working fluid has a pressure ratio of 10. The minimum and maximum temperatures in the cycle are 300 and 1250 K, respectively. Assuming a source temperature of 1550 K and a sink temperature of 310 K, find (i) exergy destruction associated with each process of the cycle, (ii) total exergy destruction of the cycle, and (iii) the second-law efficiency of the cycle.

Solution

Figure Ex. 11.8 shows the p-v diagram of the ideal Brayton cycle.

Processes 1-2 and 3-4 are isentropic compression and isentropic expansion, respectively, and hence involve no irreversibilities; therefore exergy destruction is zero in either case.

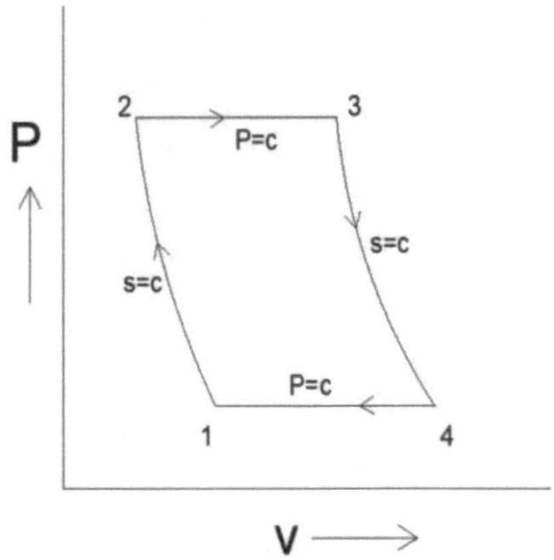

FIGURE EX. 11.8

Processes 2-3 and 4-1 are constant pressure heat addition and constant pressure heat rejection, respectively. These processes are irreversible since they involve heat transfer through a finite temperature difference.

$$\text{For process 1-2, } \frac{T_2}{T_1} = \left(\frac{P_2}{P_1}\right)^{\frac{\gamma-1}{\gamma}} \Rightarrow T_2 = 300(10)^{\frac{0.4}{1.4}} = 579.2\,\text{K}$$

$$T_1 = T_0 = 300\,\text{K}, \quad T_3 = 1250\,\text{K}$$

$$\frac{T_4}{T_3} = \left(\frac{P_4}{P_3}\right)^{\frac{\gamma-1}{\gamma}} \Rightarrow T_4 = 1250\left(\frac{1}{10}\right)^{\frac{0.4}{1.4}} = 647.45\,\text{K}$$

$$\text{Heat added}(q_1) = c_p(T_3 - T_2) = 1.005(1250 - 579.2) = 674.15\,\text{kJ/kg}$$

$$\text{Heat rejected}(q_2) = c_p(T_4 - T_1) = 1.005(647.45 - 300) = 349.19\,\text{kJ/kg}$$

From ideal gas properties of air,

s_2 at 579.2 K = 2.363 kJ/kg K and s_3 at 1250 K = 3.226 kJ/kg K

$s_1 = s_2$ and $s_3 = s_4$

For constant pressure heat addition process 2-3, $s_3 - s_2 = s_3^0 - s_2^0 - R\ln\left(\frac{P_3}{P_2}\right)$

$s_3 - s_2 = 3.226 - 2.363 - 0.287 \times \ln(1) = 0.863$ kJ/kg K (Since $P_2 = P_3$)

$$\text{For process 4-1, } s_1 - s_4 = s_2 - s_3 = -0.863\,\text{kJ/kg K}$$

i. Exergy destruction associated with each process of the cycle

$$x_{destroyed,12} = 0 \text{ and } x_{destroyed,34} = 0$$

$$x_{destroyed,23} = T_0\left((s_3 - s_2) - \frac{q_1}{T_{source}}\right) = 300\left(0.863 - \frac{674.15}{1550}\right) = 128.42\,\text{kJ/kg}$$

$$x_{destroyed,41} = T_0\left((s_1 - s_4) + \frac{q_2}{T_{sink}}\right) = 300\left(-0.863 + \frac{349.19}{310}\right) = 79.025\,kJ/kg$$

$$x_{supplied} = x_{heat,in} = \left(1 - \frac{T_0}{T_{source}}\right)q_1 = \left(1 - \frac{300}{1550}\right)674.15 = 543.67\,kJ/kg$$

ii. Total exergy destruction, $x_{destroyed} = 0 + 128.42 + 0 + 79.025 = 207.45$ kJ/kg Ans.

iii. Second – law efficiency, $\eta_{II} = 1 - \frac{x_{destroyed}}{x_{supplied}} = \left(1 - \frac{207.45}{543.67}\right) = 61.84\%$ Ans.

REVIEW QUESTIONS

11.1 Define compression ratio and pressure ratio.
11.2 What are the assumptions for air-standard cycles?
11.3 Define mean effective pressure.
11.4 Define cutoff ratio and expansion ratio.
11.5 Write down expressions for efficiencies of Stirling cycle and Ericsson cycle and explain the terms in it.
11.6 How do you compare the compression ratio range and thermal efficiency of both SI and CI engines?
11.7 Define indicated power, brake power, and mechanical efficiency.
11.8 What is air-standard efficiency?
11.9 Define bore and stroke.
11.10 What are the differences between internal combustion and external combustion engines?
11.11 What are the advantages of external combustion engines over internal combustion engines?
11.12 What are the limitations of the Stirling cycle and the Ericsson cycle?
11.13 What are assumptions for thermal efficiency of simple Brayton cycle?
11.14 What are the differences between the open and closed-cycle gas turbine power plants?
11.15 What is effectiveness of regenerator in a Brayton cycle?
11.16 What are the effects of regeneration on a Brayton cycle?
11.17 What are the effects of reheating and intercooling on a Brayton cycle?
11.18 When does a Brayton cycle approach the Ericson cycle and why?
11.19 Define effectiveness of a regenerator in the Brayton cycle.
11.20 Define back work ratio.
11.21 What are new combustion technologies for gas turbines?
11.22 What are the advantages of trapped vortex combustion (TVC)?
11.23 Why is gas turbine cycle preferred for jet propulsion?
11.24 What is thrust?
11.25 What is propulsive power?
11.26 What is propulsive efficiency?
11.27 How is aircraft propelled?

11.28 Why is axial-flow compressor preferred over centrifugal compressor in aircraft propulsion?

11.29 What is the difference between jet engine and rocket engine?

EXERCISE PROBLEMS

11.1 An air-standard Carnot cycle operates in a closed system between the temperature limits of 325 and 1050 K. The pressures before and after the isothermal compression are 160 and 320 kPa, respectively, and the net work output per cycle is 0.6 kJ. Assuming variable specific heats for air, evaluate (i) the maximum pressure in the cycle, (ii) the heat transfer to air, and (iii) the mass of air.

11.2 A Carnot cycle operates in a closed system with 0.003 kg of air. The temperature limits of the cycle are 300 and 900 K, and the minimum and maximum pressures that occur during the cycle are 20 and 2000 kPa. Assuming constant specific heats, determine the net work output per cycle.

11.3 An engine working on the Otto cycle is supplied with air at 280 kPa at 30°C. The compression ratio is 8.5. Heat supplied is 2000 kJ/kg. Calculate (i) the maximum pressure and temperature of the cycle, (ii) the cycle efficiency, and (iii) mean effective pressure. (Take $c_p = 1.005$ and $c_v = 0.718$, $R = 0.287$ kJ/kg K).

11.4 In an air-standard Diesel cycle, the compression ratio is 20 and at the beginning of isentropic compression, the temperature is 25°C and the pressure is 0.1 MPa. Heat is added until temperature at the end of isobaric process is 1500°C. Determine (i) the cutoff ratio, (ii) heat supplied, (iii) the efficiency, and (iv) the mean effective pressure.

11.5 An ideal diesel cycle with air as the working fluid has a compression ratio of 18 and the cut off ratio of 2. At the beginning of compression, the air is at 100 kPa, 27°C, and 1917 cm^3. Determine (i) the pressure and temperature at each point, (ii) the mean effective pressure, (iii) the thermal efficiency, and (iv) the work done by the cycle.

11.6 An engine of 250 mm bore and 375 mm stroke works on the Otto cycle. The clearance volume is 0.00263 m^3. The initial pressure and temperature are 1 bar and 50°C. If the maximum pressure is limited to 25 bar, find the following: (i) air-standard efficiency of the cycle and (ii) the mean effective pressure of the cycle. Assume ideal conditions.

11.7 The minimum pressure and temperature in an Otto cycle are 0.1 MPa and 27°C. The amount of heat added to the air per cycle is 1500 kJ/kg. (i) Determine the pressures and temperatures at all points of the air-standard Otto cycle and (ii) calculate the specific work and thermal efficiency of the cycle for a compression ratio of 8:1. Take $c_v = 0.718$ kJ/kg K and $\Upsilon = 1.4$.

11.8 The engine working on the Otto cycle has a volume of 0.45 m^3, pressure 1 bar, and temperature 30°C at the beginning of the compression stroke. At the end of the compression stroke, the pressure is 11 bar and 210 kJ of heat is added at isochoric process. Determine (i) pressures temperatures and volumes at salient points in the cycle, (ii) the percentage clearance, (iii) the mean effective pressure, and (iv) the ideal power developed by the engine.

11.9 An ideal Ericsson engine, using hydrogen as the working fluid, operates between temperature limits of 350 and 1700 K and pressure limits of 150 and 1500 kPa. The mass flow rate of hydrogen used in the cycle is 10 kg/s. Evaluate (i) the thermal efficiency of the cycle, (ii) the heat transfer rate in the regenerator, and (iii) the power delivered.

11.10 An ideal Stirling engine, using helium as the working fluid, operates between temperature limits of 350 and 2200 K and pressure limits of 120 kPa and 2.5 MPa. The mass flow rate of the helium used in the cycle is 5.5 kg/s. Evaluate (i) the thermal efficiency of the cycle, (ii) the amount of heat transfer in the regenerator, and (iii) the work output per cycle.

11.11 Air enters the compressor of a gas turbine operating on the Brayton cycle at 1 bar, 30°C. The pressure ratio in the cycle is 8. Calculate (i) the maximum temperature in the cycle and (ii) the efficiency of the cycle. Assume work of turbine, $W_T = 2.5W_C$ and $\Upsilon = 1.4$.

11.12 Air enters the compressor of an ideal air-standard Brayton cycle at 100 kPa, 300 K, with a volumetric flow rate of 5 m^3/s. The compressor pressure ratio is 10. The turbine inlet temperature is 1400 K. Determine (i) the thermal efficiency of the cycle, (ii) the back work ratio, and (iii) the net power developed, in kW.

11.13 In an ideal jet propulsion cycle, air enters the compressor at 1 atm and 20°C. The pressure of air leaving the compressor is 8 atm and the maximum temperature is 870°C. The air expands in the turbine to such a pressure that the turbine work is equal to the compressor work, on leaving the turbine, the air expands isentropically in a nozzle at 1 atm. (i) Determine the velocity of air leaving the nozzle and (ii) the isentropic efficiency.

11.14 In a gas turbine, the compressor is driven by high pressure turbine (HPT). The exhaust from the HPT goes to a free shaft low pressure turbine (LPT) which runs the load. The air flow rate is 20 kg/s, the maximum and minimum temperatures are 1500 and 450 K respectively. The compressor ratio is 5. Calculate (i) the pressure ratio of LPT and temperature of exhaust gasses and (ii) the efficiency of the cycle assume compressor and turbine are isentropic and take $c_p = 1$ kJ/kg K and $\Upsilon = 1.4$.

11.15 A regenerative gas turbine with intercooling and reheating operates at steady state. Air enters the compressor at 0.1 MPa and 315 K. With a mass flow rate of 5 kg/s. The pressure ratio across two stage compressor as well as turbine is 15. The intercooler and reheater operates at 0.3 MPa. At the inlets to the turbine stages, the temperature is 1500 K. The temperature at the inlet to the second compressor stage is 300 K. The efficiency of each compressor and turbine stage is 80%. The regenerator effectiveness is 80%. Determine (i) the thermal efficiency, (ii) the back work ratio, and (iii) the net power developed.

11.16 Air enters the compressor of a gas turbine plant operating on the Brayton cycle at 1 bar, 30°C. The pressure ratio in the cycle is 5. Calculate (i) the maximum temperature in the cycle and (ii) the cycle efficiency. Assume $w_t = 2.5W_c$, where w_t and w_c are the turbine and the compressor work, respectively. Take $\Upsilon = 1.4$.

11.17 An isentropic air turbine is used to supply 0.1 kg/s of air at 0.1 MN/m^2 and at 285 K to a cabin. The pressure at inlet to the turbine is 0.4 MN/m^2. Determine (i) the temperature at the turbine inlet and (ii) the power developed by the turbine. Assume $c_p = 1.0$ kJ/kg K.

11.18 A closed-cycle ideal gas turbine plant operates between temperature limits of 800°C and 30°C and produces a power of 100 kW. The plant is designed such that there is no need for regenerator. A fuel of calorific value = 45,000 kJ/kg is used. Calculate (i) the mass flow rate of air through the plant and (ii) the rate of fuel consumption. Assume $c_p = 1$ kJ/kg K and $\Upsilon = 1.4$.

11.19 Find the required air fuel ratio and efficiency of the cycle in a gas turbine cycle whose turbine and compressor efficiencies are 86% and 80%, respectively. The maximum cycle temperature is 900°C. The working fluid can be taken as air ($c_p = 1$ kJ/kg K and $\Upsilon = 1.4$), which enters the compressor at 1 bar and 30°C. The pressure ratio is 5. The fuel has a calorific value of 42,000 kJ/kg. There is loss of 10% of calorific value in the combustion chamber.

11.20 In a regenerative gas-turbine power plant with two stages of compression and two stages of expansion, the overall pressure ratio of the cycle is 8. The air enters each stage of the compressor at 298 K and each stage of the turbine at 1100 K. Assuming variable specific heats, evaluate the minimum mass flow rate of air needed to develop a net power output of 100 MW.

11.21 Air enters the compressor of a cold air-standard Brayton cycle with regeneration and reheat at 103 kPa, 310 K, with a mass flow rate of 7.2 kg/s. The compressor pressure ratio is 9, and the inlet temperature for each turbine stage is 1350 K. The pressure ratios across each turbine stage are equal. The turbine stages and compressor each have isentropic efficiencies of 80% and the regenerator effectiveness is 85%. For $k = 1.4$, determine (i) the thermal efficiency of the cycle, (ii) the back work ratio, and (iii) the net power developed, in kW.

11.22 A two-stage air compressor operates at steady state, compressing 12 m^3/min of air from 103 kPa, 298 K, to 1050 kPa. An intercooler between the two stages cools the air to 298 K at a constant pressure of 320 kPa. The compression processes are isentropic. Determine (i) the power required to run the compressor, in kW, and (ii) compare the result to the power required for isentropic compression from the same inlet state to the same final pressure.

11.23 In a turbojet engine of an aircraft, air enters at 25 kPa, 240 K, and 250 m/s. The air mass flow rate is 27 kg/s. The compressor pressure ratio is 13, the turbine inlet temperature is 1350 K, and air exits the nozzle at 25 kPa. The diffuser and nozzle processes are isentropic, the compressor and turbine each have isentropic efficiencies of 85%, and there is no pressure drop for flow through the combustor. Kinetic energy is negligible everywhere except at the diffuser inlet and the nozzle exit. On the basis of air-standard analysis, find (i) the pressures, in kPa, and temperatures, in K, at each principal state, (ii) the rate of heat addition to the air passing through the combustor, in kW, and (iii) the velocity at the nozzle exit, in m/s.

11.24 A turbojet aircraft flies with a velocity of 350 m/s at an altitude of 9000 m, where the ambient conditions are 30 kPa and 240 K. The pressure ratio across the compressor is 14, and the temperature at the turbine inlet is 1300 K. Air enters the compressor at a rate of 50 kg/s, and the jet fuel has a heating value of 42,000 kJ/kg. Assuming ideal operation for all components and constant specific heats for air at room temperature, calculate (i) the velocity of the exhaust gases, (ii) the propulsive power developed, and (iii) the rate of fuel consumption.

11.25 Air enters a turbojet engine at 5°C, at a rate of 15 kg/s and at a velocity of 280 m/s relative to the engine. Air is heated in the combustion chamber at a rate 14,500 kJ/s and it leaves the engine at 420°C. Find the thrust produced by this turbojet engine.

11.26 An ideal diesel engine has a compression ratio of 18 and uses air as the working fluid. The state of air at the beginning of the compression process is 100 kPa and 25°C. The maximum temperature in the cycle is 2000 K. Assuming the source temperature of 1500 K and the sink temperature of 310 K, find (i) exergy destruction associated with each process of the cycle and (ii) the second-law efficiency.

11.27 A gas-turbine power plant operates on the simple Brayton cycle with air as the working fluid and delivers 25 MW of power. The minimum and maximum temperatures in the cycle are 295 and 1000 K, respectively, and the pressure ratio is 8. Assuming an isentropic efficiency of 80% for the compressor and 85% for the turbine, the source temperature of 1500 K and the sink temperature of 310 K, find (i) exergy destruction associated with each process of the cycle and (ii) the second-law efficiency. Account for the variation of specific heats with temperature.

11.28 A regenerative Brayton cycle using air as the working fluid has a pressure ratio of 8. The minimum and maximum temperatures in the cycle are 300 and 1200 K. Assuming an isentropic efficiency of 75% for the compressor and 80% for the turbine and an effectiveness of 65% for the regenerator, the source temperature of 1500 K, and the sink temperature of 310 K, find (i) exergy destruction associated with each process of the cycle and (ii) the second-law efficiency.

DESIGN AND EXPERIMENT PROBLEMS

11.29 Earlier aircraft was powered by SI engine, while it is powered by gas turbines nowadays. Compare selection factors such as performance, power-to-weight ratio, space requirements, fuel cost and its availability, and environmental impact including global warming potential. Summarize your findings and prepare a report.

11.30 Design a 200-MW closed-cycle gas turbine power plant with air, nitrogen, and helium as possible working fluids that circulate through a nuclear power plant unit to absorb energy by heat transfer. Estimate key operating pressures and temperatures for each of the two working fluids of the gas turbine cycle and estimate the expected performance of the cycle. Suggest a working fluid and justify your recommendations.

12 Refrigeration Cycles

LEARNING OUTCOMES

After learning this chapter, students should be able to

- Analyze the ideal and actual vapor compression refrigeration cycles.
- Evaluate the factors involved in selecting the right refrigerant for a particular application.
- Evaluate the performance of innovative vapor compression refrigeration systems such cascade refrigeration, multistage refrigeration, and liquefaction of gases.
- Apply the thermodynamic analysis to operation of various refrigeration systems such as vapor compression, vapor absorption, and air refrigeration systems.
- Analyze the eco-friendly refrigerants and develop alternative energy sources such as solar energy for refrigeration systems.

12.1 REVERSED CARNOT CYCLE

Since Carnot cycle is a totally reversible cycle, all the processes constituted by the cycle can be reversed by reversing the directions of heat and work transfer interactions without changing the cycle, so that it becomes the reversed Carnot cycle. The cycle so developed can be an ideal cycle for both refrigerators and heat pumps and is called Carnot refrigeration cycle and Carnot heat pump cycle, respectively. Figure 12.1 shows the Carnot refrigerator operating on the reversed Carnot cycle. Figure 12.2 shows the T-s diagram of the reversed Carnot cycle. Heat Q_2 is absorbed by the refrigerant isothermally from the low-temperature reservoir maintained at T_2, the refrigerant is compressed isentropically with the help of work input, $W_{net,in}$, and heat Q_1 is rejected isothermally from the refrigerant to high-temperature reservoir maintained at T_1. The refrigerant is then condensed from vapor state to liquid state in the condenser.

The COPs of refrigerator and heat pump are expressed as

$$COP_R = \frac{Q_2}{W_{net,in}} = \frac{Q_2}{Q_1 - Q_2} \tag{12.1}$$

$$COP_{H.P} = \frac{Q_1}{W_{net,in}} = \frac{Q_1}{Q_1 - Q_2} \tag{12.2}$$

The COP of reversible (Carnot) refrigerator or heat pump can be developed by replacing the heat transfer ratio with the ratio of absolute temperatures of two reservoirs with which refrigerator or heat pump is communicating, and it is given as

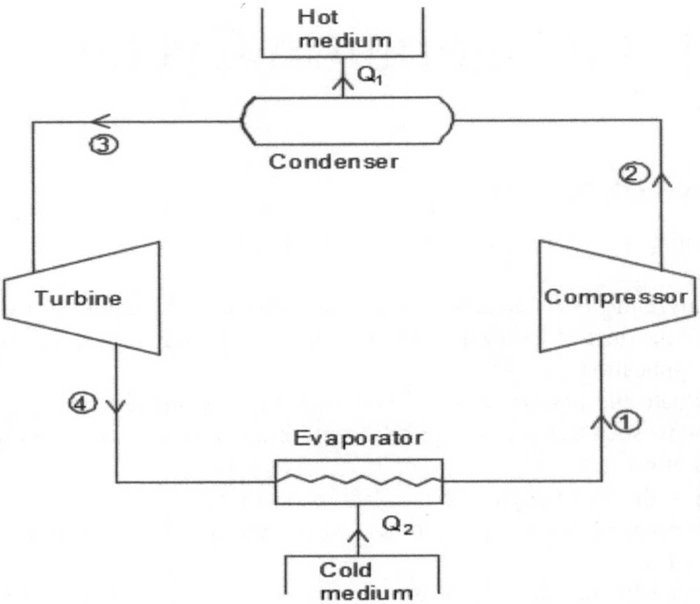

FIGURE 12.1 The Carnot refrigerator.

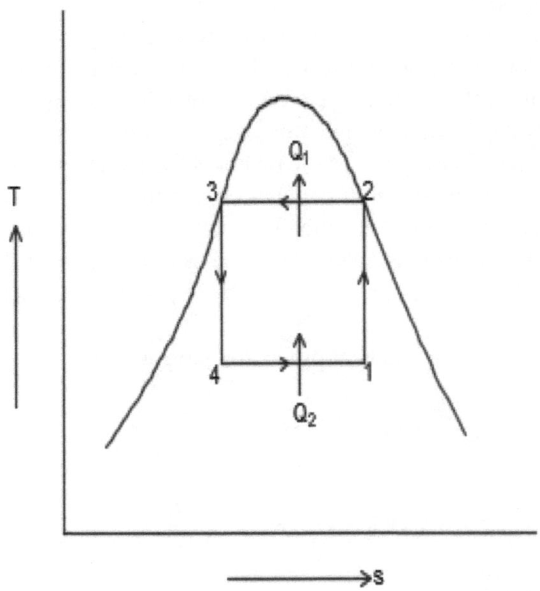

FIGURE 12.2 T-s diagram of the reversed Carnot cycle.

$$COP_{R,Rev} = \frac{T_2}{T_1 - T_2} \tag{12.3}$$

$$COP_{HP,Rev} = \frac{T_1}{T_1 - T_2} \tag{12.4}$$

Reversible refrigerator or heat pump, like the reversible heat engine, functioning between two reservoirs at temperatures T_1 and T_2 can have the highest coefficient of performance while the actual refrigerators or heat pumps functioning between the above temperatures have lower COPs. However, the reversed Carnot cycle cannot be implemented in practice due to the reason that a compressor cannot handle a two-phase (liquid–vapor) refrigerant as can be seen (process 1-2) from Figure 12.2 that the cycle is executed within the saturation region of the refrigerant. Further, if a refrigerant with a high moisture content is expanded in the turbine (process 3-4), it will damage the blades of the turbine.

12.2 REFRIGERATORS AND HEAT PUMPS

The second law of thermodynamics places limits on the performance of refrigeration and heat pump cycles as it does for power cycles. As shown in Figure 12.3, a system undergoes a cycle while exchanging heat with two thermal reservoirs at different temperatures. The energy transfers labeled on the figure are in the directions indicated by the arrows. Based on the energy conservation principle, the cycle discharges heat in the amount of Q_1 by heat transfer to the hot reservoir, equal to the sum of the heat Q_2 received by heat transfer from the cold reservoir and the net work input.

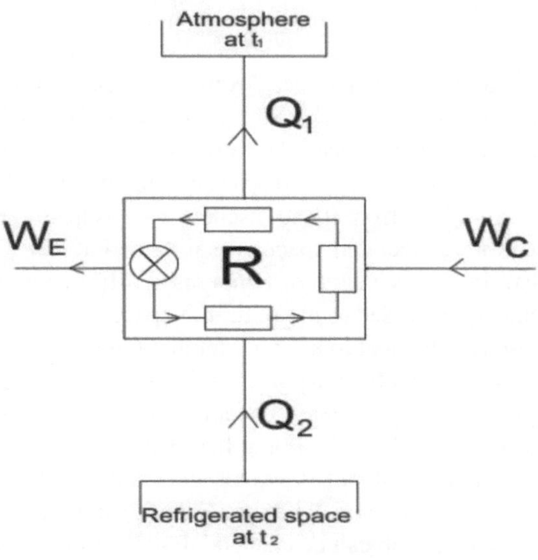

FIGURE 12.3 The refrigeration cycle.

This cycle might be a refrigeration cycle or a heat pump cycle, depending on whether its function is to remove energy Q_2 from the cold reservoir or deliver energy Q_1 to the hot reservoir.

Heat always flows in the direction of decreasing temperature, i.e., from a high-temperature body to a low-temperature one without requiring any devices. However, the reverse process, i.e., transfer of heat from a low-temperature medium to a high-temperature one, cannot occur spontaneously, and it requires special devices called refrigerators. Refrigerators are also cyclic devices like heat engines and use the refrigerant as working fluid in the refrigeration cycle. The most commonly used refrigeration cycle is the vapor compression refrigeration cycle. It comprises four main components: compressor, condenser, expansion valve, and evaporator.

The amount of heat removed from the refrigerated space is Q_2 at temperature T_2, the amount of heat rejected to the warm environment is Q_1 at temperature T_1, and $W_{net,in}$ is the net work input to the refrigerator. Q_2 and Q_1 represent magnitudes and thus are positive quantities.

The coefficient of performance (COP) of refrigeration and heat pump cycles is defined as

$$COP = \frac{\text{Desired effect}}{\text{Required input}} \quad (12.5)$$

The coefficient of performance of a refrigeration cycle (COP_R) is

$$COP_R = \frac{Q_2}{W_{net,in}} = \frac{Q_2}{Q_1 - Q_2} \quad (12.6)$$

where $W_{net,in} = W_C - W_E$.

Heat Pump

Heat pump, like refrigerator, transfers heat from a low-temperature body to a high-temperature one. Figure 12.4 shows the heat pump cycle. Though refrigerators and heat pumps operate on the same cycle, their objectives are different. The objective of a refrigerator is to keep the refrigerated space comparatively at lower temperature by continuous removal of heat from it and discharging this heat to a higher temperature medium. When the refrigerated space is maintained at low temperature, some heat will always leak into it by virtue of temperature difference, so this heat must be continuously removed in order to maintain the space at lower temperature constantly. However, discharging heat to a higher temperature medium is merely a necessary part of the operation, not the purpose. On the other hand, the objective of heat pump is to maintain a heated space at a high temperature than the surrounding environment. This is accomplished by absorbing heat from a low-temperature medium such as well water or cold outside air in winter and supplying this heat to the high-temperature medium such as a house. The same device can be operated as refrigerator and heat pump if its direction can be changed. For example, in winter conditions, a refrigerator placed in the window of a house with its door open to the cold outside air tends to do as a heat pump, since it attempts to cool the outside by absorbing heat from it and discharging this heat into the house through the coils behind it.

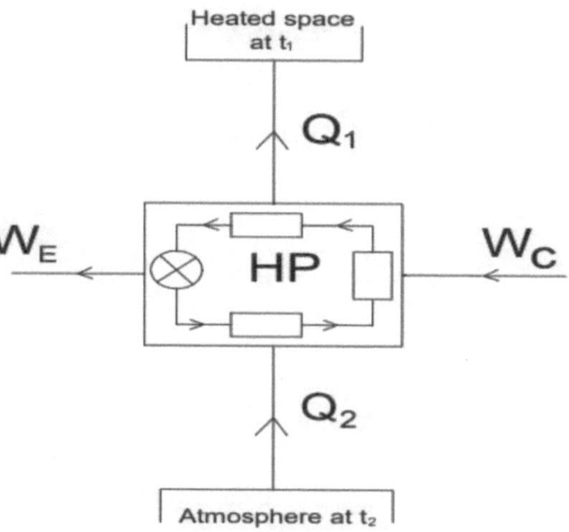

FIGURE 12.4 The heat pump cycle.

For a heat pump cycle, the coefficient of performance ($COP_{H.P}$) is

$$COP_{H.P} = \frac{Q_1}{W_{net,in}} = \frac{Q_1}{Q_1 - Q_2} \tag{12.7}$$

From Eqs. 12.6 and 12.7, it can be observed that

$$COP_{H.P} = COP_R + 1 \tag{12.8}$$

12.3 VAPOR COMPRESSION REFRIGERATION CYCLE

The limitations of the reversed Carnot cycle can be overcome by compressing the single-phase vapor and replacing the turbine with an expansion valve or capillary tube in a cycle known as vapor compression refrigeration cycle. It is a commonly used refrigeration system in domestic and commercial utility ranging from 0.5- to 200-ton capacity. In this system, the working fluid, refrigerant, readily evaporates and condenses or changes alternatively between the vapor and liquid phases without leaving the refrigeration plant. Figure 12.5 shows the schematic, T-s, and p-h diagrams of the vapor compression refrigeration system. The four basic components of the system, as shown in Figure 12.5a, are compressor, condenser, expansion valve, and evaporator.

Figure 12.6 shows the wet compression on T-s and p-h diagrams. Wet vapor enters the compressor at state 1 and is compressed reversibly and adiabatically to state 2. The vapor then flows into the condenser and is condensed to saturated liquid (state 3) at the same pressure by rejecting its heat to the surrounding cooling medium. Then, the liquid refrigerant proceeds to the throttle valve, where it is expanded to low pressure (state 4) so that temperature drops (isenthalpic expansion). For the throttling process, $h_3 = h_4$. As the refrigerant flows through the expansion valve, some of it is evaporated, and hence, a low-quality mixture emerges from the expansion valve.

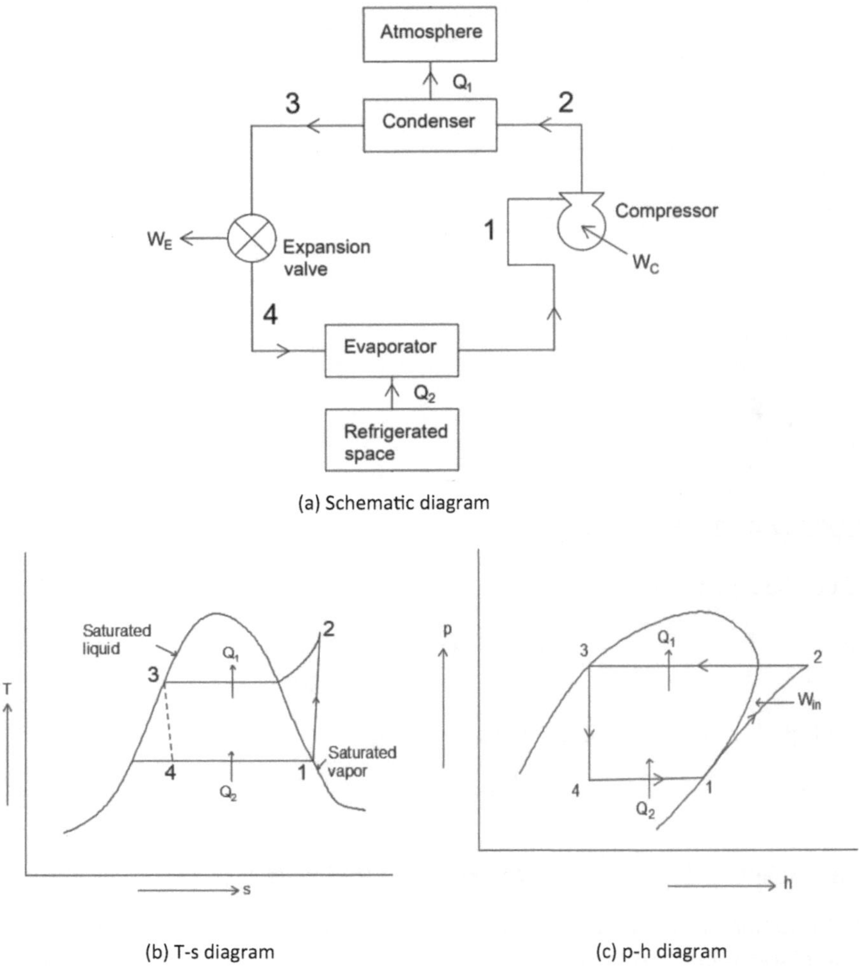

FIGURE 12.5 Vapor compression refrigeration system. (a) Schematic diagram, (b) T-s diagram, and (c) p-h diagram.

The low-pressure and low-temperature liquid refrigerant flows through an evaporator, where it absorbs heat from the space to be cooled at the same pressure and changes into vapor. The space is cooled and the vapor refrigerant enters the compressor, and the cycle repeats.

In an ideal vapor compression refrigeration system, three of the four processes are reversible while the process 3-4 is irreversible. An expansion engine is not used in the cycle as the work obtained by expanding a saturated liquid does not justify the cost of the engine. The p-h diagram is more convenient to analyze the cycle because three of the four processes that appear on it are straight lines, and further for the processes in condenser and evaporator, the heat transfer is proportional to the length of

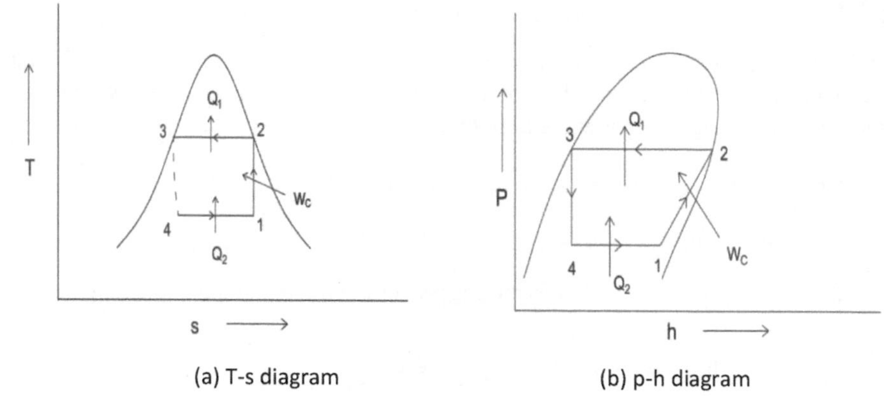

FIGURE 12.6 Vapor compression refrigeration system with wet compression. (a) T-s diagram and (b) p-h diagram.

the process paths. From the T-s diagram, it can be seen that wet compression closely resembles the reversed Carnot cycle. However, wet compression is not advisable as it reduces the performance of the compressor and causes damage to the cylinder.

Dry compression eliminates the drawback of wet compression. In dry compression, the refrigerant enters the compressor as a dry-saturated or superheated vapor. Dry compression, in particular with superheated vapor entering the compressor, makes the cycle approach the reversed Brayton cycle and, however, reduces the COP of the cycle for a given condenser and evaporator temperatures. Still, dry compression is preferred over wet compression since it improves the compressor efficiency and reduces the damage to the compressor. Also, during compression, the temperature of vapor exceeds the temperature of condensation allowing the same coolant used in condenser to be used to cool the compressor, thereby reducing the work of compression.

The vapor compression refrigeration system has some advantages compared with vapor absorption and other refrigeration systems. The refrigeration effect produced is more, it requires less refrigerant quantity per ton of refrigeration, and its COP is high. However, it entails some disadvantages, its initial cost is high, and refrigerant may be toxic and inflammable during leakage.

The vapor compression refrigeration cycle is extensively used in refrigerators, air-conditioning systems, and heat pumps. An ideal cycle consists of four processes:

1-2: Isentropic compression in a compressor
2-3: Constant-pressure heat rejection in a condenser
3-4: Throttling in an expansion device
4-1: Constant-pressure heat absorption in an evaporator

In an ideal cycle, the refrigerant enters the compressor as saturated vapor and is compressed isentropically to the condenser pressure, and there is a corresponding rise in the temperature of the refrigerant.

12.3.1 COP of Vapor Compression Refrigeration System

The COP of a vapor compression refrigeration system on a unit mass basis can be derived by considering the flow of 1 kg of refrigerant in a steady flow process, neglecting the changes in kinetic and potential energies.

$$\text{Compressor}\quad h_1 + w_c = h_2 \Rightarrow w_c = h_2 - h_1\ (\text{kJ/kg})$$

$$\text{Condenser}\quad h_2 = q_1 + h_3 \Rightarrow q_1 = h_2 - h_3\ (\text{kJ/kg})$$

$$\text{Expansion valve}\quad h_3 = h_4\ (\text{Throttling process, enthalpy remains constant})$$

$$(h_f)_{p1} = (h_f)_{p2} + x_4 (h_{fg})_{p2} \Rightarrow x_4 = \frac{(h_f)_{p1} - (h_f)_{p2}}{(h_{fg})_{p2}}$$

where x_4 is the quality of the refrigerant at the inlet to the evaporator.

$$\text{Evaporator}\quad h_4 + q_2 = h_1 \Rightarrow q_2 = h_1 - h_4\ (\text{kJ/kg})$$

Then, the COP of the vapor compression refrigeration system is

$$COP_R = \frac{\text{Desired effect}}{\text{Required input}} = \frac{q_2}{w_{net,in}} = \frac{h_1 - h_4}{h_2 - h_1} \tag{12.9}$$

where q_2 is the amount of heat removed from the cold refrigerated space per unit mass flow rate of the refrigerant, which is called refrigerating effect. If $\dot{m}$ is the mass flow rate of the refrigerant in kg/s, then the rate of heat removed from the cold refrigerated space is

$$\dot{Q}_2 = \dot{m}(h_1 - h_4)(\text{kJ/s}) = \dot{m}(h_1 - h_4) \times 3600\,(\text{kJ/h}) \tag{12.10}$$

Ton of refrigeration is defined as the rate of heat removal from the surroundings equivalent to the heat required for melting 1 tonne of ice in 1 day. One tonne is equivalent to the heat removal at the rate of [(1000 × 336)/24] 14,000 kJ/h (latent heat of fusion of ice is 336 kJ/kg).

$$1\ \text{Tonne of refrigeration} = 14{,}000\ \text{kJ/h or } 230\ \text{kJ/min or } 3.83\ \text{kJ/s}$$

$$\text{Then the capacity of the refrigerating plant} = \frac{\dot{m}(h_1 - h_4) \times 3600}{14{,}000}\ \text{tonnes} \tag{12.11}$$

$$\text{The rate of heat removal from the condenser, } \dot{Q}_1 = \dot{m}(h_2 - h_3)(\text{kJ/s}) \tag{12.12}$$

$$\text{The rate of work input required for the compressor, } \dot{W}_c = \dot{m}(h_2 - h_1)(\text{kJ/s}) \tag{12.13}$$

12.3.2 Exergy Analysis of Vapor Compression Refrigeration Cycle

It was mentioned earlier that the efficiency of all the actual power and refrigeration cycles is less than that of the ideal ones due to the irreversibilities associated with the components forming the process. In this section, exergy destruction and second-law

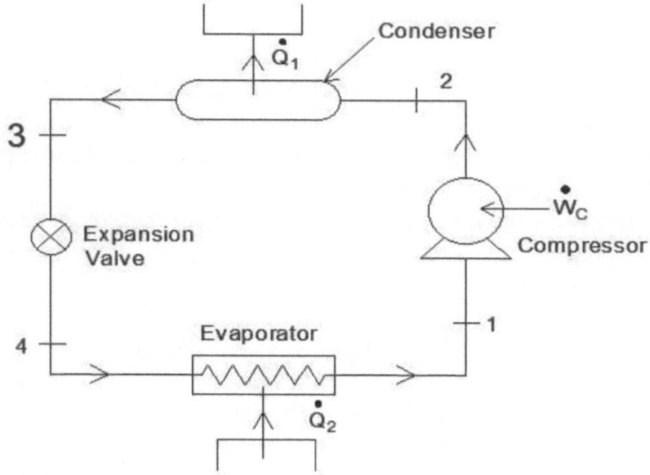

FIGURE 12.7 Vapor compression refrigeration system used for exergy analysis.

efficiency of the vapor compression refrigeration system are presented. Figure 12.7 shows the vapor compression refrigeration system used for exergy analysis.

Irreversibility is caused due to the entropy generation, and hence, the rate of exergy destruction can be calculated from the entropy generation, that is

$$\dot{X}_{des} = T_0 \dot{S}_{gen} \tag{12.14}$$

where T_0 is the temperature of the surroundings and $\dot{S}_{gen}$ is the entropy generation rate. The exergy analysis of vapor compression refrigeration system includes evaluating the exergy destruction of four main components such as compressor, condenser, expansion valve, and evaporator using Eq. 12.14.

Compressor:

$$\dot{X}_{des,12} = T_0 \dot{S}_{gen,12} = T_0 \dot{m}(s_2 - s_1) \tag{12.15}$$

where $\dot{m}$ is the mass flow rate of the refrigerant and s_1 and s_2 are the entropies at the inlet and exit of the compressor, respectively.

Condenser:

$$\dot{X}_{des,23} = T_0 \dot{S}_{gen,23} = T_0 \left[\dot{m}(s_3 - s_2) + \frac{\dot{Q}_1}{T_1} \right] \tag{12.16}$$

where $\dot{Q}_1$ is the rate of heat rejection to the warm environment and T_1 is the temperature of the high-temperature medium (for a refrigerator, T_1 is equal to T_0 temperature of the surroundings). s_2 and s_3 are the entropies at the inlet and exit of the condenser, respectively.

Expansion valve:

$$\dot{X}_{des,34} = T_0 \dot{S}_{gen,34} = T_0 \dot{m}(s_4 - s_3) \tag{12.17}$$

Evaporator:

$$\dot{X}_{des,41} = T_0\dot{S}_{gen,41} = T_0\left[\dot{m}(s_1 - s_4) - \frac{\dot{Q}_2}{T_2}\right] \tag{12.18}$$

where $\dot{Q}_2$ is the rate of heat removed from the refrigerated space and T_2 is the temperature of the low-temperature medium.

The total exergy destruction of the cycle is the sum of the exergy destructions of all the four components.

Total exergy destruction is

$$\dot{X}_{des} = \dot{X}_{des,12} + \dot{X}_{des,23} + \dot{X}_{des,34} + \dot{X}_{des,41} \tag{12.19}$$

Second-law efficiency for the cycle, η_{II}, is the ratio of minimum power required (power required on a reversible basis) to actual power required.

The COP for minimum power requirement is obtained when reversible conditions are considered:

$$COP_{R,rev} = \frac{\dot{Q}_2}{\dot{W}_{c,min}} = \left(\frac{T_2}{T_0 - T_2}\right) \rightarrow \dot{W}_{c,min} = \dot{Q}_2\left(\frac{T_0 - T_2}{T_2}\right)$$

Minimum power required is

$$\dot{W}_{c,min} = \dot{Q}_2\left(\frac{T_0 - T_2}{T_2}\right) \tag{12.20}$$

Then, the second-law efficiency is

$$\eta_{II} = \frac{\dot{W}_{c,min}}{\dot{W}_c} = 1 - \frac{\dot{X}_{des}}{\dot{W}_c} \tag{12.21}$$

12.4 REFRIGERANTS

The first refrigerant used was ether in a vapor compression machine. Later on, ethyl chloride (C_2H_5Cl) was used for some time, which was replaced by ammonia in around 1875 as ammonia has better thermodynamic properties. Its low cost and higher COPs also favored its use. Sulfur dioxide (SO_2), methyl chloride (CH_3Cl), and carbon dioxide (CO_2) also emerged as refrigerants during the same period. However, with the development of Freons, a series of fluorinated hydrocarbons, also known as fluorocarbons, derived from methane and ethane, the major breakthrough occurred in the field of refrigeration. The refrigerants that have fluorine and chlorine in their molecules will have the desired properties such as a wide range of normal boiling points and being non-toxic.

Ammonia, water, and carbon dioxide are the widely used inorganic refrigerants. Chlorofluorocarbons (CFCs), the chlorofluoro derivates of methane and ethane, are the predominantly used organic refrigerants nowadays. The halogenated chlorofluorocarbons that have hydrogen in their molecules are termed hydrochlorofluorocarbons (HCFCs), and those having hydrogen without chlorine are termed hydrofluorocarbons (HFCs) and simple hydrocarbons (HCs).

The selection of right refrigerant depends on the type of application. The refrigerants R-11, R-12, R-22, R-134a, and R-502 constitute a major portion of the refrigerant's market. However, the depletion of ozone layer with the CFCs has caused a major stir in the refrigeration industry. These CFCs, while destroying the ozone layer, allow more ultraviolet rays into the earth's surface. Thus, CFCs are responsible for global warming resulting from the greenhouse effect. R-11, R-12, and R-115 are major contributors to the greenhouse effect, while R-22 contributes partially to it. Therefore, the use of CFCs as refrigerants has been banned in many countries and phased out in some more countries. Eco-friendly refrigerants that do not contribute to the greenhouse effect have been developed. Chlorine-free R-134a is one such eco-friendly refrigerant that replaces the R-12 in applications, especially in domestic refrigerators and automobile air-conditioning.

The important parameters that play a vital role in the selection of a right refrigerant are temperatures of both the refrigerated space and the surrounding environment and the type of equipment used. Usually, a temperature difference of 5°C–10°C between the refrigerant and medium is desirable to have the reasonable heat transfer rate. For example, if the refrigerated space is to be maintained at a temperature of 5°C, it would be better if the refrigerant evaporates at –5°C.

Since the refrigerant essentially undergoes a phase change process during the heat transfer process, the pressure of the refrigerant will be the saturation pressure during the heat addition and heat rejection processes. Low pressures mean large volumes and necessitate large equipment, while high pressures mean small equipment, but it must be designed to withstand higher pressures. An important requirement in refrigeration applications is pressure, and it should be well below the critical point pressure. For very small temperature applications, a cascade refrigeration system with binary fluids can be used effectively.

The type of compressor used also has an important role to play in the selection of a right refrigerant. Usually, reciprocating compressors are preferred for low specific volumes and high pressures, while centrifugal compressors are suitable for high specific volumes. Rotary compressors lie in between the reciprocating and centrifugal compressors.

In addition to the above, the other desired properties of the refrigerant are as follows: They should be non-toxic, non-flammable, non-corrosive, and available at low cost. They should also have good chemical stability and higher enthalpy of vaporization.

12.4.1 Low–Global Warming Potential (Low-GWP) Refrigerants

Global warming potential is a relative measure of the amount of heat trapped by a particular gas, when released into the atmosphere, compared to the amount of heat trapped by an equivalent mass of carbon dioxide gas. GWP values calculated over a 100-year time interval are presented here in this section. The GWP value of carbon dioxide is defined as 1. A GWP value of 500 for a particular gas, for example, indicates that the gas would trap 500 times more heat than the equivalent mass of carbon dioxide over a 100-year time period. Ozone depletion potential (ODP) of a chemical compound is the relative amount of degradation to the ozone layer it can cause when compared with CFC-11 (trichlorofluoromethane), also known as R-11, which has an ODP of 1.0. R-11 has the maximum ODP among all the chlorocarbons since it has three chlorine atoms in the molecule, while R-22 has an ODP of 0.05.

The goal of an eventual HFC phasedown is to replace present HFC refrigerants with low-GWP alternatives; however, two HFCs, in particular HFC-32 and HFC-152a, deserve consideration as viable replacement options. HFC-32 is classified as A2L and has a GWP of 677, while HFC-152a is classified as A2 and has a GWP of 138. Though these GWP values are higher than those of other single-digit-GWP alternatives, they represent a significant improvement over most current HFC refrigerants that have GWP values between 2000 and 4000. Of the two, HFC-32 is a versatile refrigerant that is predominantly suitable for air-conditioning and heat pump applications. The use of HFC-32 has accelerated in the past 5–6 years, with at least one manufacturer having announced a switch to using HFC-32 in all successive models of residential air conditioners launched in Japan beginning in late 2012. HFC-152a has been found to be an option for replacing HFC-134a in mobile vehicle air-conditioning applications, but its A2 flammability classification poses a major obstacle to widespread adoption. HFC-152a may also be a viable replacement in commercial refrigeration applications, chillers, and industrial refrigeration. Since both HFC-32 and HFC-152a have similar efficiencies to the other widely used HFC refrigerants, implementing these alternatives would not considerably reduce system efficiency.

12.4.2 Current Low-GWP Refrigerant Options

Heat ventilation, air-conditioning, and refrigeration (HVAC&R) equipments have been primarily using high-GWP HFC refrigerants since the 1990s. To comply with the global HFC phasedown targets and proposals, the industry started developing equipment that uses low-GWP alternative refrigerants. As per those regulations, an ideal refrigerant should (i) be non-toxic, (ii) be non-flammable, (iii) have zero ozone depletion potential (ODP), (iv) have zero GWP, (v) have acceptable operating pressures, and (vi) have volumetric capacity appropriate to the application. Low-GWP HFCs include hydrocarbons, ammonia, carbon dioxide, and hydrofluoroolefins (HFOs).

Hydrocarbons: The three most viable hydrocarbon refrigerants include propane, isobutane, and propylene. These hydrocarbons have GWP values of 3, and they are classified as A3 refrigerants due to their high flammability.

Ammonia: It is classified as B2 refrigerant and has a GWP value of 0. Refrigeration systems in industrial applications often use ammonia as a refrigerant. Due to its class B toxicity rating, ammonia cannot make itself as a suitable candidate for comfort conditioning applications or indoor commercial refrigeration applications.

Carbon dioxide (CO_2): It is classified as A1 (non-flammable, non-toxic) and has a GWP of 1. CO_2 has been proved to be a viable alternative for several

TABLE 12.1
Ozone Depletion and Global Warming Potential of Refrigerants

Refrigerant Number	Type	Chemical Formula	GWP (approx.)	ODP (approx.)
R-12	CFC	CCl_2F_2	10,900	1
R-11	CFC	CCl_3F	4,750	1
R-114	CFC	$CClF_2CClF_2$	10,000	1
R-113	CFC	CCl_2FCClF_2	6,130	1
R-22	HCFC	$CHClF_2$	1,810	0.055
R-134a	HFC	CH_2FCF_3	1,430	0
R-1234yf	HFC	$CF_3CF{=}CH_2$	4	0
R-410f	HFC blend	R-32, R-125 (50/50 Weight %)	1,725	0
R-407C	HFC blend	R-32, R-125, R-134a (23/25/52 weight %)	1,526	0
R-744 (CO_2)	Natural	CO_2	1	0
R-717 (NH_3)	Natural	NH_3	0	0
R-290 (propane)	Natural	C_3H_8	10	<0 (smog)
R-50 (methane)	Natural	CH_4	25	<0 (smog)
R-600 (butane)	Natural	C_4H_{10}	10	0

applications including heat pumps, water heaters, commercial refrigerated vending machines, supermarket refrigeration, secondary expansion systems, and industrial and transport refrigeration systems. Carbon dioxide is also a technically viable option in mobile vehicle air-conditioning (MVAC) systems.

Hydrofluoroolefins (HFOs): These are emerging as the most viable alternative refrigerants. Refrigerant manufacturers have developed several HFO blends specifically to some applications. HFO-1234yf and HFO1234ze are a step ahead along in development. HFO-1234yf and HFO-1234ze are both classified as A2L and have GWP values less than 1. Moreover, the performance of HFO-1234yf is almost close to that of HFC-134a. HFO-1234yf has been extensively used outside the USA for future MVAC systems, and one automobile manufacturing company based in the USA has been dedicated to using HFO-1234yf since 2013. HFO-1234yf also shows the potential as a refrigerant in chillers and commercial refrigeration applications that are currently using HFC-134a. Table 12.1 shows the ozone depletion and global warming potential of various refrigerants.

12.5 VAPOR ABSORPTION REFRIGERATION CYCLE

Vapor compression refrigeration system requires work input to compress the vapor refrigerant from the evaporator pressure to condenser pressure. If a liquid refrigerant is pumped, instead of vapor, through the same pressure difference at the same mass flow rate, it requires less work input since the steady-flow work is directly proportional to the specific volume (vapor occupies more volume than liquid). Thus, it is possible to have a refrigerant in the liquid phase as its pressure is raised. Condensing the pure refrigerant at the evaporator pressure would fail to achieve the purpose of the refrigeration system; this difficulty can be overcome by a refrigeration system in which the refrigerant is dissolved in a liquid at the evaporator pressure, and the resulting liquid is pumped to the condenser pressure, and then, the refrigerant is separated at this pressure from the liquid. The arrangement used is the vapor absorption refrigeration system.

However, these systems are economically viable when there is a source of inexpensive thermal energy such as solar energy, geothermal energy, and waste heat from steam power plants at temperatures in the range of 100°C–200°C. In this system, the refrigerant is absorbed by the transport medium. Ammonia (NH_3)–water (H_2O) system is the prominent vapor absorption refrigeration system in which ammonia is the refrigerant and water is the absorbent (transport medium). There are other absorption refrigeration systems such as water–lithium bromide and water–lithium chloride systems in which water is the refrigerant in both of them. However, these two systems, in which water serves as the refrigerant, are suitable to the applications such as air-conditioning where the minimum temperature is above the freezing point of water.

Figure 12.8 shows the vapor absorption refrigeration system. It can be observed from the figure that the main difference between the vapor absorption and vapor compression refrigeration systems lies in that the compressor of vapor compression refrigeration system is replaced by absorber, pump, generator, regenerator, valve, and rectifier in the absorption refrigeration system. The remaining is the same as the

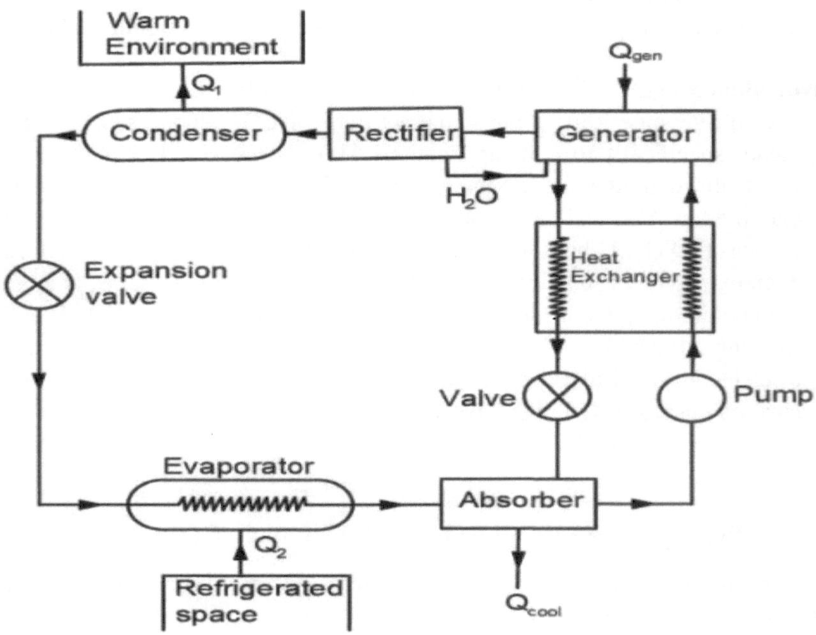

FIGURE 12.8 Vapor absorption refrigeration system.

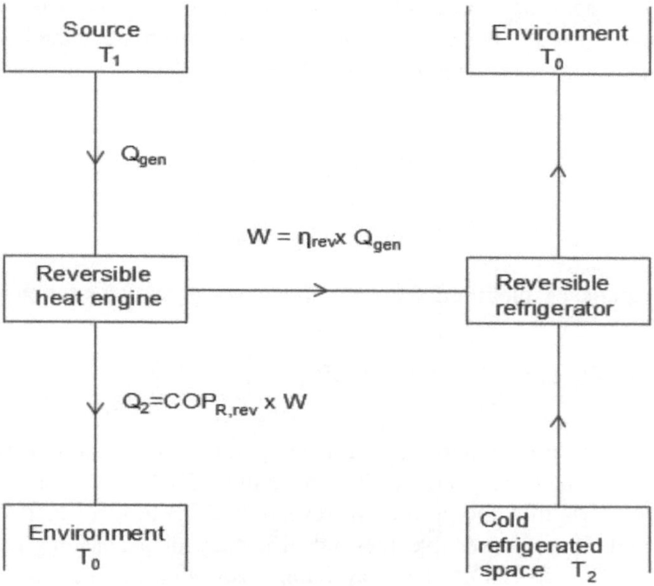

FIGURE 12.9 The maximum (reversible) COP of vapor absorption refrigeration system.

vapor compression refrigeration system. The pressure rise of NH_3 is accomplished by the use of absorber, pump, and generator or boiler as shown in the figure.

Low-pressure refrigerant vapor (NH_3) coming from the evaporator is absorbed by weak solution coming from generator in absorber (water) to form NH_3-H_2O. Water cooling is provided in the absorber to remove heat evolved due to the absorption of the refrigerant. Now, the solution in the absorber is a strong solution (rich in refrigerant). The pump draws the strong solution from the absorber and pumps it to the generator. In the generator, the strong solution is heated so that some of the solution vaporizes, and the vapor that is rich in NH_3 then passes through a rectifier where the water is separated from the vapor and sent to the generator. The high-pressure refrigerant is driven into condenser. The remaining weak solution is sent back to the absorber through a pressure-reducing valve. In the condenser, the high-pressure refrigerant vapor is converted into a high-pressure liquid refrigerant with the rejection of its heat to the cooling water. The liquid refrigerant enters into the evaporator through the expansion valve at low pressure where it absorbs latent heat from the space or products and gets converted into vapor and this vapor refrigerant is sent back to absorber, and the cycle repeats.

The COP of the vapor absorption refrigeration cycle is

$$COP_{abs} = \frac{Q_2}{Q_{gen} + w_{pump,in}} \cong \frac{Q_2}{Q_{gen}} \tag{12.22}$$

The maximum COP of the cycle is derived based on the assumption that the cycle is a reversible one; i.e., there are no irreversibilities and heat transfer is through a differential temperature difference. The reversible conditions can be achieved when the heat transfer from the source is transferred to a reversible heat engine, the work output of which is used to drive the Carnot refrigerator to remove the heat from the cold refrigerated space as shown in Figure 12.9.

$$W = \eta_{rev} \cdot Q_{gen}$$

$$Q_2 = W \times COP_{R,rev} = \eta_{rev} \cdot Q_{gen} \cdot COP_{R,rev}$$

Then, the reversible or maximum COP of vapor absorption refrigeration cycle is

$$COP_{abs,rev} = \frac{Q_2}{Q_{gen}} = \eta_{rev} \cdot COP_{R,rev} = \left(1 - \frac{T_0}{T_1}\right)\left(\frac{T_2}{T_0 - T_2}\right) \tag{12.23}$$

where T_1, T_2, and T_0 are the thermodynamic temperatures of the source, refrigerated space, and environment, respectively. The reversible COP of the vapor absorption refrigeration cycle is thus the product of reversible COP of a refrigerator working between T_0 and T_2 and reversible thermal efficiency of a heat engine operating between T_1 and T_0. The COP of actual vapor absorption refrigeration systems is always less than 1, which is lower than that determined from Eq. 12.23.

The main drawbacks of vapor absorption refrigeration systems are complexity, occupying more space, and higher cost of the equipment due to the complex structure. Therefore, they are preferred in large industrial and commercial applications. A critical analysis is necessary before deciding whether to use absorption refrigeration system, and if so, what kind of it is to be used. Certainly, the absorption refrigeration system requires comparatively less power input; for example, the power requirement of a vapor compression refrigeration system is 1000 kW (3.6×10^6 kJ/h), and if this can be replaced by an absorption refrigeration system, it requires even less than 50 kW of power, which however requires a heat input of 20×10^6 kJ/h. Therefore, the relative costs of power and heat play an important role as they vary from one location to another.

12.6 GAS CYCLE REFRIGERATION

The reversed Carnot cycle cannot be a model for actual refrigeration cycles using gaseous working substances due to the practical difficulties associated with it. If the Rankine cycle is reversed in the direction, it results in the vapor compression refrigeration cycle; similarly, if the Brayton cycle is reversed in the direction, it results in the gas refrigeration cycle. The gas refrigeration cycle, also known as the Bell-Coleman refrigeration cycle, was developed by Bell-Coleman and Light Foot by reversing Joule's cycle (i.e., all the processes described are internally reversible, and the cycle executed is the ideal gas refrigeration cycle). This cycle is used in refrigerating machines of ships to carry frozen meat, and in aircraft cooling, and with regeneration, these machines are suitable for liquefaction of gases and cryogenic applications. Figure 12.10 shows the flow and T-s diagrams of gas cycle refrigeration. The four processes of the cycle are given below:

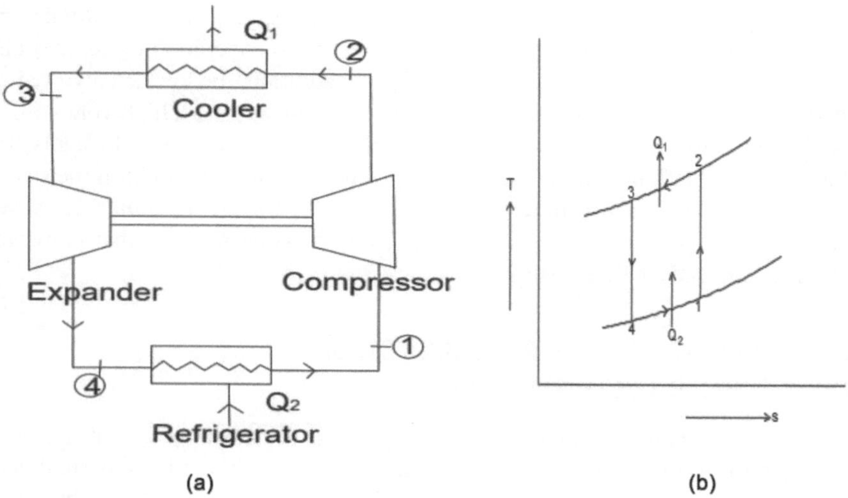

FIGURE 12.10 (a) The flow and (b) T-s diagrams of gas cycle refrigeration system.

Process 1-2: The gas is compressed isentropically from pressure P_1 to P_2 in the compressor, and consequently, the temperature increases.

Process 2-3: The high-pressure, high-temperature gas at state 2 is then cooled at constant pressure to its initial temperature by rejecting heat to the surroundings.

Process 3-4: It is an expansion process in a turbine, during which the gas temperature drops to T_4.

Process 4-1: Finally, the cold gas absorbs heat from the refrigerated space until its temperature rises to T_1.

The compression and expansion processes in actual gas refrigeration cycles deviate from the isentropic ones, and T_3 is higher than the initial temperature unless the heat exchanger is infinitely large.

The COP of the gas refrigeration cycle,

$$COP_R = \frac{q_2}{w_{net,in}} = \frac{q_2}{w_C - w_T}$$

where $q_2 = h_1 - h_4$

Turbine output, $w_T = h_3 - h_4$

Compressor input, $w_C = h_2 - h_1$

$$COP_R = \frac{h_1 - h_4}{(h_2 - h_1) - (h_3 - h_4)} \tag{12.24}$$

The main advantage of the gas refrigeration cycle is that it involves simpler and lighter components, thus making it suitable for applications where compactness is required, such as aircraft cabin cooling. Despite the advantages, there are some drawbacks with the gas refrigeration cycle. The COP of the gas refrigeration cycle is relatively lower than that of the vapor compression refrigeration cycle, and also the power requirement per unit capacity is high. In addition, the low density of air at moderate pressures necessitates either very high pressures or very high volume flow rates to maintain moderate refrigeration capacity. To overcome this difficulty, the earlier gas refrigeration systems used very high pressures in order to limit the size of the equipment; thus, they were termed *dense air refrigerating machines*. However, these machines have been replaced by vapor compression systems that employed lower pressures and smaller volumes.

12.7 INNOVATIVE VAPOR COMPRESSION REFRIGERATION SYSTEMS

The simple vapor compression refrigeration system is the most prominent refrigeration system for most of the refrigeration needs, since it is simple in construction, is less expensive, and requires less maintenance. Despite these advantages, it has some drawbacks that it is not suitable for large industrial refrigeration units in which simplicity is not the criterion but the efficiency and higher ranges of temperature are more important. It is therefore inevitable to improve the efficiency of the existing simple vapor compression refrigeration, so that they can be used advantageously for a wide range of applications. There are different alternative ways to accomplish this task. Using multistage compression is one such alternative, and operating two cycles in series, called cascading, is another approach. Apart from these methods, liquefaction of gases also proves to be a viable option.

12.7.1 Multistage Vapor Compression Refrigeration Systems

As stated earlier, simple vapor compression refrigeration system is suitable for moderate temperature ranges; however, in case of higher temperature ranges, pressure ratio becomes large, resulting in lower volumetric efficiency for the reciprocating air compressors if the cycle operates in a single stage. Moreover, high pressure ratio, in case of dry compression, also results in high discharge temperature which increases the load on the condenser. To overcome these difficulties and to improve the COP of the cycle, multistage compression with intercooling is utilized conveniently. Figure 12.11 shows the flow and p-h diagrams of the two-stage vapor compression system with a *heat exchanger*.

From Eq. 8.17 of Section 8.4, for minimum work of the two-stage compression process, the intermediate pressure, p_i, is the geometric mean of suction and delivery pressures, that is

$$p_i = \sqrt{p_1 p_2} \text{ or } \frac{p_i}{p_1} = \frac{p_2}{p_i} \tag{12.25}$$

The energy balance equation for the direct contact heat exchanger is given as

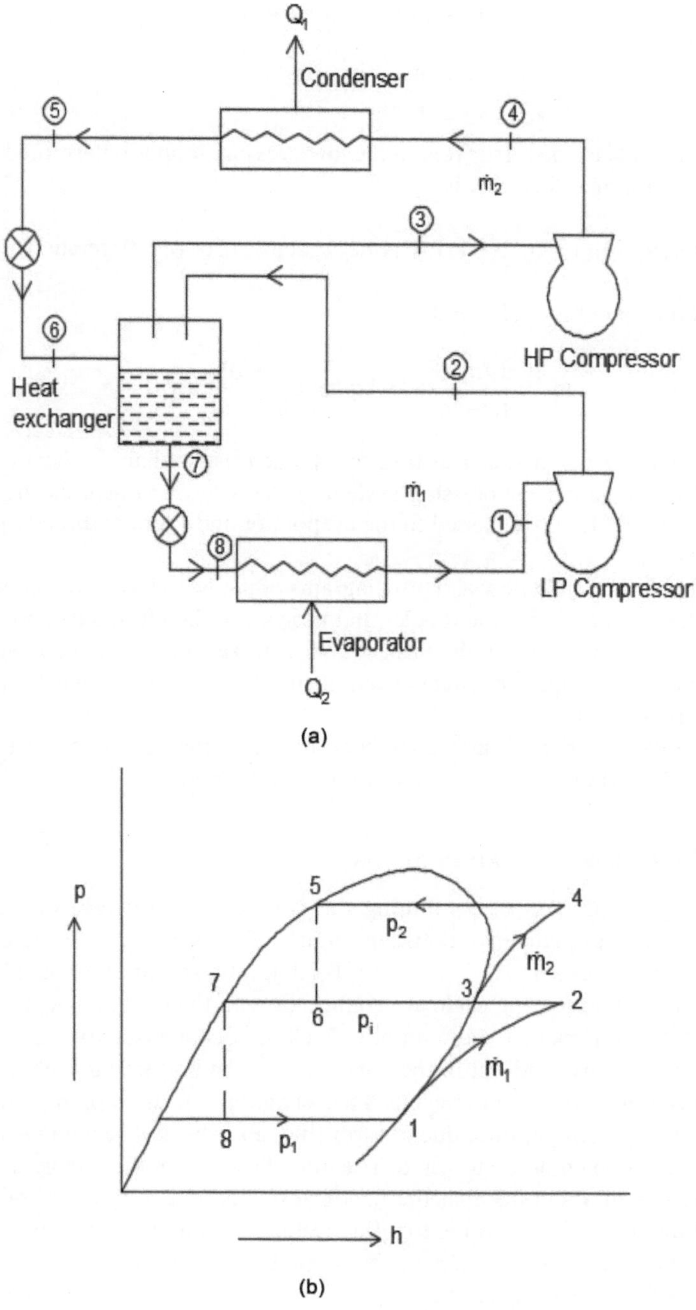

FIGURE 12.11 (a) The flow and (b) p-h diagrams of the two-stage vapor compression system.

$$\dot{m}_1 h_2 + \dot{m}_2 h_6 = \dot{m}_1 h_7 + \dot{m}_2 h_3$$

$$\frac{\dot{m}_2}{\dot{m}_1} = \frac{h_2 - h_7}{h_3 - h_6} \tag{12.26}$$

The mass flow rate of the refrigerant in the low-pressure loop is determined from the desired refrigerating effect, that is

$$\dot{m}_1 (h_1 - h_8) = 14{,}000 \times P \, (\text{P is the capacity in Ton of refrigeration})$$

The mass flow rate of the refrigerant is

$$\dot{m}_1 = \frac{14{,}000}{h_1 - h_8} \times P \, (\text{kg/h}) \, \text{or} \, \frac{3.98P}{h_1 - h_8} (\text{kg/s}) \tag{12.27}$$

The *flash chamber* has better heat transfer characteristics than the heat exchanger of the multistage vapor compression system, and also it separates the liquid from vapors so that only liquid is passed to the evaporator and vapor is directly passed to the compressor.

Figure 12.12 shows the flow and p-h diagrams of the two-stage vapor compression system with a flash chamber intercooler. In this system, the saturated vapor (state 9), being separated from liquid in the flash chamber, mixes with the superheated vapor from low-pressure compressor (state 2) and the resulting mixture (state 3) then enters the high-pressure compressor.

If a multiple number of flash chambers are used, they allow perfect pressure reduction and maximum liquid to pass through the evaporator.

12.7.2 Cascade Refrigeration System

Cascading is the process of combining the two or more refrigeration cycles in series with a heat exchanger in between. Figure 12.13 shows a two-stage cascade refrigeration cycle schematically and on a T-s diagram. In this combined cycle, the topping cycle and bottoming cycle are connected with the help of a heat exchanger that acts as an evaporator for the topping cycle and condenser for the bottoming cycle. The heat transferred out in the condenser of the bottoming cycle is utilized in the evaporator of the topping cycle. The refrigerant of the topping cycle, comparatively at lower temperature due to throttling, absorbs this heat and evaporates, and the space is cooled or refrigerated. The mass flow rate of the refrigerant in each cycle is calculated assuming that the heat transferred out in the bottoming cycle is equal to the heat utilized in the topping cycle. If $\dot{m}_1$ and $\dot{m}_2$ are respectively the mass flow rates of the refrigerant in the topping and bottoming cycles, then the ratio of mass flow rates is given as

$$\dot{m}_1 (h_5 - h_8) = \dot{m}_2 (h_2 - h_3) \Rightarrow \frac{\dot{m}_1}{\dot{m}_2} = \frac{h_2 - h_3}{h_5 - h_8} \tag{12.28}$$

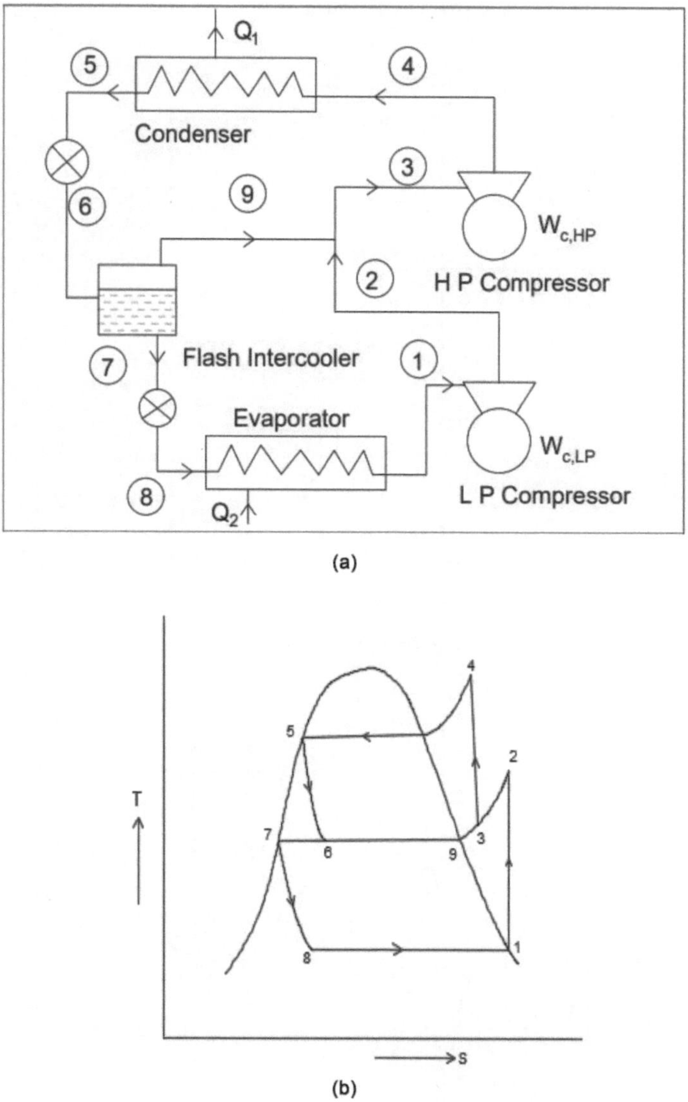

FIGURE 12.12 (a) The flow and (b) T-s diagrams of the two-stage vapor compression system with flash chamber.

The COP of the two-stage cascade refrigeration cycle is given as

$$COP_{R(cascade)} = \frac{\dot{Q}_2}{\dot{W}_{net,in}} = \frac{\dot{m}_2(h_1 - h_4)}{\dot{m}_1(h_6 - h_5) + \dot{m}_2(h_2 - h_1)} \tag{12.29}$$

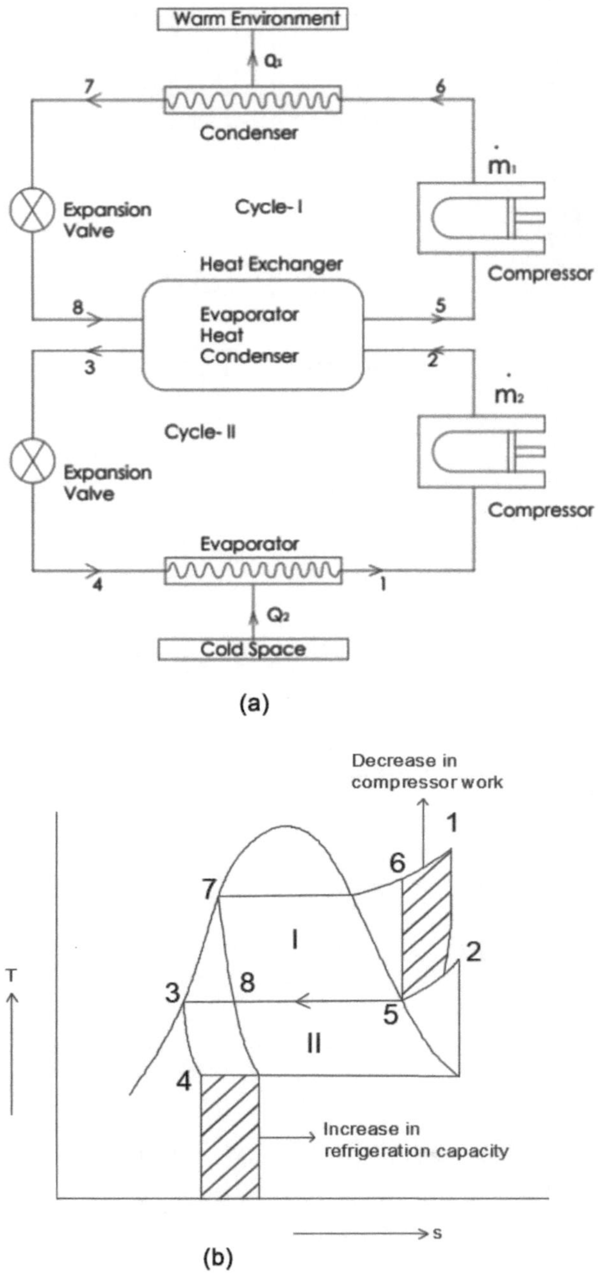

FIGURE 12.13 (a) The flow and (b) T-s diagrams of the two-stage cascade refrigeration cycle.

The major advantage with the cascade refrigeration system is that the work of compression decreases and the amount of heat removed from the refrigerated space increases, resulting in an improvement in the COP of the cycle. Further, the refrigerants of better features can be used in each cycle since there is no mixing in heat exchanger.

12.7.3 Liquefaction of Gases

Liquefaction of gases plays a vital role in engineering applications that involve temperatures below −100°C such as cryogenics (a low-temperature technology) and superconductivity. It is also used in the production of liquid propellants for rocket engines such as liquid oxygen and liquid nitrogen. The low temperatures below −100°C cannot be produced by ordinary refrigeration techniques. There are two prominent methods in which a gas can be cooled below the temperatures of that magnitude: One is by isentropic expansion through an expander and the other is by Joule–Thomson (Joule–Kelvin) expansion through a throttle valve. In the former method, temperature essentially drops, while it is not always the case with the second method. A substance exists in gas phase above its critical point only. For example, nitrogen has a critical point of −147°C; thus, it does not exist in liquid phase at atmospheric temperatures. Similarly, there are other gases such as hydrogen and helium that behave in the same way as nitrogen gas does. Therefore, the expansion of a gas through a throttle valve results in a temperature drop only when the temperature before throttling is below (maximum inversion temperature) its critical point value.

Linde–Hampson System

For the liquefaction of gas, Joule–Kelvin effect is used in this system. Figure 12.14 shows the schematic and T-s diagrams of the Linde–Hampson system. It uses a two-stage compressor with intercooling and aftercooling. The gas is compressed isothermally by a multistage compression process. First, the gas, after mixing with a portion of uncondensed gas from the previous cycle, is compressed in a low-pressure compressor, cooled in an intercooler, and compressed further in a high-pressure compressor. The high-pressure gas is cooled in an aftercooler in which a cooling agent circulates. The gas is further cooled in a regenerator with the help of uncondensed gas from the previous cycle and then throttled so that it becomes a saturated liquid–vapor mixture. The liquid is separated in a separator and collected as a desired product. The vapor is circulated through the regenerator to be used as a coolant for the high-pressure gas before it is throttled. The gas finally mixes with the fresh gas, and the cycle repeats.

12.8 ENERGY CONSERVATION IN DOMESTIC REFRIGERATORS

Among all home appliances, refrigerators are known to be the most energy-consuming devices accounting for around 30% of the total energy consumption globally. Inefficient use of energy leads to wastage of valuable resource and contributes to global warming as well. Most of the global warming effect of refrigerating systems comes from generating energy (electrical) to drive them. In 2015, the annual production of refrigerators was nearly 150 million units, with a rapid growth trend, and around 1.5 billion units are in use worldwide. Moreover, the rate of energy consumption by refrigerators is also growing at an increasing rate, particularly in developing Asia–Pacific Economic Cooperation (APEC) economies, which has a great impact on APEC energy intensity reduction goals, if there is no active support with efficient promotion of refrigerators.

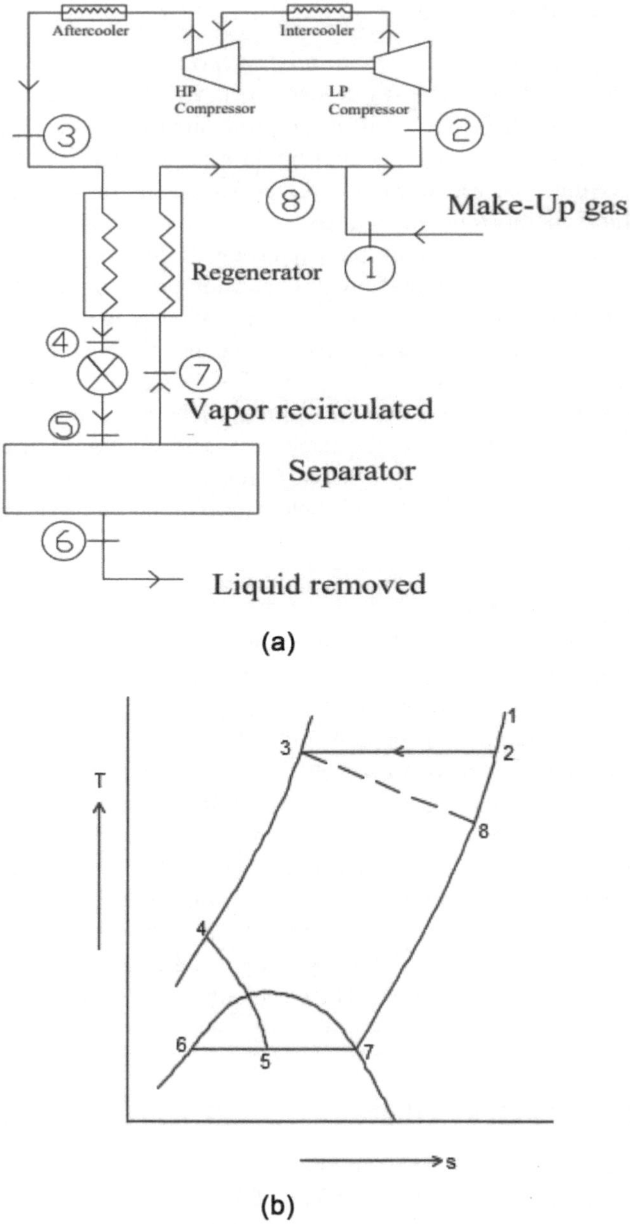

FIGURE 12.14 (a) Schematic and (b) T-s diagrams of the Linde–Hampson system of gas liquefaction.

The energy use of a refrigerator depends on the model of the refrigerator. Refrigerators with freezers on the top are known to be most energy efficient and the ones with side-mounted freezers to be less energy efficient, while the refrigerators with freezers on the bottom lie in between. Energy consumption increases with higher temperatures; that is, each 1°F variation in the ambient air temperature results in an energy consumption variation of 2.25%–2.5%. ENERGY STAR-certified refrigerators proved to be nearly 9% more energy efficient than the models that comply with the federal minimum energy efficiency standard. In the USA, nearly 190 million refrigerators and refrigerator-freezers are in use as of 2019 out of which 68 million refrigerators are about 10 years old. The energy cost per year of these old refrigerators is $5.5 billion. In addition, ENERGY STAR-certified refrigerators could also reduce the impact on the environment due to the prevention of greenhouse gas emissions.

Energy labeling plays an important role on energy savings of refrigerators, since it takes into account the factors such as the design of main components, thermal insulation, adequate thermal behavior, and alternative refrigerants, and can act as a guide to study different mechanisms. In addition to the technical characteristics of the components, there are other factors such as usage habits of the consumer and environmental conditions where the equipment is located, ambient temperature, relative humidity, and frost formation, which will influence the energy-efficient use of refrigerators.

12.8.1 Effect of Room Temperature on Energy Consumption

The information as to how the energy efficiency of refrigeration system changes under a set of operating conditions will provide a useful means of understanding the changes in energy consumption. The room temperature around the refrigerator will have a strong influence on the condensing temperature. The condenser used for the rejection of heat from the working fluid to the ambient environment will be 5–10 K warmer than the temperature of room air, and it, however, depends on the type of the condenser. Most of the refrigerators typically operate in the room temperature range of 10°C–35°C with the condensing temperatures in the range of 15°C–45°C, considerably cooler than rated conditions.

There are two primary effects of room temperature on the energy consumption of the refrigerating equipment. One is the temperature difference between compartment temperatures and the room temperature. Room temperature has two main effects on the energy consumption of refrigerating appliances. Firstly, the temperature difference between compartment temperatures and the room temperature dictates the heat gain into the appliance through the cabinet wall insulation and door seals. A second effect is that a change in room temperature affects the condensing temperature. An increase in room temperature reduces overall refrigerating system COP by increasing the difference between the evaporating and condensing temperatures. As noted previously, there are also usually fixed or variable heat loads (such as heaters, fans, and controls) that are not related to room temperature. Changes in heat gain through walls and door seals are linear and in proportion to the temperature difference between the room temperature and the compartment temperature(s). The COP of the compressor and the refrigeration system is expected to decrease linearly as the room temperature increases. An increase in room temperature increases the difference between the temperatures of evaporator and condenser leading to a reduction in the overall refrigeration system COP.

12.8.2 Effect of Thermal Load on Energy Consumption

The refrigerating capacity of a refrigerator is influenced by several factors such as the inner thermal load (mass) of the cabinet, initial temperature of food, cabinet temperature, and specific and latent heat of thermal load (water). Figure 12.15 shows how the total energy consumed by the refrigerator varies with thermal load. Since the mass is heated during the off-state of the compressor for cooling again during the on-state, energy consumption increases with an increase in thermal load. The figure shows the energy behavior for two conditions of room temperature such as 20°C and 25°C, and it can be seen that this magnitude is higher at a room temperature of 25°C when compared to that at 20°C. An increase in room temperature causes an increase in thermal gradient between surroundings and cabinet, leading to a considerable amount of heat to be transferred by conduction through the walls of the refrigerator. For instance, at 20°C, with an increase in the reference range of 0–34 kg, the energy consumption increased from 0.4 to 3.5 kWh, while it increased from 0.8 to 4.5 kWh at 25°C with an increase in reference range from 0 to 39 kg.

The energy behavior of a refrigerator is provided in Table 12.2, which shows that the time required to reach the thermal stability increases by about 4 hours when ambient temperature fluctuates from 20°C to 25°C causing the switch-on percentage of compressor to rise to 4%. This represents an increase of 0.029 kWh per operating hour. The thermal stability increases with an increase in thermal load due to the increase in ambient temperature. With the increase in thermal load, switch-on percentage and cycles decreased.

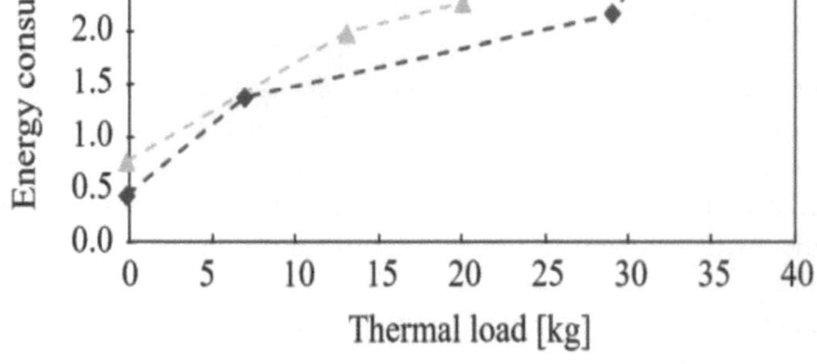

FIGURE 12.15 Energy consumption for different thermal loads. (With permission: "J.M.B Flores, D.P. Cely, M.A. Gómez-Martínez, I.H. Pérez, D.A. Rodríguez-Valderrama and Y.H. Aricapa. Thermal and energy evaluation of a domestic refrigerator under the influence of the thermal load. *Energies* 2019, 12, 400; doi: 10.3390/en12030400".)

TABLE 12.2
Energy Behavior at Different Loads and Constant Ambient Temperature

Room Temperature (°C)	Thermal Load (Kg)	Thermal Stability Time (hours)	% Switch-On	Total Energy (kWh)	Cycles (24 hours)
20	0	8	32	0.4	24
	7	21	35	1.4	21
	29	24	40	2.2	27
	32	33	37	2.8	24
	34	38	38	3.6	25
25	0	12	36	0.8	24
	13	25	39	2.0	34
	20	25	42	2.3	34
	25	38	37	3.3	26
	39	46	36	4.5	24

Source: With permission, "J.M.B Flores, D.P Cely, M.A. Gómez-Martínez, I.H Pérez, D.A. Rodríguez-Valderrama and Y.H Aricapa. Thermal and energy evaluation of a domestic refrigerator under the influence of the thermal load. *Energies* 2019, 12, 400; doi: 10.3390/en12030400".

12.8.3 Effect of Cooling of Compressor Shell with the Defrost Drips

There are certain measures toward energy conservation in refrigeration systems such as application of energy-efficient compressors, air handling units, condensers, and evaporators of high effectiveness. An effective method of energy saving is cooling of compressor shell with the defrost drips. The reduction in energy consumption not only reduces power bills for the end user of the refrigerator but also results in reduced CO_2 emissions. The quantity of defrost formation is significant while refrigeration systems are on in tropical countries where relative humidity is 70%–80% year-round. By dripping the defrost water on the compressor's shell, it could be possible to cool the compressor oil, which in turn can reduce the friction losses and winding temperature of the motor. The reduced winding temperature will reduce the compressor's ampere rating, which eventually reduces the energy consumption of the compressor.

EXAMPLE PROBLEMS

Example 12.1 A vapor compression refrigeration system uses R-12 as a refrigerant and operates between condenser and evaporator temperatures of 28°C and −10°C, respectively. The refrigeration load required is 2 kW. Determine (i) COP, (ii) swept volume of compressor, if volumetric efficiency, η_v, is 76%, and (iii) power required to drive the compressor.

Solution

From saturated tables of R-12, $p_1 = p_{sat(-10°C)} = 2.1912\,\text{bar}$

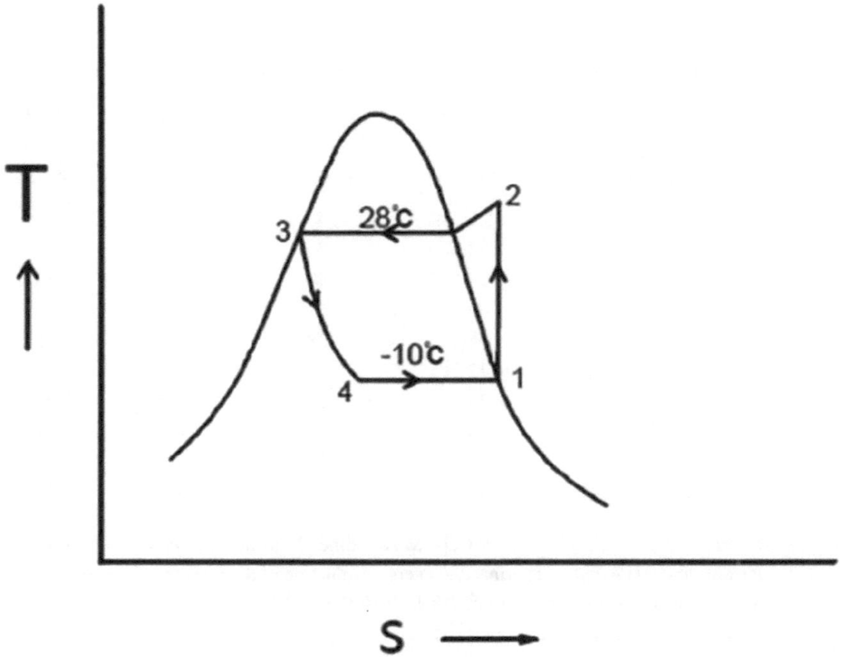

FIGURE EX. 12.1

$h_1 = 183.19$ kJ/kg, $s_1 = 0.7019$ kJ/kg K

$\nu_1 = 0.077$ m³/kg

$p_2 = p_{sat@28°C} = 7.067$ bar

$h_3 = h_4 = 62.63$ kJ/kg

At $p_2 = 7.067$ bar and $s_1 = s_2 = 0.7019$ kJ/kg K

$h_2 = 207$ kJ/kg and $t_2 = 40°$ (from interpolation)

The mass flow rate of refrigerant, $\dot{m}$, can be calculated from the refrigeration load, that is

$$\dot{m}(h_1 - h_4) = 2\,kW$$

The refrigerant flow rate, $\dot{m} = \dfrac{2}{183.19 - 62.63} = 0.016\,kg/s$

$$COP = \frac{h_1 - h_4}{h_2 - h_1} = \frac{183.19 - 62.63}{207 - 183.63} = 5.15 \qquad \text{Ans.}$$

$$\eta_v = \frac{V_a}{V_s}$$

Where $V_a = \dot{m}v_1 = 0.016 \times 0.077 = 1.23 \times 10^{-3}\,m^3/s$

$$\therefore v_s = \frac{1.23 \times 10^{-3}}{0.76} = 1.62 \times 10^{-3}\ m^3/s$$ Ans.

Power required to drive the compressor $= \dot{m}(h_2 - h_1)$

$$= 0.016 \times (207 - 183.19) = 0.3809\ kW$$ Ans.

Example 12.2 A gas refrigerating system, using air as a refrigerant, operates between −12°C and 27°C. It works on an ideal reversed Brayton cycle of pressure ratio 5 and minimum pressure 1 atm, to maintain a load of 10 tonnes. Compute (i) COP, (ii) air flow rate in kg/s, (iii) volume flow rate entering compressor in m³/s, and (iv) the maximum and minimum temperatures of the cycle.

Solution

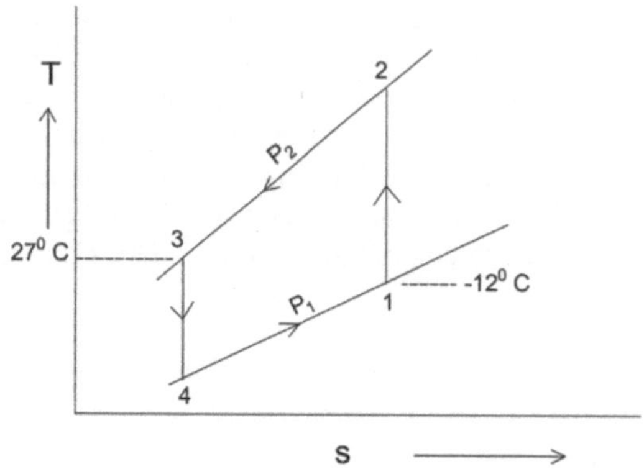

FIGURE EX. 12.2

Temperature at inlet to the compressor, $T_1 = 273 - 12 = 261$ K

Temperature at inlet to the turbine, $T_3 = 273 + 27 = 300$ K

For an ideal gas, for an isentropic process 1–2,

$$\frac{T_2}{T_1} = \left(\frac{p_2}{p_1}\right)^{\frac{\gamma-1}{\gamma}} \Rightarrow T_2 = T_1 \times \left(\frac{p_2}{p_1}\right)^{\frac{\gamma-1}{\gamma}} = 261 \times (5)^{\frac{1.4-1}{1.4}} = 412.90\,K$$

Similarly for process 3–4,

$$\frac{T_4}{T_3} = \left(\frac{p_1}{p_2}\right)^{\frac{\gamma-1}{\gamma}} \Rightarrow T_4 = 300 \times \frac{1}{(5)^{0.2857}} = 189.42\,K = -83.58°C\left(\text{since } p_1 = p_4 \text{ and } p_2 = p_3\right)$$

Refrigerating effect $= h_1 - h_4 = c_p(T_1 - T_4) = 1.005(261 - 189.42) = 71.93$ kJ/kg

Net work input, $w_{net,in} = (h_2 - h_1) - (h_3 - h_4) = c_p[(T_2 - T_1) - (T_3 - T_4)]$

$= 1.005[(412.90 - 261) - (300 - 189.42)] = 42$ kJ/kg

$$COP = \frac{\text{Refrigerating effect}}{w_{net,in}} = \frac{71.93}{42} = 1.71$$ Ans.

If $\dot{m}$ is the mass flow rate of refrigerant, then $\dot{m}(h_1 - h_4)$ is the rate of heat removal.

Again, 1 ton = 14,000 kJ/h.

$$\dot{m}(71.93) = 10 \times 14{,}000 \Rightarrow \dot{m} = 0.50\,\text{kg/s} \qquad \text{Ans.}$$

Volume flow rate, $v_a = \dot{m}v_1 = 0.4\,\text{m}^3/\text{s}$

$$\text{Where } v_1 = \frac{RT_1}{p_1} = \frac{0.287 \times 261}{100} = 0.749\,\text{m}^3/\text{kg}$$

$\therefore v_a = 0.5 \times 0.749 = 0.375$ m^3/s Ans.

Minimum temperature, $T_4 = -83.5$°C

Maximum temperature, $T_2 = 139.90$°C

Example 12.3 A food freezing system requires 20 tonnes of refrigeration at an evaporator temperature of –35°C and condenser temperature of 25°C. Refrigerant R-12 is subcooled by 4°C before entering the expansion valve, and vapor is superheated by 5°C before leaving the evaporator. A six-cylinder single-acting compressor (bore-to-stroke ratio, B/L = 1), operating at 1500 rpm, is used. Determine (i) refrigerating effect, (ii) flow rate of R-12, (iii) cylinder dimensions if volumetric efficiency is 80%, (iv) power required to drive the compressor, and (v) COP. Take $c_{pg} = 1.235$ kJ/kg K and $c_{pl} = 0.733$ kJ/kg K.

Solution

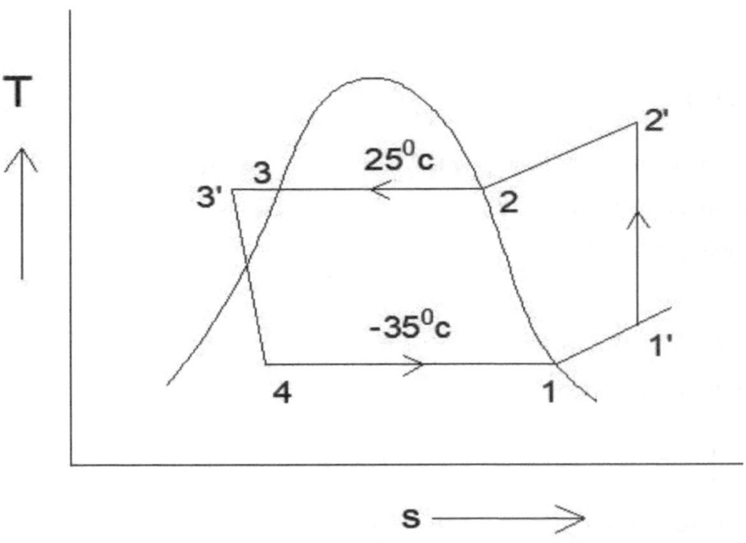

FIGURE EX. 12.3

From saturated tables of refrigerant R-12, the enthalpy and entropy values at given temperatures of 25°C and –35°C are

$$h_1(\text{at } -35°C) = h_g = 175 \text{ kJ/kg}$$

$$\nu_1 = 0.1954 \text{ m}^3/\text{kg and } s_1 = (s_g) = 0.72 \text{ kJ/kg K}$$

The enthalpy of saturated vapor, h_2 (= h_g), at 25°C = 197.73

And entropy of saturated vapor, $s_2(s_g) = 0.6868$ kJ/kg K

The enthalpy of saturated liquid, h_3, at 25°C = h_f = 59.7 kJ/kg

The vapor is superheated by 5°C before leaving the evaporator; therefore, the enthalpy of superheated vapor is

$$h_1^1 = h_1 + c_p\left(T_{sup} - T_{sat}\right) = 175 + 1.235(5) = 181.175 \text{kJ/kg}$$

And also, the entropy of superheated vapor is

$$s_1^1 = s_2^1 \Rightarrow s_1 + c_{pg}\log\left(\frac{T_1^1}{T_1}\right) = s_2 + c_{pg}\log\left(\frac{T_2^1}{T_2}\right)$$

$$= 0.72 + 1.235\log\left[\frac{-35+5+273}{-35+273}\right] = 0.6868 + 1.235\log\left(\frac{T_2^1}{298}\right)$$

$$T_2^1 = 40°C\,(\text{approx.})$$

Since the refrigerant is subcooled by 4°C before entering the expansion valve, the enthalpy of subcooled liquid refrigerant is

$$h_3^1 = h_3 + c_p\left(T_3 - T_3^1\right) = 59.7 - 0.733(4) = 56.76 \text{kJ/kg}$$

$$h_2^1 = h_2 + c_p\left(T_2^1 - T_2\right) = 197.73 + 1.235[40 - 25] = 246.54 \text{kJ/kg}$$

Refrigerating effect, $h_1 - h_4 = h_1 - h_3^1 = 175 - 56.76 = 118.24$ kJ/kg

If $\dot{m}$ is the mass flow rate of refrigerant, then $\dot{m}(h_1 - h_4)$ is the rate of heat removal.

$$\dot{m}(h_1 - h_4) = \frac{20 \times 14{,}000}{3600} \Rightarrow \dot{m} = \frac{20 \times 14{,}000}{3600} \times \frac{1}{118.24} = 0.657 \text{kg/s}$$

Volume flow rate, $v_a = \dot{m}v_1 = 0.657 \times 0.1954 = 0.1283 \text{m}^3/\text{s}$

$$\text{COP} = \frac{h_1 - h_4}{h_2^1 - h_1^1} = \frac{118.24}{65.36} = 1.809 \qquad \text{Ans.}$$

Power required to drive the compressor $= \dot{m}\left(h_2^1 - h_1^1\right) = 0.657(246.54 - 181.175) = 42.94 \text{kW}$

Volume flow rate, $v_a = \dot{m}v_1 = 0.657 \times 0.19 = 0.1283 \text{m}^3/\text{s}$

Volumetric efficiency, $\eta_v = \frac{v_a}{v_s} = 80\%$

$$v_s = \frac{0.1283}{0.8} = 0.1604 \text{m}^3/\text{s or } 9.6 \text{m}^3/\text{min}$$

Swept volume, $v_s = \frac{\pi}{4}d^2 LNn$

$$9.6 = \frac{\pi}{4}d^2 \times d \times 6 \times 1500 \rightarrow d = 11.07 \text{cm} \qquad \text{Ans.}$$

Example 12.4 In an aqua-ammonia absorption refrigeration system, heat is supplied to the generator by the steam condensed at 1.5 bar and 0.85 dry. The evaporator temperature is –20°C, and condenser temperature is 25°C. Determine the steam flow rate required when refrigeration load is 10 ton, if the cycle is operating on ideal conditions.

Solution

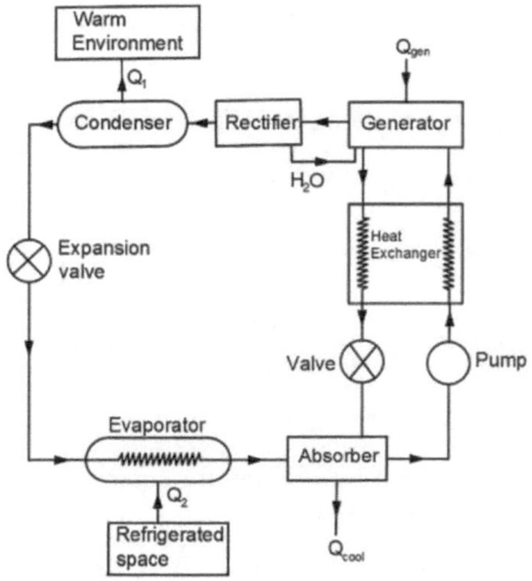

FIGURE EX. 12.4

At 1.5 bar and 0.85 dry, from steam tables:

Saturation temperature, $T_{sat} = 111.37°C$, and $h_{fg} = 2265.5$ kJ/kg

The reversible or maximum COP of vapor absorption refrigeration cycle is

$$COP_{abs,max} = \left(1 - \frac{T_0}{T_1}\right)\left(\frac{T_2}{T_0 - T_2}\right) \text{(According to Eq. 12.23)}$$

where T_1 is the generator temperature, $T_1 = T_{gen} = 111.37 + 273 = 386.7$ K

T_0 is the condenser temperature, $T_0 = 25 + 273 = 298$ K

T_2 is evaporator temperature, $T_2 = -20 + 273 = 253$ K

$$COP_{abs,max} = \left(1 - \frac{298}{386.7}\right)\left(\frac{253}{298 - 253}\right) = 1.3$$

Again for vapor absorption refrigeration cycle, $COP_{abs} = \frac{Q_2}{Q_G}$ or $COP_{abs} = \frac{\dot{Q}_2}{\dot{Q}_G}$

where $\dot{Q}_2$ is the rate of refrigeration and $\dot{Q}_G$ is the rate of heat supplied from the source.

$$\dot{Q}_G = \frac{10 \times 14000}{1.3 \times 3600} = 30\,\text{kW}$$

$$\text{Now } \dot{Q}_G = \dot{m}(h_2 - h_1)$$

Heat transferred by 1 kg of steam on condensation is

$$h_1 - h_2 = (h_f + xh_{fg}) - h_f = 0.85 \times 2226.5 = 1892.52\,\text{kJ/kg}$$

$\therefore$ The rate of steam flow required, $\dot{m} = \dfrac{\dot{Q}_G}{h_2 - h_1} = \dfrac{30}{1892.52} = 0.0158\,\text{kg/s}$ Ans.

Example 12.5 Air enters the compressor of an aircraft cooling system at 100 kPa and 0°C and is compressed to 250 kPa. The air is then cooled to 40°C in a cooler at constant pressure and expanded in a turbine to 100 kPa. Both the compressor and turbine have an isentropic efficiency of 80% each, and it is required to maintain a load of 5 tonnes. Determine (i) COP, (ii) air mass flow rate, and (iii) driving power required.

Solution

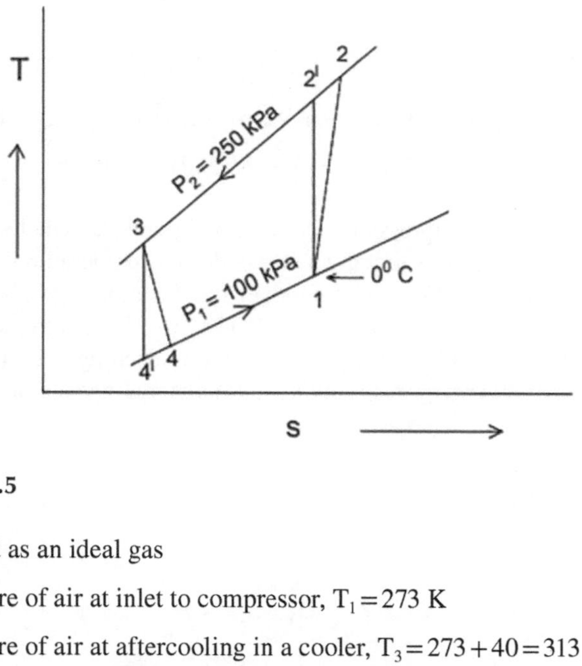

FIGURE EX. 12.5

Air is assumed as an ideal gas

The temperature of air at inlet to compressor, $T_1 = 273$ K

The temperature of air at aftercooling in a cooler, $T_3 = 273 + 40 = 313$ K

For an ideal gas, for an isentropic process 1–2,

$$\frac{T_2^1}{T_1} = \left(\frac{p_2}{p_1}\right)^{\frac{\gamma-1}{\gamma}} \Rightarrow T_2^1 = 273 \times (2.5)^{\frac{0.4}{1.4}} = 361\,\text{K}$$

$$\text{Also for an isentropic process 3–4, } \frac{T_3}{T_4^1} = \left(\frac{p_2}{p_1}\right)^{\frac{\gamma-1}{\gamma}} \Rightarrow T_4^1 = \frac{313}{(2.5)^{\frac{0.4}{1.4}}}$$

$$= 241.14\,\text{K} \left(\text{Since } p_2 = p_3 \text{ and } p_1 = p_4\right)$$

$$\eta_{comp} = \frac{T_2^1 - T_1}{T_2 - T_1} = 0.80 = \frac{361 - 273}{T_2 - 273}$$

$$\text{Then} \quad T_2 = 382\,K$$

$$\eta_{Turbine} = \frac{T_3 - T_4}{T_3 - T_4^1} \Rightarrow 0.80 = \frac{313 - T_4}{313 - 241.14}$$

$$\text{Then} \quad T_4 = 255.512\,K$$

i. $COP = \frac{Q_2}{W_{net,in}} = \frac{h_1 - h_4}{(h_2 - h_1) - (h_3 - h_4)}$

$= \frac{278 - 255.512}{(382.68 - 278) - (313 - 255.512)} = 0.48$ Ans.

ii. Mass flow rate of air $\Rightarrow \dot{m}(h_1 - h_4) = \frac{5 \times 14{,}000}{3600}$

$\dot{m}c_p(T_1 - T_4) = \frac{5 \times 14{,}000}{3600} \Rightarrow \dot{m} = \frac{19.44}{1.005(278 - 255.512)} = 0.85\,kg/s$ Ans.

iii. Driving power required $= \dot{m}\left[(h_2 - h_1) - (h_3 - h_4)\right] = \dot{m}c_p\left[(T_2 - T_1) - (T_3 - T_4)\right]$

$= 0.85 \times 1.005[(382.68 - 278) - (313 - 255.512) = 39.6\,kW$ Ans.

Example 12.6 A vapor compression refrigeration system uses R-134a as a refrigerant and operates between evaporator and condenser temperatures of −10°C and 40°C, respectively. The plant develops 5 tonnes refrigeration. The refrigerant enters the compressor at 1 bar with a flow rate of 0.08 kg/s. The isentropic efficiency of the compressor is 85%. Find (i) exergy destruction in each process, (ii) second-law efficiency, and (iii) total exergy destruction. Take surroundings temperature as 27°C.

Solution

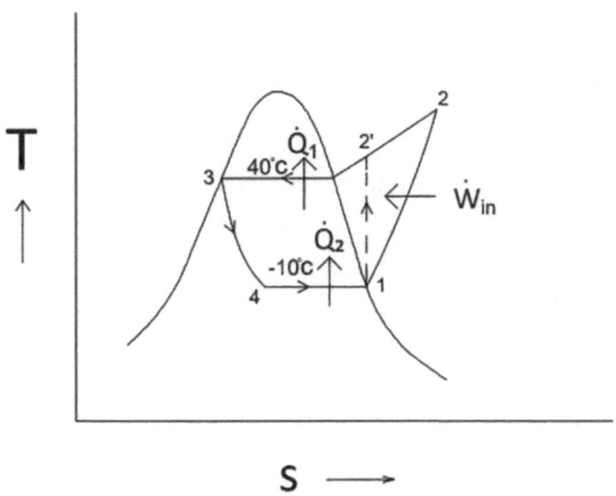

FIGURE EX. 12.6

The T-s diagram of the cycle is shown in Figure Ex. 12.6

From the properties table of R-134 a,

$T_1 = T_{sat@100kPa} = -26.4°C$

$$h_1 = 234.44\,kJ/kg, s_1 = s_2^1 = 0.9518\,kJ/kgK$$

$$P_2 = P_3 = P_{sat@40°C} = 1017.1\,kPa$$

$$\left.\begin{array}{l} P_2 = 1017.1\,kPa \\ s_2^1 = 0.9518\,kJ/kg\text{-}K \end{array}\right\} h_2^1 = 281.32\,kJ/kg$$

$$\left.\begin{array}{l} P_3 = 1017\,kPa \\ h_3 = 108\,kJ/kg \end{array}\right\} s_3 = 0.392\,kJ/kgK$$

$$\left.\begin{array}{l} P_4 = 100 \\ h_3 = h_4 = 108\,kJ/kg \end{array}\right\} s_4 = 0.437\,kJ/kgK$$

$$\text{Isentropic efficiency of compressor, } \eta_c = \frac{h_2^1 - h_1}{h_2 - h_1}$$

$$\Rightarrow 0.85 = \frac{281.32 - 234.44}{h_2 - 234.44} \Rightarrow h_2 = 289.59\,kJ/kg$$

$$\text{At}\, h_2 = 289.59\ \&\ P_2 = 1017.1\,kPa \Rightarrow s_2 = 0.973$$

$$\dot{Q}_1 = \dot{m}(h_2 - h_3) = 0.08(289.59 - 108) = 14.53\,kW$$

$$\dot{Q}_2 = \dot{m}(h_1 - h_4) = 0.08(234.44 - 108) = 10.12\,kW$$

$$\dot{W}_{in} = \dot{m}(h_2 - h_1) = 0.08(289.59 - 234.44) = 4.412\,kW$$

i. The rate of exergy destruction

$$\dot{X}_{destroyed,12} = T_0\dot{S}_{gen,12} = T_0\dot{m}(s_2 - s_1) = 300 \times 0.08 \times (0.973 - 0.9518) = 0.5088\,kW$$

$$\dot{X}_{destroyed,23} = T_0\dot{S}_{gen,23} = T_0\left[\dot{m}(s_3 - s_2) + \frac{\dot{Q}_1}{T_1}\right]$$

$$\dot{X}_{destroyed,23} = 300\left[0.08(0.392 - 0.973) + \frac{14.53}{300}\right] = 0.456\,kW$$

$$\dot{X}_{destroyed,34} = T_0\dot{S}_{gen,34} = T_0\dot{m}(s_4 - s_3) = 300 \times 0.08 \times (0.437 - 0.392) = 1.08\,kW$$

$$\dot{X}_{destroyed,41} = T_0\dot{S}_{gen,41} = T_0\left[\dot{m}(s_1 - s_4) - \frac{\dot{Q}_2}{T_2}\right]$$

$$= 300\left[0.08(0.9518 - 0.437) - \frac{10.12}{263}\right] = 0.816\,kW$$

$$\text{Minimum power input for cycle, } \dot{W}_{min,in} = \dot{Q}_2\left(\frac{T_0 - T_2}{T_2}\right)$$

$$\dot{W}_{min,in} = 10.12\left(\frac{300 - 263}{263}\right) = 1.4237\,kW$$

ii. Second law efficiency is the ratio of minimum power required, and actual power required, $\eta_{II} = \frac{\dot{W}_{min,in}}{\dot{W}_{in}} = \frac{1.4237}{4.412} = 32.27\%$ Ans.

iii. Total exergy destruction, $\dot{X}_{destroyed} = \dot{X}_{destroyed,12} + \dot{X}_{destroyed,23} + \dot{X}_{destroyed,34} + \dot{X}_{destroyed,41}$

$= 0.5088 + 0.456 + 1.08 + 0.816 = 2.8608$ kW Ans.

REVIEW QUESTIONS

12.1 Define refrigerating effect and refrigeration.
12.2 Define COP of refrigeration.
12.3 Define ton of refrigeration.
12.4 What kind of cycle is vapor compression refrigeration and why?
12.5 What is refrigeration capacity?
12.6 Which cycle is air refrigeration cycle? And name the processes in it
12.7 Why is heat pump more efficient than heating coil?
12.8 Relate COP of refrigerator and COP of heat pump.
12.9 What are the different types of refrigerants?
12.10 What is global warming potential of refrigerants?
12.11 How do you compare the vapor compression and vapor absorption refrigeration systems?
12.12 Why is subcooling and superheating done in vapor compression refrigeration?
12.13 Where do you prefer air refrigeration cycle and what are its limitations?
12.14 What is cascading refrigeration system and what are its advantages?
12.15 What is liquefaction of gases and what are its uses?
12.16 What are the factors affecting the energy efficiency of the refrigerators?
12.17 Where do you prefer multistage vapor compression systems over single-stage vapor compression systems?
12.18 Why is flash chamber preferred over heat exchanger in the multistage vapor compression system?
12.19 What are the low–global warming potential (low-GWP) refrigerants?
12.20 Derive an expression for the maximum COP of vapor absorption refrigeration system

EXERCISE PROBLEMS

12.1 A refrigerator uses R-134a as the working fluid and operates on an ideal vapor compression refrigeration cycle between 0.12 and 0.6 MPa. The mass flow rate of the refrigerant is 0.05 kg/s. Determine (i) the rate of heat removal from the refrigerated space, (ii) the power input to the compressor, and (iii) the heat rejection rate in the condenser and COP.

12.2 An ideal vapor compression refrigeration cycle, using refrigerant R-134a as the working fluid, communicates thermally with a cold region at 3°C and a warm region at 30°C. Saturated vapor enters the compressor at 3°C, and saturated liquid leaves the condenser at 30°C. The mass flow rate of the refrigerant is 0.075 kg/s. Determine (i) the compressor power in kW, (ii) the refrigeration capacity in tonnes, (iii) the coefficient of performance, and (iv) the coefficient of performance of a Carnot refrigeration cycle operating between warm and cold regions at 30°C and 3°C, respectively.

12.3 In an aqua-ammonia absorption refrigerator system, heat is supplied to the generator by condensing steam at 0.2 MPa, 90% quality. The temperature to be maintained in the refrigerator is −8°C. The ambient air is 30°C. Estimate (i) maximum COP of the refrigerator, and (ii) the steam flow rate if the actual COP is 50% of the maximum COP and the refrigeration load is 20 tonnes.

12.4 A vapor compressor heat pump with heat capacity of 500 kJ/min is driven by a power cycle with thermal efficiency of 25% for the heat pump, and R-22 is compressed from saturated vapor at −8°C to the condenser pressure of 10 bar. The isentropic efficiency of the compressor is 80%. Liquid enters the expansion valve at 10 bar and 30°C. For the power cycle, 80% of the heat rejected is transferred to the heated space. Determine (i) power input to the compressor, and (ii) evaluate the ratio of total rate that the heat is delivered to the amount of heat input to the power cycle.

12.5 A heat pump is to use an R-12 cycle to operate between outdoor air at −5°C and air in a domestic heating system at 35°C. The temperature difference in the evaporator and the condenser is 8°C. The compressor efficiency is 80%, and the compression begins with saturated vapor. The expansion begins with saturated liquid. The combined efficiency of the motor and the belt drive is 75%. Estimate (i) the electrical load in kW and (ii) the COP of the system. Assume required heat supply to the warm air is 43.6 kW.

12.6 An ideal (Carnot) refrigeration system operates between the temperature limits of −30°C and 25°C. Calculate (i) the COP of the cycle and (ii) power required from an external source to absorb 3.89 kW at high temperatures.

12.7 In a Carnot vapor refrigeration cycle, refrigerant R-22 is the working fluid for which the evaporator temperature is −25°C. Saturated vapor enters the condenser at 32°C, and saturated liquid exits at the same temperature. The mass flow rate of the refrigerant is 0.20 kg/s. Determine (i) the rate of heat transfer to the refrigerant passing through the evaporator, in kW, (ii) the net power input to the cycle, in kW, (iii) the coefficient of performance, and (iv) the refrigeration capacity, in tonnes.

12.8 A Carnot vapor refrigeration cycle operates between thermal reservoirs at 10°C and 30°C. For (i) refrigerant R-134a, (ii) CO_2, (iii) refrigerant R-22, and (iv) ammonia as the working fluids, determine the operating pressures in the condenser and evaporator, and the coefficient of performance.

12.9 A heat pump that operates on an ideal compression cycle with R-134a is used to heat a house and to maintain it at 20°C using underground water at 10°C as the heat source. The house is losing heat at the rate of 75 MJ/h. The evaporator and condenser pressures are 320 and 800 KPa, respectively.

Determine (i) power input to the heat pump and (ii) electrical power saved by using heat pump instead of a resistance heater.

12.10 Air enters the compressor of a gas refrigeration cycle at 100 kPa and 270 K. The compressor pressure ratio is 3.2, and the temperature at the turbine inlet is 310 K. The compressor and turbine both have isentropic efficiencies of 80%. Determine (i) the net work input, per unit mass of air flow, in kJ/kg, and (ii) exergy accounting of the net power input, in kJ per kg of air flowing. Let $T_0=310$ K.

12.11 A water cooler supplies chilled water at 10°C when water is supplied to it at 30°C at a rate of 0.7 l/min, while the power consumed amounts to 0.5 kW. Compare (i) COP of this refrigeration plant to ideal refrigeration plant.

12.12 The working fluid in a heat pump installation is ammonia. The ammonia after evaporation to a dry-saturated state at 2°C is compressed to a pressure of 10 bar, at which it is cooled and condensed to a saturated liquid state. It then passes through a throttle valve and returns to the evaporator. Calculate (i) COP assuming isentropic efficiency of the compressor is 0.85.

12.13 Air enters the compressor of a gas refrigeration cycle at 103 kPa and 250 K and is compressed adiabatically to 350 kPa. Air enters the turbine at 350 kPa and 300 K and expands adiabatically to 103 kPa. For the cycle, (i) determine the net work per unit mass of air flow, in kJ/kg, and the coefficient of performance if the compressor and turbine isentropic efficiencies are both 100%, (ii) plot the net work per unit mass of air flow, in kJ/kg, and determine the coefficient of performance for equal compressor and turbine isentropic efficiencies ranging from 80% to 100%.

12.14 A Carnot refrigerator requires 1.3 kW per tonne of refrigeration to maintain a region at a low temperature of −38°C. Determine (i) COP of Carnot refrigerator, (ii) higher temperature of the cycle, and (iii) heat delivered and COP.

12.15 A heat pump is used for heating the interior of the house in a cold climate. The ambient temperature is −5°C, and the desired interior temperature is 25°C. The compressor of a heat pump is to be driven by a heat engine working between 1000°C and 25°C. Treating both cycles as reversible, estimate (i) the COP of the heat pump, (ii) the efficiency of the engine, and (iii) the ratio in which heat pump and heat engine share the heating load.

12.16 An ammonia refrigerator operates between evaporating and condensing temperatures of −16°C and 50°C, respectively. The vapor is dry-saturated at the compressor inlet, the compression process is isentropic, and there is no undercooling of the condensate. Determine (i) refrigerating effect per kg, (ii) mass flow rate and power input, and (iii) COP of the cycle.

12.17 A refrigeration plant works between temperature limits of −5°C and 25°C. The working fluid ammonia has a dryness fraction of 0.6 at the entry of the compressor. If the machine has the relative efficiency of 50%, calculate the amount of ice formed during a period of 24 hours. The ice is to be formed at 0°C from water at 20°C, and 6 kg of ammonia is circulated per minute. Assume latent heat of ice is 335 kJ/kg and c_p of water is 4.187 kJ/kg K.

Temperature (K)	Liquid heat (kJ/kg)	Latent heat (kJ/kg)	Entropy (kJ/kg K)
298	298.9	1167.1	1.124
268	158.2	1280.8	0.630

12.18 Thirty tonnes of ice from 0°C is produced in a day of 24 hours by an ammonia refrigerator. The temperature range in the compressor is 298–258 K. The vapor is dry-saturated at the end of compression, and an expansion valve is used. Assume a coefficient of performance of 60% of the theoretical value, and calculate the power required to drive the compressor, in kW. Latent heat of ice is 335 kJ/kg.

Temperature (K)	Enthalpy (kJ/kg) Liquid	Enthalpy (kJ/kg) Vapor	Entropy of liquid (kJ/kg K)	Entropy of vapor (kJ/kg K)
298	100.04	1319.22	0.3473	4.4852
258	−54.56	1304.99	−2.1338	5.0585

12.19 Air at 200 kPa and 373 K is extracted from a main jet engine compressor for cabin cooling. The extracted air enters a heat exchanger where it is cooled at constant pressure to 310 K through heat transfer with the ambient. It then expands adiabatically to 95 kPa through a turbine and is discharged into the cabin. The turbine has an isentropic efficiency of 80%, and the mass flow rate of the air is 0.85 kg/s. Determine (i) the power developed by the turbine, in kW, and (ii) the rate of heat transfer from the air to the ambient, in kW.

12.20 A gas refrigeration system using air as a refrigerant is to work between −12°C and 27°C using an ideal reversed Brayton cycle of pressure ratio 8 and minimum pressure 1 atm, and to maintain a load of 10 tonnes. Find (i) the COP, (ii) air flow rate, (iii) volume flow rate entering the compressor in m^3/s, and (iv) the maximum and minimum temperatures of the cycle.

12.21 Air at 1.5 bar and 350 K is extracted from a jet engine compressor for a cabin cooling. The extracted air enters a heat exchanger where it is cooled isobarically to 300 K through heat transfer with the ambient. It then expands adiabatically to 0.90 bar through a turbine and is discharged into the cabin. The turbine has an isentropic efficiency of 70%. If the mass flow rate of air is 1 kg/s, determine (i) power developed by the turbine and (ii) the rate of heat transfer from the air to the ambient.

12.22 In a two-stage cascade refrigeration system, operating pressures are 750 and 90 kPa. Each stage operates on an ideal vapor compression refrigeration cycle with refrigerant R-22 as the working fluid. Heat rejection from the lower cycle to the upper cycle takes place in an adiabatic counterflow heat exchanger where both streams enter at about 300 kPa, and the mass flow rate of the refrigerant through the upper cycle is 0.4 kg/min. Determine (i) the mass flow rate of the refrigerant through the lower cycle, (ii) the rate of heat removal from the refrigerated space and the power input to the compressor, and (iii) the coefficient of performance of this cascade refrigerator.

12.23 In a two-stage compression refrigeration system, using refrigerant R-134a as the working fluid, operating pressures are 900 and 150 kPa. The refrigerant leaves the condenser as a saturated liquid and is throttled to a flash chamber operating at 450 kPa. The refrigerant leaving the low-pressure compressor at 450 kPa is also routed to the flash chamber. The vapor in the flash chamber is then compressed to the condenser pressure by the high-pressure compressor, and the liquid is throttled to the evaporator pressure. Determine (i) the fraction of the refrigerant that evaporates as it is throttled to the flash chamber, (ii) the rate of heat removed from the refrigerated space for a mass flow rate of 0.005 kg/min through the condenser, and (iii) the coefficient of performance. Assume that the refrigerant leaves the evaporator as saturated vapor and both compressors are isentropic.

12.24 Refrigerant R-134a enters at a rate of 0.05 kg/s into the compressor of a vapor compression refrigeration system at 150 kPa and −10°C and leaves at 1 MPa. The system rejects heat to surroundings at 25°C. The isentropic efficiency of the compressor is 80%. The refrigerant enters the throttling valve at 0.95 MPa and 35°C and leaves the evaporator as saturated vapor at −20°C Determine (i) exergy destruction associated with each of the processes of the cycle and (ii) second-law efficiency.

12.25 The vapor compression refrigeration system using refrigerant R-134a maintains the space −10°C by rejecting heat to surroundings at 25°C. The refrigerant enters the compressor of a refrigerator as superheated vapor at 100 kPa superheated by 5.5°C at a rate of 0.08 kg/s. The refrigerant is cooled in the condenser to 40°C and leaves it as a saturated liquid. Determine (i) exergy destruction associated with each of the processes of the cycle, (ii) minimum power input, and (iii) second-law efficiency.

DESIGN AND EXPERIMENT PROBLEMS

12.26 Consider the different eco-friendly refrigerants such as hydrofluorocarbons (HFCs) that include HFC-32, HFC-134a, HFC-152a, and simple hydrocarbons (HCs) that include propane (R-290), isobutene (R-600a), and propylene for refrigeration and air-conditioning systems, and evaluate their cost, availability, properties, and global warming potential. Prepare a PowerPoint presentation with your recommendations on each of these refrigerants for a particular application.

12.27 Design an experiment with complete instrument to measure the actual coefficient of performance (COP) of an industrial refrigerator, and compare it with ideal COP. Discuss the instrumentation required and measurements that need to be taken.

12.28 Design a refrigeration system to maintain the temperature in the range of –25°C to –15°C, when the surrounding temperature varies from 15°C to 30°C. Consider a vapor compression refrigeration system with the total thermal load on the storage unit not to exceed 25 kW and compressor efficiency to be in the range from 70% to 90%. Sketch the cycle of operations on a temperature–entropy (T-s) diagram to enable the determination of temperatures and pressures at the different states and also the determination of the COP value and the mass flow rate of the working fluid.

12.29 Specify the type and size in horsepower of the pump required for a 125-ton ammonia–water vapor absorption refrigeration system. The pump should handle 250 kg/min of strong solution. The generator conditions are 12 kgf/cm^2 and 1200°C, and the absorber is at 2.2 kgf/cm^2 with strong solution exiting at 400°C. The evaporator pressure is 2.4 kgf/cm^2, and the exit temperature is 40°C.

13 Thermodynamic Relations

LEARNING OUTCOMES

After learning this chapter, students should be able to

- Develop basic thermodynamic relations and express the properties that cannot be measured directly in terms of easily measurable properties.
- Develop Maxwell relations based on which they are able to develop other relations such as Clausius–Clapeyron equation.
- Apply thermodynamic property relations to ideal and real gases.
- Comprehend the thermodynamic property relations and calculate property data using fundamental thermodynamic relations.
- Develop and solve simple mathematical models of ideal gas mixtures undergoing a thermodynamic process

13.1 IMPORTANT MATHEMATICAL RELATIONS

In order to develop Maxwell relations and other relations useful in thermodynamics, it is first required to develop some important mathematical relations that will prove to be useful in the procedure.

Let us consider a variable z such that it is a continuous function of x and y:

$$z = f(x, y).$$

Then

$$dz = \left(\frac{\partial z}{\partial x}\right)_y dx + \left(\frac{\partial z}{\partial y}\right)_x dy \tag{13.1}$$

The above function can be conveniently written in the form

$$dz = Mdx + Ndy$$

where $M = \left(\frac{\partial z}{\partial x}\right)_y$ is a partial derivative of z with respect to x keeping y as constant)

and

$$N = \left(\frac{\partial z}{\partial y}\right)_x \text{ (partial derivative of z with respect to y when x is held constant).}$$

In the equation $dz = Mdx + Ndy$, if x, y and z are all point functions (the change in a property in a change of state is dependent on end states and not on the path), then the differentials are exact differentials. The following relation can be obtained:

$$\left(\frac{\partial M}{\partial y}\right)_x = \left(\frac{\partial N}{\partial x}\right)_y \tag{13.2}$$

The proof of the above equation is

$$\left(\frac{\partial M}{\partial y}\right)_x = \frac{\partial^2 z}{\partial x \partial y} \text{ and } \left(\frac{\partial N}{\partial x}\right)_y = \frac{\partial^2 z}{\partial y \partial x}$$

When point functions are involved, the order of differentiation makes no difference, and therefore

$$\frac{\partial^2 z}{\partial x \partial y} = \frac{\partial^2 z}{\partial y \partial x}$$

$$\text{Thus} \quad \left(\frac{\partial M}{\partial y}\right)_x = \left(\frac{\partial N}{\partial x}\right)_y$$

The above relation is quite useful in thermodynamics for developing Maxwell relations.

If x is a function of y and z, then $x = x(y,z)$ and

$$dx = \left(\frac{\partial x}{\partial y}\right)_x dy + \left(\frac{\partial x}{\partial z}\right)_y dz \tag{13.3}$$

Substituting the value of dz from Eq. 13.1 in Eq. 13.3

$$dx = \left(\frac{\partial x}{\partial y}\right)_z dy + \left(\frac{\partial x}{\partial z}\right)_y \left[\left(\frac{\partial z}{\partial x}\right)_y dx + \left(\frac{\partial z}{\partial y}\right)_x dy\right]$$

Rearranging the above equation, $dx = \left[\left(\frac{\partial x}{\partial y}\right)_z + \left(\frac{\partial x}{\partial z}\right)_y \left(\frac{\partial z}{\partial y}\right)_x\right] dy + \left(\frac{\partial x}{\partial z}\right)_y \left(\frac{\partial z}{\partial x}\right)_y dx$

$$\text{or} \quad dx = \left[\left(\frac{\partial x}{\partial y}\right)_z + \left(\frac{\partial x}{\partial z}\right)_y \left(\frac{\partial z}{\partial y}\right)_x\right] dy + dx$$

$$\left(\frac{\partial x}{\partial y}\right)_z + \left(\frac{\partial x}{\partial z}\right)_y \left(\frac{\partial z}{\partial y}\right)_x$$

$$\left(\frac{\partial x}{\partial y}\right)_z\left(\frac{\partial z}{\partial x}\right)_y\left(\frac{\partial y}{\partial z}\right)_x = -1 \tag{13.4}$$

Equation 13.4 is called the cyclic relation and can be applied for thermodynamic variables such as p, υ and T.

13.2 THE MAXWELL RELATIONS

In this section, we develop Maxwell relations useful for calculating the difference between the property of entropy within a single homogeneous phase—gas, liquid, or solid—assuming a simple compressible system. The equations that relate the partial derivatives of properties p, v, T, and s of a simple compressible system to each other are called *Maxwell relations.* Four Gibbs equations form the basis for the development of Maxwell relations. This can be done by utilizing the exactness of differentials of thermodynamic properties. The two of the four Gibbs equations are Eqs. 7.15 and 7.16, which are as given below:

$$Tds = du + pdv$$

$$Tds = dh - vdp$$

The other two equations are derived based on the Helmholtz function 'a' and Gibbs function 'g' given as

$$a = u - Ts \tag{13.5}$$

$$g = h - Ts \tag{13.6}$$

Differentiating Eqs. 13.5 and 13.6 yields

$$da = du - Tds - sdT$$

$$dg = dh - Tds - sdT$$

Simplification of the above two equations using Eqs. 7.15 and 7.16 yields the other two Gibbs equations for simple compressible systems as given below:

$$da = -sdT - pdv \tag{13.7}$$

$$dg = -sdT + vdp \tag{13.8}$$

It can be observed from the four Gibbs equations that they are in the form

$$dz = Mdx + Ndy$$

with $\left(\frac{\partial M}{\partial y}\right)_x = \left(\frac{\partial N}{\partial x}\right)_y$

Now properties u, h, a, and g are point functions and have exact differentials. Equation 7.25 can be applied to each of the four Gibbs equations:

$$\left(\frac{\partial T}{\partial v}\right)_s = -\left(\frac{\partial p}{\partial s}\right)_v \quad (13.9)$$

$$\left(\frac{\partial T}{\partial p}\right)_s = -\left(\frac{\partial v}{\partial s}\right)_p \quad (13.10)$$

$$\left(\frac{\partial s}{\partial v}\right)_T = \left(\frac{\partial p}{\partial T}\right)_v \quad (13.11)$$

$$\left(\frac{\partial s}{\partial p}\right)_T = -\left(\frac{\partial v}{\partial T}\right)_p \quad (13.12)$$

Eqs. 13.9–13.12 are called Maxwell relations. These relations are much useful in thermodynamics as they form the basis for the determination of change in entropy, which cannot be done by other means, by simply measuring the changes in properties p, v, and T.

13.3 CLAUSIUS–CLAPEYRON EQUATION

Maxwell relations form the basis for the development of other important relations that play a vital role in thermodynamics. The Clapeyron equation is one such relation that is useful to evaluate the change in enthalpy during a phase change process such as vaporization, sublimation, or melting at a constant temperature from pressure-specific volume–temperature data pertaining to the phase change.

To derive the Clapeyron equation, let us consider the third Maxwell relation given in (Eq. 13.11)

$$\left(\frac{\partial s}{\partial v}\right)_T = \left(\frac{\partial p}{\partial T}\right)_v.$$

During phase change, when the temperature is held constant, the pressure is independent of specific volume and the quantity $\left(\frac{\partial p}{\partial T}\right)_v$ is determined by the temperature and can be given as

$$\left(\frac{\partial p}{\partial T}\right)_v = \left(\frac{dp}{dT}\right)_{sat} \quad (13.13)$$

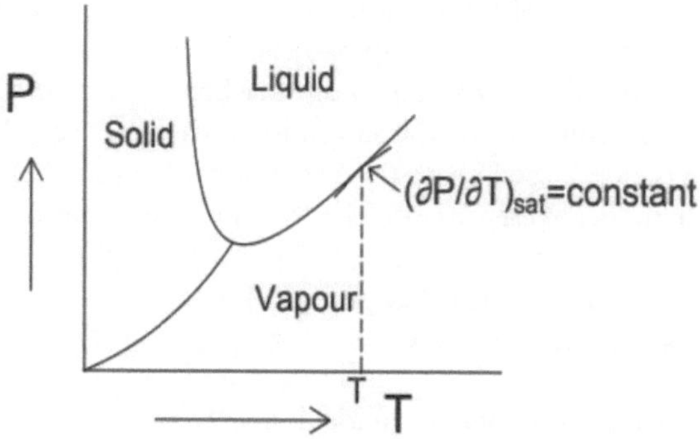

FIGURE 13.1 The p-T diagram on which the slope of the saturation curve is constant at T=constant.

where the subscript "sat" specifies that the derivative is the slope of the saturation pressure–temperature curve at the point determined by the temperature held constant during the phase change as shown in Figure 13.1.
Combining the two equations Eqs. 13.11 and 13.13,

$$\left(\frac{\partial s}{\partial v}\right)_T = \left(\frac{dp}{dT}\right)_{sat} \tag{13.14}$$

Integrating Eq 13.14 gives

$$s_g - s_f = \left(\frac{dp}{dT}\right)_{sat} v_g - v_f \tag{13.15}$$

Rearranging Eq. 13.15 gives

$$\left(\frac{dp}{dT}\right)_{sat} = \frac{s_g - s_f}{v_g - v_f} \tag{13.16}$$

During the phase change process, the pressure is also constant as the temperature is constant (since pressure is dependent on temperature alone), and therefore from the equation Tds = dh − vdp,

$$\int_f^g dh = \int_f^g T ds \Rightarrow h_{fg} = Ts_{fg} (\text{when } vdp = 0)$$

Substituting the above in Eq. 13.16 gives

$$\left(\frac{dp}{dT}\right)_{sat} = \frac{h_g - h_f}{T(v_g - v_f)} \tag{13.17}$$

Equation 13.17 is the *Clapeyron equation* and is useful in determining the enthalpy change of vaporization as it is possible to experimentally determine the slope of the vapor pressure as a function of temperature and also specific volumes of saturated liquid and saturated vapor at a given temperature.

A special case of the Clapeyron equation is when it involves the vapor phase occurring at low temperatures when the saturation pressure is very low. In this case, the specific volume of condensed phase (υ_f) is not only much smaller when compared to that of vapor (υ_g) but is also represented by the ideal gas equation of the state very closely. With these approximations, Eq. 13.17 can be simplified to be applied for liquid–vapor and solid–vapor phase changes.

At very low pressures, when $v_g \gg v_f \Rightarrow v_{fg} \cong v_g$

Again, when the vapor is treated as an ideal gas, $v_g = \dfrac{RT}{p}$.

Substituting these approximations into Eq. 13.17, we have

$$\left(\frac{dp}{dT}\right)_{sat} = \frac{h_g - h_f}{RT^2/p}$$

$$\left(\frac{dp}{p}\right)_{sat} = \frac{h_{fg}}{R}\left(\frac{dT}{T^2}\right)_{sat} \tag{13.18}$$

At very low temperatures, h_{fg} does not change significantly with temperature and hence assumed to be constant. Then Eq. 13.18 can be integrated over a range of temperatures to determine the saturation pressure at a temperature.

$$\ln\left(\frac{p_2}{p_1}\right)_{sat} = \frac{h_{fg}}{R}\left(\frac{T_2 - T_1}{T_1 T_2}\right) \tag{13.19}$$

Equation 13.19 is called the Clausius–Clapeyron equation and is useful for evaluating the variation of saturation pressure with temperature. A similar expression applies for the case of sublimation (solid–vapor region) as well by replacing the enthalpy of vaporization h_{fg} with enthalpy of sublimation h_{ig}.

13.4 THE JOULE–THOMSON COEFFICIENT

The useful property of substances defined as a partial derivative is the Joule–Thomson coefficient, which is used to describe the temperature behavior of a fluid during a throttling (constant enthalpy) process. Whenever a fluid is allowed to pass through a narrow cross-section such as a porous plug or a capillary tube, its pressure drops. The enthalpy of the fluid remains constant during this throttling process. There may be a large drop in fluid temperature during throttling. Refrigerators and air-conditioners operate on this principle. However, during throttling, this is not always the case as the temperature of the fluid may remain constant, or it may even increase. The Joule–Thomson coefficient μ_J is defined as

$$\mu_J = \left(\frac{\partial T}{\partial p}\right)_h \tag{13.20}$$

The Joule–Thomson coefficient of a liquid or gas can be evaluated experimentally by allowing the liquid or gas to expand steadily through a porous plug as shown in Figure 13.2. In this process, the gas or fluid enters the apparatus at a given temperature T_1 and pressure p_1 and expands through the plug to a reduced pressure p_2, which is controlled by an outlet valve. The downstream temperature (T_2) is measured. The apparatus is designed in such a way that the gas undergoes a throttling process as it expands from 1 to 2. Accordingly, the exit state fixed by p_2 and T_2 will have the fixed value of specific enthalpy, i.e., $h_2 = h_1$. Repeating the experiment for different sets of inlet pressure and temperature, it is possible to construct several isenthalpic curves (h = constant) on a T-p diagram for a given substance as shown in Figure 13.3.

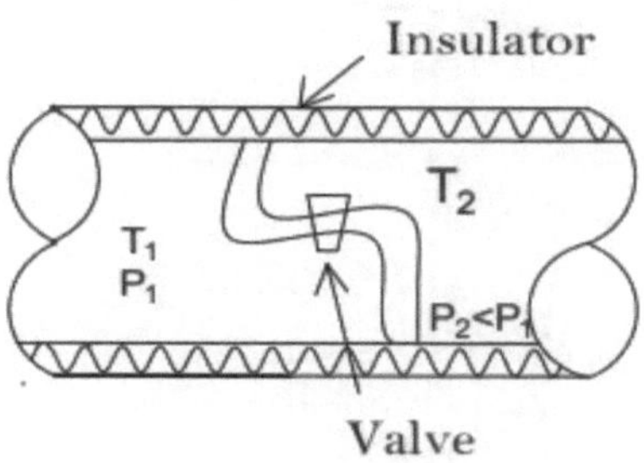

FIGURE 13.2 The throttling process through a porous plug.

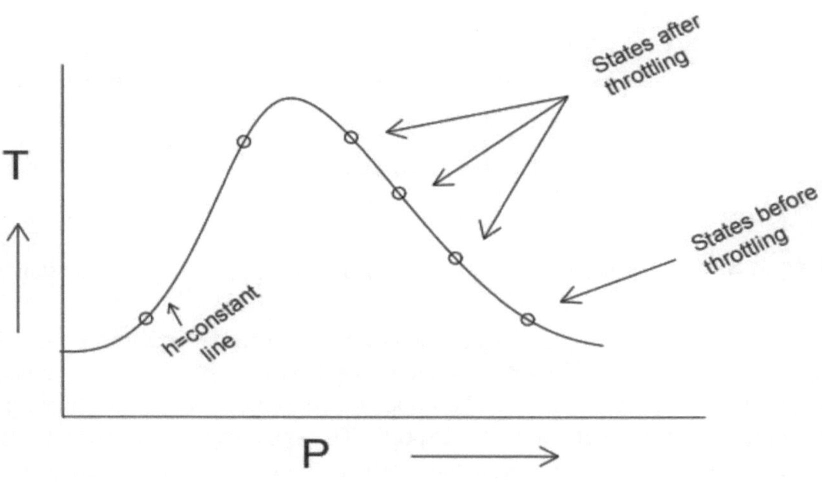

FIGURE 13.3 The constant enthalpy (isenthalpic) line on a T-p diagram.

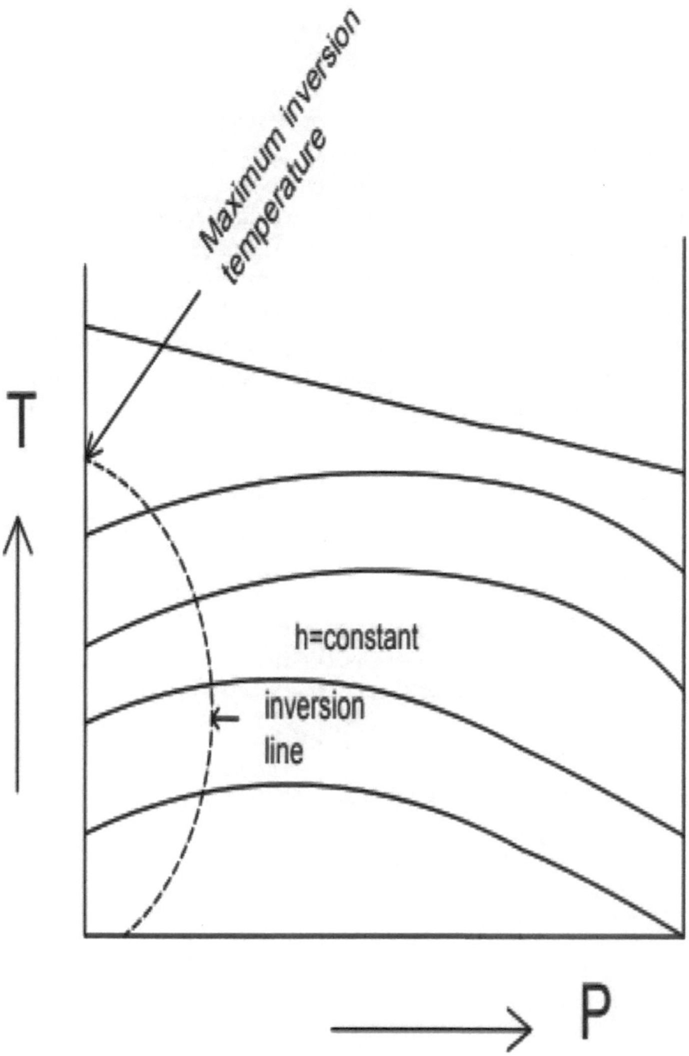

FIGURE 13.4 The T-p diagram showing the constant enthalpy lines of a pure substance.

An isenthalpic curve is the locus of the points representing equilibrium states of the same specific enthalpy. Figure 13.4 shows the T-p diagram showing the constant enthalpy lines of a pure substance. The slope of a constant enthalpy line at any point is the Joule–Thomson coefficient at that point. The slope may be positive, negative, or zero. The states with the zero Joule–Thomson coefficient are called inversion states, and the constant enthalpy line joining the zero coefficient is called an inversion line. Throttling of a gas along a constant enthalpy line will always take place in the direction of decreasing pressure. However, for isenthalpic curves having an inversion state, the temperature at the exit of the apparatus may be greater than, equal to, or less than the initial temperature, depending on the exit pressure specified. For those states that are to the right of an inversion state, the value of the Joule–Thomson coefficient is negative. For these states, the temperature increases as the pressure at the exit of the apparatus decreases. At states to the left of an inversion line, the value of the Joule–Thomson coefficient is positive and the temperature decreases as the pressure at the exit of the device decreases. This can be used conveniently in systems meant for liquefaction of the gases.

13.5 GENERAL RELATIONS FOR CHANGES IN ENTHALPY, INTERNAL ENERGY, AND ENTROPY

13.5.1 Change in Enthalpy

Specific heat at constant pressure, $c_p = \left(\frac{\partial h}{\partial T}\right)_p$

Also from Eq. 7.16 $Tds = dh - \nu dp$.

The general relation for change in enthalpy of a pure substance can be derived from the enthalpy is a function of temperature and pressure for a pure substance,

$$h = h(T, p), \text{then}$$

$$dh = \left(\frac{\partial h}{\partial T}\right)_p dT + \left(\frac{\partial h}{\partial p}\right)_T dp$$

$$dh = c_p dT + \left(\frac{\partial h}{\partial p}\right)_T dp \tag{13.21}$$

Similarly, when entropy is a function of temperature and pressure, $s = s(T,p)$,

$$ds = \left(\frac{\partial s}{\partial T}\right)_p dT + \left(\frac{\partial s}{\partial p}\right)_T dp \tag{13.22}$$

Substituting Eq. 13.22 in $dh = Tds + \upsilon dp$

$$dh = T\left[\left(\frac{\partial s}{\partial T}\right)_p dT + \left(\frac{\partial s}{\partial p}\right)_T dp\right] + \mathrm{v}dp$$

On rearranging the above equation $dh = T\left(\frac{\partial s}{\partial T}\right)_p dT + \left(\mathrm{v} + T\left(\frac{\partial s}{\partial p}\right)_T\right)dp$ (13.23)

Comparing Eqs 13.21 and13.23,

$$c_p = T\left(\frac{\partial s}{\partial T}\right)_p \tag{13.24}$$

$$\text{and}\left(\frac{\partial h}{\partial p}\right)_T = v + T\left(\frac{\partial s}{\partial p}\right)_T \tag{13.25}$$

Equation 13.25 can be rearranged using the fourth Maxwell relation $\left(\frac{\partial s}{\partial p}\right)_T = -\left(\frac{\partial v}{\partial T}\right)_p$, which gives

$$\left(\frac{\partial h}{\partial p}\right)_T = v - T\left(\frac{\partial v}{\partial T}\right)_p \tag{13.26}$$

Substituting Eq. 13.26 in Eq. 13.21 gives

$$dh = c_p dT + \left[v - T\left(\frac{\partial v}{\partial T}\right)_p\right]dp \tag{13.27}$$

Equation 13.27 can be integrated to give the change in enthalpy associated with a change of state, that is,

$$h_2 - h_1 = \int_1^2 c_p\, dT + \int_1^2 \left[v - T\left(\frac{\partial v}{\partial T}\right)_p\right]dp$$

$$\text{or}\quad h_2 - h_1 = \int_{T_1}^{T_2} c_p\, dT + \int_{p_1}^{p_2} \left[v - T\left(\frac{\partial v}{\partial T}\right)_p\right]dp \tag{13.28}$$

13.5.2 Change in Internal Energy

Change in internal energy can also be derived in a similar way by assuming that the internal energy is a function of temperature and volume, that is,

$$u = u(T, \upsilon)$$

$$du = \left(\frac{\partial u}{\partial T}\right)_v dT + \left(\frac{\partial u}{\partial v}\right)_T dv$$

Specific heat at constant volume, $c_v = \left(\frac{\partial u}{\partial T}\right)_v$.

Then

$$du = c_v dT + \left(\frac{\partial u}{\partial v}\right)_T dv \tag{13.29}$$

Again, as entropy is a function of temperature and volume, that is, $s = s(T, \upsilon)$,

$$ds = \left(\frac{\partial s}{\partial T}\right)_v dT + \left(\frac{\partial s}{\partial v}\right)_T dv \tag{13.30}$$

From Eq. 7.15, Tds = du + pdν or du = Tds − pdυ

Substituting Eq. 13.30 in du = Tds − pdυ gives

$$du = T\left(\frac{\partial s}{\partial T}\right)_v dT + \left[T\left(\frac{\partial s}{\partial v}\right)_T - p\right]dv \tag{13.31}$$

Comparing Eqs. 13.29 and 13.31,

$$c_v = T\left(\frac{\partial s}{\partial T}\right)_v \tag{13.32}$$

$$\left(\frac{\partial u}{\partial v}\right)_T = T\left(\frac{\partial s}{\partial v}\right)_T - p \tag{13.33}$$

Equation 13.33 can be rearranged using the third Maxwell relation $\left(\frac{\partial s}{\partial v}\right)_T = \left(\frac{\partial p}{\partial T}\right)_v$, which gives

$$\left(\frac{\partial u}{\partial v}\right)_T = T\left(\frac{\partial p}{\partial T}\right)_v - p \tag{13.34}$$

Therefore $du = c_v dT + \left[T\left(\frac{\partial p}{\partial T}\right)_v - p\right]dv$.

Equation 13.34 can be integrated to give the change in internal energy associated with a change of state, that is,

$$u_2 - u_1 = \int_{T_1}^{T_2} c_v\, dT + \int_{v_1}^{v_2}\left[T\left(\frac{\partial p}{\partial T}\right)_v - p\right]dv \tag{13.35}$$

13.5.3 Change in Entropy

Change in entropy can be derived by assuming that the entropy is a function of temperature and pressure, that is,

$$s = s(T, p)$$

Then

$$ds = \left(\frac{\partial s}{\partial T}\right)_p dT + \left(\frac{\partial s}{\partial p}\right)_T dp \tag{13.36}$$

Replacing the first partial derivative of Eq. 13.36 by the equation $c_p = T\left(\frac{\partial s}{\partial T}\right)_p$ and the second partial derivative by the fourth Maxwell relation gives

$$ds = c_p \frac{dT}{T} - \left(\frac{\partial v}{\partial T}\right)_p dp \tag{13.37a}$$

Similarly by assuming that the entropy is a function of temperature and specific volume, that is,

$$s = s(T, \upsilon),$$

it can be shown that

$$ds = c_v \frac{dT}{T} + \left(\frac{\partial p}{\partial T}\right)_v dv \tag{13.37b}$$

Equation 13.37a can be integrated to give the change in entropy associated with a change of state, that is,

$$s_2 - s_1 = \int_1^2 c_p \frac{dT}{T} - \int_1^2 \left(\frac{\partial v}{\partial T}\right)_p dp$$

13.6 SPECIFIC HEAT RELATIONS

When entropy is a function of temperature and volume, $s = s(T,\upsilon)$,

$$ds = \left(\frac{\partial s}{\partial T}\right)_v dT + \left(\frac{\partial s}{\partial v}\right)_T dv$$

Multiplying the above equation by T gives

$$Tds = T\left(\frac{\partial s}{\partial T}\right)_v dT + T\left(\frac{\partial s}{\partial v}\right)_T dv \tag{13.38}$$

Comparing Eq. 13.38 with $Tds = du + pd\upsilon$,

$$Tds = c_v dT + T\left(\frac{\partial p}{\partial T}\right)_v dv \tag{13.39}$$

Since from Eq. 13.32, $c_v = T\left(\frac{\partial s}{\partial T}\right)_v$

and from the third Maxwell relation $\left(\frac{\partial s}{\partial v}\right)_T = \left(\frac{\partial p}{\partial T}\right)_v$

Equation 13.39 is first Tds equation

When entropy is a function of temperature and pressure, $s = s(T,p)$,

$$ds = \left(\frac{\partial s}{\partial T}\right)_p dT + \left(\frac{\partial s}{\partial p}\right)_T dp$$

Multiplying the above equation by T gives

$$Tds = T\left(\frac{\partial s}{\partial T}\right)_p dT + T\left(\frac{\partial s}{\partial p}\right)_T dp \qquad (13.40)$$

Comparing Eq. 13.40 with $Tds = dh - \nu dp$,

$$Tds = c_p dT - T\left(\frac{\partial v}{\partial T}\right)_p dp \qquad (13.41)$$

Since from Eq. 13.24 $c_p = T\left(\frac{\partial s}{\partial T}\right)_p$ and from the fourth Maxwell relation $\left(\frac{\partial s}{\partial p}\right)_T = -\left(\frac{\partial v}{\partial T}\right)_p$,

Equation 13.41 is the second Tds equation.

Now equating Eqs. 13.39 and 13.41

$$Tds = c_p dT - T\left(\frac{\partial v}{\partial T}\right)_p dp = c_v dT + T\left(\frac{\partial p}{\partial T}\right)_v dv$$

$$(c_p - c_v)dT = T\left(\frac{\partial p}{\partial T}\right)_v dv + T\left(\frac{\partial v}{\partial T}\right)_p dp$$

$$dT = \frac{T\left(\frac{\partial p}{\partial T}\right)_v}{(c_p - c_v)} dv + \frac{T\left(\frac{\partial v}{\partial T}\right)_p}{(c_p - c_v)} dp \qquad (13.42)$$

When temperature is a function of volume and pressure, $T = T(\upsilon, P)$,

$$dT = \left(\frac{\partial T}{\partial v}\right)_p dv + \left(\frac{\partial T}{\partial p}\right)_v dp \qquad (13.43)$$

Comparing Eqs. 13.42 and 13.43,

$$\frac{T\left(\frac{\partial p}{\partial T}\right)_v}{(c_p - c_v)} = \left(\frac{\partial T}{\partial v}\right)_p \text{ and } \frac{T\left(\frac{\partial v}{\partial T}\right)_p}{(c_p - c_v)} = \left(\frac{\partial T}{\partial p}\right)_v$$

Rearranging the first equation above, $c_p - c_v = T\left(\frac{\partial p}{\partial T}\right)_v \left(\frac{\partial v}{\partial T}\right)_p$

$$\left(\frac{\partial p}{\partial T}\right)_v \left(\frac{\partial T}{\partial v}\right)_p \left(\frac{\partial v}{\partial p}\right)_T = -1 (\text{Cyclic relation Eq. 13.4})$$

Then

$$c_p - c_v = -T\left(\frac{\partial v}{\partial T}\right)_p^2 \left(\frac{\partial p}{\partial v}\right)_T \tag{13.44}$$

For an ideal gas, $p\upsilon = RT \rightarrow \left(\frac{\partial v}{\partial T}\right)_p = \frac{R}{p} = \frac{v}{T}$

$$p = \frac{RT}{v} \rightarrow \left(\frac{\partial p}{\partial v}\right)_T = -\frac{RT}{v^2}$$

Substituting the above two in Eq. 13.44, we get $c_p - c_\upsilon = R$

Equation 13.44 can also be expressed in terms of isothermal compressibility (α), and coefficient of volumetric expansion (β) defined as the change in volume with temperature when pressure is held constant.

$$\alpha = -\frac{1}{v}\left(\frac{\partial v}{\partial p}\right)_T \text{ and } \beta = \frac{1}{v}\left(\frac{\partial v}{\partial T}\right)_p$$

Then

$$c_p - c_v = \frac{Tv\beta^2}{\alpha} \tag{13.45}$$

EXAMPLE PROBLEMS

Example 13.1 Verify the fourth Maxwell relation for the refrigerant R-134a at 30°C and 600 kPa.

Solution The fourth Maxwell relation is $\left(\frac{\partial s}{\partial p}\right)_T = -\left(\frac{\partial v}{\partial T}\right)_p$.

According to this relation, the change in entropy with pressure at constant temperature is equal to the negative change in specific volume with temperature when pressure is constant. This problem can be solved by replacing the differential quantities in the above equation with the finite quantities from the property tables.

$$\left(\frac{\Delta s}{\Delta p}\right)_{T=30^\circ C} \cong -\left(\frac{\Delta v}{\Delta T}\right)_{p=600 kPa}$$

$$\left[\frac{s_{700kPa} - s_{500kPa}}{(700-500)kPa}\right]_{T=30^\circ C} \cong -\left[\frac{v_{40^\circ C} - v_{20^\circ C}}{(40-20)^\circ C}\right]_{p=600kPa}$$

$$= \left[\frac{0.9313 - 0.9713}{200}\right] = -\left[\frac{0.0378 - 0.0343}{20}\right]$$

$$= -0.00023 = -0.0002$$

Since the two values obtained are in close agreement with each other, the refrigerant R-134a satisfies the fourth Maxwell relation at the specified state.

Example 13.2 Evaluate the enthalpy of vaporization of water at 50°C using the Clapeyron equation and compare it with the value from property tables.

Solution The enthalpy of vaporization from the Clapeyron equation is

$$h_{fg} = Tv_{fg}\left(\frac{dp}{dT}\right)_{sat}$$

From steam tables at 50°C, $v_f = 0.001012\,m^3/kg, v_g = 12.026\,m^3/kg$

$$\therefore v_{fg} = v_g - v_f = 12.025\,m^3/kg$$

$$\left(\frac{dp}{dT}\right)_{sat@50°C} \cong \left(\frac{\Delta p}{\Delta T}\right)_{sat@50°C} = \frac{p_{sat@55°C} - p_{sat@45°C}}{55 - 45}$$

$$= \frac{15.763 - 9.595}{10} = 0.616\,kPa/K$$

Then $h_{fg} = 323 \times 12.025 \times 0.616 = 2392.59\,kJ/kg$

The value of h_{fg} from steam tables is 2382.0 kJ/kg

The difference between two values is around 0.4%.

Example 13.3 Develop an expression for the entropy change of a gas that follows van der Waals equation of state.

Solution Van der Waals equation of state is $\left(p + \frac{a}{v^2}\right)(v - b) = RT$

Rearranging the above equation $p = \frac{RT}{(v - b)} - \frac{a}{v^2}$

Then $\left(\frac{\partial p}{\partial T}\right)_v = \frac{R}{(v - b)}$

Let us assume that the entropy is a function of temperature and specific volume, that is,

$$s = s(T, v)$$

Then it follows that $ds = \left(\frac{\partial s}{\partial T}\right)_v dT + \left(\frac{\partial s}{\partial v}\right)_T dv$

$c_v = T\left(\frac{\partial s}{\partial T}\right)_v$ and from third Maxwell relation $\left(\frac{\partial s}{\partial v}\right)_T = \left(\frac{\partial p}{\partial T}\right)_v$

Entropy change is $ds = c_v \frac{dT}{T} + \left(\frac{\partial p}{\partial T}\right)_v dv$

Substituting $\left(\frac{\partial p}{\partial T}\right)_v = \frac{R}{(v-b)}$ in entropy change equation above

$$ds = \frac{c_v}{T}dT + \frac{R}{(v-b)}dv$$

Integrating the above equation gives $s_2 - s_1 = \int_1^2 c_v \frac{dT}{T} + R\int_1^2 \left(\frac{1}{(v-b)}\right)dv$

or $s_2 - s_1 = c_v \ln\frac{T_2}{T_1} + R\log(v-b) + c_1$.

Example 13.4 Determine the saturation pressure of the refrigerant R-134a at –50°C.

Solution The Clausius–Clapeyron equation, given below, is used for determining the saturation pressure of the refrigerant R-134a,

$$\ln\left(\frac{p_2}{p_1}\right)_{sat} = \frac{h_{fg}}{R}\left(\frac{T_2 - T_1}{T_1 T_2}\right)$$

This equation is solved with the limits $T_2 = -40°C + 273 = 233$ K and

$T_1 = -50°C + 273 = 223$ K. At –40°C from R-134a tables, $h_{fg} = 225.86$ kJ/kg, and $p_2 = 51.25$ kPa, and the R value for R-134a is 0.08149 kJ/kg K.

$$\ln\left(\frac{51.25}{p_1}\right) = \frac{225.86}{0.08149}\left(\frac{233-223}{233\times 223}\right) = 30.09\,\text{kPa}$$

Saturation pressure of the refrigerant R-134a at –50°C is 30.09 kPa.

REVIEW QUESTIONS

13.1 For an ideal gas, what is the value of the Joule–Thomson coefficient?
13.2 What are the approximations involved in the Clapeyron–Clausius equation?
13.3 Define the inversion line and the maximum inversion temperature.
13.4 Does the Joule–Thomson coefficient of a substance change with temperature at a fixed pressure?
13.5 What is the significance of Maxwell relations?
13.6 What is the significance of Clapeyron equation?
13.7 What is Joule–Thomson coefficient?
13.8 Define isothermal compressibility and coefficient of volumetric expansion.
13.9 What is a throttling process and what is its use?
13.10 How does the temperature change in an adiabatic throttling process? Explain with respect to the inversion line.
13.11 Explain how you would achieve cooling from substances such as hydrogen whose maximum inversion temperature is –67°C.
13.12 What is the significance of Clapeyron–Clausius equation?

EXERCISE PROBLEMS

13.1 Estimate the Joule–Thomson coefficient of steam at (i) 20 bar and 2500°C and (ii) 50 bar and 500°C.

13.2 What will happen if the temperature of the steam is throttled slightly from 10 bar and 250°C?

13.3 Derive the expressions for changes in (i) enthalpy, (ii) internal energy, and (iii) entropy of an ideal gas that follows the Berthelot equation of state.

13.4 The vapor pressure of mercury at 390 and 394 K is observed as 0.922 and 0.946 mm of mercury, respectively. Determine the latent heat of vaporization of mercury at 392 K.

13.5 Determine the Joule–Thomson coefficient for a gas that follows van der Waals equation of state $\left(p+\frac{a}{v^2}\right)(v-b)=RT$.

13.6 Determine the enthalpy of vaporization of R-134a at 25°C using (i) the Clapeyron equation and (ii) the Clausius–Clapeyron equation. Compare these results with the tabulated data.

13.7 Determine for a gas obeying the equation of state, $p(\upsilon-b)=RT$, where b is a positive constant, whether the temperature can be reduced in a Joule–Thomson expansion?

13.8 Using Maxwell relations, develop a relation for $\left(\frac{\partial s}{\partial p}\right)_T$ for a gas whose equation of state is $\left(p+\frac{a}{v^2}\right)(v-b)=RT$.

13.9 Develop expressions for the volume expansivity and the isothermal compressibility for (i) an ideal gas, (ii) a gas whose equation of state is $p(\upsilon-b)=RT$, and (iii) a gas obeying the van der Waals equation.

13.10 A gas is described by $v=\frac{RT}{p}-\frac{a}{T}+b$, where a and b are constants. Obtain the expressions for (i) the temperatures at the Joule–Thomson inversion states and (ii) c_p-c_v.

13.11 Determine the change in internal energy of carbon dioxide when it undergoes a change of state from 101.32 kPa, 25°C to 500 kPa, 250°C using (i) the equation of state $p(\upsilon-b)=RT$, where the constant $b=0.072$ m^3/kg and (ii) ideal gas equation of state.

13.12 Determine the volume expansivity (α), and the isothermal compressibility (β) for water at 30°C, 30 bar and at 250°C, 180 bar using steam tables.

13.13 Obtain the relationship between c_p and c_υ for a gas that obeys (i) the equation of state $p(\upsilon-b)=RT$ and (ii) $\left(p+\frac{a}{v^2}\right)(v-b)=RT$.

13.14 Determine the change in entropy of hydrogen when it undergoes a change of state from 101.32 kPa, 20°C to 500 kPa, 300°C using (i) the equation of state $p(\upsilon-a)=RT$, where constant $b=0.01$ m^3/kg and (ii) ideal gas equation of state.

13.15 Develop an expression for the variation in temperature with pressure in a constant-entropy process, $(\partial T/\partial P)_s$, that includes only the properties P–v–T and the specific heat, c_p.

DESIGN AND EXPERIMENT PROBLEMS

13.16 It is required to store compressed natural gas (CNG) used as a fuel for automobile engines, at storage pressures up to 200 kgf/cm^2, with a total mass of 75 kg, with suitable material to provide lightweight, economical, and safe onboard storage. The storage vessels should have the potential of storing CNG for 100–125 miles of long distance. Specify the size and number of cylinders required to meet the above design constraints.

14 Psychrometry

LEARNING OUTCOMES

After learning this chapter, students should be able to

- Define and differentiate between dry air and atmospheric air
- Calculate the properties of atmospheric air such as specific and relative humidity, dew-point temperature, and wet-bulb temperature
- Use psychrometric charts and estimate various essential properties related to psychrometry and processes
- Demonstrate an understanding of thermal comfort conditions with respect to temperature and humidity, human clothing and activities and its impact on human comfort, productivity, and health
- Develop generalized psychrometrics of moist air and apply them to air-conditioning processes
- Perform the mass and energy balance calculations on various air-conditioning processes

14.1 PROPERTIES OF ATMOSPHERIC AIR

The content of water vapor in atmospheric air plays a vital role in comfort air-conditioning. Therefore, the knowledge of fundamental laws of gaseous mixture for a thorough understanding of psychrometry, which is a study of properties of air and water vapor, is important. *Dry air* is a mixture of nitrogen, oxygen, and in small quantities several other gases such as carbon dioxide, helium, argon, and neon. The volumetric composition of air is 21% oxygen and 79% nitrogen, and the molecular weight of dry air is approximately equal to 29. *Moist air* is a mixture of dry air and water vapor. Atmospheric air that has no moisture content is termed dry air. The amount of water vapor in the air keeps changing due to the condensation and evaporation from rivers, oceans, lakes, etc. The simultaneous control of temperature and moisture plays a significant role in human comfort and air-conditioning.

The mixture of air and water vapor at a given temperature is said to be *saturated* when it contains the maximum amount of water vapor that it can hold. If the temperature of mixture of dry air and water vapor is above the saturation temperature of water vapor, then the vapor is *superheated vapor.*

Dalton's Law of Partial Pressures

According to Dalton's law of partial pressures, the total pressure of a mixture of ideal gases is equal to the sum of the partial pressures exerted by each constituent gas when it occupies the volume of the mixture at the mixture temperature. For the moist air, as per this law

$$p = p_a + p_w \tag{14.1}$$

where p is the total pressure of moist air and p_a and p_w are the partial pressures of dry air and water vapor, respectively.

14.1.1 Specific Humidity and Relative Humidity

Specific humidity (ω) is defined as the mass of water vapor per unit mass of dry air.

$$\omega = \frac{m_w}{m_a} = \frac{p_w v/R_w T}{p_a v/R_a T} = \frac{p_w/R_w}{p_a/R_a}$$

$$\omega = \frac{0.622 p_w}{p - p_w} \tag{14.2}$$

where m_a and m_w are mass of dry air and water vapor, respectively. Specific humidity is maximum when air is saturated at a given temperature.

Relative Humidity

The amount of moisture air holds relative to the maximum amount of moisture air can hold at the same temperature is called *relative humidity.* It is the ratio of partial pressure of water in a mixture to the saturation pressure of pure water at the same temperature. If some water is sprayed into unsaturated air in a container, water evaporates, resulting in an increase in moisture content of air, and water pressure will increase. The evaporation will continue till air becomes saturated at that temperature; thereafter, no more evaporation will take place. For saturated air, relative humidity is 100%.

$$\phi = \frac{m_w}{m_s} = \frac{p_w v/R_w T}{p_s v/R_s T} = \frac{p_w}{p_s} \tag{14.3}$$

In the above equation, p_s is the saturation pressure of the water vapor at the mixture temperature.

The relation between relative and specific humidities can be expressed by substituting the value of p_w from Eq. 14.2 into Eq. 14.3, that is

$$\phi = \frac{\omega p_a}{0.622} \times \frac{1}{p_s} = 1.6\omega \frac{p_a}{p_s} \tag{14.4}$$

Relative humidity is an important parameter when compared to specific humidity in comfort air-conditioning as it implies the absorption capacity of air. If the initial relative humidity is less, air absorbs more moisture.

Enthalpy of Atmospheric Air

The enthalpy of atmospheric air is the sum of the enthalpies of dry air and water vapor associated with dry air and is given as

Enthalpy of moist air = enthalpy of 1 kg of dry air

+ enthalpy of water vapor per kg of dry air

$$h = h_a + w \cdot h\omega$$

$$\text{or} \quad h = h_a + \omega \cdot h_g \left(\text{Since } h\omega = h_g\right) \tag{14.5}$$

In Eq. 14.5, h_a is the enthalpy of 1 kg of dry air and w hω is the enthalpy of water vapor per kg of dry air.

Degree of Saturation (μ)

The ratio of mass of water vapor per unit mass of dry air to mass of water vapor per unit mass of dry air saturated at the same temperature is called *degree of saturation,* that is

$$\mu = \frac{\omega}{\omega_s} = \frac{0.622\dfrac{p_\omega}{p - p_\omega}}{0.622\dfrac{p_s}{p - p_s}}$$

Simplification of the above equation yields

$$\mu = \frac{p_\omega}{p_s} \cdot \frac{p - p_s}{p - p_\omega} \tag{14.6}$$

The value of μ varies from 0 to 1.

14.1.2 Dew-Point Temperature

An important feature of the behavior of moist air is that partial condensation of the water vapor takes place when the temperature is reduced. There are several situations in which this kind of happening is usually encountered. The formation of dew on grass is a familiar example, and the condensation of vapor on windowpanes and on pipes carrying cold water is another example.

Dew-point temperature, t_{dp}, is defined as the temperature at which condensation begins when the air is cooled at constant pressure; it is the saturation temperature of water corresponding to vapor pressure. Figure 14.1 shows this; when the mixture is cooled at constant pressure, the partial pressure of the vapor remains constant, until point 2 is reached. Suppose that the temperature of the gas–vapor mixture and the partial pressure of vapor in the mixture are in such a way that the vapor is superheated initially at state 1. If any further drop in temperature, condensation begins. The temperature at state 2 is the dew-point temperature. Dew-point temperature is equal to the saturation temperature at the partial pressure of water vapor in the mixture (p_w), which is given as

$$T_{dp} = T_{sat@p_w} \tag{14.7}$$

14.1.3 Wet-Bulb and Dry-Bulb Temperatures

Psychrometer is a device universally used to measure the humidity of air–water vapor mixtures. *Wet-bulb temperature (WBT)* is defined as the temperature measured by the thermometer when its bulb is covered with a cotton wick that is saturated with water and air is blown over the wick. The ordinary temperature of atmospheric air is usually referred to as *dry-bulb temperature (DBT).*

Wet-bulb temperature measurement is basically complex due to the reason that if the air–water vapor mixture is not saturated, it results in evaporation of some of the water in the wick, which in turn diffuses into the surrounding air, thereby cooling the water in the wick. Consequently, the temperature of water drops, causing the heat transfer from both the air and the thermometer to water. A steady state is reached

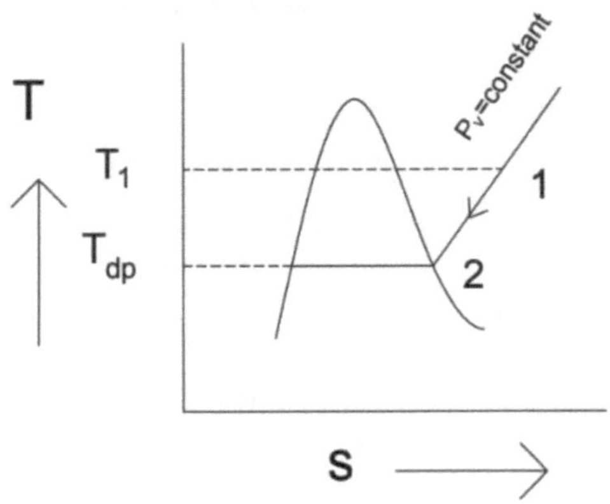

FIGURE 14.1 Constant-pressure cooling of moist air.

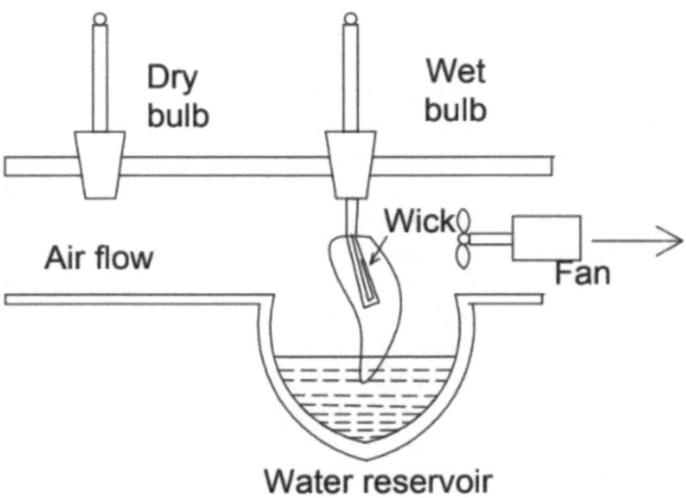

FIGURE 14.2 Dry-bulb and wet-bulb temperature measurement.

as the heat loss from water by evaporation equals the heat gain from the air and the water temperature stabilizes. The temperature recorded by the thermometer at this point is called wet-bulb temperature. Figure 14.2 shows a continuous psychrometer with a fan for drawing air over the bulbs of both the thermometers.

Although the adiabatic saturation temperature and wet-bulb temperature are often treated as the same, they are not the same. The wet-bulb temperature measured by a psychrometer is dependent on heat and mass flow rates and not on thermodynamic equilibrium properties. However, for air–water vapor mixtures, at atmospheric pressure both are assumed to be equal.

14.2 ADIABATIC SATURATION

The specific humidity of air–water vapor mixture can be determined if we know the values of the mixture, pressure, temperature, and adiabatic saturation temperature. The adiabatic saturation process is used for determining the specific humidity of air–water vapor mixture (unsaturated). In this process, an air–vapor mixture comes in contact with a body of water in an insulated chamber. Some of the water will evaporate in doing so, and temperature of air–vapor mixture decreases and moisture content of mixture increases; this is because some part of latent heat of vaporization of air is absorbed by water that evaporates. The adiabatic saturation process is shown schematically and on T-s diagram in Figures 14.3 and 14.4, respectively. If the mixture that exits the chamber is saturated and if the process is adiabatic, then the temperature of the mixture leaving is adiabatic saturation temperature. To make this a steady-state process, the makeup water at adiabatic saturation temperature (T_2) is added at a rate at which it evaporates, while pressure is assumed to be constant. It is important to note that in an adiabatic saturation process, the adiabatic saturation temperature and the temperature of the air–vapor mixture that exits the chamber are dependent on the pressure, temperature, and relative humidity of the entering mixture and the exit pressure. Therefore, measuring the pressure and temperature

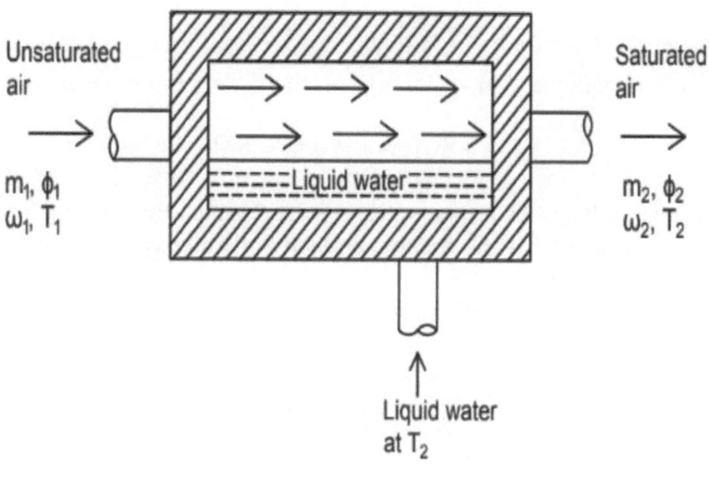

FIGURE 14.3 Adiabatic saturation process.

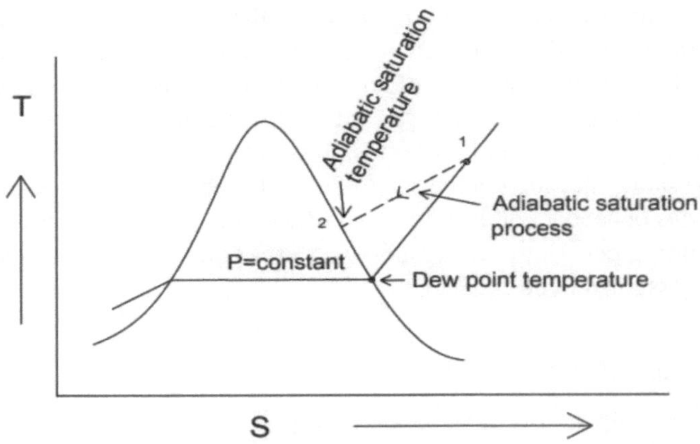

FIGURE 14.4 T-s diagram showing the adiabatic saturation process.

of the mixture entering and leaving the adiabatic saturator, the specific and relative humidity of air–water vapor mixture entering can be determined.

Conservation of mass and energy can be used to establish the relationship among these properties. The continuity equation (mass balance) for air and water is

$$\dot{m}_{a_1} = \dot{m}_{a_2}$$

$$\dot{m}_{w_1} + \dot{m}_f = \dot{m}_{w_2} \left(\text{where } \dot{m}_f \text{ is the rate of evaporation}\right)$$

$$\dot{m}_a\omega_1 + \dot{m}_f = \dot{m}_a\omega_2 \quad \left(\text{where } \omega = \frac{m_w}{m_a}\right)$$

The mass flow rate of water vapor removed is equal to the rate of evaporation, that is

$$\dot{m}_f = \dot{m}_a(\omega_2 - \omega_1)$$

The energy conservation equation (energy balance) for air and water is

$$\dot{m}_a h_1 + \dot{m}_a(\omega_2 - \omega_1)h_{f_2} = \dot{m}_a h_2 \tag{14.8}$$

where h_{f_2} is the enthalpy of liquid water at state 2, since the water is introduced at temperature T_2, and h_w is the specific enthalpy of water vapor in air. And also, $h_{w_2} = h_{g2}$.

Dividing Eq. 14.8 by $\dot{m}_a$ results in the following:

$$h_1 + (\omega_2 - \omega_1)h_{f_2} = h_2$$

$$\omega_1 = \frac{h_2 - h_1 + \omega_2 h_{fg2}}{h_{w_1} - h_{f_2}} \tag{14.9}$$

$$\text{or} \qquad \omega_1 = \frac{c_p(T_2 - T_1) + \omega_2 h_{fg2}}{h_{w_1} - h_{f_2}}$$

At state 2, the air is saturated with water vapor so that $p_{w_2} = p_{s_2}$ and also $\phi_2 = 1.0$.
Then, from Eq. 14.4

$$\omega_2 = \frac{m_{w2}}{m_a} = \frac{0.622 p_{s2}}{p - p_{s2}} \tag{14.10}$$

14.3 PSYCHROMETRIC CHART

It is a graphical representation of properties of air–water vapor mixtures. It is a plot of dry-bulb temperature as abscissa and specific humidity (humidity ratio) as ordinate. The relative humidity, wet-bulb temperature, specific volume, and mixture enthalpy per unit mass of dry air are parameters. Vapor pressure is also considered as ordinate in some charts since at a fixed total pressure, there is one-to-one correspondence between specific humidity and vapor pressure. Curves of constant relative humidity are shown on psychrometric charts and on the left end of the chart, there is a 100% relative humidity curve called the saturation line. Psychrometric charts are used in air-conditioning applications. Figure 14.5 shows the psychrometric chart.

The state of atmospheric air at a specified pressure is specified by two independent intensive properties. The other properties can be calculated from the relations shown in the previous sections. The enthalpy of air–vapor mixture is given per kg of dry air on chart, with an assumption that the enthalpy of dry air is zero at −20°C,

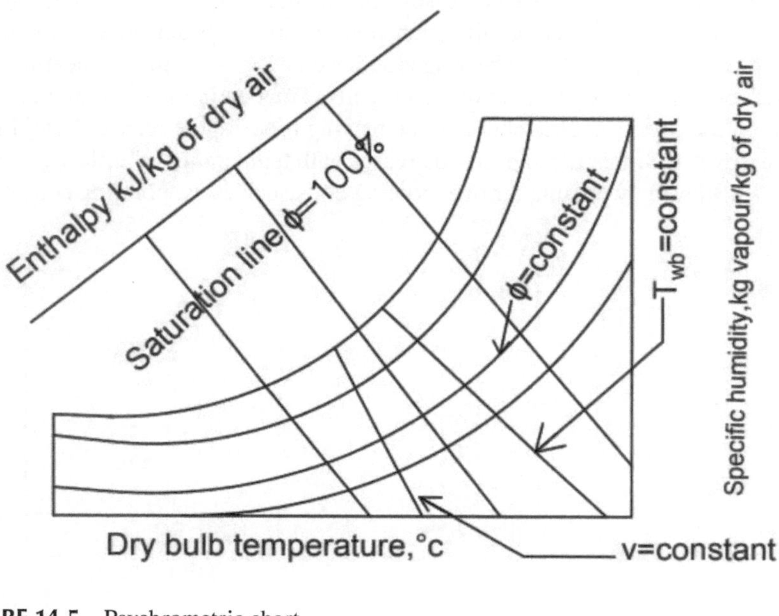

FIGURE 14.5 Psychrometric chart.

and that of vapor is taken from steam tables, with an assumption that enthalpy of saturated liquid is zero at 0°C. The constant–wet-bulb temperature line represents the adiabatic saturation process and also coincides with constant-enthalpy line.

If an unsaturated air–vapor mixture comes in contact with a body of water in an insulated chamber, some of the water will evaporate, resulting in an increase in specific humidity of air and a decrease in temperature of air–vapor mixture. Both the air and water are cooled as evaporation takes place, and the process continues until the energy transferred from air to the water is equal to the energy required to vaporize the water. There exists a thermal equilibrium at this point with respect to air, water, and water vapor, and air becomes saturated. This equilibrium temperature is termed adiabatic saturation temperature or thermodynamic wet-bulb temperature.

14.4 AIR-CONDITIONING PROCESSES

Air-conditioning is basically simultaneous control of temperature and moisture so that the conditioned space is maintained at the desired temperature and humidity. Air-conditioning usually involves two or more of the processes such as simple heating, simple cooling, humidification, and dehumidification to make the conditions human comfort. For example, winter air-conditioning requires heating and humidification of air, while summer air-conditioning requires cooling and dehumidification. Some of the air-conditioning processes are discussed in the next sections.

14.4.1 Sensible Heating and Cooling

Heating or cooling of air without addition or subtraction of moisture is termed sensible heating or cooling. In both these processes, specific humidity of air remains constant with no humidification or dehumidification, and the processes appear as a horizontal line on psychrometric chart. The heating is achieved by passing air over heating coils such as electric resistance heaters or steam coils. The heating and cooling are represented on the psychrometric chart. In the heating process, the relative humidity of air decreases as the moisture capacity increases with temperature. Similarly, sensible cooling is achieved by passing air over cooling coils such as evaporating coils of the

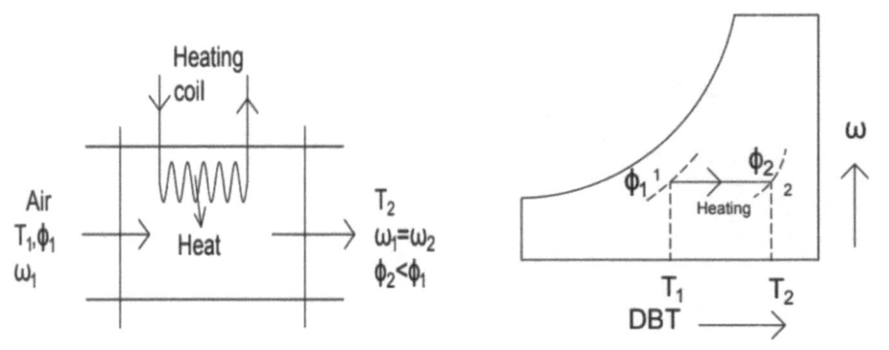

FIGURE 14.6 Sensible heating process. (a)Schematic diagram, (b) Psychrometric chart.

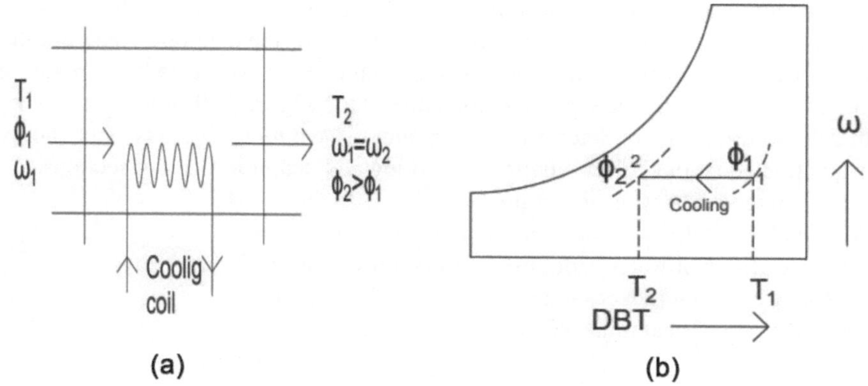

FIGURE 14.7 Sensible cooling process. (a)Schematic diagram, (b) Psychrometric chart.

refrigeration cycle. In this, the relative humidity of air increases and dry-bulb temperature decreases. Figures 14.6 and 14.7 show both the sensible heating and cooling processes schematically and on the psychrometric chart, respectively.

The rate of heat transfer is given as

$$\dot{Q} = \dot{m}_a (h_2 - h_1) \quad (14.11)$$

where h_1 and h_2 are the enthalpies per kg of dry air at the inlet and exit of the heating or cooling coils, respectively and $\dot{m}_a$ is the mass flow rate of air.

14.4.2 Heating with Humidification

The addition of moisture to air is called humidification. Usually, steam injection can be used for increasing the specific humidity of air. Steam used for this should essentially be dry and saturated, and none of the steam is condensed during mixing. Also it is not possible to use the steam below 100°C since the steam is sprayed through the nozzles that the lowest possible enthalpy carried with the steam is the total heat of steam at 100°C. Figure 14.8 shows the heating with humidification. To determine the rate of heat supplied, mass and energy conservation equations can be applied on heating and humidification sections.

Mass balance for dry air $\dot{m}_{a_1} = \dot{m}_{a_2} = \dot{m}_a$

Mass balance for water vapor $\dot{m}_{a_1}\omega_1 = \dot{m}_{a_2}\omega_2 \quad (\text{since}\, \omega_1 = \omega_2)$

First-law equation reduces to $\dot{m}_a h_1 + \dot{Q}_{in} = \dot{m}_a h_2$

Then the rate of heat supplied is

$$\dot{Q}_{in} = \dot{m}_a (h_2 - h_1) \quad (14.12)$$

14.4.3 Cooling with Dehumidification

The removal of water vapor from air is termed *dehumidification*. The simple cooling results in an increase in relative humidity; however, higher levels of relative humidity cause uncomfortable. Hence, the removal of some moisture (dehumidification) is required, and it is possible only if the air is cooled below the dew-point temperature of air. For effective dehumidification, it is essential that the coil surface temperature be maintained below the dew-point temperature of air. Figure 14.9 shows the cooling with dehumidification. To determine the amount of heat removed, mass and energy conservation equations can be applied on cooling and dehumidification sections.

Mass balance for dry air $\dot{m}_{a_1} = \dot{m}_{a_2} = \dot{m}_a$

Mass balance for water vapor $\dot{m}_{a_1}\omega_1 = \dot{m}_{a_2}\omega_2 + \dot{m}_w$

Then, the mass of water vapor (moisture) removed $\dot{m}_w = \dot{m}_a(\omega_1 - \omega_2)$

First-law equation reduces to $\dot{m}_a h_1 = \dot{m}_a h_2 + \dot{Q}_{out} + \dot{m}_w h_w$

Then the rate of heat removed is

$$\dot{Q}_{out} = \dot{m}_a(h_1 - h_2) - \dot{m}_w h_w \quad (14.13)$$

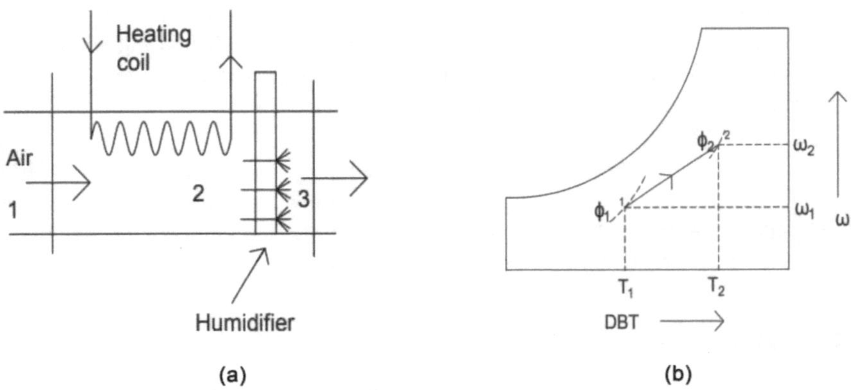

FIGURE 14.8 Heating with humidification. (a)Schematic diagram, (b) Psychrometric chart.

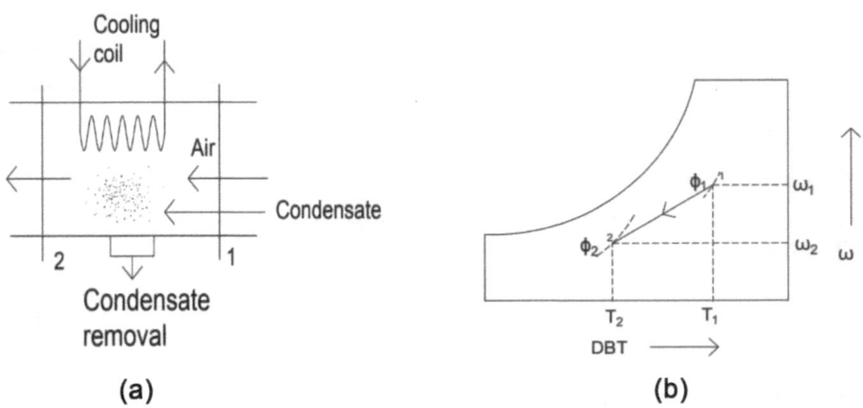

FIGURE 14.9 Cooling with dehumidification. (a)Schematic diagram, (b) Psychrometric chart.

14.4.4 Evaporative Cooling

Evaporative cooling is based on the principle that when water evaporates, the latent heat of vaporization is absorbed from the water body and surrounding air cooling both the water and air. This approach is quite old for cooling the water. In this, a large quantity of water is made to flow through a spray chamber. The air–vapor mixture is allowed to pass through the spray so that a part of circulating water evaporates. The specific humidity of air increases, which is in proportion to the quantity. Figure 14.10 shows the evaporative cooling process schematically and on T-s diagram.

The evaporative cooling process follows the line of constant–wet-bulb temperature line on the psychrometric chart. Therefore, the enthalpy remains constant in this process since the constant–wet-bulb temperature lines are nearly parallel to constant-enthalpy lines on the chart.

Mass balance for dry air $\dot{m}_{a_1} = \dot{m}_{a_2} = \dot{m}_a$

Mass balance for water vapor $\dot{m}_{a_1}\omega_1 + \dot{m}_w = \dot{m}_{a_2}\omega_2$

$$\text{where} \qquad \dot{m}_w = \dot{m}_a(\omega_2 - \omega_1)$$

First-law equation reduces to

$$\dot{m}_a h_1 + \dot{m}_a(\omega_2 - \omega_1)h_f = \dot{m}_a h_2$$

Rearranging the above equation, $\dot{m}_a(h_1 - h_2) + \dot{m}_a(\omega_2 - \omega_1)h_f = 0$

Dividing by $\dot{m}_a$, and rearranging, we get

$$h_1 - \omega_1 h_f = h_2 - \omega_2 h_f \tag{14.14}$$

The evaporative cooling process is widely used in cooling towers to cool the hot water below the dry-bulb temperature of air. However, it is not possible to cool the water below the wet-bulb temperature.

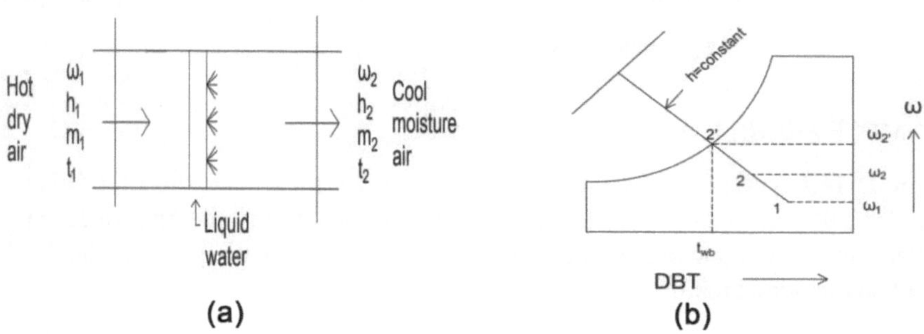

FIGURE 14.10 Evaporative cooling. (a)Schematic diagram, (b) Psychrometric chart.

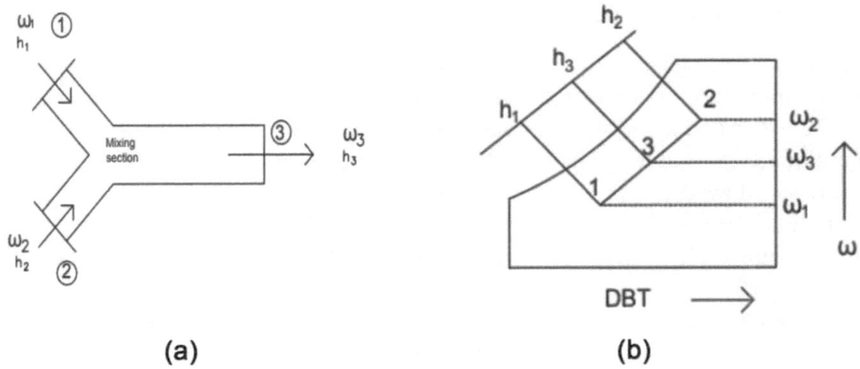

FIGURE 14.11 Adiabatic mixing of two airstreams. (a)Schematic diagram, (b) Psychrometric chart.

14.4.5 Adiabatic Mixing of Airstreams

When the two airstreams of different psychrometric properties are mixed together, then the properties of the mixture can be evaluated by balancing the enthalpies and mass of water vapor in the mixed air. Figure 14.11 shows the adiabatic mixing of two airstreams.

Mass balance of dry air $\dot{m}_{a_1} + \dot{m}_{a_2} = (\dot{m}_{a_1} + \dot{m}_{a_2})a_3$

Mass balance of water vapor $\omega_1\dot{m}_{a_1} + \omega_2\dot{m}_{a_2} = (\dot{m}_{a_1} + \dot{m}_{a_2})\omega_3$

Energy balance $\Rightarrow \dot{m}_{a_1}h_1 + \dot{m}_{a_2}h_2 = (\dot{m}_1 + \dot{m}_2)h_3$

$$\dot{m}_{a_1}(\omega_1 - \omega_3) = \dot{m}_{a_2}(\omega_3 - \omega_2)$$

$$\Rightarrow \frac{\dot{m}_{a_1}}{\dot{m}_{a_2}} = \frac{\omega_3 - \omega_2}{\omega_1 - \omega_3} \tag{14.15}$$

$$\text{and} \quad \frac{\dot{m}_{a_1}}{\dot{m}_{a_2}} = \frac{h_3 - h_2}{h_1 - h_3} \tag{14.16}$$

The specific humidity and enthalpy scale are linear on the psychometric chart, so that the final state lies on the straight line forming the points 1 and 2 and mixture point divides the line into two parts in the ratio of $\frac{\dot{m}_{a_1}}{\dot{m}_{a_2}} = \frac{\text{Distance 2-3}}{\text{Distance 3-1}}$

EXAMPLE PROBLEMS

Example 14.1 Air at a steady rate of 49.5 m³/min, 1 atm, 15°C and 60% RH is first heated to 20°C in a heating section and then humidified by introducing water vapor.

Determine i) the rate of heat transferred to the air in the heating section and ii) the rate of steam added to air.

Solution

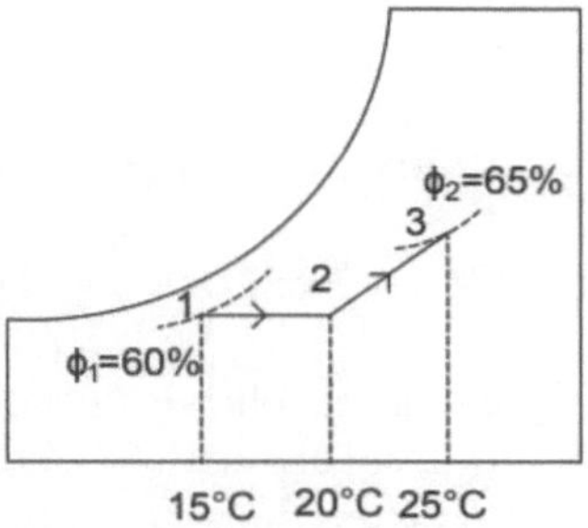

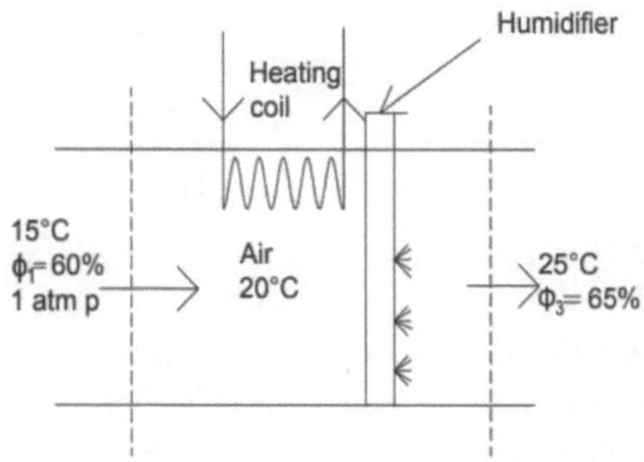

FIGURE EX. 14.1

Assumptions: 1. Dry air and water vapor are considered as ideal gas

2. Changes in kinetic and potential energies are neglected

From psychrometric chart (Refer Fig.Ex.14.1)

$$c_p = 1.005\,\text{kJ/kgK} \qquad R = 0.287\text{kJ/kgK and } v_1 = 0.825\,\text{m}^3/\text{kg}$$

$$h_1 = 31.5\,\text{kJ/kg of dry air} \quad h_2 = 37\,\text{kJ/kg of dry air}$$

$$\omega_1 = \omega_2 = 0.00653\,\text{kg vapor/kg dry air}$$

$$\omega_3 = 0.01308\,\text{kg vapor/kg dry air}$$

$$p_{sat} \text{ of water at } 15°\text{C} \Rightarrow 1.757\,\text{kPa}$$

$$p_{sat} \text{ of water at } 25°\text{C} \Rightarrow 3.1698\,\text{kPa}$$

Enthalpy of saturated water vapor at 15°C $\Rightarrow h_{g_1} = 2528.3\,\text{kJ/kg of dry air}$

Enthalpy of saturated water vapor at 20°C $\Rightarrow h_{g_2} = 2538.1\,\text{kJ/kg of dry air}$

$$\varnothing = \frac{p_{w_1}}{p_{s_1}} \Rightarrow p_{w_1} = 0.6 \times 1.757 = 1.05\,\text{kPa}$$

$$p_{a_1} = p_1 - p_{w_1} = 101.325 - 1.0542 = 100.27\,\text{kPa}$$

$$\omega_1 = \frac{0.622 p_{w_1}}{p_1 - p_{w_1}} = \frac{0.622 \times 1.0542}{100.27} = 0.00653$$

$$h = h_a + h_w \omega \quad \text{and also } \omega_1 = \omega_2$$

$$h_1 = c_p T_1 + \omega_1 h_{g_1} = 1.005 \times 15 + 0.00653 \times 2528.3 = 31.584\,\text{kJ/kg dry air}$$

$$h_2 = c_p T_2 + \omega_2 h_{g_2} = 1.005 \times 20 + 0.00653 \times 2538.1 = 36.67\,\text{kJ/kg dry air}$$

i. The rate of heat transferred to the air in the heating section

$$\therefore \dot{Q}_{in} = \dot{m}_a (h_2 - h_1) = 1(36.67 - 31.584) = 5.108\,\text{kJ/s}$$

$$\text{where } \dot{m}_a = \frac{V_1}{v_1} = \frac{49.5}{0.825 \times 60} = 1\,\text{kJ/s}$$

ii. The rate of steam added to air can be found from mass balance for water in the humidifying section

$$\dot{m}_{a_2}\omega_2 + \dot{m}_w = \dot{m}_{a_3}\omega_3$$

$$\omega_3 = \frac{0.622\phi_3 p_{g3}}{p_3 - \phi_3 p_{g3}} = \frac{0.622(0.65)(3.1698)}{101.325 - 0.65 \times 3.1698} = 0.01308\,\text{kg vapor/kg of dry air}$$

The rate of steam added to air, $\dot{m}_w = \dot{m}_a(\omega_3 - \omega_2) = 1(0.01308 - 0.00653) = 0.0065\,\text{kg/s}$

Example 14.2 An air-conditioning system is to be designed for the following conditions: outdoor: 25°C dbt and 70% RH; required indoor: 20°C dbt and 65% RH; coil dew-point temp: 12°C; and amount of air circulated: 10 m³/s. The required condition is achieved first by cooling, dehumidification, and then heating. Determine (i) the capacity of cooling coil, (ii) the capacity of heating coil, and (iii) the rate of water vapor removed.

Solution

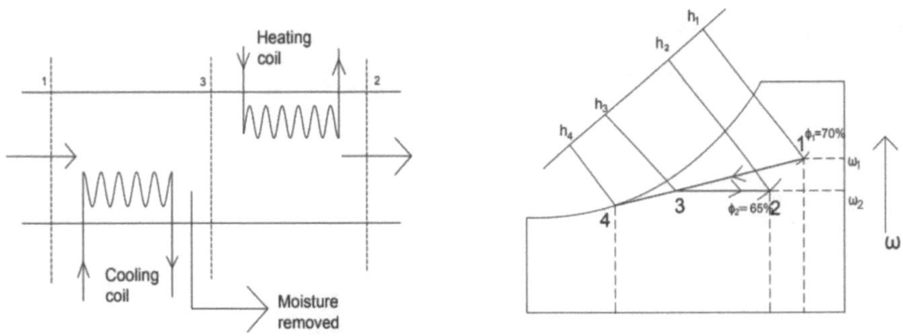

FIGURE EX. 14.2

The property values are taken from the psychrometric chart (refer to Figure Ex. 14.2)

$h_1 = 61$ kJ/kg of dry air $\quad h_2 = 45$ kJ/kg of dry air

$h_3 = 38$ kJ/kg of dry air $\quad h_4 = 35$ kJ/kg of dry air

Volume flow rate, $\dot{V} = 10\,\text{m}^3/\text{s}$ $\quad \nu_1 = 0.865\,\text{m}^3/\text{kg}$ dry air

$\omega_1 = 0.014$ kg vapor/kg dry air

$\omega_2 = \omega_3 = 0.009$ kg vapor/kg dry air

$$\therefore \text{Mass flow rate of air}, \dot{m}_a = \frac{\dot{V}}{\nu} = \frac{10}{0.865} = 11.56\,\text{kg/s}$$

i. Cooling coil capacity $= \dot{m}_a(h_1 - h_3) = 11.56(61 - 38)$

$$= 265.88\,\text{kJ/s} = \frac{265.58 \times 3600}{14{,}000} = 68\,\text{Tonnes}$$

ii. Capacity of heating coil = $\dot{m}_a (h_2 - h_3)$

$= 11.56\ (45 - 38) = 80.92\,\text{kW}$

iii. The rate of water vapor removed = $\dot{m}_a (\omega_1 - \omega_3)$

$= 11.56\ (0.014 - 0.009) = 0.0578\ \text{kg/s}$

Example 14.3 A room of $100\,m^3$ contains air at 27°C and 101.325 kPa at a relative humidity of 60%. Estimate (i) partial pressure of dry air, (ii) specific humidity, (iii) enthalpy of dry air, and (iv) mass of dry air and water vapor.

Solution

From steam tables, for water at 27°C, $p_{sat} = 3.422\,\text{kPa}$, and $h_{fg} = 2551\ \text{kJ/kg}$

i. Partial pressure of dry air, $p_a = p - p_w = 101.325 - 2.053 = 99.27\,\text{kPa}$

Relative humidity, $\phi = \dfrac{p_w}{p_{sat}} \Rightarrow p_w = 0.60 \times 3.422 = 2.053\,\text{kPa}$

ii. Specific humidity, $\omega = 0.622 \times \dfrac{p_w}{p - p_w} = 0.622 \times \dfrac{2.053}{99.27} = 0.0129\,\text{kg vapor/kg of dry air}$

iii. The enthalpy of air–vapor mixture is $G_h = G_{ha} + mh_w$

$$\text{or} \qquad h = h_a + \omega\, h_w$$

$$= 1.005 \times 27 + 0.129 \times 2551 = 59.95\ \text{kJ/kg}$$

iv. Mass of dry air and water vapor:

Since the dry air and water vapor both occupy the entire room, the volume of each gas is equal to the volume of the room:

$$V_w = V_a = 100\,m^3$$

$$p_a\, V_a = m_a\, R_a T$$

Mass of dry air, $\quad m_a = \dfrac{99.27 \times 100}{0.287 \times 300} = 115.3\,\text{kg}$

Mass of water vapor, $m_w = \dfrac{p_w V_w}{R_w T} = \dfrac{2.053 \times 100}{0.4619 \times 300} = 1.4816\,\text{kg}$

Example 14.4 A room contains air at 101.32 kPa, 30°C, and 50% relative humidity. Using the psychometric chart, evaluate (i) enthalpy, (ii) specific volume of air, (iii) specific humidity, and (iv) wet-bulb and dew-point temperatures.

Solution

i. The enthalpy of air per unit mass of dry air can be determined by drawing a line parallel to h = constant lines on the psychrometric chart (refer to Figure Ex. 14.4), which gives

$$h = 65\ \text{kJ/kg of dry air}$$

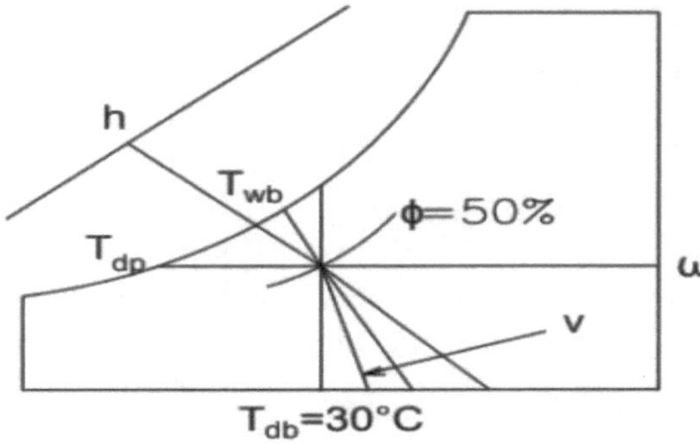

FIGURE EX. 14.4

ii. The specific volume per unit mass of dry air can be determined by taking the distances between a specified state and ν = constant lines on both the sides of the point, which gives

$$\nu = 0.89\,m^3/kg$$

iii. Specific humidity is found by drawing a horizontal line from the specified point to the right until it intersects with the vertical axis, which gives

$$\omega = 0.014\,\text{kg vapor/kg of dry air}$$

iv. Wet-bulb temperature is determined by drawing a line parallel to T_{wb} = constant lines from the specified state until it intersects the saturation line, which gives

$$T_{wb} = 23°C$$

Dew-point temperature is determined by drawing a horizontal line from the specified point to left until it intersects the saturation line, which gives

$$T_{dp} = 19°C$$

Example 14.5 If two streams of air having temperatures 20°C and 25°C, pressure of both 101.32 kPa, relative humidities 25% and 75%, and flow rates 20 and 25 m³/min, respectively, are mixed adiabatically and pressure is maintained 100 kPa, determine the specific humidity.

Solution

$$\omega_1 = 0.622\frac{p_{w_1}}{p - p_{w_1}} \text{ and } \qquad p = p_a + p_w$$

From steam tables, at 20°C, $P_{sat} = P_{s1} = 2.339\,kPa$

at 25°C, $P_{sat} = P_{s2} = 3.169\,kPa$

$$\phi_1 = 25\% = \frac{p_{w_1}}{p_{s_1}} \Rightarrow p_{w_1} = 0.25 \times 2.339 = 0.585\,kPa$$

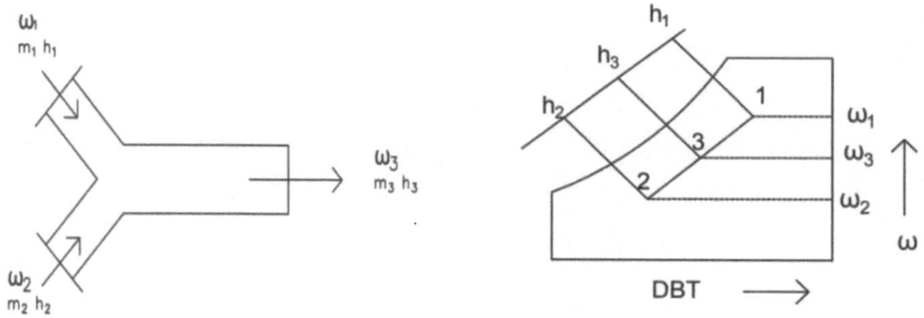

FIGURE EX. 14.5

$$\therefore \omega_1 = 0.622 \times \frac{0.585}{101.325 - 0.585} = 0.0036\,\text{kg vapor/kg of dry air}$$

$$\omega_2 = 0.622 \times \frac{p_{w2}}{p - p_{w2}}$$

$$\Phi_2 = 75\% = \frac{p_{w2}}{p_{s2}} \Rightarrow p_{w_2} = 0.75 \times 3.16 = 2.377\,\text{kPa}$$

$$\therefore \omega_2 = 0.622 \times \frac{2.377}{101.325 - 2.377} = 0.0149\,\text{kg/kg of dry air}$$

The continuity equation for water vapor is $\omega_1 m_{a_1} + \omega_2 m_{a_2} = \omega_3 (m_{a_1} + m_{a_2})$

$$m_{a_1} = \frac{P_{a_1} v_1}{R_a T_1} = \frac{(101.325 - 0.585) \times 20}{0.287 \times 293} = 23.96\,\text{kg/min where } P_{a_1} = P - P_{w_1}$$

$$m_{a_2} = \frac{P_{a_2} v_2}{R_a T_2} = \frac{(101.325 - 2.377) \times 20}{0.287 \times 298} = 28.92\,\text{kg/min where } P_{a_2} = P - P_{w_2}$$

$$\therefore \omega_3 = \frac{0.0036 \times 23.96 + 0.0149 \times 28.92}{23.96 + 28.92} = 0.0098\,\text{kg/kg of dry air}$$

Example 14.6 Air–water vapor mixture at 15°C and 60% RH at a pressure of 101.32 kPa is heated at constant pressure to a temperature of 32°C. Determine (i) the initial and final specific humidities of mixture, (ii) the final relative humidity, (iii) the dew-point temperature, and (iv) the amount of heat transferred per kg of dry air. If the initial mixture is cooled isobarically to 2°C, calculate the amount of water vapor condensed.

Solution

Point 1 is located on the psychometric chart at t_{db} of 15°C and $\phi_1 = 60\%$.

Point 2 is set by drawing a horizontal line from 1 and a vertical line from t_{db} of 32°C.

From the psychrometric chart (refer to Figure Ex. 14.6)

$\omega_1 = \omega_2 = 0.00625$ kg/kg of dry air

$\omega_3 = 0.0048$ kg/kg of dry air

$h_1 = 32.6$ kJ/kg of dry air $h_2 = 48$ kJ/kg of dry air

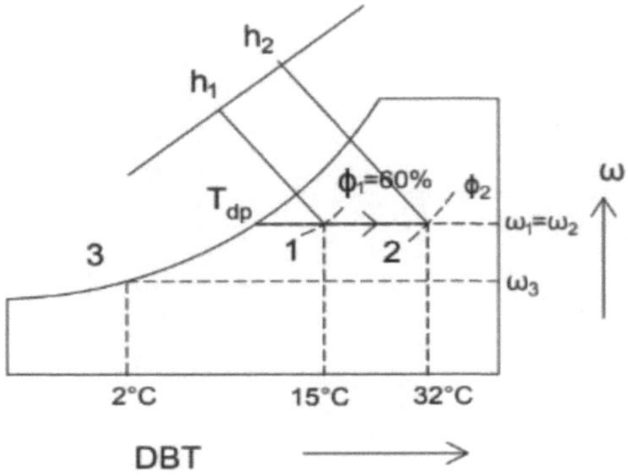

FIGURE EX. 14.6

$t_{db} = 8°C$ $\phi_2 = 24.5\%$

Heat transferred $Q = h_2 - h_1 = 48 - 32.6 = 15.4$ kJ/kg of dry air

Mass of vapor condensed $= \omega_1 - \omega_3 = 0.00625 - 0.0048$

$= 0.0014$ kg/kg of dry air

Example 14.7 Atmospheric air at 27°C and 65% relative humidity is brought to 23°C and 45% RH by means of cooling followed by heating. The barometric pressure is 101 kPa, and the mass flow rate of atmospheric air including moisture is 85 kg/h. Determine (i) the mass flow rate of water vapor removed per hour and (ii) heat removed in the cooler per hour.

Solution

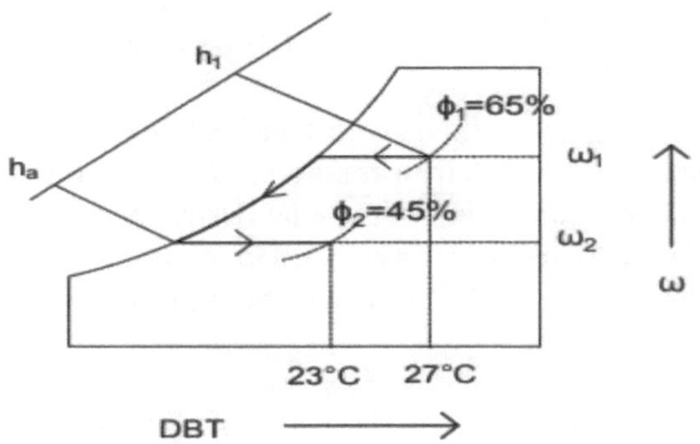

FIGURE EX. 14.7

From the psychrometric chart (refer to Figure Ex. 14.7)

At 27°C and 65% relative humidity, $h_1 = 65$ kJ/kg of dry air

$\omega_1 = 0.017$ kg/kg of dry air

At 23°C and 45% relative humidity, $h_a = 30$ kJ/kg of dry air

$$\omega_2 = 0.008 \text{ kg/kg of dry air}$$

$$\text{Mass flow rate of air}: \dot{m}_a(1+\omega_1) = 85$$

$$= (1+0.017)\dot{m}_a = 85 \Rightarrow \dot{m}_a = 83.58 \text{ kg/hr}$$

i. Mass flow rate of water vapor removed

$$\dot{m}_w = (\omega_1 - \omega_2) = 0.017 - 0.008 = 0.009 \text{ kg/kg of dry air}$$

Mass flow rate of water vapor removed per hour $= \dot{m}_a(\omega_1 - \omega_2)$

$$= 83.58 \times 0.009 = 0.7522 \text{ kg/h}$$

ii. The rate of heat removed in the cooler per hour

$$= \dot{m}_a(h_1 - h_a)$$

$$= 83.58(65 - 30) = 2925.30 \text{ kJ/h}$$

REVIEW QUESTIONS

14.1 What is vapor pressure?
14.2 What is specific humidity?
14.3 What is relative humidity?
14.4 How will (i) the specific humidity and (ii) the relative humidity of the air contained in a well-sealed room change as it is heated?
14.5 When are dry-bulb and dew-point temperatures identical?
14.6 What is dew-point temperature? Explain dew formation
14.7 What is the purpose of psychrometric chart?
14.8 Distinguish between relative humidity and specific humidity
14.9 What is the difference between dry air and atmospheric air?
14.10 What are saturated air and unsaturated air?
14.11 What is adiabatic saturation process?
14.12 What is degree of saturation?
14.13 What is air-conditioning?
14.14 What are the comfort conditions for human beings?
14.15 What is cooling and dehumidification and when do you prefer?
14.16 What is heating and humidification and when do you prefer?
14.17 What are sensible heating and sensible cooling?
14.18 What is evaporative cooling? Explain it with an example
14.19 Why does a sensible heating or cooling line appear as a horizontal line on the psychrometric chart?
14.20 How does relative humidity change during sensible heating and sensible cooling processes?

EXERCISE PROBLEMS

14.1 The air in a room has a pressure of 101 kPa, a dry-bulb temperature of 22°C, and a wet-bulb temperature of 14°C. Using the psychrometric chart, determine (i) the specific humidity, (ii) the relative humidity, (iii) the enthalpy (in kJ/kg dry air), (iv) the dew-point temperature, and (v) the specific volume of the air (in m^3/kg dry air).

14.2 Air at 30°C, 1 bar, and 50% relative humidity enters an insulated chamber operating at steady state with a mass flow rate of 3 kg/min and mixes with a saturated moist airstream entering at 5°C and 1 bar with a mass flow rate of 5 kg/min. A single mixed stream exits at 1 bar. Determine (i) the relative humidity and temperature, in 8°C, of the exiting stream.

14.3 Air enters a compressor operating at steady state at 45°C, 0.9 bar, and 70% relative humidity with a volumetric flow rate of 1 m^3/s. The moist air exits the compressor at 200°C and 1.5 bar. Assuming the compressor is well insulated, determine (i) the relative humidity at the exit, (ii) the power input, in kW, and (iii) the rate of entropy production, in kW/K.

14.4 A flow of moist air at 101.32 kPa, 30°C, and 50% relative humidity is cooled to 10°C in a constant-pressure device. Find the humidity ratio of the inlet and exit flow, and the heat transfer in the device per kg dry air.

14.5 A large room contains moist air at 25°C and 100 kPa. The partial pressure of water vapor is 1.45 kPa. Determine (i) the relative humidity, (ii) the humidity ratio, in kg (vapor) per kg (dry air), (iii) the dew-point temperature, in 8°C, and (iv) the mass of dry air, in kg, if the mass of water vapor is 10 kg.

14.6 Moist air enters a duct at 10°C and 80% relative humidity and a volumetric flow rate of 150 m^3/min. The mixture is heated as it flows through the duct and exits at 30°C. No moisture is added or removed, and the mixture pressure remains approximately constant at 1 bar. For steady-state operation, determine (i) the rate of heat transfer, in kJ/min, and (ii) the relative humidity at the exit. Changes in kinetic and potential energy can be ignored.

14.7 An air-conditioning system is to take in air at 100 kPa, 30°C, and 60% relative humidity and deliver it at 20°C and 40% relative humidity. The air flows first over the cooling coils, where it is cooled and dehumidified, and then over the resistance heating wires, where it is heated to the desired temperature. Assuming that the condensate is removed from the cooling section at 8°C, determine (i) the temperature of air before it enters the heating section, (ii) the amount of heat removed in the cooling section, and (iii) the amount of heat transferred in the heating section, both in kJ/kg dry air.

14.8 Air enters a compressor operating at steady state at 50°C, 0.9 bar, and 70% relative humidity with a volumetric flow rate of 0.8 m^3/s. The moist air exits the compressor at 195°C and 1.5 bar. Assuming the compressor is well insulated, determine (i) the relative humidity at the exit, (ii) the power input, in kW, and (iii) the rate of entropy production, in kW/K.

14.9 The requirements for an air-conditioned room are air at 18°C, 100 kPa, and 60% RH at a flow rate of 25 m^3/min. The conditioner takes in air at 100 kPa and 75% RH, cools it to adjust the moisture content, and reheats it to room temperature. Assuming the fan before the cooler absorbs the 0.45 kW and the condensate is discharged at the temperature to which the air is cooled, determine (i) the temperature to which the air is cooled and (ii) thermal loading on both the cooler and the heater.

14.10 The discharge moist air from a clothes dryer is at 35°C and 75% relative humidity. The flow is guided through a pipe up through the roof and a vent to the atmosphere. Due to heat transfer in the pipe, the flow is cooled to 20°C by the time it reaches the vent. Find the humidity ratio in the flow out of the clothes dryer and at the vent. Find the heat transfer and any amount of liquid that may be forming per kg dry air for the flow.

14.11 A fixed amount of air initially at 52°C, 1 atm, and 10% relative humidity is cooled at constant pressure to 15°C. Using the psychrometric chart, determine whether condensation occurs. If so, evaluate the amount of water condensed, in kg per kg of dry air. If there is no condensation, determine the relative humidity at the final state.

14.12 Ambient air at 100 kPa, 30°C, and 40% relative humidity goes through a constant-pressure heat exchanger as a steady flow. In one case, it is heated to 45°C, and in another case, it is cooled until it reaches saturation. For both the cases, find the exit relative humidity and the amount of heat transfer per kg of dry air.

14.13 The mixture of air–water vapor enters a heater–humidifier system at 7°C, 1.03 bar, and 60% RH. Liquid water at a flow rate of 1.5 kg/min is sprayed into the mixture, while dry air flow rate is 50 kg/min. The mixture leaves the unit at 27°C and 1.03 bar. Determine (i) the rate of heat transfer to the unit and (ii) the relative humidity at outlet.

14.14 An automobile air conditioner uses refrigerant R-134a as the cooling fluid. The evaporator operates at 250 kPa gauge and the condenser operates at 15 bar gauge. The compressor requires a power input of 5.6 kW and has an isentropic efficiency of 80%. Atmospheric air at 24°C and 45% relative humidity enters the evaporator and leaves at 7°C and 80% relative humidity. Determine the volume flow rate of the atmospheric air entering the evaporator of the air conditioner, in m^3/min.

14.15 Ambient air is at a condition of 100 kPa, 35°C, and 50% relative humidity. A steady stream of air at 100 kPa, 23°C, and 70% relative humidity is to be produced by first cooling a flow of ambient air to an appropriate temperature to condense out the proper amount of water and then mix this stream adiabatically with another flow under ambient conditions. What is the ratio of the two flow rates? To what temperature must the first stream be cooled?

14.16 A 180 m^3/min of air with a dry-bulb temperature of 10°C and a wet-bulb temperature of 6°C is continuously mixed with 450 m^3/min of air with a dry-bulb temperature of 30°C and a relative humidity of 60%. The mixing chamber is at atmospheric pressure and is electrically heated with a power consumption of 4 kW. For the resulting mixture, determine (i) the dry-bulb temperature, (ii) the wet-bulb temperature, (iii) the dew-point temperature, and (iv) the relative humidity.

14.17 Air at 35°C, 1 atm, and 50% relative humidity enters a dehumidifier operating at steady state. Saturated moist air and condensate exit in separate streams, each at 15°C. Neglecting kinetic and potential energy effects, (i) determine the heat transfer from the moist air, in kJ per kg of dry air, and (ii) determine the amount of water condensed, in kg per kg of dry air, and (iii) check your answers using data from the psychrometric chart.

14.18 Air enters an evaporative cooler at 100 kPa, 28°C, and 30% relative humidity at a rate of 0.08 m3/s, and it leaves with a relative humidity of 80%. Determine (i) the exit temperature of the air and (ii) the required rate of water supply to the evaporative cooler.

14.19 A flow of moist air at 45°C and 15% relative humidity with a flow rate of 15 kg/min dry air is mixed with a flow of moist air at 27°C and absolute humidity of 0.0165 with a rate of 20 kg/min dry air. The mixing takes place in an air duct at 100 kPa, and there is no significant heat transfer. After the mixing, there is heat transfer to a final temperature of 27°C. Find the temperature and relative humidity after mixing. Find the heat transfer and the final exit relative humidity.

14.20 Outside atmospheric air with a dry-bulb temperature of 30°C and a wet-bulb temperature of 20°C is to be passed through an air-conditioning device so that it enters a house at 15°C and 35% relative humidity. The process consists of two steps: First, the air passes over a cooling coil, where it is cooled below its dew-point temperature and the water condenses out until the desired humidity ratio is reached. Then, the air is passed over a reheating coil until its temperature reaches 15°C. Determine (i) the amount of water removed per kg of dry air passing through the device, (ii) the heat removed by the cooling coil in kJ/kg of dry air, and (iii) the heat added by the reheating coil in kJ/kg of dry air.

14.21 Atmospheric air can be dehumidified by cooling the air at constant total pressure until the moisture condenses out. Suppose that air with a humidity ratio of 0.008 kg water per kg of dry air must be achieved by cooling incoming atmospheric air with a dry-bulb temperature of 23°C and a wet-bulb temperature of 18°C. Determine (i) to what temperature the incoming air–water vapor mixture must be cooled to achieve a humidity ratio of 0.008 kg water per kg dry air and (ii) how much water must be removed per kg of dry air to achieve this state.

DESIGN AND EXPERIMENT PROBLEMS

14.22 Dual-fuel heat pumps use both natural gas and electricity. At lower temperatures, gas is used for increased efficiency, while at higher temperatures, electricity is advantageous. Design an air-conditioning system with advanced heat ventilation and air-conditioning (HVAC) technology concepts to work on dual-fuel heat pumps with the features such as controls, advanced monitoring systems, building automation systems, and smart thermostats.

14.23 Electric clothes dryers consume a part of residential electricity consumption. Available electric clothes dryers today are based on either electric resistance (low-cost but energy-inefficient) or vapor compression (energy-efficient but high-cost). Thermoelectric dryers have the potential to alleviate the disadvantages of both through a low-cost, energy-efficient solution. Design an energy-efficient and low-cost thermoelectric dryer with cooling capacity of 1 kW and to heat air from 35°C to 60°C.

14.24 Study the air-conditioning system in your college seminar hall in terms of the parameters such as occupant comfort, potential impact on productivity, and energy requirements, and propose a thermally driven air-conditioning system that is based on solar power supplemented by natural gas. Compare the energy requirements, cost of the energy, comfort conditions, and maintenance cost of the existing and proposed systems. Draw your conclusions based on your study.

14.25 Design a solar-powered evaporative cooling system of 1 m^3 capacity to increase the shelf life of stored chilli based on the principle of evaporative cooling and increasing the relative humidity (RH) in the preservation chamber. The storage system should be designed in such a way that the jute pad should trap the moisture by water flowing through a series of perforated pipes from the reservoir located at the top of the storage system. The average cooling efficiency is required to be 80%, the temperature in the system should range from 5°C to 10°C, and the relative humidity in the cooling chamber should be 85%. The design should also consider the capacity of battery charger.

14.26 Compare a commercial air-conditioning system in your locale installed way back in the 1990s with the newly installed one. Evaluate the efficiency of the systems in terms of comfort level provided, operating and maintenance costs, global warming potential of the refrigerants used, and other relevant issues. Assess whether running the older system is feasible, and recommend upgradation or full system replacement in comparison with the new unit. Prepare a report of your findings.

15 Chemical Potential of Ideal Fermi and Bose Gases

LEARNING OUTCOMES

After learning this chapter, students should be able to

- Form a sound base for the development of the concepts related to chemical thermodynamics such as chemical potential and fugacity
- Apply Gibbs function in chemical equilibrium
- Derive the fugacity and fugacity coefficient for real gases
- Evaluate the properties of ideal Fermi–Dirac and Bose–Einstein gases
- Understand the low-temperature behavior of physical systems

15.1 INTRODUCTION

Rudolf Clausius (1822–1888) in 1865 published *The Mechanical Theory of Heat*, in which he proposed that the principles of thermochemistry (e.g., heat evolved in combustion reactions) can conveniently be applied to principles of thermodynamics. Founded by the work of Clausius, Josiah Willard Gibbs (1839–1903), during 1973–1976, established the relations for the evaluation of thermodynamic equilibrium of chemical reactions and their tendencies of occurrence. For his pioneering contributions, Gibbs is recognized as the father of chemical thermodynamics. His contributions led to the development of a unified body of thermodynamic theorems based on the principles introduced by others such as Sadi Carnot and Clausius. Chemical thermodynamics deals with the interrelation of heat and work with chemical reactions or with physical changes of state under the limitations of the laws of thermodynamics. It applies mathematical methods to solve the chemical questions and evaluates various thermodynamic properties by using laboratory measurements.

15.2 CHEMICAL POTENTIAL AND FUGACITY

The concept of chemical potential was introduced by Gibbs long back. Gibbs established the mathematical beauty of thermodynamics by articulating the fundamental equation of thermodynamics of a system. The Gibbs inventions, i.e., the formation of the fundamental equation and introduction of chemical potential, are the starting points that led to the development of chemical thermodynamics and showed the path to apply thermodynamics to materials science and engineering.

For a simple compressible system of fixed mass with uniform temperature and pressure throughout, in the absence of motion and gravity of overall system, the first law of thermodynamics in differential form is $\delta Q = dU + \delta W$.

When the system involves PdV work only, then

$$\delta Q = dU + PdV \tag{15.1}$$

The entropy balance equation in differential form when the temperature is uniform with position throughout the system, based on Eq. 7.34, is

$$dS = \frac{\delta Q}{T} + S_{gen} \tag{15.2}$$

We can eliminate δQ from both Eqs. 15.1 and 15.2:

$$TdS - dU - PdV = TS_{gen} \tag{15.3}$$

According to the entropy principle, entropy of an isolated system (adiabatic closed system) during a process always increases. Entropy is generated in all actual processes and conserved only in the absence of irreversibilities. Hence, Eq. 15.3 puts a limitation on the direction of processes. The only processes allowed are those for which $S_{gen} \geq 0$. Equation 15.3 thus becomes

$$TdS - dU - PdV \geq 0 \tag{15.4}$$

The *total Gibbs function*, G, is given by

$$G = H - TS \tag{15.5}$$

Molar Gibbs function, g, is given by $g = h - Ts$.

The differential of Gibbs function is $dG = dH - TdS - SdT$,

where $H = U + PV$ and $dH = dU + PdV + VdP$.

Then, $dG = (dU + PdV + VdP) - TdS - SdT$

$$dG - VdP + SdT = -(TdS - dU - PdV) \tag{15.6}$$

The right-hand side of Eq. 15.6 is the same as Eq. 15.4 except the negative sign:

$$dG - VdP + SdT \leq 0 \tag{15.7}$$

In Eq. 15.7, due to the negative sign considered above, inequality reverses the direction. From Eq. 15.7, for any process that occurs at a specified temperature and pressure, it can be concluded that

$$(dG)_{T,P} \leq 0 \tag{15.8}$$

A chemical reaction at a given temperature and pressure will essentially proceed in the direction of a reducing Gibbs function. Therefore, any chemical reaction, at a definite temperature and pressure, proceeding in the direction of increasing Gibbs function is a violation of second law of thermodynamics. A reacting system in a change of pressure or temperature will assume a different equilibrium state that is characterized by minimum Gibbs function at that pressure or temperature.

Gibbs free energy or specific Gibbs function (g) is a measure of the maximum available work that can be derived from any system at constant temperature and pressure. The total Gibbs function G is a thermodynamic state function that depends on the conditions that are imposed on the system such as electrical, magnetic, and gravitational fields and temperature and pressure but not on the past history of the system. Since it is not possible to determine the absolute values of G, differential change in G of a pure substance based on the equation $dh = vdP + Tds$ is given as

$$dG = VdP - SdT \tag{15.9}$$

Let 1, 2, 3…k substances of a mixture are contained by a system. The total Gibbs function for a mixture is dependent not only on two independent intensive properties but also on the composition, given as $G = G(P, T, n_1, n_2, n_3 \ldots, n_k)$. In the differential form, we can write the total Gibbs function for a mixture as

$$dG = \left(\frac{\partial G}{\partial P}\right)_{T,n_i} dP + \left(\frac{\partial G}{\partial T}\right)_{P,n_i} dT + \sum_{i=1}^{k}\left(\frac{\partial G}{\partial n_i}\right)_{P,T,n_j} dn_i \tag{15.10}$$

where the subscript i indicates any substance while the subscript j indicates any other substance except the one whose number of moles varies.

Comparing Eqs. 15.9 and 15.10,

$$dG = VdP - SdT + \sum_{i=1}^{k}\left(\frac{\partial G}{\partial n_i}\right)_{P,T,n_j} dn_i \tag{15.11}$$

$G = H - TS$ and $H = U + PV \rightarrow G = U + PV - TS$.

Then, $dG = dU + PdV + VdP - TdS - SdT$.

$$dU = TdS - PdV + \sum_{i=1}^{k}\left(\frac{\partial G}{\partial n_i}\right)_{P,T,n_j} dn_i \tag{15.12}$$

Similarly, we can write the expressions for changes in other thermodynamic potentials dA and dH:

$$dA = -SdT - PdV + \sum_{i=1}^{k}\left(\frac{\partial G}{\partial n_i}\right)_{T,V,n_j} dn_i \tag{15.13}$$

$$dH = TdS + VdP + \sum_{i=1}^{k} \left(\frac{\partial G}{\partial n_i} \right)_{S,P,n_j} dn_i \tag{15.14}$$

Similarly, for a mixture (system of variable composition), the internal energy depends on S, V, and number of moles of different constituents of the system, n_1, n_2, $n_3 \ldots$, n_k, and is given as

$$U = U(S, V, n_1, n_2, n_3 \ldots, n_k)$$

In the differential form, we can write the total internal energy for a mixture as

$$dU = \left(\frac{\partial U}{\partial S} \right)_{V,n_i} dS + \left(\frac{\partial U}{\partial V} \right)_{S,n_i} dV + \sum_{i=1}^{k} \left(\frac{\partial U}{\partial n_i} \right)_{S,V,n_j} dn_i \tag{15.15}$$

and

$$dU = TdS - PdV \tag{15.16}$$

Comparing Eqs. 15.15 and 15.16,

$$dU = TdS - PdV + \sum_{i=1}^{k} \left(\frac{\partial U}{\partial n_i} \right)_{S,V,n_j} dn_i \tag{15.17}$$

Comparing Eqs. 15.2 and 15.17,

$$\mu_i = \left(\frac{\partial G}{\partial n_i} \right)_{P,T,n_j} = \left(\frac{\partial U}{\partial n_i} \right)_{S,V,n_j} \tag{15.18}$$

where μ_i is the chemical potential of the ith component of the mixture. Chemical potential is equal to the Gibbs function for a pure substance. The *chemical potential* of a chemical substance, μ, is analogous to electrical potential, gravitational potential, thermal potential, and mechanical potential. It can be defined as the chemical energy (U_c) possessed by 1 mol of the substance, which is given as

$$\mu = \frac{U_c}{n} \tag{15.19}$$

where n is the number of moles of the substance.

Chemical potential is a thermodynamics concept that plays an important role particularly in the field of materials science and also others including physics, chemistry, biology, and chemical engineering. Thermodynamic properties of materials can be obtained at a specified temperature and pressure from the known value of chemical potential. It is also useful in the determination of stability of chemical species, compounds, and solutions at a constant temperature and pressure. In addition, chemical potential also plays a vital role in finding out the tendency of chemical substances to react chemically, to transform to new physical states, etc. However, its use is confined due to its nature of being difficult to understand and confusing. Another reason may be the lack of a unique unit associated with chemical potential.

The *standard chemical potential,* μ^0, of a pure substance in a given phase and at a given temperature is the chemical potential of the substance when it is in the standard state of the phase at this temperature and standard pressure, p^0.

It cannot be possible to evaluate the absolute value of μ at a specified temperature and pressure and μ^0 at the same temperature. However, we can evaluate the difference $\mu - \mu^0$.

From Eq. 15.9, $dG = VdP - SdT$. Dividing both sides of this equation by n results in the total differential of the chemical potential with the same independent variables as given below

$$d\mu = -S_m dT + V_m dP \tag{15.20}$$

By considering the coefficients on the right-hand side of Eq. 15.20 as partial derivatives, we can write the expressions for a pure substance:

$$\left(\frac{d\mu}{dT}\right)_P = -S_m \tag{15.21}$$

and

$$\left(\frac{d\mu}{dP}\right)_T = V_m \tag{15.22}$$

Equation 15.22 shows that chemical potential increases with an increase in pressure for an isothermal process.

We can integrate $d\mu = V_m dP$ (by setting dT equal to zero) from the standard state at pressure P^0 to the experimental state at pressure P^I as given below:

$$\mu\left(P^I\right) - \mu^0 = \int_{P^0}^{P^I} V_m \, dP \tag{15.23}$$

Figure 15.1 shows the chemical potential as a function of pressure when temperature is constant for the real gas and the same gas when it behaves as an ideal gas, shown as solid curve and dotted curve, respectively. Point 1 is the gas standard state while point 2 is the state of the real gas at p^I. The fugacity f (p^I) of the real gas at p^I is equal to the pressure of the ideal gas when its chemical potential is equal to the real gas, indicated as point 3.

An ideal gas is said to be in its standard state at a specified temperature when its pressure is the standard pressure. It can be possible to derive the relation between chemical potential of an ideal gas and its pressure and standard chemical potential at the same temperature by setting V_m equal to RT/P in Eq. 15.23:

$$\mu\left(P^I\right) - \mu^0 = \int_{P^0}^{P^I} \left(\frac{RT}{P}\right) dP = RT\ln\left(P^I/P^0\right) \tag{15.24}$$

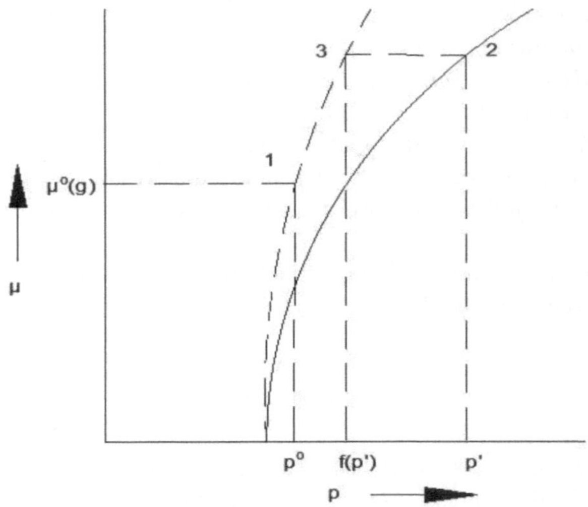

FIGURE 15.1 Chemical potential as a function of pressure when temperature is constant.

Now we can write the general relation for μ as a function of P for both ideal and real gases:

$$\mu = \mu^0(g) + RT \ln \frac{P}{P^0} \text{(pure ideal gas at constant T)} \quad (15.25)$$

$$\mu = \mu^0(g) + RT \ln \frac{f}{P^0} \text{(pure gas)} \quad (15.26)$$

The ‘f’ in Eq. 15.26 is called fugacity of the real gas.

Chemical potential is quantified by *fugacity*. If a chemical with two different fugacities is kept in two different compartments that are in contact, the flow of chemical potential takes place from the compartment of higher fugacity to the compartment of lower fugacity. This will be continued until there is an equilibrium—i.e., the fugacities are equal—at which net transfer is zero. Fugacity can be effectively used for the calculation of chemical equilibrium of real gases based on the condition that the chemical potential of the reactants is equal to the chemical potential of the products. Fugacity has pressure dimensions and is a kind of effective pressure. Fugacity is the pressure that an imaginary gas will have so that it is the chemical potential at a specified temperature to be the same as the chemical potential of the real gas. Fugacity is equal to its pressure when the gas is an ideal gas. In chemical thermodynamics, fugacity can be used to compute the chemical equilibrium constant of a real gas, which is an effective partial pressure of a real gas that replaces mechanical partial pressure. Fugacity has the dimensions of pressure. This effective partial pressure is equal to the pressure of an ideal gas that has the same temperature and molar Gibbs free energy as the real gas.

The *fugacity coefficient*, ϕ, is defined by

$$\phi = \frac{f}{P} \tag{15.27}$$

The fugacity coefficient for a pure gas at P^I (pressure at experimental state) is given by

$$\ln \phi\left(P^I\right) = \int_0^{P^I} \left(\frac{V_m}{RT} - \frac{1}{P}\right) dP \tag{15.28}$$

15.3 CHEMICAL POTENTIAL AND THERMAL RADIATION

Radiation is basically an electromagnetic wave phenomenon consisting of photons that do not interact with each other. It is defined as the propagation of a collection of particles (photons) or the propagation of electromagnetic waves. It is the emission and transmission of energy in the form of waves such as gamma rays, X-rays, and ultraviolet (UV) and infrared rays through the space or a medium. The properties of photons are evaluated by interaction with the matter, emitters, and absorbers. *Thermal radiation* is the one that is caused by and affects the temperature of the matter. It is in the spectrum extending from 0.1 to 100 μm including a portion of UV and all of the visible and infrared radiation. The intensity and the spectrum of thermal radiation are a function of wavelength and emitter's temperature. Nonthermal radiation, on the other hand, is identified by its high intensity, and the emitter's temperature in this case has less significance. The thermodynamic dealing of radiation by Kirchhoff, Wien, Rayleigh, and Planck led to the development of well-known Planck's blackbody radiation formula.

Chemical potential of photons is used to account for the difference between thermal and nonthermal radiation. Planck's law for thermal radiation can be conveniently used to describe the radiation of any kind that has the real temperature of the emitting material. The thermal radiation can be well defined by a vanishing chemical potential μ_γ of photons that results from the non-conservation of photon number N_γ. The energy changes in a system dU, according to the Gibbs, are related to the changes in extensive variables. We can apply this to photons in a cavity:

$$dU = TdS - pdV + \mu_\gamma dN_\gamma \tag{15.29}$$

where S is entropy, p pressure, V volume, and T temperature. The equilibria of the system are characterized by minimal energy. For the chemical equilibrium, at constant entropy and volume $\mu_\gamma\ dN_\gamma = 0$. However, due to the non-conservation of photon number, $dN_\gamma \neq 0$, and hence in equilibrium, chemical potential must be zero ($\mu_\gamma = 0$). The vanishing chemical potential is assumed to be a general property of the photons and is true to nonthermal radiation as well.

15.4 PROPERTIES OF IDEAL FERMI–DIRAC AND BOSE–EINSTEIN GASES

Classical thermodynamics is based on the continuum macroscopic approach that disregards the recognition of the existence of atoms and molecules while statistical thermodynamics analyzes the molecular-level interactions by using standard statistical methods and is related to the microscopic approach.

There are four basic attributes of statistical thermodynamics. First, it can be used to explain certain apparent discontinuities in physical behavior, such as superconductivity. Second, it can be used to extend classical thermodynamic results into regions where the continuum hypothesis is no longer valid, as in the case of rarefied gases. Third, it can often provide a molecular interpretation of physical phenomena that are observed at the macroscopic level but originate at the molecular level (such as fluid viscosity). Fourth and perhaps most important, it can function as a tool to provide accurate equations of state that describe the behavior of non-measurable thermodynamic properties, such as internal energy, enthalpy, and entropy, as a function of measurable properties, such as pressure, temperature, and density, without resorting to experimental measurements.

Maxwell–Boltzmann statistics is also known as a classical case, while Bose–Einstein statistics and the Fermi statistics are quantum-mechanical cases.

The knowledge of chemical potential is vital for the calculation of properties of quantum gases in Bose–Einstein and Fermi–Dirac distribution functions. And these properties are the basis for understanding of a wider range of physical systems such as electrons in metals, helium liquids, and trapped gas systems at low-temperatures. Thus, deriving expressions for chemical potentials as functions of temperature using Bose–Einstein and Fermi–Dirac statistics, it can be possible to calculate chemical potentials of monatomic gases across the temperature region in which quantum degeneracy effect is typically present.

The point at which phase transition takes place in case of an ideal gas of Bose atoms is dependent on temperature along the chemical potential. Chemical potential is zero at the transition point, and it continues to be so with a further decrease in temperature. And at this transition point, there appears a gas component with a macroscopically large number of particles called Bose–Einstein condensate (BEC) in one quantum-mechanical state identified by the lowest energy level of the system in the boson distribution function. In this distribution function, Bose-condensate component is mathematically described by a generalized function (Dirac delta function). All the gas particles must be in Bose-condensate state of zero temperature.

In bosons, a greater number of particles are allowed in one-particle state, while only one particle is allowed in a single one-particle state, and thus, there exists only one permissible state of the system. For an ideal gas of Fermi atoms, on temperature scale, absolute zero is the only singularity. Fermi energy, that is, the energy of the highest populated one-particle state of the system, is defined to introduce a physically meaningful energy scale at zero temperature. When the temperature approaches Fermi energy, quantum properties of the fermion gas are apparent. Fermi distribution function is a generalized function in the limit of zero temperature, and according to this function, the energy of any Fermi particle of the gas at zero temperature does not exceed the Fermi energy.

Let us choose the well-known distribution of an average number of particles over quantum states with energy ε to describe the ideal monatomic gases:

$$f(\varepsilon, T) = \frac{1}{e^{(\varepsilon-\mu)/T} \pm 1} \tag{15.30}$$

In Eq. 15.30, both the signs '+' and '–' specify respectively the Fermi–Dirac and Bose–Einstein statistics and chemical potential is usually the function of temperature T and is expressed in energy units.

Now we can develop the relation for determining the chemical potential as a function of temperature from the relation for the total number (N) of gas particles by integrating Eq. 15.30 over the phase space.

$$N = \frac{gV}{(2\pi h)^3} \int_0^{\alpha} \frac{4\pi p^2 dp}{e^{(\varepsilon-\mu)/T} \pm 1} \tag{15.31}$$

where g is the degree of the degeneracy of the energy states of the gas particles and V is the volume of the system of the space. If we specifically consider standard energy dependence on the momentum in accordance with the dispersion law for nonrelativistic monatomic ideal gas, the resulting relation is $\varepsilon(p) = p^2/m$ (m is the mass of the particles).

For a Bose gas with the density $n = N/V$, the Bose condensation temperature, that is, the temperature (in energy units) at which chemical potential equals the energy of the lowest energy state, is given by

$$T_B = \frac{1}{2m} \left(\frac{4\pi^2 h^3 n}{\Gamma(3/2)\zeta(3/2) g} \right)^{2/3} \tag{15.32}$$

For a Fermi gas, the temperature, that is, the energy of the highest populated state of the system at zero temperature known as Fermi energy, is given by

$$T_F = \frac{1}{2m} \left(\frac{6\pi^2 h^3 n}{g} \right)^{2/3} \tag{15.33}$$

In Eq. 15.32, at $T = 0$, $\mu_F = \varepsilon_F$, meaning that at zero temperature, chemical potential equals Fermi energy. The integral of the right-hand side of Eq. 15.29 can be made dimensionless by writing

$$\int_0^{\alpha} \frac{\sqrt{x}dx}{e^{-v/t} e^x \pm 1} = a \pm t^{-3/2} \tag{15.34}$$

In Eq. 15.34, for Bose gas, $t = T/T_B$, $v = \mu/T_B$, and $a_- = \Gamma(3/2)\zeta(3/2)$.

And for Fermi gas, $t = T/T_F$, $\nu = \mu/T_F$, and $a_+ = 2/3$.

The solution of the parametric equation, Eq. 15.34, which defines the chemical potential as a function of temperature at a fixed density of particle, can be used to describe the thermodynamics of quantum gases.

15.5 BOSE AND FERMI FUGACITY

For a quantum gas, after eliminating the fugacity f, the equation of state is given as

$$\frac{P}{kT} = \frac{1}{V} \ln \Phi (T,V,f) \tag{15.35}$$

$$n = \frac{1}{V} f \frac{\partial}{\partial f} \ln \Phi (T,V,f) \tag{15.36}$$

In Eqs. 15.35 and 15.36, P is pressure, V is volume, n is particle number density, and Φ is the grand potential a quantity used in statistical mechanics, in particular for irreversible processes. We can obtain the equation of state for isobaric processes by solving $f = f(P, T)$ from Eq. 15.35 and substituting in Eq. 15.36:

$$n = \frac{1}{V} f \frac{\partial}{\partial f} \ln \Phi (T,V,f)\Big|_{f=f(P,T)} \tag{15.37}$$

Similarly for isochoric processes, solving $f = f(n, T)$ from Eq. 15.36 and substituting in Eq. 15.35, we can obtain the equation of state:

$$n = \frac{1}{V} f \frac{\partial}{\partial f} \ln \Phi (T,V,f)\Big|_{f=f(n,T)} \tag{15.38}$$

Fugacity f can only be obtained approximately in most of the cases such as high-temperature and low-density and low-temperature high-density limits. Only in some cases such as classical ideal gases and two-dimensional ideal quantum gases, the explicit expression for fugacity f can be obtained by solving Eq. 15.35 or 15.36.

15.6 LOW-TEMPERATURE BEHAVIOR OF PHYSICAL SYSTEMS

With the advents in cryogenics, it can be possible to develop and study the systems at very low temperatures near to absolute zero. Low-temperature systems exhibit quantum phenomena such as quantum Hall effect and superconductivity which are essentially useful in a diverse field of applications that include precision measurements, fast digital electronics, and nuclear magnetic resonance (NMR) spectroscopy. NMR is extensively used to study molecular physics and to determine the structure of organic molecules in solution, both crystalline and non-crystalline materials. Low-temperature systems also play a vital role in the fundamental research of particle physics. At low temperatures, the thermodynamic behavior of superfluids is influenced largely by thermal excitation of phonons. As a result, at low temperatures, chemical potential increases with T, while at high temperatures, it decreases with T in the ideal gas regime.

The systems that are of particular interest for modern physicists are dense collections of spin-1/2 particles known as fermions whose dynamics are governed by Wolfgang Pauli's exclusion principle. The notable among them are electrons in heavy atoms, in metals, and in white dwarf stars, and the protons and neutrons in nuclei and neutron stars. The specific forces between particles are of less significance than the general structure in accordance with the exclusion principle.

At low temperatures, the particles have much less energy and less quantum states are available, and the occupation rate of each state increases. At high temperatures, in contrast, the particles have much high energy and have many quantum states; consequently, the chance of every quantum state being occupied is very less ($\ll$1) and the exclusion principle is of less importance. It can be possible to distinguish between fermions and bosons at the quantum level by cooling the atoms to quite low temperatures. At the quantum level, both the fermions and bosons differ from each other. According to Wolfgang Pauli's exclusion principle, it is not possible for identical fermions to occupy the same quantum state at the same time, while bosons can share the quantum states. However, this difference can be recognized when the boson or fermion gases are cooled to ultra-low temperatures at which there is more chance for individual quantum states to be occupied. At these low temperatures, there is a tendency of formation of Bose–Einstein condensate as bosons can easily occupy the single quantum state while fermions tend to occupy the energy states from the lowest up with one particle per quantum state as shown in Figure 15.2.

At high temperatures, both the fermions and bosons spread out over several states with much less than one atom per each state. In ultra-low-temperature regime, the fermions don't undergo a sudden phase transition, and instead of it, the quantum behavior appears to develop as the fermion gas is cooled below the Fermi temperature $T_F = E_F/k_B$, in which E_F and k_B are Fermi energy and Boltzmann's constant, respectively. Fermi energy is the energy of the highest filled state. The value of Fermi temperature, T_F, is less than 1 μK for atomic gases, which is identified as a crossover from the classical to the quantum regime.

Quantum Degeneracy

Quantum degeneracy is defined as a regime in which de Broglie wavelength of particles is comparable with that of spacing between the particles. The first experimental recognition of Bose–Einstein condensation in dilute gases took place in 1995. The credit for this achievement goes to the combined efforts of Eric Cornell, Carl Wieman, and Wolfgang Ketterle who shared Nobel Prize in Physics in 2001. In the later years, the degenerate Fermi gas was recognized by Deborah Jin and Brian DeMarco in 1999. From then, it took nearly 20 years of time to discover the quantum degeneracy in polar molecules.

15.6.1 Fermi Low-Temperature Expansions

The quantum statistics discovered by Fermi and Dirac incorporating the Pauli's exclusion principle enables us to understand precisely several physical phenomena in terms of ideal Fermi gas (IFG) in a broad range of particle density values starting

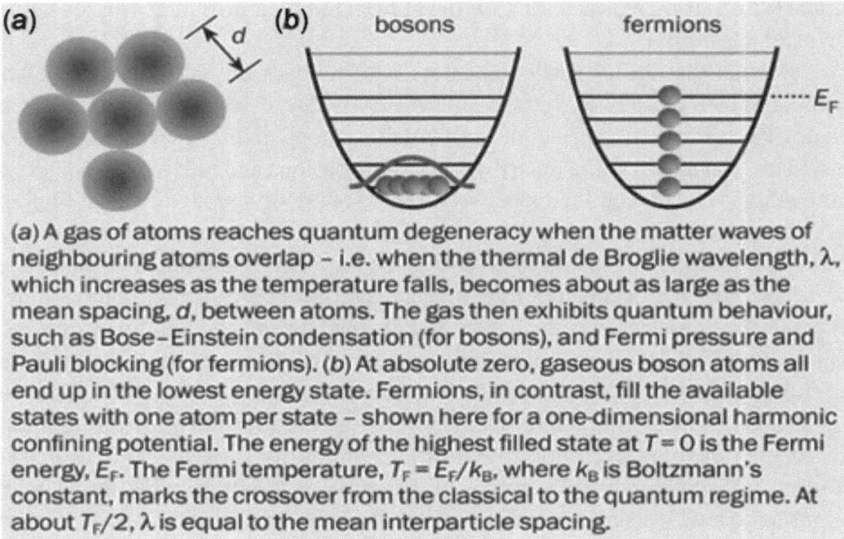

FIGURE 15.2 The low-temperature behavior of fermions and bosons. (a) A gas of atoms reaches quantum degeneracy when the matter waves of neighboring atoms overlap—i.e., when the thermal de Broglie wavelength, λ, which increases as the temperature falls, becomes about as large as the mean spacing, d, between atoms. The gas then exhibits quantum behavior, such as Bose–Einstein condensation (for bosons), and Fermi pressure and Pauli blocking (for fermions). (b) At absolute zero, gaseous boson atoms all end up in the lowest energy state. Fermions, in contrast, fill the available states with one atom per state—shown here for a one-dimensional harmonic confining potential. The energy of the highest filled state at $T=0$ is the Fermi energy, E_F. The Fermi temperature, $T_F=E_F/k_B$, where k_B is Boltzmann's constant, marks the crossover from the classical to the quantum regime. At about $T_F/2$, λ is equal to the mean interparticle spacing. (With permission: *Physics World*, IOP Publishing, "A Fermi gas of atoms" *Low-Temperature Physics.* https://physicsworld.com/a/a-fermi-gas-of-atoms/.)

from astrophysical scales to subnuclear ones. In contrast to the ideal Bose gas (IBG) that has a setback of Bose–Einstein condensation in three dimensions, IFG exhibits a smooth thermodynamic behavior as a function of both temperature and particle density.

15.6.2 Bose Low-Temperature Expansions

The chemical potential shows a distinctive T^2 behavior at a low temperature which is due to the thermal excitation of phonons similar to that which takes place in superfluids. In Fermi gas, the chemical potential shows a maximum in the neighborhood of superfluid critical temperature. The chemical potential shows always a decreasing trend with T at high temperatures; hence, T^2 increase shown by the chemical potential at a low temperature can be the main reason for the non-monotonic behavior as a function of T. The calculation of coefficient of T^2 law using Lieb–Liniger results for the sound velocity is useful for determining the interacting 1D Bose gas.

The coefficient in T^2 expansion of the chemical potential can be demonstrated by the dependence of sound velocity on zero-temperature density.

REVIEW QUESTIONS

15.1 What is chemical potential?
15.2 What is fugacity?
15.3 What are the applications of chemical potential?
15.4 Define Gibbs free energy
15.5 What are the applications of low-temperature systems?
15.6 What is Bose ionic grand canonical potential?
15.7 Write down the equations of pressure, density, and internal energy for an ideal Bose gas
15.8 What is Bose fugacity? Write down its general expression
15.9 Write down the corrections along with their equations in classical limit of Bose gas
15.10 What is Bose–Einstein condensation? Write down its equation
15.11 What are the properties that scale with Bose–Einstein condensation?
15.12 What is standard chemical potential?
15.13 What is fermionic grand canonical potential?
15.14 Write down the equations of pressure and density for an ideal Fermi gas
15.15 What is Fermi fugacity? Write down its general expression
15.16 What is a degenerated Fermi gas?
15.17 What are the properties that scale with low-temperature expansion of Fermi gas?

16 Irreversible Thermodynamics

LEARNING OUTCOMES

After learning this chapter, students should be able to

- Understand the concept of equilibrium and non-equilibrium thermodynamics
- Demonstrate the knowledge of coupled phenomena
- Evaluate the entropy flow and entropy production
- Apply irreversible thermodynamics to thermoelectricity
- Demonstrate the knowledge of Onsager's reciprocal relations in irreversible thermodynamics
- Apply Onsager's reciprocal relations to connect thermodynamics, transport theory, and statistical mechanics

16.1 NEW CONCEPTS BASED ON THE SECOND LAW OF THERMODYNAMICS

Thermodynamics is essentially a vital part of science, and it is a science of development. The advancements and achievements in thermodynamics play a key role in natural sciences, social sciences, and philosophy. Hence, the scientists across the fields should have the fundamental knowledge of classical and modern thermodynamics. The modern formulation of the second law of thermodynamics provides the means of establishing the relations for entropy production. The three most widely accepted mathematical expressions of the second law of thermodynamics for different systems as per the developments in thermodynamics during the 19th century and in the beginning of the 20th century are as follows:

i. Principle of entropy increase for an isolated macroscopic system

$$(dS)_{iso} \geq 0 \tag{16.1}$$

where dS_{iso} refers to the change in entropy of an isolated system

ii. The decrease of Gibbs free energy principle for isothermal isobaric macroscopic systems

$$(dG)_{T,P} \leq 0 \tag{16.2}$$

where $(dG)_{T,P}$ refers to the Gibbs free energy change of isothermal and isobaric macroscopic systems

iii. Positive entropy production principle for any macroscopic system

$$d_iS \geq 0 \tag{16.3}$$

The change in entropy dS is divided into two parts: One is part of entropy production of the system that is part of entropy change due to irreversible processes inside the system d_iS, and the other is part of entropy flow associated with the exchange of energy with the surroundings d_eS:

$$dS = d_iS + d_eS \tag{16.4}$$

The relation between entropy production and Gibbs free energy change under isothermal conditions is

$$(dG)_{T,P} = -Td_iS \tag{16.5}$$

The principle of positive entropy production, besides being suitable for open, closed, and isolated systems, is also suitable for isothermal and isobaric systems and non-isothermal and non-isobaric systems. Thus, the positive entropy production principle is recognized as general mathematical expression of the second law of thermodynamics. Moreover, the entropy production of a system is connected directly with the internal irreversible processes of the system. Entropy production of a system is the sum of entropy productions of internal irreversible processes. This is another advantage with the positive entropy production principle.

16.2 AN OVERVIEW OF EQUILIBRIUM AND NON-EQUILIBRIUM THERMODYNAMICS

A system is said to be in thermodynamic equilibrium when there are no noticeable changes macroscopically and the system is isolated from its surroundings. One of the primary requirements for equilibrium is the uniform temperature throughout the system or each part of the system in thermal contact. If this condition is not met, heat transfer takes place spontaneously from one location to another when the system is isolated. Another requirement is the absence of unbalanced forces between parts of the system. The system will be in thermal and mechanical equilibrium if the above two conditions are met, but it does not ensure the complete equilibrium. A process occurs involving a chemical reaction, a transfer of mass between phases, or both. Equilibrium thermodynamics is the study of transformations of matter and energy in systems with the theory based on thermodynamic equilibrium. Equilibrium is a state of balance in which potentials or driving forces within a system are in exact balance. Non-equilibrium thermodynamics, on the other hand, deals with the physical systems that are not in thermodynamic equilibrium. Non-equilibrium thermodynamics can be effectively used to describe the biological processes that involve protein folding/unfolding and transport through the membranes.

The second law of thermodynamics is often misused to describe that life's order, as in the case of DNA, cells, and proteins, cannot emerge by chance. The entropy either remains constant or increases in an isolated system (the whole universe). In open systems that are characterized by fluxes of energy and matter, however, order can emerge until there is a sufficient entropy increase of the surrounding system so that the total entropy from the two parts of the system together increases. Moreover, the second law cannot predict how fast a system can approach equilibrium except at the stationary non-equilibrium steady state. Carnot efficiency, under this situation, puts a limitation on the rate of entropy production by the heat flux into the system. This qualitative nature of the second law makes it extremely difficult to predict dynamical thermodynamic systems.

Irreversible thermodynamics is an approach favored in physics which connects the thermodynamics and transport phenomenon. This approach does not provide any additional physical insight into the relationship between the rates of mass and energy interactions and local thermodynamic properties of the system penetrated by these interactions. However, it can be possible to determine many coefficients of the newly developed empirical relations just like the thermal conductivity coefficient determined from Fourier's law. This approach is quite useful to analyze the coupled transport phenomena, where two or more transport processes coexist and influence one another. In addition, this approach is the integrated and analytically compact treatment of all irreversible-flow phenomena.

16.3 LOCAL EQUILIBRIUM THERMODYNAMICS

Though the non-equilibrium thermodynamics can be potentially applied to the broad range of situation pertaining to the fields such as physics, chemistry, and engineering, currently its formulation focuses only on the equilibrium systems that are based on local equilibrium assumptions. Thermodynamics that defines the systems in equilibrium is of interest to the engineers and technologists. This can be attributed to the fact that almost all the systems are locally in thermodynamic equilibrium. Thermodynamics forms the basis for the understanding of energy interactions with the surroundings and the rules that the macroscopic properties of the systems at equilibrium follow. These rules do not apply when we talk of the systems outside equilibrium, and it is not possible to define clearly the above quantities. However, with certain assumptions that the system is composed of several subsystems, a local equilibrium hypothesis can be applied to those away from but close to equilibrium.

16.4 COUPLED PHENOMENA

The study of coupled processes is the core of classical theory of non-equilibrium thermodynamics. The theoretical framework of non-equilibrium thermodynamics can be conveniently applied to coupled transport processes. However, these coupled transport processes were discovered prior to the development of this theoretical framework. For example, thermoelectric effects such as those discovered by Seebeck and Peltier respectively in 1821 and 1835 came into existence before even the Ohm's law was formulated quantitatively. The thermodiffusion was first observed by Ludwig in 1856 before the discovery of Dufour's effect (1872) and Soret's effect (1879). *Dufour's effect* is the energy flux caused by a mass concentration gradient resulting from the coupled effect of irreversible processes and is the reciprocal phenomenon to the Soret effect. The *Soret effect*, also called thermal diffusion, is the appearance of a component flux as a result of a thermal gradient. Thermal diffusion in porous medium has been an active area of research due to its wide range of practical applications. Despite the fact that these concepts are quite old, they have recently gained research interest for practical applications because of the recent advancements in materials science, high-power lasers, and optimization of energy generation.

Usually, physical phenomena are considered independent of each other as long as there are no mutual effects. However, if the presence of one physical phenomenon induces one or more other physical phenomena to occur simultaneously, then the physical phenomenon is considered as a coupled phenomenon. The transport of heat energy through a system induces a flow of electrical energy and vice versa is an example of such coupled phenomenon termed thermoelectric effect. Energy is a conserved quantity, and hence, some of the thermal energy is converted directly into electric energy, the conversion being quite low in this case. However, the reverse process, i.e., the conversion of electrical energy to thermal energy, is 100% efficient.

This coupled phenomenon can be a future technology in direct energy conversion processes. Currently, it is widely used in sensors to produce low-level electrical signals that are proportional to the magnitude of the other phenomenon that is present in that. For example, thermocouple is a device used as a temperature sensor. The voltage produced by thermoelectric effect is proportional to the local temperature that can be measured by thermocouple. If we think of reversing this thermoelectric process that is the application of a voltage to the leads of a thermocouple, it will generate a heating or cooling effect in the thermocouple junction. Thermoelectric cooling can be a primary source of cooling in industries where the space is a criterion.

Due to the inherent irreversibilities of the energy conversion processes, energy dissipation takes place within the system, resulting in the loss of efficiency, and hence, the efficiency of every energy conversion processes is always less than 100%. The overall energy conversion efficiency is the product of all the energy conversion efficiencies involved in the system. Power plant is the best example that involves various efficiencies for generating electric power, and its overall efficiency is expressed as given by Eq. 3.63b.

$$\eta_{\text{overall}} = \eta_{\text{combustion}} \times \eta_{\text{thermal}} \times \eta_{\text{generator}} \tag{16.6}$$

The overall energy conversion efficiency is very less since the energy conversion efficiency of each parameter in Eq. 16.6 is less than 100%. Therefore, there is a renewed interest nowadays to develop the direct energy conversion technologies that can potentially compete with conventional indirect energy conversion technologies.

16.5 ONSAGER'S RECIPROCAL RELATIONS

The classical thermodynamics is basically founded on reversible process and equilibrium states, and hence, it is the primary limitation of classical thermodynamics in dealing with the macroscopic behavior of processes since true equilibrium is achieved in a limited number of cases only. Irreversible thermodynamics plays a vital role in determining many of the coefficients related to the newly developed correlations. The irreversible thermodynamics developed by Lars Onsager (1903–1976) is one such approach that can effectively describe the coupled transport phenomenon involving two or more transport processes. L. Onsager of Yale University and I. Prigogine of Universite Libre de Bruxelles (ULB) were awarded Nobel prizes in chemistry for their outstanding contributions in non-equilibrium thermodynamics in the years 1968 and 1977, respectively. Their pioneering contributions are a big breakthrough in the history of non-equilibrium thermodynamics and modern thermodynamics as well. The credit of proposing the first general relation in non-equilibrium thermodynamics goes to Onsager who developed reciprocity relation in 1931, and it is coincident with the term classical thermodynamics that also emerged in the same year. Thus, the year 1931 can be marked as the milestone year for modern thermodynamics.

Onsager in 1931, by using fluctuation theory, identified reciprocal relations among transport coefficients and established the reciprocal relations that connect the thermodynamics, transport theory, and statistical mechanics. Statistical mechanics is an appropriate filed for analyzing the systems near equilibrium and introducing so many elementary properties of equilibrium correlations. Onsager applied statistical methods to equilibrium fluctuations that evolve in accordance with the laws that govern macroscopic variations.

When a system is close to equilibrium, a general theory based on linear relations between forces and flows could be formulated. The Onsager reciprocal relations apply well if the irreversible fluxes are linear functions of driving forces and later become non-linear functions of the state variables. Let us consider an adiabatic system that is expressed by the fluctuations a_i ($i = 1, 2, 3\ldots$) of a set of variables with respect to their equilibrium. At equilibrium, the entropy of a system reaches its maximum (S_m), and change in entropy, $\Delta S = S - S_m$. We can express this as a quadratic equation:

$$\Delta S = -\frac{1}{2}\sum_{i,k}^{n} b_{ik} a_i a_k \quad \text{where } b_{ik} = \frac{-\partial^2 S}{\partial a_i \, \partial a_k} \tag{16.7}$$

where b_{ik} is a positive definite.

Ludwig Boltzmann (1844–1906) established a relation between entropy and probability given as

$$S = k_B \ln W \tag{16.8}$$

where k_B is Boltzmann's constant, and its value is 1.38×10^{-23} J/K. W is the number of microscopic states corresponding to the macroscopic state, and it is thermodynamic probability. Albert Einstein (1879–1955) derived a relation for the probability of a fluctuation given as

$$P(\Delta S) = Z^{-1}\exp(\Delta S/k_B) \tag{16.9}$$

where ΔS is the change in entropy corresponding to the fluctuation from the state of equilibrium and Z is a normalization constant. While in Eq. 16.8, the probability of a state is fundamental quantity and we can derive entropy from it, the opposite is true in Eq. 16.9 in which entropy is a fundamental quantity and we can derive probability of a function from it.

The probability density function according to Boltzmann's entropy hypothesis is given as

$$f(a_i \ldots a_n) = f(0\ldots0)\exp(\Delta S/k_B) \tag{16.10}$$

Onsager assumed conjugated variables as linear combinations of a_i:

$$X_i = k_B \frac{\partial \ln f}{\partial a_i} = \frac{\partial \Delta s}{\partial a_i} = -\sum_k b_{ik} a_k \tag{16.11}$$

where X_i are the variables conjugated to a_i.

The Proof of Reciprocal Relations

From Eq. 16.11,

$$\left(a_i X_j\right) = -k_B \delta_{ij}\ \delta_{ij} = 0 \text{ if } i \neq j, \delta_{ii} = 1 \tag{16.12}$$

According to the principle of detailed balance, if a_i has a value of $a_i(t)$ at time t and if at time $t+\tau$, then a correlated variable a_j has a value of $a_i(t+\tau)$. Then, there occurs a time-reversed transition given as

$$\left(a_i(t)a_j(t+\tau)\right) = \left(a_j(t)a_i(t+\tau)\right) \tag{16.13}$$

In the region close to equilibrium (not far away from equilibrium), the variables a_i obey linear macroscopic equations expressed as

$$\frac{da_i(t)}{dt} = -\sum_j M_{ij} a_j(t) = \sum_k L_{ik} X_k(t) \tag{16.14}$$

Onsager coefficients L_{ik} can be defined as

$$L_{ik} = \sum_j M_{ij} b_{jk}^{-1} \tag{16.15}$$

Onsager assumed that the fluctuations result in the mean according to macroscopic laws:

$$a_j(t+\tau) = a_j(t) + \tau \Sigma L_{jk} X_k(t) \tag{16.16}$$

The principle of detailed balance given in Eq. 16.13 results in reciprocal relations:

$$L_{ij} = L_{ji} \tag{16.17}$$

It is important to note that when L_{ij} are functions of B, then the reciprocal relations in the presence of a magnetic field B will take the form

$$L_{ik}(B) = L_{ki}(-B) \tag{16.18}$$

16.6 ENTROPY AND ENTROPY PRODUCTION

There are certain misconceptions as far as entropy is concerned. According to the second law of thermodynamics, entropy of a closed system is non-decreeing. However, in the quantum physics, the second-law discussion leads to exploring different opportunities and raising different issues resulting in the quantum-engineered devices. There are some kinds of speculations for about the last 150 years that the universal extremal principles such as the maximum entropy production principle (MEPP) proposed by Ziegler and Paltridge determine what happens in nature. MEPP can be used to explain several phenomena such as Rayleigh–Benard convection, transition from laminar to turbulent flow in pipes, flow regimes in plasma physics, and eco-systems.

Onsager relations form the basis of linear non-equilibrium thermodynamics that can describe the non-equilibrium processes that occur in nature. Onsager theoretically generalized the empirical laws developed G. Ohm, A. Fick, and J. Fourier and proposed a linear relationship between thermodynamic fluxes such as heat flux and thermodynamic forces such as temperature gradient given as

$$J_i = \sum_k L_{ik} X_k \tag{16.19}$$

where J_i and X_i are respectively thermodynamic fluxes and thermodynamic forces. The reciprocal relation for the kinetic coefficients L_{ik} is grounded as

$$L_{ik} = L_{ki} \tag{16.20}$$

Equations 16.19 and 16.20 are useful in solving the energy, momentum, and mass transfer equations. However, Eq. 16.19 may not give correct results in the case of some non-equilibrium processes such as chemical reactions. In such cases, non-linear relationship between thermodynamic fluxes and thermodynamic forces can be of use. Now the relation between entropy production density σ and J_i and X_i is established as

$$\sigma = \Sigma_i X_i J_i \tag{16.21}$$

The quantity entropy production plays an important role in non-equilibrium thermodynamics as entropy does in equilibrium thermodynamics. There are two prominent principles associated with entropy production: (i) Prigogine's and Ziegler's principle and (ii) maximum entropy production principle (MEPP).

Prigogine's and Ziegler's Principle

This principle constitutes two statements: The first one is proposed by I. Prigogine during 1945–1947 and the second by H. Ziegler in 1963. The statement of Prigogine is called the principle of minimum entropy production. The minimum entropy production is the necessary and sufficient condition required for the stationary state of non-equilibrium system, provided the main equations of the linear non-equilibrium thermodynamics (Eqs. 16.1 and 16.2) are realized in a system and a part of the total number of thermodynamic forces X_i are maintained constant. However, there appear two drawbacks for this principle: First, it applies to linear non-equilibrium thermodynamics only, and second, it does not provide any constructive information as a theorem and adds nothing new. Laws of conservation and Eqs. 16.1 and 16.2 can be more conveniently used to solve the problems than using this principle. Despite the above drawbacks, biologists and philosophers still focus their attention toward this principle.

The Ziegler statement is maximum entropy production and seems to be an antipode to the Prigogine's statement. True thermodynamic fluxes J_i with the presence of thermodynamic forces X_i satisfying the side condition of Eq. 16.19 give the entropy production density, $\sigma(J)$. We can express this mathematically with the help of Lagrange multiplier, μ, as given below:

$$\Sigma_J \left[\sigma(J_K) - \mu\left(\sigma(J_K) - \Sigma_i X_i J_i\right)\right]_X = 0 \tag{16.22}$$

From Eq. 16.19, we can develop the relationship between forces and fluxes (both linear and non-linear) by varying the entropy production in terms of thermodynamic fluxes with constant forces:

$$X_i = \frac{\sigma(J)}{\Sigma_i\left(J_i\, \partial\sigma/\partial J_i\right)} \partial\sigma/\partial J_k \tag{16.23}$$

For $\sigma = \sum R_{ik} J_i J_k$, we can obtain the main equations of linear non-equilibrium thermodynamics given in Eqs. 16.7 and 16.8 from Eq. 16.19, meaning that the Prigogine theorem can be developed from Eqs. 16.7 and 16.8 with the introduction of additional restrictions. This makes the Prigogine theorem much narrower than the Ziegler theorem.

Maximum Entropy Production Principle (MEPP)

Based on the above discussion, the maximum entropy production principle seems to be a common statement compared to the minimum entropy production principle. MEPP, unlike Prigogine's principle, has not been a popular approach, since Ziegler's principle is solely based on the theoretical viewpoint in the area of non-equilibrium thermodynamics and focused mainly on solving the problems of plasticity. Moreover, there are other principles in non-equilibrium thermodynamics for solving non-linear problems, and the theory of plasticity on the other hand has various other principles that can effectively solve the problems. However, Ziegler's method forms the basis for the development of similar principles by various researchers, and hence, it invites special focus. Some of these methods are presented in the following sections.

M. Kohler and J. Ziman developed a variational method of transport theory for solving the linearized Boltzmann equation used in the study of the transfer in gases, metals, and semiconductors. According to their theory, the velocity distribution function for non-equilibrium gas systems is such that the entropy production density is a maximum at preset gradients of the temperature, the concentration, and the mean velocity. This statement is generalized later so that it is valid for both the quantum systems and relatively dense systems.

M. Berthelot stated that if a system involves several chemical reactions without the aid of an external energy, it results in a reaction with the largest amount of heat release. In a large number of chemical processes that occur at low temperatures, the rate of heat produced is proportional to entropy production and it is in accordance with the maximum entropy production principle.

D. Temkin, J. Kirkaldy, and E. Ben-Jacob assumed that, in the case of fixed supersaturation/supercooling, the growing dendrite will select the maximum possible rate. The entropy production density of the system under consideration is proportional to the squared growth rate of the crystal. MEPP plays a crucial role in non-equilibrium thermodynamics to demonstrate the direction of biological and social evolution.

The maximum entropy production principle allows us to see the surrounding world from the same perspective without dividing it into the animate and inanimate. There are the simplest and relatively well-studied physical and chemical processes satisfying the principle at the lowest levels of this world. The formation of the higher levels, the construction of which we witness and which are a part, also takes place according to this principle of the non-equilibrium thermodynamics.

16.7 LINEAR PHENOMENOLOGICAL EQUATIONS

The linear relations between fluxes (X_i) and flows (J_i) can be formulated when a system is close to equilibrium. The fluxes will be zero when all the driving forces are zero (at equilibrium). Though the fluxes are driven by forces, they are not entirely determined by them. The presence of a catalyst is another factor that also can cause the flow. The presence of a catalyst can influence the corresponding flow and the rate of chemical reaction for a fixed value of affinity. According to this cross-coupling effect, the particular flow J_i is driven not only by its conjugate force X_i but also by the remaining driving forces, given as

$$J_i = J_i(X_0, X_1 \ldots X_n) \tag{16.24}$$

At equilibrium, $J_i = (J_i)_{x_0=x_1=\cdots=x_n=0} = 0$

Expanding J_i about the equilibrium state using Taylor's series function results in

$$J_i = (J_i)_{x_0=x_1=\cdots=x_n=0} + \frac{\partial J_i}{\partial X_0} X_0 + \cdots \frac{\partial J_i}{\partial X_n} X_n + \text{higher-order terms} \, (i = 0,1,2,\ldots n) \tag{16.25}$$

The primary coefficients can be defined as $L_{11} = \frac{\partial J_1}{\partial X_1}$ and $L_{22} = \frac{\partial J_2}{\partial X_2}$.

The secondary coefficients can be defined as $L_{12} = \frac{\partial J_1}{\partial X_2}$ and $L_{21} = \frac{\partial J_2}{\partial X_1}$.

$$J_i = \sum_{k=0}^{n} L_{ik} X_k \, (i = 0,1,2\ldots n) \tag{16.26}$$

The entropy production rate, for the conditions under which linear phenomenological relations are valid, takes the form

$$\dot{s}_{gen} = \sum_{ik} L_{ik} X_i X_k \geq 0 \tag{16.27}$$

where $L_{ik} = \frac{\partial J_i}{\partial X_k}$ (coefficients L_{ik} are called phenomenological coefficients).

Equation 16.27 defines the local entropy generation rate, conjugate pairs of flows, and driving forces. If only two flows are present and the conjugate pairs of flows and forces are (J_0, X_0) and (J_1, X_1), then the linear relations can be

$$J_0 = L_{11} X_1 + L_{12} X_2 \tag{16.28}$$

$$J_1 = L_{21} X_1 + L_{22} X_2 \tag{16.29}$$

In case of pure heat transfer through an isotropic medium, the driving force is $X_0 = \frac{\partial}{\partial x}\left(T^{-1}\right)$ and the flow is J_0. The Fourier law of heat conduction can be used to define thermal conductivity k:

$$J_0 = -k \frac{\partial T}{\partial x} \text{ where q is the heat flux } \left(J_0 = q\right) \tag{16.30}$$

Then, Eq. 16.28 becomes $\Rightarrow J_0 = L_{11} \frac{\partial}{\partial x}\left(T^{-1}\right)$ and $L_{11} = T^2 k$.

16.8 THERMOELECTRIC PHENOMENA

Lord Kelvin (William Thomson (1824–1907)) in 1854 proposed a quasi-thermodynamic method of analysis for establishing the relationships among the phenomenon such as thermoelectric effect. Although Kelvin himself announced that his method was not fully justifiable, experimental results confirmed the validity of the proposed method that was successfully applied by Helmholtz to the theory of electrolytic cells and later by Eastman to the Soret effect. In addition, this method was also extended to galvano- and thermo-magnetic effects by Bridgman. Onsager developed a set of reciprocal relations and proved them to be effective in treating the symmetry in the mutual interference of the two or more of the simultaneously occurring irreversible processes of a system. Onsager concluded that the reciprocal relations form the basis for Kelvin relations. The phenomenon of thermoelectric effect occurs due to this mutual interference of the two or more of the simultaneously occurring irreversible processes, that is, when the simultaneous flow of electric current and heat current takes place in a system.

In the case of simultaneous flow of heat and electrical energy in a system, the entropy generation rate can be

$$\dot{s}_{gen} = J_Q X_Q + J_E X_E \tag{16.31}$$

where $J_Q = L_{qq} X_q + L_{qe} X_E$ and $J_E = L_{eq} X_q + L_{ee} X_E$

Then, Eq. 16.31 becomes

$$\dot{s}_{gen} = L_{qq} X_Q + L_{qe} X_Q X_E + L_{eq} X_Q X_E + L_{ee} X_E \tag{16.32}$$

According to Onsager's reciprocity relations, $L_{eq} = L_{qe}$.

$$\dot{s}_{gen} = L_{qq} X_Q + 2L_{qe} X_E X_Q + L_{ee} X_E > 0 \tag{16.33}$$

From Eq. 16.33, it can be concluded that there occurs the heat flow even in the absence of temperature gradient, provided that the coefficient L_{qe} is nonzero and there is a voltage source X_E within the system. Similarly, there exists a flow of electrical energy J_E without any voltage source, provided that the L_{eq} is nonzero and heat transport J_Q occurs. This coupled effect leads to the development of thermoelectric technology.

16.8.1 Seebeck Effect

When two dissimilar metal wires are joined together to form a closed loop and one of the junctions is heated, an electromotive force (EMF) is generated. This is called *Seebeck effect*, and the circuit is called thermoelectric circuit since it incorporates both thermal and electrical effects. Thermoelectric power generation is based on the Seebeck effect. The EMF generated in the circuit is of the order 10^{-5} V/K of temperature difference. Thomas Johann Seebeck (1770–1831) in 1821 observed the generation of an EMF in a thermocouple circuit by heating one junction of a bimetallic couple and cooling the other. For thermoelectric phenomena, the semiconductors are an important area of research, since a thermocouple made up of semiconductors is capable of producing large electromotive potentials and can be used for converting heat into electricity.

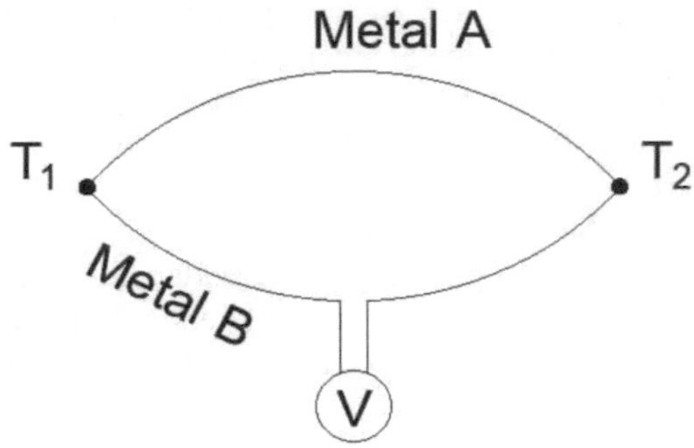

FIGURE 16.1 Seebeck effect.

Let us consider a thermocouple (materials A and B) with its junctions maintained at different temperatures ($T_2 > T_1$) as shown in Figure 16.1. A voltmeter is inserted in the thermocouple in such a way that it does not allow any flow of electricity, and it, however, offers no resistance to the flow of heat. Heating one of the junctions of the bimetallic closed loop, an electromotive force (EMF) is generated, which is directly proportional to the temperature of the junction, and this effect leads to relative Seebeck coefficient given as

$$\alpha_{AB} = -\lim_{\Delta T \to 0}\left(\frac{\Delta\phi_{AB}}{\Delta T}\right)_{I=0} \tag{16.34}$$

where $\Delta\phi_{AB}(\phi_A - \phi_B)$ is the potential difference termed Seebeck voltage. α_{AB} is the negative value of slope of open-circuit voltage–temperature relationship for the pair of conductors. The thermoelectric power of the thermocouple (α_{AB}) is defined as the change in voltage per a unit change in temperature difference. The sign of α_{AB} is considered positive if the voltage increment is such that it drives the current from conductor A to B at the hot junction.

16.8.2 Peltier Effect

When a small current is passed through the junction of two dissimilar metal wires, the junction cools. This is called *Peltier effect.* Thermoelectric refrigeration is based on the Peltier effect. The heat current in the circuit is 10^{-5} J/s A. Jean Charles Athanase Peltier (1785–1845), in 1834, found that when an electric current is passed through a bimetallic circuit, heat absorption takes place at one junction and rejection at other junction, as shown in Figure 16.2. The heat flow per unit current in isothermal circuit is called Peltier heat, q_{Pe}, which is given by

$$\dot{Q}_P = \frac{\text{Heat added or removed}}{I} \tag{16.35}$$

The Peltier coefficient is

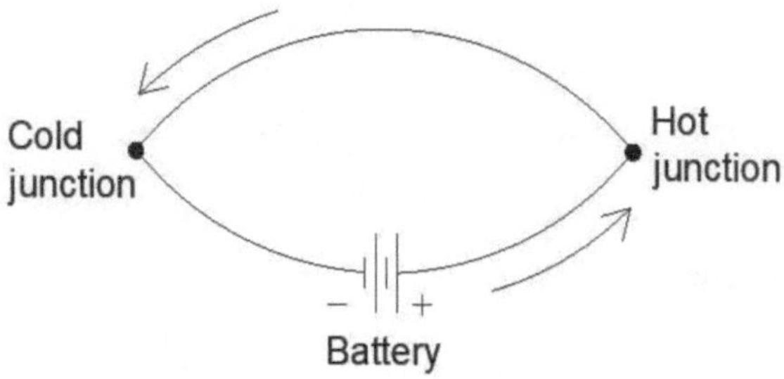

FIGURE 16.2 Peltier effect.

$$\pi_{AB} = \pi_A - \pi_B = \frac{\dot{Q}_B - \dot{Q}_A}{I} \tag{16.36}$$

$\dot{Q}_A$ and $\dot{Q}_B$ are the Peltier heating or cooling rates of the conductors A and B.

The Peltier coefficient is defined as the heat that must be supplied to the function when unit electric current passes from conductor A to B. The Peltier coefficient has dimensions of electromotive force and is sometimes referred to as Peltier e.m.f.

The use of semiconductors makes it possible to achieve rapid heating or cooling with the Peltier effect. It can be possible to maintain larger temperature differences as high as 70°C between hot and cold junctions.

16.8.3 Joule Effect

When an electric current is passed through a homogeneous isothermal conductor, it produces an internal heating of the conductor independent of the direction of current flow. This is called Joule effect, discovered by James Prescott Joule (1818–1889) in 1841. It is shown in Figure 16.3. This effect results in Joule heating formula given as

$$\dot{Q}_J = k_J I^2 R \tag{16.37}$$

where $\dot{Q}_J$ is the Joule heating rate, R is the electrical resistivity of conductor, I is the electric current, and k_J is the Joule coefficient.

Since k_J in Eq. 16.37 is just the unit conversion factor, it can be eliminated as given below

$$\dot{Q}_J = I^2 R \tag{16.38}$$

16.8.4 Kelvin Effect

When an electric current is passed through a homogeneous conductor in which a temperature gradient exists, heating or cooling of conductor takes place depending on the current flow direction relative to temperature gradient. This is called *Kelvin effect* in the honor of Lord Kelvin (1824–1907) who discovered this in 1854. The Kelvin effect is shown in Figure 16.4. The Kelvin effect can be achieved by heating

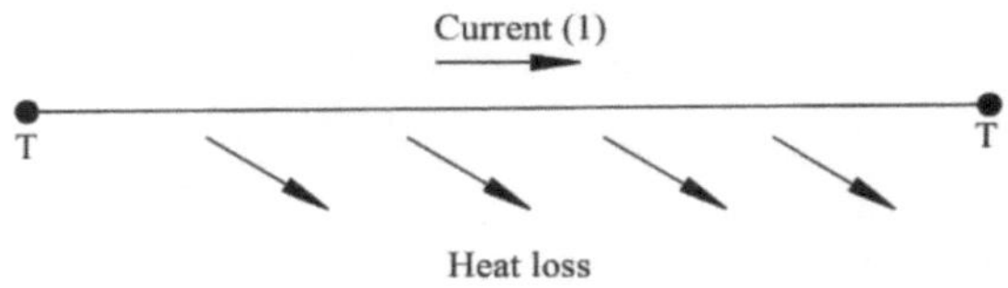

FIGURE 16.3 Joule effect.

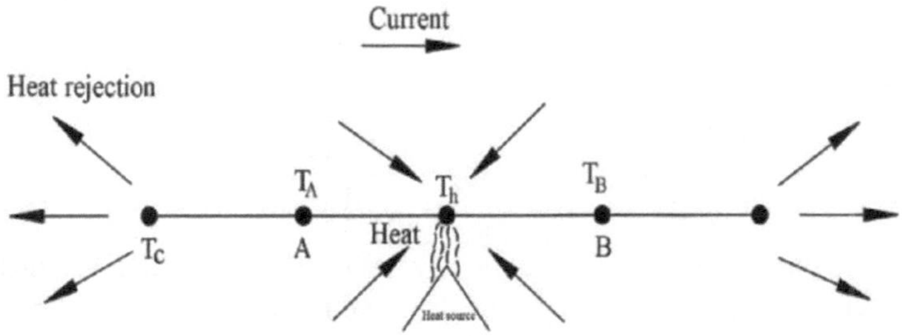

FIGURE 16.4 Kelvin effect.

the center of a uniform wire and cooling its ends and passing a current through it. If the temperature of two points A and B (T_A and T_B) is measured, it can be observed that $T_A \neq T_B$. The electric current distributes the temperature profile in the wire which results in Kelvin coefficient.

$$\tau = \lim_{\Delta T \to 0} \frac{\dot{Q}}{1(\Delta T)} \tag{16.39}$$

$J_E = {}^{1}\!/_{A}$ is the electric current density.

The Kelvin coefficient is related to the Seebeck coefficient as

$$\tau_{AB} = -T\left(\frac{d\alpha_{AB}}{dT}\right) \tag{16.40}$$

The Kelvin coefficient is related to the Peltier coefficient as

$$\pi_{AB} = T\alpha_{AB} \tag{16.41}$$

where T is the absolute temperature.

16.9 THERMODYNAMIC FORCES AND THERMODYNAMIC VELOCITIES

Irreversible processes are assumed as thermodynamic forces driving the thermodynamic flows; i.e., thermodynamic flows result from thermodynamic forces. This can be explained with some examples. To cause an irreversible flow of heat, thermodynamic force required is temperature gradient; similarly, to cause the flow of matter, the driving force required is concentration gradient. The irreversible increase in entropy in the case of the free expansion of a gas is given by

$$d_iS = \frac{P_{gas} - P_{piston}}{T} dV \tag{16.42}$$

where $P_{gas} - P_{piston}/T$ refers to the thermodynamic force and dV/dt the corresponding flow. The term $(P_{gas} - P_{piston})dV$ refers to the uncompensated heat of Clausius. De Donder introduced the concept of affinity to establish the new formalism for the second law of thermodynamics by incorporating uncompensated heat of Clausius. This approach can be useful to define the affinity in chemical reactions, thereby writing the entropy production equation of the reaction more elegantly as the product of thermodynamic force and thermodynamic flow.

16.10 STATIONARY STATES, FLUCTUATIONS, AND STABILITY

Fluctuations in thermodynamic quantities such as temperature, concentration, and partial molar volume are caused due to random motion of molecules, and the state of a system is subject to constant agitations due to its interaction with the exterior. Despite all these fluctuations and agitations, the state of equilibrium must retain its stability. For an isolated system, at equilibrium, entropy reaches its maximum value and it can be reduced by fluctuations only. As long as fluctuations grow, the state of the system cannot be in equilibrium. In response to these fluctuations, the irreversible processes help the system regain its state of equilibrium. For isolated systems, the stability theory can be developed based on the considerations that the total energy U, volume V, and molar amounts N_k are constants.

Thermal Stability

Let us consider an arbitrary isolated system in which a fluctuation occurs in a small part of it. Due to the fluctuation, there is a flow of energy, δU, from one part to the other, resulting in a small temperature fluctuation, δT, in the smaller part. The total entropy and internal energy of the system are as follows:

$$S = S_1 + S_2 \text{ and } U = U_1 + U_2$$

Let us consider S = S(U,V) and the expand entropy as a power series in these parameters:

$$\delta_1 S = \left(\frac{\partial S}{\partial U_1}\right)_V \delta U_1 + \left(\frac{\partial S}{\partial U_2}\right)_V \delta U_2 + \left(\frac{\partial S}{\partial V_1}\right)_U \delta V_1 + \left(\frac{\partial S}{\partial V_2}\right)_U \delta V_2 \tag{16.43}$$

Comparing Eq. 16.43 with TdS = dU + pdV,

$$\left(\frac{\partial S}{\partial U}\right)_V = \frac{1}{T} \text{ and } \left(\frac{\partial S}{\partial V}\right)_U = \frac{P}{T}$$

$$\delta_1 S = \left(\frac{1}{T_1}\right)\delta U_1 + \left(\frac{1}{T_2}\right)\delta U_2 + \left(\frac{P_1}{T_1}\right)\delta V_1 + \left(\frac{P_2}{T_2}\right)\delta V_2$$

$$\text{(Neglecting the second order terms)} \tag{16.44}$$

Since the total energy of the system remains the same and volume remains constant,

$$\delta U_1 = -\delta U_2 \text{ and } \delta V_1 = -\delta V_2$$

$$\text{At equilibrium, } \delta_1 S = 0, T_1 = T_2 \text{ and } p_1 = p_2$$

Let us consider $U = U(S,V)$ and the expand internal energy as a power series in these parameters:

$$\delta U = \left(\frac{\partial U}{\partial S}\right)_V \delta S + \frac{1}{2}\left(\frac{\partial^2 U}{\partial S^2}\right)_V (\delta S)^2 + \left(\frac{\partial U}{\partial V}\right)_S \delta V + \frac{1}{2}\left(\frac{\partial^2 U}{\partial V^2}\right)_S (\delta V)^2 + \left(\frac{\partial^2 U}{\partial V \cdot \partial S}\right)\partial V \cdot \partial S + \cdots$$

$$= T\delta S - p\delta V + \frac{1}{2}\left(\frac{\partial^2 U}{\partial S^2}\right)_V (\delta S)^2 + \left(\frac{\partial U}{\partial V}\right)_S \delta V + \frac{1}{2}\left(\frac{\partial^2 U}{\partial V^2}\right)_S (\delta V)^2 + \left(\frac{\partial^2 U}{\partial V \cdot \partial S}\right)\partial V \cdot \partial S + \cdots \tag{16.45}$$

For the stability of the system, $\delta U + p\delta V - T\delta S > 0$, and it is valid if the following conditions are met:

$$\left(\frac{\partial^2 U}{\partial S^2}\right)_V > 0, \left(\frac{\partial^2 U}{\partial V^2}\right)_S > 0 \text{ and } \frac{\partial^2 U}{\partial V \cdot \partial S} > 0$$

From the equation, $\left(\frac{\partial S}{\partial U}\right)_V = \frac{1}{T} \rightarrow \left(\frac{\partial U}{\partial S}\right)_V = T.$

By substituting $\partial U = C_v\,\partial T$ in the above equation, $\left(\frac{C_v\,\partial T}{\partial S}\right)_V = T \rightarrow \left(\frac{\partial T}{\partial S}\right)_V = \frac{T}{C_v}$

$$\frac{T}{C_v} > 0 \rightarrow C_v > 0 \tag{16.46}$$

Equation 16.46 is known as the condition for *thermal stability.*

Mechanical Stability

The Helmholtz function is

$$A = U - TS \tag{16.47}$$

Differentiating Eq. 16.6, $dA = dU - TdS - SdT$

$$TdS = dU + PdV \rightarrow dU - TdS = -PdV$$

$$\therefore dA = -PdV - SdT \tag{16.48}$$

Let us choose A = A(T,V) and expand δA:

$$\delta A = -P\delta V - S\delta T + \frac{1}{2}\left(\frac{\partial^2 A}{\partial V^2}\right)_T (\delta V)^2 + \frac{1}{2}\left(\frac{\partial^2 A}{\partial T^2}\right)_S (\delta T)^2 + \frac{\partial^2 A}{\partial V \cdot \partial T} \cdot \delta V \cdot \delta T + \cdots$$

For stability, $\delta A + P\delta V + S\delta T > 0$, provided that $\left(\frac{\partial^2 A}{\partial V^2}\right)_T > 0$ and $\left(\frac{\partial A}{\partial V}\right)_T = -P$.

Differentiation of $\left(\frac{\partial A}{\partial V}\right)_T = -P$ gives $\left(\frac{\partial^2 A}{\partial V^2}\right)_T = -\left(\frac{\partial P}{\partial V}\right)_T$.

To satisfy the above equation,

$$\left(\frac{\partial P}{\partial V}\right)_T < 0 \tag{16.49}$$

Eq. 16.49 is known as the condition for *mechanical stability.*

REVIEW QUESTIONS

16.1 What are the differences between equilibrium and non-equilibrium thermodynamics approaches?
16.2 What is the significance of Onsager's reciprocal relations?
16.3 What are local intensive quantities?
16.4 What is irreversible thermodynamics?
16.5 Give the equation for quasi-low-temperature behavior of specific heat?
16.6 What is threshold of quasi-low-temperature regime?
16.7 Give typical examples of dissipative and non-dissipative thermodynamics
16.8 What is Peltier effect and what is its significance?
16.9 What is Seebeck effect and what is its significance?
16.10 What is the role of irreversible thermodynamics in thermoelectricity?
16.11 What is the condition for mechanical stability?
16.12 What is the condition for thermal stability?
16.13 What is the significance of the maximum entropy production principle?
16.14 What are the reasons for fluctuations in thermodynamic quantities such as temperature and concentration?
16.15 What are the limitations of classical thermodynamics?
16.16 What is the role of irreversible thermodynamics in dealing with non-equilibrium processes?
16.17 Explain how Onsager's reciprocal relations connect thermodynamics, transport theory, and statistical mechanics
16.18 What is the role of the principle of positive entropy production in the context of modern formulation of the second law of thermodynamics?
16.19 What are coupled phenomena? Give some examples

REVIEW QUESTIONS

References

1. Cengel Y.A and Boles M.A. *Thermodynamics an Engineering Approach*. 7th Edition. Singapore: McGraw-Hill Education, 2011.
2. Nag P.K. *Engineering Thermodynamics*. 6th Edition. India: McGraw-Hill Education, 2018.
3. Jones J.B and Dugan R.E. *Engineering Thermodynamics*. India: Prentice Hall, 1996.
4. Sonntag R.E, Borgnakke C and Van Wylen G.J. *Fundamentals of Thermodynamics*. 5th Edition. Wiley India Edition, 2003.
5. Moran M.S and Shapiro H.N. *Fundamentals of Engineering Thermodynamics*. New York: John Wiley & Sons, 1988.
6. Bejan A. *Advanced Engineering Thermodynamics*. 2nd Edition. New York: Wiley, 1997.
7. Modi P.N and Seth S.M. *Hydraulics and Fluid Mechanics Including Hydraulics Machines*. 20th Edition. Standard Book House Delhi.
8. Som S.K and Biswas G. *Introduction to Fluid Mechanics and Fluid Machines*. 2nd Edition. India: McGraw Hills, 2011.
9. Nakayama Y and Boucher R.F. *Introduction to Fluid Mechanics*. Jordan Hill, Oxford: Butterworth-Heinemann Linacre House.
10. Holman J.P. *Heat Transfer*. 8th Edition. New York: McGraw-Hill, Inc. 2007.
11. Özışık M.N. *Heat Transfer a Basic Approach*. McGraw-Hill Book Company, 1985.
12. Black W.J and Hartley J.G. *Thermodynamics*. New York: Harper & Row, 1985.
13. Keszei E. *Chemical Thermodynamics*. Berlin, Heidelberg: Springer-Verlag, 2012. doi: 10.1007/978-3-642–19864-9_2.
14. Singhal B.L. *Engineering Thermodynamics*. India: Macmillan Publishers India Ltd, 2020.
15. Zhu G, Liu J, Fu J and Wang S. A combined organic Rankine cycle with double modes used for internal combustion engine waste heat recovery. *Journal of Engineering for Gas Turbines and Power* 139 p. 112804-1, 2017.
16. Ganesan V. *Internal Combustion Engine Fundamentals*. 4th Edition. New Delhi: McGraw-Hill, 2012.
17. Heywood J.B. *Internal Combustion Engine Fundamentals*. McGraw-Hill, 2018.
18. Ganesan V. *Gas Turbines*. 3rd Edition. New Delhi: Tata McGraw-Hill, 2010.
19. Arora C.P. *Refrigerating and Air-Conditioning*. 3rd Edition. New Delhi: Tata McGraw-Hill, 2008.
20. *ASHRAE Handbook of Fundamentals*. SI version. Atlanta, GA: American Society of Heating, Refrigerating, and Air-Conditioning Engineers, Inc., 1993.
21. *ASHRAE Handbook of Refrigeration*. SI version. Atlanta, GA: American Society of Heating, Refrigerating, and Air-Conditioning Engineers, Inc., 1994.
22. Goetzler W, Sutherland T, Rassi M and Burgos J. Research & Development Roadmap for Next-Generation Low Global Warming Potential Refrigerants, Prepared by Navigant Consulting, Inc. U.S Department of Energy, November 2014. Available electronically at www.osti.gov/home.
23. Njoku I.H, Oko C.O.C and Ofodu J.C. Performance evaluation of a combined cycle power plant integrated with organic Rankine cycle and absorption refrigeration system. *Cogent Engineering* 5 (1) pp. 1–30, 2018.
24. Ersayin E and Ozgener L. Performance analysis of combined cycle power plants: A case study. *Renewable and Sustainable Energy Reviews* 43 pp. 832–842, 2015.

25. Boyce M.P. Performance testing of combined cycle power plant. In: *Handbook for Cogeneration and Combined Cycle Power Plants*, M.P. Boyce (ed.), Chapter 13. ASME Press. doi: 10.1115/1.859537.ch13.
26. Tiwari A.K, Hasan M.M and Islam M. Effect of operating parameters on the performance of combined cycle power plant. *Open Access Scientific Reports* 1 (7) 1:351, 2012. doi: 10.4172/scientificreports.351.
27. Yüksek I and Karadayi T.T. Energy-Efficient Building Design in the Context of Building Life Cycle, 2017. https://www.intechopen.com/books/energy-efficient-buildings/energy-efficient-building-design-in-the-context-of-building-life-cycle. doi: 10.5772/66670.
28. Flores J.M.B, Cely D.P, Gómez-Martínez M.A, Pérez I.H, Rodríguez-Valderrama D.A and Aricapa Y.H. Thermal and energy evaluation of a domestic refrigerator under the influence of the thermal load. *Energies* 12 p. 400, 2019. doi: 10.3390/en12030400.
29. Harrington L, Aye L and Fuller B. Impact of room temperature on energy consumption of household refrigerators: Lessons from analysis of field and laboratory data. *Applied Energy* 211 pp. 346–357, 2018.
30. Khosravy el Hossaini M. *Review of the New Combustion Technologies in Modern Gas Turbines*, 2013. Book chapter, doi: 10.5772/54403.
31. Liu Y, Sun X, Sethi V, Nalianda D, Li Y.-G and Wang L. Review of modern low emissions combustion technologies for aero gas turbine engines. *Progress in Aerospace Sciences* 94 pp. 12–45, 2017. www.elsevier.com/locate/paerosci. doi: 10.1016/j.paerosci.2017.08.001.
32. Li X, Liu T and Chen L. Thermodynamic performance analysis of an improved two-stage organic Rankine cycle. *Energies* 11 p. 2864, 2017. doi: 10.3390/en11112864.
33. Kanoglu M and Dincer I. Performance assessment of cogeneration plants. *Energy Conversion and Management* 50 pp. 76–81, 2009.
34. Cowan B. On the chemical potential of ideal Fermi and Bose gases. *Journal of Low Temperature Physics* 197 pp. 412–444, 2019. doi: 10.1007/s10909-019-02228-0.
35. Majoumerd M.M, Raas H, Jana K, De S and Assadi M. Coal quality effects on the performance of an IGCC power plant with CO_2 capture in India. *Energy Procedia* 114 pp. 6478–6489, 2017.
36. Martyushev L.M. Entropy and entropy production: Old misconceptions and new breakthroughs. *Entropy* 15 pp. 1152–1170, 2013. doi: 10.3390/e15041152.
37. Kondepudi D and Prigogine I. *Modern Thermodynamics from Heat Engines to Dissipative Structures*. 2nd Edition. Wiley.
38. A Fermi Gas of Atoms. *Low-Temperature Physics*, 04 April 2002. https://physicsworld.com/a/a-fermi-gas-of-atoms/.
39. Callen H.B. The application of Onsager's reciprocal relations to thermoelectric, thermomagnetic, and galvanomagnetic effects. *Physical Review* 73 (11) pp. 1349–1358, 1948.
40. Balmer R.T. *Modern Engineering Thermodynamics*. Academic Press is an imprint of Elsevier, Amsterdam, Netherlands.

Index

Note: **Bold** page numbers refer to tables and *italic* page numbers refer to figures.

I

J

K

L

M

N

O

P

Q

R

S

T

U

V

W

Z